GENETIC ANALYSIS OF COMPLEX DISEASES

GENETIC ANALYSIS OF COMPLEX DISEASES

SECOND EDITION

Jonathan L. Haines PhD
Center for Human Genetics Research
Department of Molecular Physiology and Biophysics
Vanderbilt University Medical Center
Nashville, Tennessee

Margaret Pericak-Vance PhD
Center for Human Genetics, Section of Medical Genetics
Duke University Medical Center
Durham, North Carolina

WILEY-LISS

A JOHN WILEY & SONS, INC., PUBLICATION

For general information on our other products and services or for technical support, please contact our
Customer Care Department within the United States at (800) 762-2974, outside the United States at
(317) 572-3993 or fax (317) 572-4002.

Wiley also publishes its books in a variety of electronic formats. Some content that appears in print may
not be available in electronic formats. For more information about Wiley products, visit our web site at
www.wiley.com.

Library of Congress Cataloging-in-Publication Data is available.

ISBN-10: 0-471-08952-4
ISBN-13: 978-0-471-08952-0

Printed in the United States of America

10 9 8 7 6 5 4 3 2

CONTENTS

Foreword   **xv**

Preface   **xvii**

Contributors   **xix**

1. Basic Concepts in Genetics and Linkage Analysis   **1**
   Elizabeth C. Melvin and Marcy C. Speer

   Introduction   1

   Historical Contributions   2
      Segregation and Linkage Analysis   2
      Hardy Weinberg Equilibrium   5

   DNA, Genes, and Chromosomes   5
      Structure of DNA   5
      Genes and Alleles   9
      Genes and Chromosomes   10

   Inheritance Patterns in Mendelian Disease   13

   Genetic Changes Associated with Disease/Trait Phenotypes   14
      Point Mutations   14
      Deletion/Insertion Mutations   17
      Novel Mechanisms of Mutation: Unstable DNA and
        Trinucleotide Repeats   18

   Susceptibility Versus Causative Genes   19

   Genes, Mitosis, and Meiosis   23
      When Genes and Chromosomes Segregate Abnormally   25

   Ordering and Spacing of Loci by Mapping Techniques   26
      Physical Mapping   26
      Genetic Mapping   29

Interference and Genetic Mapping   30
Meiotic Breakpoint Mapping   31
Disease Gene Discovery   31
Information Content in a Pedigree   41
Disease Gene Localization   42
Extensions to Complex Disease   45

Summary   45

References   46

2. Defining Disease Phenotypes   **51**
Arthur S. Aylsworth

Introduction   51

Exceptions to Traditional Mendelian Inheritance Patterns   52
Pseudodominant Transmission of a Recessive   53
Pseudorecessive Transmission of a Dominant   54
Mosaicism   55
Mitochondrial Inheritance   56
Incomplete Penetrance and Variable Expressivity   58
Genomic Imprinting   61
Phenocopies and Other Environmentally Related Effects   63

Heterogeneity   64
Genetic Heterogeneity   64
Phenotypic Heterogeneity   65

Complex Inheritance   67
Polygenic and Multifactorial Models   67
Role of Environment   70
Role of Chance in Phenotype Expression   70

Phenotype Definition   71
Classification of Disease   71
Nonsyndromic Phenotypes   72
Syndromic Phenotypes   72
Associations and Syndromes of Unknown Cause   73
Importance of Chromosomal Rearrangements in Mapping   74
Qualitative (Discontinuous) and Quantitative (Continuous) Traits   74

Defining Phenotypes for Analysis of Complex Genetic Disorders   75
Select Most Biologically Meaningful Phenotype   75
Partition Phenotype or Dataset by Cause and Associated Pathology   75
Summary: Approach to Phenotype Definition   80
Resources for Information about Clinical Genetics and
  Phenotype Definition   82

References   82

3. Determining Genetic Component of a Disease   **91**
Allison Ashley-Koch

Introduction   91

Study Design   92
    Selecting a Study Population   93
    Ascertainment   94

Approaches to Determining the Genetic Component of a Disease   99
    Cosegregation with Chromosomal Abnormalities and
      Other Genetic Disorders   100
    Familial Aggregation   101
    Twin and Adoption Studies   104
    Recurrence Risk in Relatives of Affected Individuals   105
    Heritability   107
    Segregation Analysis   108

Summary   110

References   111

4. Patient and Family Participation in Genetic Research Studies   **117**
Chantelle Wolpert, Amy Baryk Crunk, and Susan Estabrooks Hahn

Introduction   117

Step 1: Preparing to Initiate a Family Study   118
    Confidentiality   118
    Certificate of Confidentiality   119
    Need for a Family Studies Director   119
    Working with Human Subjects   122

Step 2: Ascertainment of Families for Studies   124
    Family Recruitment   124
    Informed Consent and Family Participation   128

Step 3: Data Collection   131
    Confirmation of Diagnosis   131
    Art of Field Studies   132
    Special Issues in Family Studies   133

Step 4: Family Follow-Up   135
    Need for Additional Medical Services   135
    Duty to Recontact Research Participants   136
    Maintaining Contact with Participants   137
    Guidelines for Releasing Genetic Information   137
    Genetic Testing of Children   139
    Genetic Discrimination   139
    DNA Banking   141

Future Considerations   142

Appendix   142

References   148

5. Collection of Biological Samples for DNA Analysis   **153**
  Jeffery M. Vance

Establishing Goals of Collection   153

Types of DNA Sample Collection   153
  Venipuncture (Blood)   153
  Buccal Samples   155
  Dried Blood   156
  Tissue   156

DNA Extraction and Processing   157
  Blood   157
  Quantitation   157
  Tissue Culture   159
  Buccal Brushes   160
  Dried Blood Cards   161
  Fixed Tissue   161
  Whole-Genome Amplification   161

Sample Management   162

Informed Consent/Security   164

References   164

6. Methods of Genotyping   **167**
  Jeffery M. Vance

Brief Historical Review of Markers Used for Genotyping   167
  Restriction Fragment Length Polymorphisms   167
  Variable Number of Tandem Repeat Markers   168
  Short Tandem Repeats or Microsatellites   168
  Single-Nucleotide Polymorphisms   168

Sources of Markers   168
  Restriction Fragment Length Polymorphisms   169
  Microsatellites   169
  Single-Nucleotide Polymorphisms   171

PCR and Genotyping   171
  Laboratory and Methodology Optimization   171
  Optimization of Reagents   172
  "I Can't Read a Marker, What Should I Do?"   173

Marker Separation   175
  Manual or Nonsequencer Genotyping   175
  Loading Variants   176
  DNA Pooling and Homozygosity Mapping   177

Detection Methods   178
  Radioactive Methods (^{32}P or ^{33}P)   178
  Silver Stain   178
  Fluorescence   179

SNP Detection   181
  DNA Array or "Chip"   181
  Oligonucleotide Ligation Assay   181
  Fluorescent Polarization   182
  Taqman   182
  Single-Base-Pair Extension   184
  Pyrosequencing   184
  Matrix-Assisted Laser Desorption/Ionization Time-of-Flight
    Spectrometry   184
  Invader and PCR-Invader Assays   184
  Single-Strand Conformational Polymorphism   186
  Denaturing High-Pressure Liquid
    Chromatography   186

Data Management   186
  Objectivity   187
  Genotype Integrity   187
  Scoring   187
  Standards   187
  Quality Control   188

References   189

7. Data Analysis Issues in Expression Profiling   193
  Simon Lin and Michael Hauser

Introduction   193

Serial Analysis of Gene Expression   194
  Analysis of SAGE Libraries   195
  Microarray Analysis   196
  Data Preparation   197
  Expression Data Matrix   198
  Dimension Reduction of Features   198
  Measures of Similarity between Objects   200
  Unsupervised Machine Learning: Clustering   201
  Supervised Machine Learning   204

Data Visualization 207
Other Types of Gene Expression Data Analysis 207
Biological Applications of Expression Profiling 209

References 212

8. Information Management **219**
Carol Haynes and Colette Blach

Information Planning 220
Needs Assessment 220

Information Flow 222

Plan Logical Database Model 223

Hardware and Software Requirements 225
Software Selection 226
System Administration 226
Database Administration 226

Database Implementation 227
Conversion 227
Performance Tuning 228
Data Integrity 228

User Interfaces 231

Security 231
Transmission Security 231
System Security 233
Patient Confidentiality 233

Pedigree Plotting and Data Manipulation
Software 234

Summary 235

9. Quantitative Trait Linkage Analysis **237**
Jason H. Moore

Introduction to Quantitative Traits 237

Genetic Architecture 238

Study Design 240

Haseman–Elston Regression 240

Multipoint IBD Method 242

Variance Component Linkage Analysis 243

Nonparametric Methods 246

Future Directions   247

Summary   249

References   250

10. Advanced Parametric Linkage Analysis   **255**
Silke Schmidt

Two-Point Analysis   256
    Example of LOD Score Calculation and Interpretation   259

Effects of Misspecified Model Parameters in LOD Score Analysis   260
    Impact of Misspecified Disease Allele Frequency   261
    Impact of Misspecified Mode of Inheritance   262
    Impact of Misspecified Disease Penetrances   263
    Impact of Misspecified Marker Allele Frequency   264

Control of Scoring Errors   265

Genetic Heterogeneity   266

Multipoint Analysis   269

Practical Approaches for Model-Based Linkage Analysis
    of Complex Traits   273
    Affecteds-Only Analysis   274
    Maximized Maximum LOD Score   275
    Heterogeneity LOD   275
    MFLINK   276

Summary   277

References   277

11. Nonparametric Linkage Analysis   **283**
Elizabeth R. Hauser, Jonathan Haines, and David E. Goldgar

Introduction   283

Background and Historical Framework   284

Identity by State and Identity by Descent   286

Measures of Familiality   289
    Qualitative Traits   289
    Measuring Genetic Effects in Quantitative Traits   293
    Summary of Basic Concepts   295

Methods for Nonparametric Linkage Analysis   295
    Tests for Linkage Using Affected Sibling Pairs (ASPs)   295
    Methods Incorporating Affected Relative Pairs   301
    Power Analysis and Experimental Design Considerations for
        Qualitative Traits   311

Nonparametric Quantitative Trait Linkage Analysis   314
Power and Sampling Considerations for Mapping
  Quantitative Trait Loci   316

Examples of Application of Sibpair Methods for Mapping
  Complex Traits   318

Additional Considerations in Nonparametric Linkage Analysis   319
  WPC Analysis   319

Software Available for Nonparametric Linkage Analysis   322

Summary   323

References   323

12. Linkage Disequilibrium and Association Analysis   **329**
Eden R. Martin

Introduction   329

Linkage Disequilibrium   330
  Measures of Allelic Association   330
  Causes of Allelic Association   331

Mapping Genes Using Linkage Disequilibrium   334

Tests for Association   335
  Case–Control Tests   335
  Family-Based Tests of Association   340

Analysis of Haplotype Data   345

Association Tests for Quantitative Traits   347

Association and Genomic Screening   347

Special Populations   348

Summary   349

References   349

13. Sample Size and Power   **355**
Yi-Ju Li, Susan Shao, and Marcy Speer

Introduction   355

Power Studies for Linkage Analysis:
  Mendelian Disease   358
  Information Content of Pedigrees   358
  Computer Simulation Methods   359

Definitions for Power Assessments   363

Power Studies for Linkage Analysis: Complex Disease   365
    Discrete Traits   367
    Quantitative Traits   373

Power Studies for Association Analysis   376
    Transmission/Disequilibrium Test for Discrete Traits   378
    Transmission/Disequilibrium Test for Quantitative Traits   380
    Case–Control Study Design   380
    DNA Pooling   381
    Genomic Screening Strategies for Association Studies   381
    Simulation of Linkage and Association Program   382

Summary   383
    Appendix 13.1: Example of Monte Carlo Simulation Assuming
      That Trait and Marker Loci Are Unlinked to Each Other   384
    Appendix 13.2: Example LOD Score Results for Pedigree
      in Figure 13.2   385
    Appendix 13.3: Example of Simulation of Genetic Marker
      Genotypes Conditional on Trait Phenotypes Allowing for
      Complete and Reduced Penetrance   386

References   393

14. Complex Genetic Interactions   **397**
William K. Scott and Joellen M. Schildkraut

Introduction   397

Evidence for Complex Genetic Interactions Genetic
  Heterogeneity   398
    Genetic Heterogeneity   398
    Gene–Gene Interaction (Epistasis)   399
    Gene–Environment Interaction   400

Analytic Approaches to Detection of Complex Interactions   401
    Segregation Analysis   402
    Linkage Analysis   402
    Association Analysis   406
    Potential Biases   414

Conclusion   415

References   415

15. Genomics and Bioinformatics   **423**
Judith E. Stenger and Simon G. Gregory

Introduction   423
    Era of the Genome   423

Mapping the Human Genome   424
   Genetic Mapping   425
   Radiation Hybrid Mapping   427
   Physical Mapping   428
   Public Data Repositories and Genome Browsers 432

Single-Nucleotide Polymorphisms   434
   SNP Discovery   435
   Utilizing SNPs   436
   Computational SNP Resources   437

Model Organisms   438

Identifying Candidate Genes by Genomic Convergence   439

De Novo Annotation of Genes   440
   Software Suites   441
   Online Sequence Analysis Resources   441
   Understanding Molecular Mechanisms of Disease 442
   Assigning Gene Function   442

Looking Beyond Genome Sequence   444
   Other Databases   445

Summary   446

References   448

16. Designing a Study for Identifying. Genes in Complex Traits   455
Jonathan L. Haines and Margaret A. Pericak-Vance

Introduction   455
   Components of a Disease Gene Discovery Study 457
   Define Phenotype   459
   Develop Study Design   460
   Analysis   463
   Follow-Up   464

Keys to a Successful Study   465
   Foster Interaction of Necessary Expertise   465

Develop Careful Study Design   466

References   467

Index   469

■■■■■ FOREWORD

The promise of the Human Genome Project to revolutionize human genetics has been fulfilled. The generation of the consensus sequence, the massive improvements in technology, particularly for sequencing and genotyping, and the current push to characterize single-nucleotide polymorphisms (SNPs) have radically changed the approach toward finding genes that influence human disease. Thousands of simple Mendelian traits have now been mapped, and in the majority of cases the responsible gene has been cloned.

From a historical perspective, the evolution of mapping techniques began with Morton in 1955 with his now classic paper "Sequential Tests for the Detection of Linkage." The method was amenable only to sibships, though with much tedious calculation, large families could be analyzed. The next milestone occurred in 1971, when Elston and Stewart published their efficient algorithm to determine the likelihood of a pedigree, which became the basis of the well-known computer program LIPED written by Ott.

The weak link was the number of markers available, on the order of 30–40 blood groups and serum proteins, which were tedious to genotype and in general were not very polymorphic. The 1980 publication by Botstein and his colleagues, which advocated the use of restriction fragment length polymorphisms (RFLPs) as markers for the construction of genetic linkage maps, ushered in a new era. The pace of human gene mapping increased exponentially. Though RFLPs were a major advance in genotyping, they were overshadowed by the discovery by Weber and May in 1989 of microsatellites, which are both abundant and highly polymorphic. Most recently, the cost and efficiency breakthroughs in high-throughput genotyping of SNPs give us an unparalleled view of the variation in a person's genome.

Geneticists have realized for many years that most of the common disorders that affect humans have a major genetic component. Examples are manic depression and schizophrenia, type 2 diabetes, osteoporosis, and hypertension. With the advent of dense maps, gene mappers felt that they could, in theory at least, unravel the genes responsible for a major component of these disorders. Many hurdles needed to be surmounted, however. The use of large multigeneration families in early linkage studies of bipolar disorder and schizophrenia produced what were later found to be spurious linkages. It soon became clear that new approaches were necessary if mapping genes involved in complex disorders was to be successful. A number of theoretical population geneticists soon rallied to the task, leading to a number of innovative new approaches of mapping genes for complex human diseases. New or modified approaches began appearing in numerous scientific journals.

These approaches have been assimilated into this book. The editors have assembled experts in the field to provide scientists with a comprehensive guide to human disease gene discovery in complex diseases in one volume. Most publications on gene mapping in complex disorders tend to emphasize the analytic techniques. This volume, on the other hand, covers all areas, not just the statistical methodology. These include the overall design, the clinical phenotype and subphenotypes, ascertainment of families and cases, computer software, data analysis, and interpretation. The decision not to include the detailed theoretical background for the linkage analysis will be seen as a boon to most of us, who are mainly interested in the application of the techniques. This volume will, hopefully, allow researchers to avoid the many pitfalls involved in the genetic study of complex disease. All major areas are covered and should be studied before a researcher embarks on such studies. All of us should be grateful to the contributors for providing in a straightforward readable fashion all the key elements involved in finding genes for common/complex diseases.

I take great pleasure in writing a foreword to this revised volume. As one who has been involved in human gene mapping from its early beginnings, it is especially gratifying to see two of my former students produce a comprehensive and up-to-date volume in this rapidly advancing field. They allow me and, hopefully, many others to understand the entire process involved in the mapping of complex human disease genes.

P. MICHAEL CONNEALLY

Distinguished Professor
Department of Molecular and Medical Genetics
Indiana University Medical School
Indianapolis, IN

When we wrote the first edition of this book in 1998, we could not have imagined the incredibly rapid advances that would take place in human genetics over the next seven years. In that time, the human genome sequence has gone from a dream to a reality. Genotyping has gone from relying primarily on microsatellite markers to utilizing single-nucleotide polymorphisms (SNPs) to great effect. Our capacity to generate genotypic data has exploded from hundreds of genotypes per day to hundreds of thousands per day. The HapMap Project has begun, initiating an ongoing effort to characterize linkage disequilibrium across the genome. The tools to characterize whole-genome gene expression through microarrays have emerged, along with a growing ability to characterize the resulting proteins in a systematic way. Bioinformatic tools have changed dramatically as well, racing to keep up with the massive amounts of data we can now generate.

Because of these advances, it is more important than ever to outline the application of all these data and methods to the dissection of human traits. Thus, the goal of this book is the same as in the first edition: to capture the state of the art in the emerging field of gene discovery in common and genetically complex traits. While we cannot hope to provide a comprehensive review of all the background, methods, and designs in the field, we have tried to include the most common and useful approaches currently available. The book has been reorganized and expanded so that we can cover the use of quantitative traits, bioinformatics, and explicit testing for complex interactions in more detail.

It is still important to outline the advantages of identifying genes underlying human traits. The potential benefits of this sort of research are many and include the following factors:

1. Identifying the genes underlying a disorder can provide insight into the pathogenesis of the trait.
2. Discovered genes may serve as direct targets for successful therapeutic intervention.
3. Characterizing the risk of disease or response to intervention related to specific genetic variability will lead to more accurate diagnosis and prognosis. This type of genetically personalized care may be the first widespread application of genetics in medicine.
4. Controlling for known genetic susceptibility improves our ability to identify and characterize additional genes, other risk factors, and gene–gene and gene–environment interactions.

This second edition continues to be modeled on the original four-day course, "Analysis of Complex Human Disease," that many of the authors have helped to teach over the past 12 years. This book is intended to provide a comprehensive overview of gene discovery in common and genetically complex traits without going too deeply into the statistical and experimental details. We hope to provide the interested reader with an understanding of the entire genetic dissection process and a guide to the often-difficult aspect of experimental design. We hope the reader will gain a healthy respect for the process, learning, as we have, that there are as many ways to study common and genetically complex traits as there are genes underlying them.

It would not have been possible to put together this book without substantial help from many individuals, not the least of whom are all our coauthors. We would like to thank Tangela Lauderdale and especially Kate Lewis for their help in putting together and formatting the manuscript. We also thank Luna Han and Thom Moore, our editors at John Wiley, for their patience with the inevitable delays that occurred as we worked to bring our book up to date with the rapid and exciting changes taking place in the field.

JONATHAN L. HAINES, Ph.D.

MARGARET A. PERICAK-VANCE, Ph.D.

CONTRIBUTORS

Allison Ashley-Koch, Section of Medical Genetics, Center for Human Genetics, Duke University Medical Center, Durham, North Carolina

Arthur S. Aylsworth, Division of Genetics and Metabolism, Department of Pediatrics and the Neuroscience Research Center, The School of Medicine, The University of North Carolina at Chapel Hill, Chapel Hill, North Carolina

Amy Baryk Crunch, Center for Human Genetics Research, Vanderbilt University Medical Center, Nashville, Tennessee

Colette Blach, Section of Medical Genetics, Center for Human Genetics, Duke University Medical Center, Durham, North Carolina

Susan Estabrooks Hahn, Section of Medical Genetics, Center for Human Genetics, Duke University Medical Center, Durham, North Carolina

David E. Goldgar, Unit of Genetic Epidemiology, International Agency for Research on Cancer, Lyon, France

Simon G. Gregory, Section of Medical Genetics, Center for Human Genetics, Duke University Medical Center, Durham, North Carolina

Jonathan Haines, Department of Molecular Physiology and Biophysics, Program in Human Genetics, Vanderbilt University School of Medicine, Nashville, Tennessee

Carol Haynes, Section of Medical Genetics, Center for Human Genetics, Duke University Medical Center, Durham, North Carolina

Elizabeth R. Hauser, Section of Medical Genetics, Center for Human Genetics, Duke University Medical Center, Durham, North Carolina

Michael Hauser, Section of Medical Genetics, Center for Human Genetics, Duke University Medical Center, Durham, North Carolina

Yi-Ju Li, Section of Medical Genetics, Center for Human Genetics, Duke University Medical Center, Durham, North Carolina

Simon Lin, Section of Bioinformatics, Robert H. Lurie Comprehensive Cancer Center, Northwestern University, Chicago, Illinois

Eden R. Martin, Section of Medical Genetics, Center for Human Genetics, Duke University Medical Center, Durham, North Carolina

Elizabeth C. Melvin, Section of Medical Genetics, Center for Human Genetics, Duke University Medical Center, Durham, North Carolina

Jason H. Moore, Computational Genetics Laboratory, Dartmouth Medical School, Lebanon New Hampshire

Margaret A. Pericak-Vance, Section of Medical Genetics, Center for Human Genetics, Duke University Medical Center, Durham, North Carolina

Joellen M. Schildkraut, Community Family Medicine, Duke University Comprehensive Cancer Center, Durham, North Carolina

Silke Schmidt, Section of Medical Genetics, Center for Human Genetics, Duke University Medical Center, Durham, North Carolina

William K. Scott, Section of Medical Genetics, Center for Human Genetics, Duke University Medical Center, Durham, North Carolina

Susan Shao, SAS, Cary, North Carolina

Marcy C. Speer, Section of Medical Genetics, Center for Human Genetics, Duke University Medical Center, Durham, North Carolina

Judith E. Stenger, Section of Medical Genetics, Center for Human Genetics, Duke University Medical Center, Durham, North Carolina

Jeffery M. Vance, Division of Neurology, Department of Medicine, Duke University Medical Center, Durham, North Carolina

Chantelle Wolpert, Section of Medical Genetics, Center for Human Genetics, Duke University Medical Center, Durham, North Carolina

CHAPTER 1

Basic Concepts in Genetics and Linkage Analysis

ELIZABETH C. MELVIN and MARCY C. SPEER

This chapter explores the underpinnings for observational and experimental genetics. Concepts ranging from laws of Mendelian inheritance through molecular and chromosomal aspects of deoxyribonucleic acid (DNA) structure and function are defined; their ultimate utilization in linkage mapping of simple Mendelian disease and common and genetically complex disease is presented. The chapter concludes with clinical examples of the various types of DNA mutation and their implications for human disease.

INTRODUCTION

For centuries, the hereditary basis of human disease has fascinated both scientists and the general public. The Talmud gives behavioral proscriptions regarding circumcision in sons born after a male sibling who died of a bleeding disorder, suggesting the ancient Hebrews knew of hemophilia; nursing students in Britain in the 1600s tracked the recurrence of spina bifida in families; and questions as to whether Abraham Lincoln and certain celebrity sports figures had Marfan syndrome sometimes arise in casual dinner conversation.

In many respects, the study of the genetic factors in disease today remains, as it has for centuries, dependent on careful description of human pedigree data in which patterns of transmission from parent to offspring are characterized. For example, Gregor Mendel provided the groundwork for the study of human genetics by carefully constructing quantified observations of the frequency of variable characteristics in the pea plant. The importance of detailed pedigree analysis was exemplified recently in the delineation of patterns of transmission of the fragile X syndrome: the most common genetic cause of mental retardation. Careful documentation of pedigrees

Genetic Analysis of Complex Diseases, Second Edition, Edited by Jonathan L. Haines and Margaret Pericak-Vance

from families with more than one person with fragile X syndrome led to description of the aptly named Sherman paradox (Sherman et al., 1985), in which different recurrence risks for relatives of various types were described. From this, the complicated workings of unstable DNA harbored in expanding trinucleotide repeats were later elucidated (e.g., Fu et al., 1991; Burke et al., 1994).

HISTORICAL CONTRIBUTIONS

Segregation and Linkage Analysis

In 1865, Gregor Mendel, an Austrian monk (Fig. 1.1), published his findings on the inheritance of a series of traits in the pea plant, including seed texture (round or wrinkled), seed color (yellow or green), and plant height (tall or short). He described three properties of heritable factors that explained his quantified observations of these scorable (discontinuous or qualitative) traits. The first property was unit inheritance, which is now considered the basis for defining the gene. He hypothesized that a factor was transmitted from parent to offspring in an unchanged form. Such a factor produced an observable trait. This idea represented a radical departure

Figure 1.1. Mendel's garden at the old monastery at Brno. On the right is the door to the Mendel museum, which contains exhibits celebrating his life and research. It is located in what used to be the monastery refectory (dining hall). Mendel's apartment window(s) overlooked the garden. The garden is planted in ornamental red and white flowers, with the first two (farthest from the camera) labeled P (parental generation), the next single red one is labeled F_1, the next row of four (three red and one white) is labeled F_2, and the next nine labeled F_3. At the far left, under a tree, is a large statue of Mendel that once stood in the town square just outside the garden gates. The square is called Mendelovaplatz. (Photo, taken in 1986, courtesy of Arthur S. Aylsworth.)

from the scientific thinking at the time, which suggested parental characteristics were blended in the offspring.

Mendel also described the behavior of factors controlling observable traits, such as flower color or plant height as a single unit. He proposed that these factors were transmitted ("segregated") independently and with equal frequency to germ cells (egg and sperm), and this observation is referred to as Mendel's first law. In experiments for a variety of traits, Mendel crossbred the offspring (the F_1 generation) of two phenotypically different, pure-breeding parental strains with one another. The offspring of these matings (the F_2 generation) expressed the grandparental traits in a 3 : 1 ratio (Fig. 1.2). Serendipitously, some of the traits Mendel had chosen to study were simple dominant traits, such that the presence of one factor was sufficient to express the trait, with the other trait being recessive (expressed only in the absence of the dominant factor); later work showed that the factor defining a characteristic need not be dominant to the other. For instance, if each factor contributes to the trait equally, as in *codominant* systems, then three different classes of offspring from the same cross described above are possible: the two parental traits and a third intermediate trait. These classes occur in proportions parental to intermediate to parental of 1 : 2 : 1.

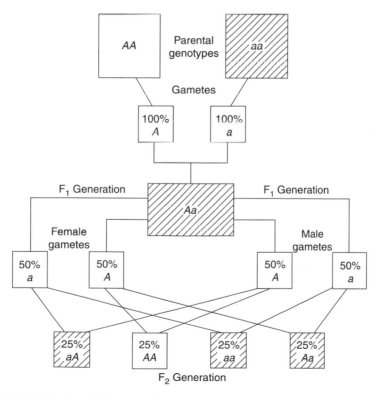

Figure 1.2. Principles of Mendel's first law of segregation of heritable characters for a dominant trait.

Mendel extended his observations from the transmission of a single trait from parent to offspring to the interaction of two traits. Mendel's law of independent assortment, also referred to as Mendel's second law, predicts that factors controlling different traits will segregate to offspring independently from one another. For instance, seed texture will segregate to offspring independently of seed color. In one experiment Mendel crossed pure-breeding round, green seed plants to pure-breeding wrinkled, yellow seed plants. Since round is dominant to wrinkled and green is dominant to yellow, the resulting seeds (F_1 generation) yielded entirely round, green seed plants. These F_1 plants were then crossed with one another, confirming the predictions from his theory of independent assortment: The seed (offspring) types in the F_2 generation were nine round/green, three round/yellow and three wrinkled/green, one wrinkled/yellow (Fig. 1.3). Any observed departure from these expected ratios using identical parental crossing strategies suggests the two traits fail to segregate independently and may be physically linked. We will

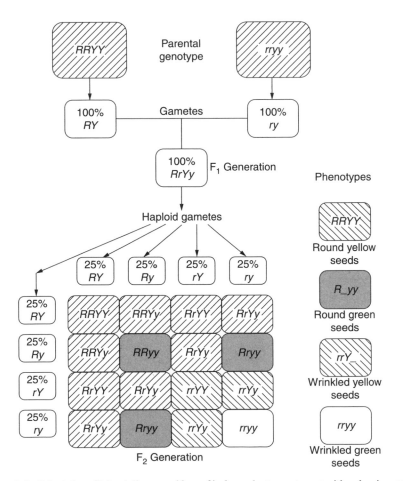

Figure 1.3. Principles of Mendel's second law of independent assortment with a dominant trait.

describe later how the failure of two traits to segregate independently can be exploited to find genes and diagnose genetic disorders.

Mendel's observations remained largely obscure until the early 1900s, when they were independently rediscovered by plant geneticists and by Sir Archibald Garrod, who was studying the human hereditary disorder alkaptonuria (Garrod, 1902). Garrod's work provided the basis of our understanding of alleles and genetic linkage. Mendel's observations remain one of the most important contributions of critical descriptive science in the history of genetics.

Hardy–Weinberg Equilibrium

Another historical landmark in genetics occurred in the early 1900s as evolutionary biologists attempted to explain why the frequency of a dominant trait or disease in the population did not increase until, over many generations, everyone in the population was affected. The answer to this question was provided independently by Hardy (1908) and Weinberg (1908), who predicted the behavior of alleles in a population using the binomial theorem. Their proof, now called the Hardy–Weinberg theorem, shows that in a large, randomly mating population, trait (genotypic) frequencies for autosomal traits will achieve and remain in a state of equilibrium after one generation. Several evolutionary forces can alter equilibrium frequencies, including selection for or against a phenotype, migration into or out of a population, new mutation, and genetic drift. For a sex-linked trait, the attainment of equilibrium will require more than one generation.

Specifically, in a two-allele autosomal system with alleles A and a (having frequencies p and q, respectively), $p + q = 1$. The Hardy–Weinberg theorem predicts the frequencies of genotypes AA, Aa, and aa are p^2, $2pq$, and q^2. Various manipulations of these algebraic formulas allow many useful calculations, such as carrier frequencies of diseases, disease prevalence, and gross estimates of penetrance. Some example applications of the Hardy–Weinberg theorem are shown in Table 1.1.

DNA, GENES, AND CHROMOSOMES

Structure of DNA

When Mendel described his genetic factor, he did not know what the underlying biological factor was. It was 90 years later when the actual genetic molecule was identified. Mendel's fundamental unit of inheritance is termed the gene. A gene contains the information for synthesizing proteins necessary for human development, cellular and organ structure, and biological function. Deoxyribonucleic acid is the molecule that comprises the gene and encodes information for synthesizing both proteins and RNA (ribonucleic acid). Deoxyribonucleic acid is present in the nucleus of virtually every cell in the body. It is made up of three components: a sugar, a phosphate, and a base. In DNA, the sugar is deoxyribose, while in RNA the sugar is ribose. The four bases in DNA are the pyrimidines adenine (A) and

TABLE 1.1. Useful Applications of Hardy–Weinberg Theory

Recall that $p + q = 1$ and $p^2 + 2pq = q^2 = 1$.

Example 1. Cystic fibrosis (CF), an autosomal recessive disease, has an incidence of $\frac{1}{400}$. What is the frequency of CF carriers in the general population?

The frequency of the CF allele (q) is calculated as $\sqrt{\frac{1}{400}} = \frac{1}{20}$.

The frequency of CF carriers is calculated as $2pq = 2\left(\frac{19}{20}\right)\left(\frac{1}{20}\right) = 0.095$.

Example 2. The frequency of the allele (q) for an autosomal dominant disorder is $\frac{1}{100}$. What is the frequency of the disease itself in the population?

Since the frequency of the disease allele is $\frac{1}{100}$, the frequency of the normal allele $(p) = 1 - \frac{1}{100} = \frac{99}{100}$.

Since the disease is dominant, both heterozygous carriers and homozygous individuals are affected with the disease:

$$q^2 + 2pq = \left(\frac{1}{100}\right)^2 + 2\left(\frac{99}{100}\right)\left(\frac{1}{100}\right) = 0.0199$$

Example 3. An autosomal dominant disorder with incomplete penetrate (f) has a population prevalence of $\frac{16}{1000}$. If the allele frequency for the normal allele p is 0.99, what is the estimated penetrance of the disease allele?

Since $p = 0.99$, then $q = 0.01$.

As in the Example 2, both heterozygous and homozygous gene carriers are affected (assuming no difference in penetrance) between homozygotes and heterozygotes. Therefore,

$$f(q^2) + f(2pq) = 0.016$$

$$f(q^2 + 2pq) = 0.016$$

$$f(0.0199) = 0.016$$

$$f = 0.804$$

---

guanine (G) and the purines cytosine (C) and thymine (T). A DNA sequence is often described as an ordered list of bases, each represented by the first letter of its name (e.g., ACTGAAACTTGATT). A nucleoside is a molecule made of a base and a sugar; a nucleotide is made by adding a phosphate to a nucleoside. A single strand of DNA is a polynucleotide, consisting of nucleotides bonded together.

A single strand of DNA is, however, unstable. The double-helical nature of DNA, which confers stability to the molecule, was hypothesized in 1953 by J. D. Watson and F. H. C. Crick. Their cohesive theory of the structure of DNA accounted for some of the previously identified properties of DNA. A fascinating account of the internecine struggles in science surrounding this discovery was provided later by Watson (1968).

Specifically, Watson and Crick postulated that DNA is a double-stranded structure and that the two strands of DNA are arranged in an antiparallel orientation. In the central portion of the molecule, hydrogen bonds link a base with its complement, such that a purine always bonds with a pyrimidine (e.g., adenine always bonds with thymine and guanine always bonds with cytosine). The conformation of the

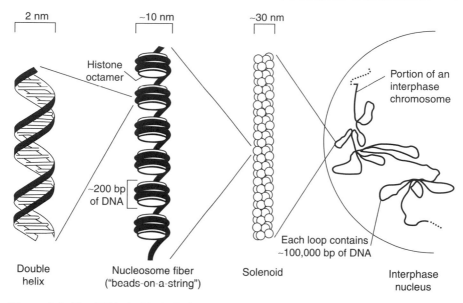

Figure 1.4. The DNA double helix is packaged and condensed in several different forms. (Reprinted by permission from Thompson et al., eds., *Genetics in Medicine*, 5th ed., W. B. Saunders Company, Philadelphia, PA, 1991.)

resultant molecule is the double helix, which undergoes several levels of compacting to fit within the cell (Fig. 1.4).

The sequence of DNA bases represents a code for synthesizing proteins. The fundamental unit of this genetic code is termed a codon, which consists of three nucleotides. Since there are four different nucleotides (one made with each of the four bases) and a codon is made of three nucleotides, there are $4^3 = 64$ different codons. However, these 64 codons specify only 20 different amino acids, which are the building blocks of proteins. Thus, the genetic code is degenerate: Different codons may code for the same amino acid. In addition, some codons act as punctuation. For instance, one specific codon in a string of DNA signals the molecular code "interpreter" to start, and then the reading of the DNA strand proceeds in three base-pair chunks; several other specific codons signal the reading process to cease. These reading signals are called start and stop codons, respectively.

Not all of the DNA in a cell actually codes for a protein product; in fact, the vast majority of the DNA sequence does not carry the information for protein formation. Within a gene, exons are the portions utilized (transcribed) to make proteins. Introns are the sequences between exons that are not transcribed. The size and number of introns and exons vary dramatically between genes (Fig. 1.5).

The central dogma of genetics is that the utilization of DNA is unidirectional such that DNA → RNA → protein (Fig. 1.6). Specifically genes are encoded in the DNA. Then, in the nucleus, messenger RNA (mRNA) is transcribed (produced) from the DNA. Subsequently mRNA undergoes a series of posttranscriptional modifications: The introns are spliced out, a cap is added at the 5′ end of the

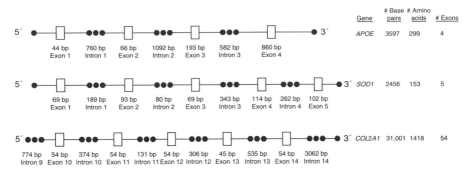

Gene	# Base pairs	# Amino acids	# Exons
APOE	3597	299	4
SOD1	2456	153	5
COL2A1	31,001	1418	54

Figure 1.5. Intron and exon sizes vary between genes.

molecule, and a string of adenylate residues (poly-A tail) is added to the 3′ end. The mRNA is then transported out of the nucleus into the cytoplasm, where it is translated into protein by means of cellular machinery called the *ribosomes*. Many excellent resources describe the very complicated process of transcription and translation (e.g., Strachan and Read, 1996).

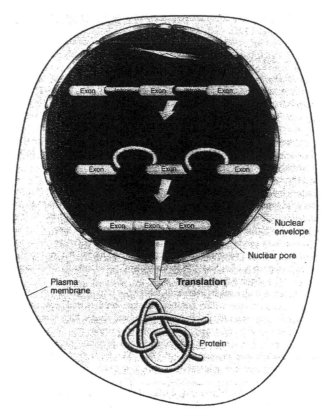

Figure 1.6. Central dogma of genetics: DNA → RNA → protein. (Reprinted by permission from Jorde et al., eds., *Medical Genetics*, C. W. Mosby, St. Louis, MO, 1995.)

Genes and Alleles

The physical site or location of a gene is called its locus. At any particular gene site, or locus, there can exist different forms of the gene, called alleles. Except on the sex chromosomes of males, an individual has two alleles at each locus. These alleles are analogous to the factors identified in the 1800s by Mendel. Homozygosity is defined as the presence of two alleles that are indistinguishable from one another. In heterozygotes, the two alleles can be distinguished from another. Males with a normal chromosome complement are hemizygous for all X chromosome loci, since they have only one copy of the X chromosome.

The difference between two alleles may be as subtle as a single base-pair change, such as the thymine-to-alanine substitution that alters the B chain of hemoglobin A from its wild type to its hemoglobin sickle cell state. Some base-pair changes have no deleterious effect on the function of the gene; nevertheless, these functionally neutral changes in the DNA still represent different forms of a gene. Alternatively, allelic differences can be as extensive as large, multicodon deletions, such as those observed in Duchenne muscular dystrophy. Any locus having two or more alleles, each with a frequency of at least 1% in the general population, is considered to be polymorphic (i.e., having many forms).

Differences in alleles can be scored via laboratory testing. The ability to score allele differences accurately within families, between families, and between laboratories is critically important for linkage analysis in both simple Mendelian and genetically complex common disorders. Allele scoring strategies may be as simple as the presence (+) or absence (−) of a deletion or point mutation or as complicated as assessing the allele size in base pairs of DNA. The latter application is common when highly polymorphic microsatellite repeat markers are used in linkage analysis.

A measure frequently utilized to quantitate the extent of polymorphism of a gene or marker system is the heterozygosity value H, which is calculated as

$$H = 1.0 - \sum_{i=1}^{n} p_i^2$$

where n is the number of alleles at the locus and p is the frequency of the ith allele at the locus. For example, the heterozygosity value of a three-allele marker with frequencies of allele $1 = 0.25$, allele $2 = 0.30$, and allele $3 = 0.45$ is calculated as $1 - [(0.25 \times 0.25) + (0.30 \times 0.30) + (0.45 \times 0.45)] = 0.645$. This measure can be interpreted as the probability that an individual randomly selected from the general population will be heterozygous at the locus. Polymorphic loci or genetic markers are critically important to linkage analysis because they allow each individual in a family a high probability of being heterozygous for the locus. The investigator may then be able to deduce the parental origin of each allele and identify recombinant and nonrecombinant gametes, which is important for linkage analysis, as described later.

In general, genetic marker systems with high heterozygosities are desirable because there is a high probability that an individual in a genetic linkage analysis will be heterozygous for that marker and thus be likely to contribute information to a genetic linkage study. However, recent emphasis has been on utilizing single-nucleotide polymorphisms (SNPs) as genetic markers, despite their limited heterozygosity, because of their ubiquity in the genome.

The PIC (or polymorphism information content) is a modification of the heterozygosity measure that subtracts from the H value an additional probability that an individual in a linkage analysis does not contribute information to the study. Formally, the PIC value subtracts from the H value the probability of obtaining a heterozygous offspring from an intercross mating (i.e., when a mother whose marker genotype is $\frac{1}{3}$ and a father whose marker genotype is $\frac{1}{3}$ have an offspring whose marker genotype is also $\frac{1}{3}$, it is impossible to tell whether the *1* allele came from the mother or from the father; thus, this combination contributes essentially no information to the linkage analysis).

Genes and Chromosomes

Genes are organized as linear structures called chromosomes, with many thousands of genes on each chromosome. Each chromosome has distinguishable sites that aid in cell division and in the maintenance of chromosome integrity. The centromere is visualized as the central constriction on a chromosome and it separates the p (short) and the q (long) arms from one another. The centromere enables correct segregation of the duplicated chromosomal material during meiosis and mitosis. Telomeres are present at both ends of the chromosome and are required for stability of the chromosomal unit.

Using appropriate staining techniques, the chromosomes in a cell can be analyzed under the microscope following cell culture and the arrest of cell division at metaphase (when the chromosomes have duplicated and condensed). At this stage of the cell cycle, a chromosome has two double-stranded DNA molecules. Together, the strands are called *sister chromatids*. The sister chromatids are held together by the centromere. Photographs are magnified and the chromosomes are arranged into a karyotype. The normal human chromosome complement consists of 46 chromosomes arranged in 23 pairs, with one member of each pair inherited from each parent (Fig. 1.7). The first 22 pairs, called autosomes, are arranged according to size and are the same in males and females. The pair of sex chromosomes generally predicts an individual's gender. Most females have two X chromosomes, while males have one X inherited from the mother and one Y chromosome inherited from the father. Therefore, the gender of an individual is determined by the father.

Because two copies of each chromosome are present in a normal somatic (body) cell, the human organism is diploid. In contrast, egg and sperm cells have haploid chromosomal complements, consisting of a single member of each chromosome pair. The correct number of chromosomes in the normal human cell was finally established in 1956, three years after the double-helical structure of DNA was described, when Tjio and Levan (1956) demonstrated unequivocally that the chromosomal complement is 46.

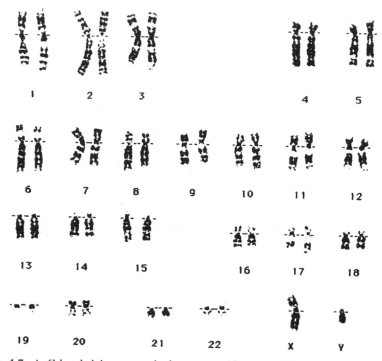

Figure 1.7. A G-banded human male karyotype. (Courtesy of Mazin Qumsiyeh, Duke University Medical Center, Durham, NC.)

Regions of chromosomes are defined by patterns of alternating light and dark regions called bands, which become apparent after a chemical treatment has been applied. One of the most common types of banding process, called Giemsa or G banding, involves digesting the chromosomes with trypsin and then staining with a Giemsa dye. G banding identifies late-replicating regions of DNA; these are the dark bands. Other chemical processes will produce different banding patterns and identify unique types of DNA.

A specific genetic locus can then be defined quite precisely along a chromosome, such as the gene *FRAXA* (fragile X syndrome), which is located on the X chromosome at band q27.3. Alternatively, its localization may be specified as an interval flanked by two genetic markers. Any two loci that occur on the same chromosome are considered to be syntenic or physically linked. Two genes may be syntenic yet far enough apart on the chromosome to segregate independently from one another. Thus, two syntenic genes may be genetically unlinked. Two syntenic genes that fail to be transmitted to gametes independently from one another are genetically linked (Fig. 1.8). The location of two loci on the same arm of the chromosome is specified by their positions relative to each other and to the centromere. The gene closer to the centromere is termed centromeric or proximal to the other; similarly, the gene further from the centromere is distal or telomeric to the other (Fig. 1.9).

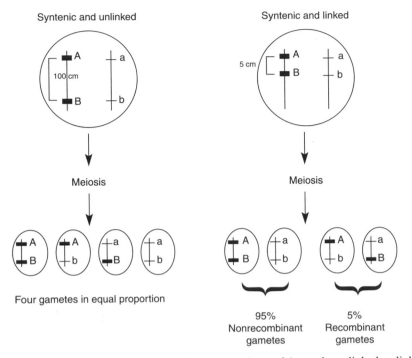

Figure 1.8. Genes that are on the same chromosome (syntenic) may be unlinked or linked.

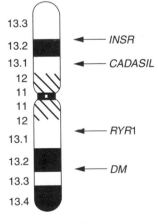

Chromosome 19

Figure 1.9. The myotonic dystrophy (*DM*) and insulin receptor (*INSR*) genes are distal (telomeric) to the ryanodine receptor 1 and *CADASIL*, respectively; *RYR1* and *CADASIL* are proximal (centromeric) to *DM* and *INSR*, respectively.

The X and Y chromosomes are very different in their genetic composition except for an area at the distal end of the p arm of each, termed the pseudoautosomal region. The pseudoautosomal region behaves similarly to the autosomes during meiosis by allowing for segregation of the sex chromosomes. Just proximal to the pseudoautosomal region on the Y chromosome are the SRY (sex-determining region on the Y) and TDF (testes-determining factor) genes, which are critical for the normal development of male reproductive organs. When crossing over extends past the boundary of the pseudoautosomal region and includes one or both of these genes, sexual development will most likely be adversely affected. For instance, the rare occurrences of chromosomally XX males and XY females are due to such aberrant crossing over.

INHERITANCE PATTERNS IN MENDELIAN DISEASE

Alleles whose loci are on an autosome can be transmitted in a dominant or recessive (or codominant) fashion; similarly, alleles having loci on the X chromosome are expressed and transmitted as either X-linked recessive or X-linked dominant disorders. These well-known patterns of inheritance are shown in Figure 1.10.

The hallmark of dominant inheritance, regardless of whether the underlying gene is located on an autosome or on an X chromosome, is that only a single allele is necessary for expression of the phenotype. In an autosomal recessive trait, two copies of a trait allele must be present for it to be expressed. In most cases, it is correctly assumed that each parent of an offspring with the trait carries a recessive

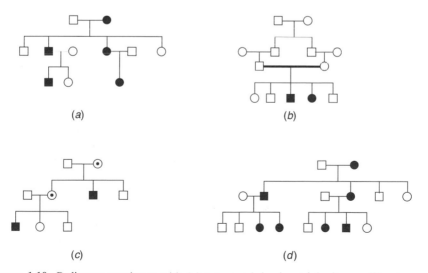

Figure 1.10. Pedigrees consistent with (*a*) autosomal dominant inheritance, (*b*) autosomal recessive inheritance, (*c*) X-linked recessive inheritance, and (*d*) X-linked dominant inheritance. Here and elsewhere squares indicate males, circles indicate females, open symbols indicate unaffected individuals, and solid symbols indicate affected individuals.

allele at the locus. Rarely, an individual expressing an autosomal recessive disorder has inherited both abnormal recessive alleles from one parent; this phenomenon is called uniparental disomy.

To express an X-linked recessive disorder, a male needs only one abnormal allele; a female usually needs two abnormal alleles (one on each X chromosome) and this is generally rare. Thus, mostly males are affected. For X-linked dominant traits, both men and women can be affected since only one copy of the trait allele is necessary for phenotypic expression. Because females have two X chromosomes, twice as many females are affected as males.

Alleles located on the Y chromosome are transmitted from affected males to all sons, and in each case the son's Y-linked phenotype will be identical to that of the father; daughters of males with a Y-linked trait will not inherit the trait, since they receive their father's X chromosome. Very few expressed genes have been localized to the Y chromosome. Characteristics of each of the different Mendelian inheritance patterns are summarized in Table 1.2.

GENETIC CHANGES ASSOCIATED WITH DISEASE/TRAIT PHENOTYPES

Alterations or mutations in the genetic code can be neutral, beneficial, or deleterious. Changes in the genetic code can lead to trait and/or disease phenotypes; the pathology can be the result of either loss or gain of function of the gene product. Such changes can occur in a number of different ways.

Point Mutations

A point mutation is defined as an alteration in a single base pair in a stretch of DNA, thereby changing the 3-bp codon. Since the genetic code is degenerate, many such changes do not necessarily alter the resulting amino acid; however, if the single-base-pair change leads to the substitution of one amino acid for another, the result can be devastating. Point mutations can be classified as transition mutations (a purine → purine or a pyrimidine → pyrimidine) or as the less common transversion mutations (purine → pyrimidine or a pyrimidine → purine). In general, transitions are less likely than transversion mutations to change the resulting amino acid. Five effects of point mutations have been defined: Synonymous or silent mutations are single-base-pair changes in the DNA that do not affect the resultant amino acid; nonsense mutations result in a premature stop codon, leading to a polypeptide of reduced length; missense mutations lead to the substitution of one amino acid for another; splice site mutations affect the correct processing of the mRNA strand by eliminating a signal for the excision of an intron; and mutations in regulatory genes alter the amount of material produced. Several examples of point mutations in human diseases are discussed in the sections that follow.

TABLE 1.2. Hallmarks of Mendelian Inheritance Patterns of Different Types

Inheritance Pattern	Examples	Gender Differences in Proportion of Affecteds?	Transmission Features	Recurrence Risks	Prevalence in Population	Other Critical Features
Autosomal dominant	Marfan syndrome; neurofibromatosis; myotonic dystrophy	No	Transmitted from affected parent to affected offspring (vertical transmission); male-to-male transmission	For each offspring of affected parent, risk to child to inherit disease gene is 50%	$p^2 + 2pq$	Reduced penetrance frequent; for a "true dominant," individuals heterozygous for trait allele are no more severe than individuals homozygous for trait allele
Autosomal recessive	Sickle cell anemia; cystic fibrosis	No	Carrier parents generally unaffected	For parents who have one affected child, risk for each subsequent child is 25%	q^2	Consanguinity frequent
Sex-linked recessive	Duchenne muscular dystrophy; hemophilia	Males more frequently affected; carrier females generally unaffected; rare cases of nonrandom X inactivation can lead to affected females	Gene transmitted from unaffected carrier mother to affected son; no male-to-male transmission	Carrier mother has 25% chance to have affected son and 25% chance to have carrier daughter; all daughters of affected males are carriers and no sons of affected males are affected	Females: q^2; males: q	Females affected in rare cases of nonrandom X inactivation

(continued)

TABLE 1.2. *Continued*

Inheritance Pattern	Examples	Gender Differences in Proportion of Affecteds?	Transmission Features	Recurrence Risks	Prevalence in Population	Other Critical Features
Sex-linked dominant	Hypophosphatemic rickets; fragile X syndrome	No	Vertical transmission from mothers to both sons and daughters; fathers transmit to daughters only; no male-to-male transmission	50% of offspring of affected mothers are affected (unless mother is homozygous for disease allele); all daughters of affected males are affected and no sons of affected males are affected	Females: $p^2 + 2pq$; males: p	Females affected three times more frequently than males
Y linked	Genes *SRY* and *TDF*, important in sex determination, are on the Y chromosome; no known diseases are located on Y	Yes; only males would express trait	Exclusively male-to-male transmission	All sons of affected males are affected; no daughters of affected males are affected	Females: 0; males: q	Male-determining genes are located just proximal to pseudoautosomal region on Y chromosome; faulty recombination in pseudoautosomal region can lead to errors in sex determination
Autosomal codominant	MN blood group; microsatellite repeat markers	No	Each allele confers measurable component to phenotype	Varies according to mating type	Genotypes expected to occur in Hardy–Weinberg proportions of p^2, $2pq$, and q^2	

Amyotrophic Lateral Sclerosis. Approximately 10–15% of amyotrophic lateral sclerosis (ALS) patients have a positive family history consistent with autosomal dominant inheritance. This rapidly progressive neurodegenerative disorder has an average age of onset in the mid-40s and is usually fatal within a few years after onset. The gene responsible for about 15–20% of familial ALS cases has been identified as the cytosolic form of superoxide dismutase (Cu, Zn SOD) at 21q22.1. To date, 38 different point mutations have been identified. A recent study of clinical correlations associated with different mutations within the *SOD1* gene (Juneja et al., 1997) found evidence for a significantly faster rate of progression (1.0 ± 0.4 year vs. 5.1 ± 5.1 years) in patients with the most common mutation, a transition mutation in which a thymine at codon 4 is substituted for a cytosine. This substitution changes the resultant amino acid from an alanine to a valine.

Sickle Cell Anemia. Sickle cell anemia, an autosomal recessive disorder with a carrier frequency in African Americans of approximately $\frac{1}{12}$, is a classic example of a point mutation leading to disease. Affected patients have the familiar phenotype of chronic anemia, sickle cell crises, and debilitating pain. Sickle cell anemia results from a single nucleotide substitution of an adenine to a thymine at position 6 in the B chain of hemoglobin. This changes the resultant amino acid from glutamine to valine. Interestingly, the carrier state for sickle cell may lead to a selective advantage in certain environments: Carriers have a resistance to malaria that is useful in tropical climes.

Achondroplasia. Achondroplasia, the most common type of short-limbed dwarfism, is an autosomal dominant disorder. About 85% of cases are the result of a new mutation. It has been observed that the rate of new dominant mutations increases with advancing paternal age (Penrose, 1955; Stoll et al., 1982). Achondroplasia is now known to result from mutations in the fibroblast growth receptor 3 gene (*FGFR3*), located on chromosome 4p16.3. Interestingly, over 95% of the mutations are the identical G-to-A transition at nucleotide 1138 on the paternal allele (Rousseau et al., 1994; Shiang et al., 1994; Bellus et al., 1995). Other mutations in *FGFR3* are also responsible for hypochondroplasia and thanatophoric dysplasia, types of dwarfism that are clinically distinct from achondroplasia.

Deletion/Insertion Mutations

Another class of mutations involves the deletion or insertion of DNA into an existing sequence. Deletions or insertions may be as small as 1 bp or they may involve one or many exons or even the entire gene. Even single-base-pair deletions or insertions can have devastating effects, frequently by altering the reading frame of the DNA strand.

Neurofibromatosis. Neurofibromatosis type 1 (NF1) is an autosomal dominant disorder with variable expression; penetrance of the disorder is high, and some clinicians consider penetrance of this gene to be complete. The most common phenotypic manifestations are multiple café-au-lait spots and peripheral neurofibromatous skin tumors. Approximately 50% of all cases are due to new mutations, most frequently on the paternally inherited allele.

The gene for NF1, coding for a protein called neurofibromin, is located at chromosome 17q11.2 and has been cloned. Its function is thought to involve tumor suppression (DeClue et al., 1992). To date, only a fraction of the mutations responsible for NF1 have been identified. Those identified include entire deletions of the gene (Wu et al., 1995), insertion mutations, and small and large deletions. Single-base-pair mutations leading to premature stop codons (Valero et al., 1994) and deletions, both leading to the production of a truncated protein, account for the majority of NF1 mutations.

Duchenne and Becker Forms of Muscular Dystrophy. Duchenne muscular dystrophy (DMD) is a severe, childhood-onset X-linked muscular dystrophy; Becker muscular dystrophy (BMD) is its allelic, clinically milder variant. Boys with DMD develop normally for the first few years of life, after which rapidly progressive muscle deterioration becomes obvious. Affected males lose the ability to walk by age 10–12 years. The eventual loss of muscle strength in the cardiac and respiratory muscles leads to death in early adulthood. The gene coding for the protein dystrophin, which is abnormal in DMD/BMD, has been cloned (Koenig et al., 1988). Approximately two-thirds of mutations in this very large gene have been identified; the majority are deletions and duplications, although point mutations have also been identified. Correlations of the clinical phenotype with the molecular mutation have been complicated in DMD/BMD. The general deletions and point mutations leading to alterations in the reading frame of the DNA molecule (frame-shift mutations) are more severe than those that do not alter the reading frame (in-frame mutations).

Cystic Fibrosis. Cystic fibrosis (CF), an autosomal recessive disorder, is the most common hereditary disease among Caucasians, with a carrier frequency of between $\frac{1}{20}$ and $\frac{1}{30}$. The function of the pancreas, lungs, and sweat glands, among other organ systems, is affected. In American Caucasians, a single mutation called ΔF508 accounts for about 70% of the abnormal CF alleles. Three base pairs (codon 508) are deleted, and the resulting amino acid sequence is missing a phenylalanine. Over 900 other deleterious mutations have been identified throughout the world. The frequency of specific mutations differs among populations.

Novel Mechanisms of Mutation: Unstable DNA and Trinucleotide Repeats

Dynamic mutations, or unstable DNA, have received considerable attention of late. Some loci of the genome have variable numbers of dinucleotide or trinucleotide repeats. Most are not associated with expressed genes but can be exploited as markers, since they are highly polymorphic. A few loci with trinucleotide repeats are near or within genes, and by expansion beyond a certain threshold, these disrupt gene expression and cause disease. To date, 11 disorders (7 autosomal and 4 X linked) have been shown to be the result of expansion of these unstable triplet repeats (Table 1.3).

The phenotypes of myotonic dystrophy, Huntington disease, and Machado–Joseph disease, among others, are associated with anticipation, a clinical phenomenon in which disease severity worsens in each successive generation (see also Chapter 2). Because disease expression can be quite variable and difficult to measure, age of onset is frequently utilized as an analogue of severity, and anticipation is then observed as a decreasing mean age of onset with each passing generation. Since the discovery that expanding trinucleotide repeats may explain anticipation, investigators have reported clinical evidence for anticipation in many various disorders, including bipolar affective disorder (McInnis et al., 1993), limb–girdle muscular dystrophy (Speer et al., 1998), familial spastic paraplegia (Raskind et al., 1997; Scott et al., 1997), and facioscapulohumeral muscular dystrophy (Tawil et al., 1996; Zatz et al., 1995). None of these disorders is proven to be caused by trinucleotide repeat expansions, and elucidation of their underlying defect will shed additional light on the phenomenon of anticipation.

SUSCEPTIBILITY VERSUS CAUSATIVE GENES

As the study of common and genetically complex human diseases identifies the significant contribution of heredity in their development, it is likely that more genes or genetic risk factors will be found to affect susceptibility to disease rather than the more traditionally considered causative genes. Historically, one of the most widely investigated examples of susceptibility loci is the human leukocyte antigen (HLA) system on the p arm of chromosome 6. Specific HLA antigens have been associated with various human diseases; for instance, the Bw47 antigen confers an 80–150-fold increased risk for congenital adrenal hyperplasia; the B27 antigen confers an 80–100-fold increased risk for ankylosing spondylitis; and the DR2 antigen confers a 30–100-fold increased risk for narcolepsy, a 3-fold increased risk for systemic lupus erythematosus, and a 4-fold increased risk for multiple sclerosis.

A recent and well-characterized example of a susceptibility locus is that of the apolipoprotein E (*APOE*) gene and Alzheimer's disease (AD). The *APOE* gene on chromosome 19 has three different alleles, scored as *2*, *3*, and *4*, which occur with frequency 6, 78, and 16% in most European populations, respectively (e.g., Saunders et al., 1993). These alleles differ in their DNA sequence by only one base at codons 112 and/or 158 (Fig. 1.11). The *APOE 4* allele increases risk and decreases age of onset in familial and sporadic late-onset AD and early-onset sporadic AD. The *2* allele has been shown to be protective to some extent for risk to develop AD (Corder et al., 1994, 1995a,b; Farrer et al., 1997). Interestingly, the *4* allele has been shown to exist at lower frequency in the Indiana Amish, at least partially explaining the decreased frequency of AD in this inbred population (Pericak-Vance et al., 1996). It is important to note that for *APOE* and AD, the *4* allele is not by itself sufficient or necessary for the development of AD but has

TABLE 1.3. Salient Features of Known Human Trinucleotide Repeat Diseases

Condition	Gene Symbol	Chromosome Location	Repeat Type	Repeat Localization	Repeat Number Abnormal Range	Clinical Features
Fragile X syndrome	*FRAXA*	Xq27.3	CGG	5′ Untranslated region (premutations in range of 52–200)	200–1000	Moderate to severe mental retardation, macroorchidism, large ears, and prominent jaw; *FRAXA* accounts for about one-half of all X-linked mental retardation
Fragile site mental retardation-2	*FRAXE*	Xq28	GCC	?	200–1000	Similar to fragile X syndrome phenotypically; cytogenetic evidence for Xq fragile site; negative for expansion in *FRAXA*
Fragile site F	*FRAXF*	Xq28	(GCCGTC)$_n$ (GCC)$_n$	?	300–500	Cytogenetic evidence for Xq fragile site without molecular expansion at *FRAXA* or *FRAXE*; whether aberrant phenotype is associated with expansion at *FRAXF* is uncertain.

Disease	Gene	Location	Repeat		Number	Comments
Fragile site 16q22	*FRA16A*	16q22	CCG	?	1000–2000	Expansion is molecular explanation for cytogenetic observation of fragile site at 16q22; expansion has been associated with infertility and spontaneous abortions
Kennedy spinal and bulbar muscular atrophy (SBMA)	*SBMA*	Xq11–q12	CAG	Open reading frame	40–52	Caused by defect in androgen receptor gene, *SBMA* usually presents in midlife with bulbar signs and facial fasciculations
Huntington disease (HD)	*HD*	4p16.3	CAG	Open reading frame	37–100	Caused by a defect in *Huntingtin* gene, *HD* is characterized by choreiform movements, rigidity, and dementia
Spinocerebellar ataxia, type 1	*SCA1*	6p23	CAG	Open reading frame	<100	Autosomal dominant ataxia with onset in 30s; upper motor neuron signs and extensor planton responses

(continued)

TABLE 1.3. *Continued*

Condition	Gene Symbol	Chromosome Location	Repeat Type	Repeat Localization	Repeat Number Abnormal Range	Clinical Features
Dentatorubropallidoluysian atrophy, Haw River syndrome	*DRPLA*	12p13.31	CAG	Open reading frame	<100	Myoclonus epilepsy, dementia, ataxia, and choreoathetosis transmitted as autosomal dominant
Machado–Joseph (spinocerebellar ataxia type 3)	*MJD*; *SCA3*	14q24.3–q31	CAG	Open reading frame	61–84	Ataxia with onset usually in 40s; frequent dystonia and facial fasciculations
Myotonic dystrophy	*DM*	19q13.2–q13.3	CTG	3′ Untranslated region	200–4000	Myotonia, ptosis, characteristic cataracts, testicular atrophy, and frontal balding
Spinocerebellar ataxia type 2	*SCA2*	12q24.1	CAG	5′ Coding region	36–59	Abnormalities of balance due to cerebellar dysfunction or pathology; clinically identical to *SCA1*

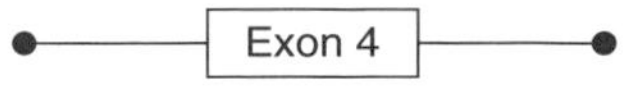

	Exon 4	

	Codon 112		Codon 158	
APOE allele	Sequence	Amino acid	Sequence	Amino acid
2	UGC	Cysteine	UGC	Cysteine
3	UGC	Cysteine	CGC	Arginine
4	CGC	Arginine	CGC	Arginine

Figure 1.11. Single base-pair changes in exon 4 of *APOE* define the *2*, *3*, and *4* alleles at this locus. (Modified from M. A. Pericak-Vance and J. L. Haines, *Trends Genet* 11, 1995.)

been shown to be associated with increased susceptibility to AD. The underlying biological mechanism for this observation has not yet been precisely identified.

The most frequent successes to date in the localization of genes underlying disease linkage analysis have been with diseases whose mode of inheritance is known (as illustrated above). These disorders are often highly or completely penetrant and are due to a defect in a single gene, yet these Mendelian disorders are often relatively rare in the population. However, some of the most common and deadly diseases of society such as cardiovascular disease and obesity have significant genetic components. These diseases are termed "complex" because they are likely due to the interaction of multiple factors, both environmental and genetic. Susceptibility genes for such complex disorders are substantially harder to identify than genes responsible for Mendelian disorders.

GENES, MITOSIS, AND MEIOSIS

A cell's ability to reproduce itself is critical to the survival of an organism. This cell duplication process, utilized by somatic cells, is called mitosis. Similarly, an organism's ability to reproduce itself is critical to the survival of the species. In sexual organisms, the reproductive process involves the union of gametes (sperm and egg cells), which are haploid. Meiosis is the process by which these haploid gametes are formed and is the biological basis of linkage analysis.

Meiosis consists of two parts: meiosis I and meiosis II. In meiosis I, which is called a reduction division stage, each chromosome in a cell is replicated to yield two sets of duplicated homologous chromosomes. During meiosis I, physical contact between chromatids may occur, resulting in the formation of chiasmata. Chiasmata are thought to represent the process of crossing over or recombination, in which an exchange of DNA between two (of the four) chromatids occurs (Fig. 1.12). A chiasma occurs at least once per chromosome pair. Thus, a parental haplotype (the arrangement of many alleles along a chromosome) may not remain intact upon transmission to an offspring. When two loci are unlinked to one another, the recombination fraction (θ) between them is 0.50. The upper limit for observed recombination between two unlinked loci is set at 50% because the frequency with which odd numbers of recombination events between a pair of loci occur

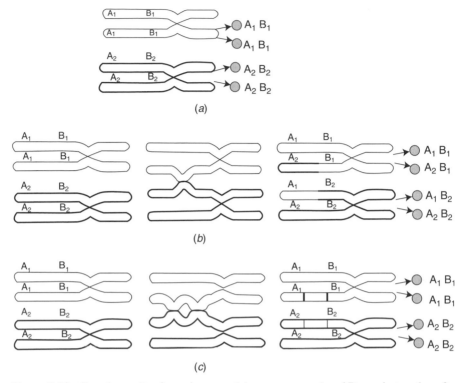

Figure 1.12. Genetic results of crossing over: (*a*) no crossover: A and B remain together after meiosis; (*b*) crossover between A and B results in a recombination (A and B are inherited together on a chromosome and A and B are inherited together on another chromosome); (*c*) double crossover between A and B results in no recombination of alleles. (Reprinted by permission from Jorde et al., eds., *Medical Genetics*, C. W. Mosby, St. Louis, MO, 1995.)

should equal the frequency with which even numbers of recombination events occur; when an even number of recombination events occurs between two loci, the resultant gametes appear to be nonrecombinant and hence these recombination events are unobserved.

Following crossing over, at least two of the four chromatids become unique, unlike those of the parent. The cellular division process that occurs ensures that one paternal homologue and one maternal homologue are transmitted to each of two diploid daughter cells. This cell division marks the end of meiosis I.

The process of genetic recombination helps to preserve genetic variability within a species by allowing for virtually limitless combinations of alleles in the transmission from parent to offspring. Estimates of genetic recombination can also predict distance between two loci: The closer two loci are to one another, the less chance for recombination between them. The frequency of recombination is not uniform through the genome. Some areas of some chromosomes have increased rates of recombination (hot spots), while others have reduced rates of recombination (cold

TABLE 1.4. Differences between Meiosis and Mitosis

	Meiosis	Mitosis
Purpose	Produce gametes; ensure or produce genetic variability through recombination	Replace somatic cells
Location	Gonads	Body cells
Number of cell divisions per cycle	Two: meiosis I and meiosis II (latter identical to mitosis)	One
Chromosome number in daughter cells	Halved from parental complement to produce gametes; resultant cells are haploid	Identical to parental complement; resultant cells are diploid
Recombination	Occurs in diplotene of prophase I of meiosis I	Occurs rarely, usually result of abnormality

spots). For instance, recombination frequencies may vary between sexes or may vary depending on whether the loci are at the telomere or centromere of the chromosome.

The second phase of meiosis is identical to a mitotic (somatic cell) division, in which genetic material is transmitted equally, identically, and without recombination to daughter cells. However, in contrast to a mitotic division, which yields two identical diploid daughter cells, the end result of the entire meiotic process in sperm cells is four haploid daughter cells with chromosomal haplotypes different from those originally present in the parent; in egg cells, the final outcome is a single haploid daughter cell, with the remainder of the genetic material lost because of the formation of nonviable polar bodies.

The fundamental differences between meiosis and mitosis are summarized in Table 1.4.

When Genes and Chromosomes Segregate Abnormally

Failure of meiosis at either phase (meiosis I or II) is termed nondisjunction and leads to aneuploidy, or abnormal chromosomal complements. The most well known aneuploidy is Down syndrome, caused by an extra copy of chromosome 21. Down syndrome is often called trisomy 21 because patients have a total of 47 chromosomes, with three copies of chromosome 21. A monosomy, or the absence of a second member of a chromosome pair, is rarely viable. A noted exception is Turner syndrome, in which a female has a total of 45 chromosomes, only one of which is an X chromosome.

Triploidy and tetraploidy are the terms for the presence of one or two entire extra sets of chromosomes, leading to a total of 69 or 92 chromosomes. These anomalies, which usually are inviable in humans, are due to errors in fertilization such as dispermy (two sperm fertilizing an ovum) or failure of the ovum's polar body to separate.

Segregation distortion, a phenomenon so far observed only rarely in humans, is characterized by a departure from the 50 : 50 segregation ratio expected from normal meiosis. Specifically, one allele at a locus is transmitted to the gamete more than 50% of the time. Segregation distortion, also termed meiotic drive, has been described in many experimental systems. Myotonic dystrophy (Beri et al., 1994), an autosomal dominant muscular dystrophy, was among the first human disease alleles suggesting preferential transmission of an allele to offspring.

ORDERING AND SPACING OF LOCI BY MAPPING TECHNIQUES

The segregation of loci in meiosis provides the opportunity for assessment of Mendel's law of independent assortment. When two loci are unlinked to one another ($\theta = 0.50$), this law holds true; however, the law is violated when two loci are linked ($\theta < 0.50$) to one another such that the transmission of one is not independent of transmission of the other. Estimating the distance between linked loci by assessing the frequency of recombination between them allows the development of an order of the markers relative to one another.

Once a gross localization for a disease or trait locus has been identified, either through linkage analysis or from some other approach (e.g., clues from chromosomal rearrangements), it is necessary to home in on the actual gene. This process is always complicated, but it can be simplified by the use of mapping resources, many of which were developed as a direct result of the Human Genome Initiative. No single mapping resource is best for all situations. Regardless of the mapping approach utilized, the resultant locus order along a chromosome should be identical.

Genetic maps order polymorphic markers by specifying the amount of recombination between markers, whereas physical maps quantify the distances among markers in terms of the number of base pairs of DNA. For small recombination fractions (usually <10–12%), the estimate of the recombination fraction provides a very rough estimate of the physical distance. In general, 1% recombination corresponds to one crossover per 100 meioses and is equivalent to about a million base pairs of DNA and is defined as one centimorgan (cM). Physical measurements of DNA are often described in terms of thousands of kilobases (10 kb of DNA is equivalent to 10,000 bp). A specific type of physical map, the radiation hybrid (RH) map (see below), also allows quantification of the length of a segment of DNA. The RH map distance is measured in centirays (cR), and on average throughout the genome 1 cR is equivalent to about 30,000 bp of DNA, although this estimate varies according to radiation dose. Estimates of distance from physical and genetic maps of the identical region may vary dramatically (Table 1.5) throughout the genome. A summary of the characteristics of different types of maps is shown in Table 1.6.

Physical Mapping

The purpose of a physical map of the genome is identical to that of a genetic map: to order pieces of DNA and, subsequently, genes. However, the materials utilized and

TABLE 1.5. Estimated Physical and Genetic Lengths of Selected Chromosomes

Chromosome	Physical Length (Mb)[a]	Genetic Length (cM)[a]	Length (cR)[b,c]
1	263	305	7894 (31.4)
2	255	271	6973 (34.4)
3	214	237	7785 (25.9)
4	203	244	2867 (24.4)
5	194	224	5611 (32.6)
6	183	207	6095 (28.3)
7	171	178	6606 (24.3)
8	155	172	3996 (36.6)
9	145	146	4513 (30.0)
10	144	181	5423 (25.1)
11	144	150	4858 (27.9)
12	143	160	5002 (26.9)
13	114	130	3306 (27.8)
14	109	122	3513 (25.0)
15	106	154	2822 (29.8)
16	98	157	2735 (34.4)
17	92	208	3039 (28.6)
18	85	143	2977 (26.8)
19	67	148	2122 (29.6)
20	72	122	2010 (33.8)
21	50	114	1562 (23.7)
22	56	81	1522 (26.9)
X	164	220	3644 (42.5)

[a]From Morton (1991).
[b]Numbers in parenthesis give kilobases per centiray.
[c]Data extracted from Stanford Human Genome Center (see Appendix for website).

the average resolutions of various mapping methods differ. Some of the oldest available physical maps of the genome are restriction maps, which identify sites at which an enzyme cuts (digests) a specific sequence of DNA. Contig maps are developed by cloning pieces of DNA into vectors, such as yeast artificial chromosomes (YACs), bacterial artificial chromosomes (BACs), or cosmids, and then ordering them by their overlapping sequences. Sequence-tagged site (STS) maps utilize unique stretches of DNA to identify particular clones. Expressed sequence tag (EST) maps order the coding stretches of DNA. The DNA sequence maps order specific stretches of DNA at the level of single base pairs. Radiation hybrid maps are developed by exposing DNA to high doses of radiation, thereby breaking the DNA into small pieces. The frequency with which particular markers are retained in a piece of DNA is scored, and this provides an estimate of the relative order and distance between markers in centirays. As with genetic maps, the basis of the RH maps is statistical: The relative order of a marker is determined in a manner analogous to calculating LOD (logarithm of the odds of linkage) scores

TABLE 1.6. Summary Characteristics of Genome Maps of Selected Types

Map Type	Measurement of Distance	Material Needed	Caveats and Notes
Genetic maps	Recombination frequency (θ) or centimorgan (cM); 1% recombination approximately equal to 1 cM	Reference pedigrees; polymorphic markers	Extremely sensitive to genotyping error; usually not useful in areas <2 cM
Meiotic breakpoint maps	Not applicable	Reference pedigrees; polymorphic markers	Good mechanism for minimizing genotyping to determine marker order using statistical techniques designed to minimize number of recombination events
Radiation hybrid maps	Centiray (cR): 1 cR represents 1% breakage between two markers	Somatic cell hybrid panel; markers not necessarily polymorphic	Maps developed using statistical techniques that assess frequency of chromosome breakage
Sequence-tagged site (STS) maps	Not applicable	Clones from which STSs are derived must be ordered	Resulting maps have landmarks assayed by polymerase chain reaction (PCR), but markers are often not polymorphic
Restriction maps	Tens of thousands of base pairs of DNA; 1000 bp of DNA is termed a kilobase	Genomic DNA	
Expressed sequence tag (EST) maps	Not applicable	Genomic DNA	

(Falk, 1991; Lange and Boehnke, 1992) and is reported in terms of odds in favor of a placement relative to another placement. All these maps were generated as steps in the process of obtaining the human genome sequence, which is the ultimate physical map. Even when declared completed, there will be some small regions of the genome that are difficult to sequence completely. The various physical maps will still be helpful in spanning these regions.

Genetic Mapping

The study of human inherited disease has benefited from numerous experiments in other organisms. Although mapping in humans has a relatively recent history, the idea of a linear arrangement of genes on a chromosome was first proposed in 1911 by T. H. Morgan from his work with the fruit fly *Drosophila melanogaster*. The possibility of a genetic map was first formally investigated by A. H. Sturtevant, who ordered five markers on the X chromosome in *Drosophila* and then estimated the relative spacing among them.

In experimental organisms, genetic mapping of loci involves counting the number of recombinant and nonrecombinant offspring of selected matings (Table 1.7).

TABLE 1.7. Example Development of Genetic Map Using Four Linked Loci, *A*, *B*, *C*, and *D*, Scored in 100 Offspring[a]

A. Score Recombination Events between All Pairwise Combinations of Four Loci

Loci Scored	Number of Recombinants	Frequency of Recombination
A–B	10	0.10
A–C	3	0.03
A–D	15	0.15
B–C	7	0.07
B–D	5	0.05
C–D	12	0.12

B. Process

1. Determine which two loci have the highest frequency of recombination between them; these two loci are the farthest apart on the map.
2. Fit the other loci into the map like pieces of a puzzle.

C. Resulting Genetic Map

Percent recombination loci order	| 0.03 | 0.07 | 0.05 |
	A C B D

[a]This example demonstrates no evidence for interference (see text). Positive interference would be manifested as a decrease in overall map length from what would be expected by adding the pairwise distances. In this example, if $(A–C) + (C–B) + (B–D) > A–D$, positive interference may be present.

Genetic mapping in humans is usually more complicated than in experimental organisms for many reasons, including our inability to design specific matings of individuals, which limits the unequivocal assignment of recombinants and nonrecombinants. Therefore, maps of markers in humans are developed by means of one of several statistical algorithms used in computer programs such as CRIMAP and MAPMAKER (Lander and Green, 1987), CLINKAGE and MULTIMAP (Matise et al., 1994), and MAP-O-MAT (Matise and Gitlin, 1999). Genetic maps can assume equal recombination between males and females or can allow for sex-specific differences in recombination since it has been well established that there are substantial differences in recombination frequencies between men and women; on average, the female map is twice as long as the male map (Li et al., 1998). These maps are generally produced utilizing a single set of reference pedigrees, such as the those developed by the Centre d'Étude du Polymorphisme Humain (CEPH) (Dausset et al., 1990), which are mostly comprised of three-generation pedigrees with a large number of offspring (average 8.5). Both sets of maternal and paternal grandparents are usually available, so linkage phase frequently can be established. The collection of CEPH pedigrees, in its entirety, consists of more than 60 pedigrees and includes more than 600 individuals; DNA from this valuable resource is available through the Coriell Institute for Medical Research. The complexity of the underlying statistical methods used to generate genetic maps renders them sensitive to marker genotyping errors, particularly in small intervals (Buetow, 1991), and these maps are less useful in small regions of less than about 2 cM. While marker order is usually correct, genotyping errors can result in falsely inflated estimates of map distances.

Interference and Genetic Mapping

Another factor complicating genetic mapping is interference, where the probability of a crossover in a given chromosomal region is influenced by the presence of an already existing crossover. In positive interference, the presence of one crossover in a region decreases the probability that another crossover will occur nearby. Negative interference, the opposite of positive interference, implies the formation of a second crossover in a region is made more likely by the presence of a first crossover. Most documented interference has been positive, but some reports of negative interference exist in experimental organisms. Interference is very difficult to measure in humans because exceedingly large sample sizes, usually on the order of 300–1000 fully informative meiotic events, are required to detect it (Weeks et al., 1994).

The investigation of interference is important because accurate modeling of interference will provide better estimates of true genetic map length and intermarker distances and more accurate mapping of trait loci. Interference (I) is frequently measured in terms of the coefficient of coincidence (c.c.) in genetic crosses where three separate linked markers can be scored. The coefficient of coincidence is the ratio of the observed number of double crossovers to the expected number of double crossovers assuming no interference. When $I > 0$, interference is present (positive); when $I < 0$, interference is negative; when $I = 0$, there is no evidence for interference and recombination fractions across intervals are additive.

For example, assume three loci whose order is *A–B–C*. If the distance between *A* and *B* is 10 cM and between *B* and *C* is 5 cM, when $I = 0$, the distance between *A* and *C* is 15 cM. As noted earlier, the frequency of recombination in humans is generally decreased near the centromeric region of chromosomes, tends to be greater near the telomeric regions, and is increased in females when compared to males. It should be further noted that genetic map distance is not tied directly to physical map distance.

Several mathematical formulas have been developed to account for interference in predicting an additive measure of genetic map distance from recombination frequencies in human linkage studies. These mapping functions include those developed by Kosambi (1944), Rao (Watzke et al., 1977), and Haldane (1919). Haldane's map function assumes the absence of interference, while Kosambi's map function assumes interference is large at small genetic distances but decreases as the genetic distance between two loci increases. A program in the LINKAGE utility package, MAPFUN, translates recombination frequencies into map distances and vice versa under a variety of mapping functions. This package, along with a comprehensive listing of other available linkage analysis programs, is available at http://www.linkage.rockefeller.edu.

Meiotic Breakpoint Mapping

Meiotic breakpoint maps, an outgrowth of genetic maps, are graphical descriptions of critical, confirmed recombination events within reference pedigrees (Fig. 1.13).

CEPH meiotic breakpoint panel

Family	ID	D5S818	D5S804	D5S816	D5S812
1331	9	○	●	●	●
1331	10	●	○	○	○
1332	3	○	●	□	□
1332	12	○	□	●	●
1333	4	○	□	●	□
1333	9	○	□	●	□
1344	5	○	□	●	□
1347	6	○	○	○	●
1347	10	○	○	○	●
1362	9	□	○	○	●
1362	11	□	○	●	●
884	3	○	○	○	●
884	11	○	○	○	●

Figure 1.13. Sample meiotic breakpoint map from data at the CHLC. Shaded circles indicate markers that recombine with those having open circles. Squares indicate marker is either uninformative or not genotyped. To test whether a marker is located between D5S818 and D5S804, an investigator would need to genotype individuals 9 and 10 in family 1331 and 3 in 1332, in addition to their parents. In typical practice, the laboratory is blinded to which individuals are recombinants to avoid potential bias. Consequently, two or three nonrecombinant siblings for each known recombinant individual are also genotyped.

Once a meiotic breakpoint map for a region has been developed, a marker whose location is nearby can be genotyped in a subset of pedigrees in which critical recombination events have occurred. Limiting the genotyping efforts to a small number of specific individuals within pedigrees greatly increases genetic mapping efficiency.

Disease Gene Discovery

Disease gene discovery is greatly facilitated by the availability of dense genetic maps. Linkage analysis for the localization of disease genes boils down to the "simple" idea of counting recombinants and nonrecombinants, but in humans this process is complicated for a variety of reasons. For example, the generation time is long in humans such that large, multigenerational pedigrees in which a disease or trait is segregating are rare; scientists cannot dictate matings or exposures, nor can they require participation of specific individuals in a study. Thus, the process of linkage analysis in humans requires a statistical framework in which various hypotheses about the linkage of a trait locus and marker locus can be considered. How far apart are the disease and marker, and how certain is the conclusion of linkage?

Linkage analysis has traditionally been performed using either a parametric approach or a nonparametric approach. Parametric approaches require the assumption of a genetic model and the specification of various parameters such as disease allele frequency and penetrance. In contrast, nonparametric methods do not require specification of a genetic model and thus do not suffer from the potentially hazardous effects of model misspecification. When the genetic model for a disease is clearly known, parametric- (likelihood-) based methods are more powerful than the nonparametric approaches.

Example 1: Linkage Analysis in Pedigree with Unlinked Marker. For illustrative purposes, consider the pedigree in Figure 1.14a. In this pedigree, shaded individuals are affected with disease and unshaded individuals are unaffected. Results for a genetic marker are shown underneath each individual. In pedigrees in which the genetic model can be deduced with a high degree of certainty, linkage analysis can be broken down into five steps.

It should be emphasized that this example is a simplified version of an extremely complicated process.

Step 1: State Components of Genetic Model. For parametric linkage analysis, the genetic model must be specified. Components of the genetic model include the inheritance pattern of the disease locus (autosomal or sex linked; dominant, recessive, or codominant), disease allele frequency and penetrance, and frequency of phenocopies and new mutation. Rough estimates of the disease allele frequency and penetrance can often be obtained from the literature or from computer databases such as Online Mendelian Inheritance in Man (http://www3.ncbi.nlm.nih.gov/Omim/); estimates of the rate of phenocopies and new mutation are frequently guesses, included as a nuisance parameter in some cases to allow for the fact that these can exist. Linkage analysis using LOD scores is relatively robust to modest

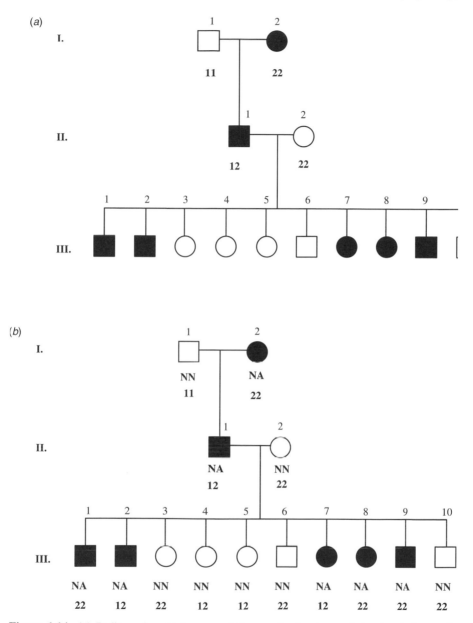

Figure 1.14. (*a*) Pedigree in which a rare, fully penetrant autosomal dominant disease is segregating. Genotypes are shown beneath pedigree symbols. (*b*) Putative disease genotype listed beneath pedigree figure. (*c*) Assignment of disease-associated haplotypes. Results on the left of the bar were transmitted from the male parent and those on the right of the bar were transmitted from the female parent. In the third generation, meioses are scored as recombinant (R) or nonrecombinant (NR). Since the number of recombinant meioses is equal to the number of nonrecombinant meioses, this pedigree is consistent with nonlinkage between the disease and marker loci.

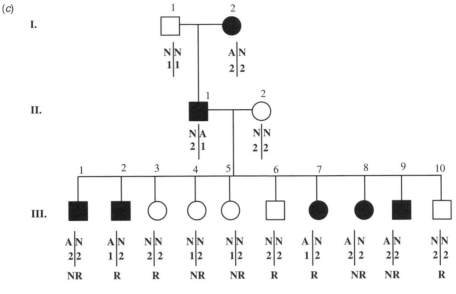

Figure 1.14. *Continued.*

misspecification of the disease allele frequency and penetrance, but misspecification of whether the disease is dominant or recessive can lead to incorrect conclusions of linkage or nonlinkage (Clerget-Darpoux et al., 1986). In this example, the disease allele will be assumed to be rare and to function in an autosomal dominant fashion with complete penetrance; the disease locus will be assumed to have two alleles, *N* (for normal or wild type) and *A* (for affected or disease). In addition, mutation and phenocopies are assumed to be absent. These assumptions allow substantial simplification of the problem, as outlined further below.

In addition to specifying the parameters of the disease locus, frequencies for the alleles at the marker locus are required. In pedigrees in which genotypes are missing in founding individuals (due either to an unsampled individual or to laboratory complications), the misspecification of allele frequencies can have substantial impact, leading to incorrect conclusions of linkage and nonlinkage and biased estimates of the recombination fraction (Ott, 1992; Knowles et al., 1992). Other factors can also wreak havoc in a linkage analysis. Such factors include linkage heterogeneity and scoring errors in the pedigree structure, diagnostic status, or marker genotyping.

Step 2: Assign Underlying Disease Genotypes Given Information in Genetic Model (Fig. 1.14*b*). In preparation for scoring recombinant and nonrecombinant individuals in a pedigree, the relationship between genotype and phenotype as defined by the genetic model can be used to assign the underlying genotype of pedigree members at the disease locus. The assumption of complete penetrance of the disease allele allows all unaffected individuals in the pedigree to be assigned a disease genotype of *NN*. Since the disease allele is assumed rare, the disease genotype for affected individuals can be assigned a disease genotype of *AN*. In other words, since the disease allele is rare, the chance that an affected individual is homozygous for the disease allele is so small that, for the purposes of this example, it can be considered to be zero.

Again, it is important to emphasize that this example is a simplified version; most linkage analysis is performed by computer analysis that allows the consideration of the small probability that a founder in such a pedigree (e.g., individual I-1) is homozygous for the disease allele. When a computer program performs this analysis, it assigns probabilities for genotypes *AA* and *AN* in individuals like I-1 by using user-specified information on disease allele frequencies.

Step 3: Determine Putative Linkage Phase. Individual II-1 has inherited the disease trait together with marker allele *2* from his affected mother. Thus, the *A* allele at the disease locus and the *2* allele at the marker locus were inherited in the gamete transmitted to II-1. There are two mutually exclusive hypotheses to consider: The null hypothesis is that the disease and marker loci are unlinked to one another. If the loci are genetically unlinked, there will be an approximately equal number of recombinant and nonrecombinant gametes among the offspring of II-1. The alternate hypothesis is that the disease and marker loci are linked to one another. If the loci are genetically linked, there will be more noncombinants than recombinants among the offspring ("meiotic events") of II-1. The basic idea is to consider these two competing hypotheses of linkage versus nonlinkage and determine which hypothesis best describes the available data from the pedigree.

Thus, the putative linkage phase (the disease allele "segregates" with marker allele *2*) has been established, and this phase can be tested in subsequent generations.

Step 4: Score Meiotic Events as Recombinant or Nonrecombinant. For this mating type, there are four possible gametes from the affected parent II-1: *N1*, *N2*, *A1*, and *A2*. Based on the putative linkage phase assigned above in step 3, gametes *A2* and *N1* are recombinant. Thus, all affected offspring of II-1 and II-2 who have inherited marker allele *2* from their father will be scored as nonrecombinant for the disease and marker; affected offspring who have inherited the *1* allele will be scored as recombinant for the disease and marker. Similar reasoning applies to the unaffected offspring, except that the unaffected offspring who have inherited allele *1* are nonrecombinant and those who have inherited allele *2* are recombinant.

In this pedigree (Fig. 1.14c), five offspring of II-1 are recombinant and five are nonrecombinant. Thus, out of 10 scorable meiotic events, the number of recombinant gametes is equal to the number of nonrecombinant gametes. These data are consistent with the hypothesis of nonlinkage between the disease and marker loci.

One frequent question arises—why is the transmission of the *2* allele from the affected grandmother to the affected son not counted as a meiotic event? Or, why are there not 11 instead of 10 meioses in this pedigree? The answer is that we do not know the linkage phase in individual II-1; we are just using the transmission from his affected mother to him to determine our hypothesis about what the linkage phase would be *if* the disease and marker loci were linked to one another.

Step 5: Calculate and Interpret LOD Scores. The pedigree data are used to consider both the hypothesis of linkage between the disease and marker locus and nonlinkage between the disease and marker locus. Morton (1955) suggests a likelihood ratio approach in which the likelihood of the pedigree and marker data is calculated under the null hypothesis that assumes free recombination ($\theta = 0.50$) between the disease and marker loci, where θ represents the recombination fraction, and then compared to the likelihood of the hypothesis of linkage between the disease and marker loci. This likelihood of the pedigree data under the hypothesis of linkage between the disease and marker loci is calculated at various increments of $\theta < \frac{1}{2}$ within the range of allowable values (0.00–0.49), representing unique subhypotheses of linkage between the disease and marker loci. The likelihood ratio (LR) is constructed as $L(\text{pedigree} | \theta = x)/L(\text{pedigree} | \theta = 0.50)$, where x is some value of θ.

The likelihood of observing the pedigree data is just $\Theta^R(1 - \Theta)^{NR}$, where R and NR are the number of recombinants for the two phases and N is the total number of scored offspring ($R + NR = N$); this likelihood is the numerator in the ratio of likelihoods. The denominator is the likelihood assuming that the two loci are unlinked (i.e., when $\Theta = 0.5$). Thus, the ratio of the likelihoods of the two hypotheses is constructed as

$$LR = \frac{\theta^R(1 - \theta)^{NR}}{(0.5)^R(0.5)^{NR}}$$

and reduces to

$$LR = \frac{\theta^R(1 - \theta)^{NR}}{0.5^N}$$

Typically, the base-10 logarithm of this ratio is taken to obtain a LOD score [Note: $z(\theta)$ is sometimes used to denote a LOD score]. A LOD score of 3 indicates odds of $10^3 : 1$ (1000 : 1) in favor of linkage and is considered to be conclusive evidence for linkage between two markers (or a marker locus and disease locus) in most cases. A LOD score of -2 indicates odds of $10^{-2} : 1$ (0.01 : 1) in favor of linkage, or more commonly, odds of 100 : 1 *against* linkage of the two markers. A LOD score of -2 or less is considered to be conclusive evidence that the two markers under study are unlinked at the specified recombination fraction. It is important to note that this "exclusion" of a disease gene from a region is only valid under the specific set of assumptions made for the analysis. For example, if an analysis were made assuming an autosomal dominant trait but the trait was actually autosomal recessive, it could be falsely excluded from a region. The LOD scores between the values of -2 and 3 are considered inconclusive and warrant additional study. A study is "expanded" by rendering the currently available family data more informative (i.e., testing the family with different or more informative markers) or by increasing the number of families under study. A LOD score should always be considered in conjunction with its respective estimate of Θ.

It is important to note that using a LOD score of 3.0 (odds 1000 : 1 in favor of linkage) as a test statistic does not equate with a type 1 error rate (α) of 0.001, which is much more stringent than the standardly used type I error rate of 0.05 in most statistical analyses. Because of the prior probability of linkage between two traits, the true p-value associated with a LOD score of 3.0 is approximately 0.04,

TABLE 1.8. LOD Scores for Pedigrees in Examples 1, 2, and 3

	$\theta = 0.01$	$\theta = 0.05$	$\theta = 0.10$	$\theta = 0.150$	$\theta = 0.20$	$\theta = 0.30$	$\theta = 0.40$
Example 1	−7.01	−3.61	−2.22	−1.46	−0.97	−0.39	−0.09
Example 2	0.97	1.51	1.60	1.55	1.44	1.09	0.62
Example 3	0.99	1.21	1.30	1.25	1.14	0.79	0.33

consistent with a false-positive rate of 1 in 25. In other words, when evidence for linkage between two loci is declared significant at a LOD score of 3.0, there is a 4% chance it is a spurious positive result (for more detail, see Ott, 1999).

Table 1.8, Example 1, shows the two-point LOD scores for the marker at a variety of hypotheses about the estimate of the recombination fraction between the disease and marker locus. In this example, the highest LOD score is -0.09 at $\theta = 0.40$ and at no value of θ is the LOD score positive, let alone ≥ 3.0, so this pedigree demonstrates no evidence in favor of linkage between the disease and marker loci. However, the pedigree *does* provide important information about where the disease locus is *not* located. Visual inspection of the LOD score data suggests that the value of θ at which the LOD score exceeds -2.0 is between 0.10 and 0.15, so approximately 13 cM on either side of the marker locus can be excluded as harboring the disease gene, for a total exclusion of 26 cM as a result of typing this marker!

Several well-tested computer programs including the LINKAGE computer package (Lathrop et al., 1984), FASTLINK (Schaffer et al., 1994, Cottingham et al., 1993), VITESSE (O'Connell and Weeks, 1995), Genehunter (Kruglyak et al., 1996), and Allegro (Gudbjartsson et al., 1999) are often utilized for calculation of LOD scores.

Example 2. Linkage Analysis in Pedigree with Linked Marker. Consider the pedigree in Figure 1.15a.

Step 1: State Components of Genetic Model. Assume the genetic model is the same as in Example 1 (rare, autosomal dominant disease allele with complete penetrance, no mutation, no phenocopies).

Step 2: Assign Underlying Disease Genotypes Given Information in Genetic Model. The assignment of disease genotypes to pedigree members is the same as in Example 1 (Fig.1.15b). Unaffected individuals have an underlying disease genotype of *NN* and affected individuals have a disease genotype of *AN*.

Step 3: Determine Putative Linkage Phase. The assignment of putative linkage phase is identical to that in Example 1, so that the disease allele is transmitted in the same gamete as marker allele 2.

Step 4: Score Meiotic Events as Recombinant or Nonrecombinant (Fig. 1.15c). In this example, all five of the affected children have inherited marker allele 2 from their affected father and are thus nonrecombinant. In addition, four of the five unaffected children have inherited marker allele *1* from their affected father. These four offspring, too, are nonrecombinant with respect to the disease and marker loci. Individual III-6, however, is unaffected and has inherited marker allele

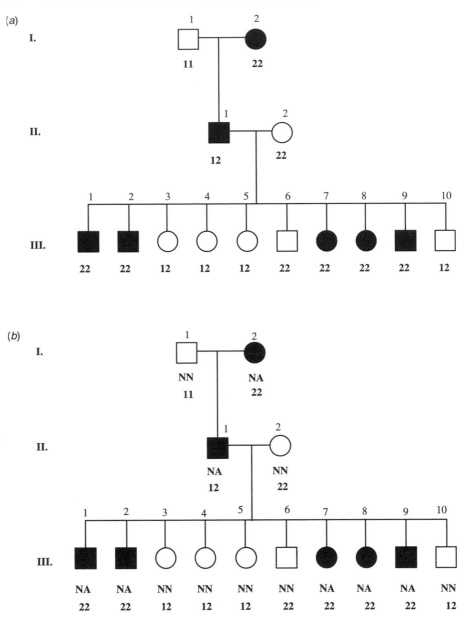

Figure 1.15. (*a*) Pedigree in which a rare, fully penetrant autosomal dominant disease is segregating. Genotypes are shown beneath pedigree symbols. (*b*) Putative disease genotype listed beneath pedigree figure. (*c*) Assignment of disease-associated haplotypes. Results on the left of the bar were transmitted from the male parent and those on the right of the bar were transmitted from the female parent. In the third generation, meioses are scored as recombinant (R) or nonrecombinant (NR). Since the number of nonrecombinant gametes is larger than the number of recombinant gametes, this pedigree is consistent with linkage between the disease and marker loci.

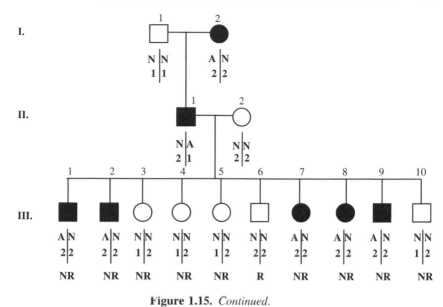

Figure 1.15. *Continued.*

2 from his affected father. This individual is scored as a recombinant individual (a recombinant meiosis). This nonrandom segregation of the disease allele and a marker allele within a pedigree is consistent with linkage between the disease and marker locus.

Step 5: Calculate and Interpret LOD Scores. The LOD scores for this pedigree at a variety of hypotheses about the recombination fraction between the disease and marker loci are shown in Table 1.8, Example 2. No LOD score at any tested recombination fraction is higher than 3.0, so this pedigree does not provide significant evidence in favor of linkage between the disease and marker locus. The highest LOD score in this pedigree is 1.6 so that more data from additional pedigrees need to be generated in order to interpret the results. However, this LOD score is probably "interesting" enough that it would indeed warrant such follow-up. Here, it is critical to note that the recombination event occurs in an unaffected individual. If the assumption of complete penetrance of the disease allele is incorrect, the assignment of this meiotic event as a "recombinant" could be in error. Thus, it is incumbent on the investigator to weigh carefully information about disease gene localization gleaned from unaffected family members.

The maximum-likelihood estimate (MLE) of the recombination fraction is that value of θ at which the LOD score is largest. It is this value of the LOD score (the highest LOD score) with which the determination of statistical significance is made. In this case, the maximum LOD score is 1.60, which occurs when $\theta = 0.10$, and so 0.10 is the MLE for θ in this example (note that this result is identical to the direct estimate for θ obtained earlier). This result does not meet the established criteria of odds $\geq 1000 : 1$ (a LOD score ≥ 3) for concluding evidence for linkage, yet it is interesting in and of itself and perhaps merits further investigation by the genotyping of additional tightly linked markers or additional families.

An intuitive, direct estimate of the recombination frequency (usually designated as θ), which is related to the genetic distance between the disease and marker locus, is just the proportion of recombinant meioses counted among total meioses scored. In this example, the direct estimate of the recombination frequency, based on the observation of 1 recombinant out of 10 scorable meioses, is $\frac{1}{10}$, or 10%. This direct approach is possible in this example because these are phase-known events so that recombinants and nonrecombinants can be counted.

When significant evidence for linkage is found, a formal presentation of results of a linkage analysis is incomplete without an indication of a support interval for the MLE of θ. A description of the calculation of these support intervals, usually performed using the "one LOD score down" method, can be found in either Ott (1999) or Conneally et al. (1985). In this approach, all values of the recombination fraction that fall within one LOD score below the highest attained LOD score are considered within the support interval.

Example 3. Linkage Phase Unknown. Next consider an example in which the grandparents are unavailable (Fig. 1.16). In this pedigree, the linkage phase cannot be established with certainty; however, only one of two linkage phases is possible. Under phase 1, the disease allele is segregating with the *1* allele. Alternatively, the disease allele could be segregating with the *2* allele (phase 2). Under each of these alternate scenarios, it can be determined whether or not the offspring—these 10 scorable meioses—are recombinant or nonrecombinant gametes. In this example the LOD scores are calculated a little differently:

$$\text{LOD} = \log 10 \left[\frac{1}{2} \left(\underbrace{\frac{(\theta^R (1 - \theta)^{NR})}{(\theta = 0.5)^N}}_{\text{phase 1}} + \underbrace{\frac{(\theta^R (1 - \theta)^{NR})}{(\theta = 0.5)^N}}_{\text{phase 2}} \right) \right]$$

where *NR* and *R* are the number of nonrecombinants for the two phases.

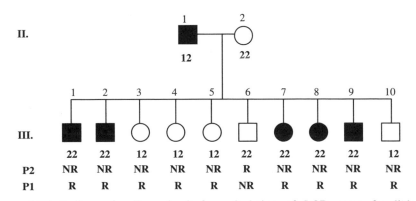

Figure 1.16. Pedigree for Example 3 for calculation of LOD score for linkage-phase-unknown pedigree.

Note that the uncertainty associated with the unknown linkage phase reduces the overall linkage information. The highest LOD score obtained in this pedigree is 1.3 (Table 1.7), as opposed to the highest LOD score of 1.6 in the earlier pedigree. Although this may not seem like a substantial loss of information, it does influence the results when working with pedigree samples that are large enough to give evidence in favor of linkage. If that much information is lost on a series of 10 pedigrees, a substantial amount of potential linkage information has been lost.

Additional Notes. Note that in the pedigree in Figure 1.15*c*, the 2 allele is segregating with all affected individuals and with only one of five unaffected individuals. One might characterize this observation as an "association" between the 2 allele and the disease allele within this family. By definition, two traits that are linked to one another will show an association *within* a family, but the associated allele may vary between families linked to the same locus.

Information Content in a Pedigree

For the meiotic events of a parent to be scorable for linkage analysis, the parent must be heterozygous at both loci (disease and marker) under consideration (Fig. 1.17). The family studied in the above examples is large and relatively atypical for a human pedigree since all family members are available for sampling. How much information is required to obtain significant results from a linkage analysis? The answer to this question can be complicated and highly dependent on the inheritance pattern of the disease (i.e., dominant, recessive, or sex linked), the penetrance of the disease, whether sporadic cases of the disease (phenocopies) may be present, and other factors. As a guideline, each scorable, phase-known meiotic event (with no recombination between the trait and marker locus) contributes about 0.30 to the LOD score. Table 1.9 gives some additional information about the contribution of each meiotic event to the LOD score at a variety of recombination fractions. Thus, at a minimum, 10 phase-known meiotic events demonstrating no recombination between the trait and marker locus are required to obtain a LOD score of 3.0. In general, human pedigrees have few phase-known meiotic events since some individuals may be unable or unwilling to participate in the study. Computer simulation studies should be performed to assess the power of an available dataset to detect linkage under an assumed genetic model prior to initiating a screen of the entire genome to detect linkage. Several programs including SIMLINK and SLINK are available for performing these power studies (Boehnke, 1986; Ploughman and Boehnke, 1989; Ott, 1989; Weeks et al., 1990).

Disease Gene Localization

Once linkage between a disease locus and marker locus is established via two-point linkage analysis, the next step is to identify the smallest region of the genome that should contain the disease gene, the minimum-candidate region (MCR). Two approaches to disease gene localization are generally used: multipoint linkage analysis and haplotype analysis.

Multipoint linkage analysis is a statistical technique using available genetic maps in which the order of markers is relative to one another and the distances between

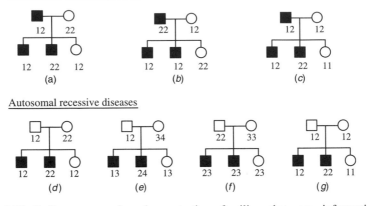

Figure 1.17. Pedigree examples demonstrating families that are informative and noninformative for linkage analysis. Both the autosomal dominant and autosomal recessive disease are assumed to be fully penetrant. Shaded individuals are affected with the disease, unshaded individuals are unaffected, and stippled individuals are asymptomatic gene carriers (in the recessive disease). Marker results are indicated beneath each individual. *Autosomal dominant diseases:* The affected parent, who is by definition homozygous at the disease locus, must also be heterozygous at the marker locus. (*a*) Fully informative pedigree: All three offspring can be scored for linkage analysis. (*b*) Uninformative pedigree: Cannot tell which 2 allele is transmitted with the disease allele. (*c*) Partially informative pedigree (an "intercross"): Homozygous offspring of this mating contribute significantly to linkage analysis. *Autosomal recessive diseases:* For pedigrees to be fully informative, both parents, who are heterozygous at the disease locus by virtue of the fact that they have at least one affected child with a recessive disease, must also be heterozygous for different alleles at the marker locus. (*d*) Fully informative pedigree: Both paternal and maternal gametes can be scored with respect to disease and marker locus. (*e*) Partially informative pedigree: Only paternally contributed gametes contribute to linkage analysis. (*f*) Uninformative pedigree: Both parents are homozygous at the marker locus. (*g*) Partially informative pedigree: Homozygous offspring contribute significantly to the linkage analysis.

TABLE 1.9. Number of Phase-Known, Fully Informative Meioses Needed to Detect Linkage at Various Values of Recombination Fraction θ

θ	Number of Informative Meioses Needed for LOD = 3	Expected LOD per Meiosis
0.00	10	0.30
0.01	11	0.28
0.02	12	0.26
0.05	14	0.21
0.10	19	0.16
0.20	36	0.08
0.25	53	0.06

the markers are known. Multipoint linkage analysis allows the simultaneous consideration of genotypes from multiple linked loci. This technique is useful for localization of disease genes between two markers and for maximizing the informativeness of a series of markers. In a general sense, given a series of markers of known location, order, and spacing, the likelihoods of the pedigree data are sequentially calculated for the disease gene to be at any position within the known map of markers. The multipoint LOD scores are typically graphed as in Figure 1.18. The x axis represents the genetic distance between the markers and the y axis represents the LOD score. In this example the most likely location for the disease is in the 7-cM interval bounded by IL9 and D5S178. In this interval, the LOD score reaches its highest value of approximately 23.5. The next most likely interval is the interval outside IL9, where the maximum LOD score reaches approximately 21.0. Note that the interval between D5S178 and D5S210 has a LOD score that maximizes at approximately 18. This interval between D5S178 and D5S210 is outside the approximate 99% confidence interval for disease gene localization because it does not fall with the three-LOD unit support interval.

Haplotype analysis is a tool for ordering alleles on a chromosome. It complements multipoint linkage analysis in that it provides visual confirmation of statistical testing. Given the order of the genetic markers on a chromosome, we can establish with varying degrees of certainty the alleles that were passed from parent to offspring in any specific gamete by haplotype analysis and identify critical recombination events that flank the MCR. Haplotype analysis can be done by hand or with the use of a computer program such as SIMWALK2 (Weeks et al., 1995). Through haplotype analysis in Mendelian disease, critical recombination events may allow us to identify the upper and lower bounds of the disease gene interval. The process of haplotyping involves identifying which alleles were transmitted in the same

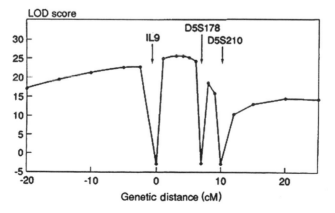

Figure 1.18. The most likely location for the disease gene is in the 7-cM interval bounded by IL9 and D5S178. The maximum LOD score is 23.50. Narrow support intervals for disease gene localization are calculated by subtracting 3 from the maximum LOD score; all locations within this three-LOD unit are within the approximate 99% confidence interval for disease gene localization.

gamete from each parent and identifying where recombination events between markers and/or between markers and the disease gene are located. The main goal is to identify markers that flank the disease gene. In the presence of reduced penetrance, recombination events that occur in affected individuals tend to be associated with a higher degree of certainty than recombination events that involve unaffected individuals because an unaffected individual could either be a non–gene carrier or a non–penetrant carrier. The confidence with which a recombination event defines a boundary of the MCR then is heavily dependent on the penetrance.

An example of haplotype analysis is shown in Figure 1.19. In the pedigree on the left, the unordered genotypes for three markers are shown underneath the pedigree symbols. The markers are arranged linearly and are tightly linked to one another such that 2 cM with intermarker distances. The task in haplotype analysis is to assign the allelic combinations that were transmitted in the gamete from parent to offspring. In the pedigree on the left one can see that from the affected grandfather the gamete with the combination of alleles *3/4/1* was transmitted to the affected son. The haplotype with the combination of alleles *2/3/2* was transmitted from the unaffected grandmother to the affected son. The haplotype *3/4/1* is now considered to represent the linkage phase that we test in the third generation. This haplotype was transmitted in its complete state from the affected father to his first affected son, individual 5. The unaffected individual 6 inherited the grand maternal combination of alleles at the three loci. Both individuals 5 and 6 represent nonrecombinant gametes. Individual 7 inherited the disease-associated haplotype from markers *1* and *2* but inherited the grand maternal derived allele *2* at marker *3*.

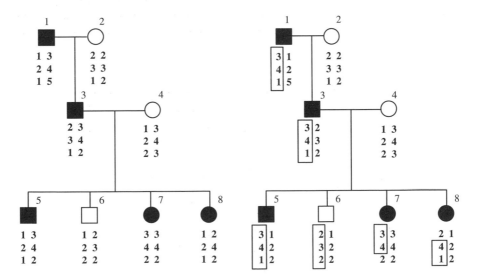

Figure 1.19. Pedigree on the left shows unordered genotypes underneath pedigree symbol for three linked markers. Pedigree on the right shows ordered genotypes, with alleles transmitted from father on the left. Alleles included in the disease-associated haplotype are in the box. See text for details.

The combination of markers that individual 7 inherited demonstrates that a recombination event between the disease gene and the marker loci occurred distal to marker 2. The affected daughter, 8, inherited the disease-associated haplotype for markers 2 and 3. She inherited the grand maternal allele from her affected father at marker 1 so that a recombination event occurred between the disease and marker loci proximal to marker 2. These two recombination events suggest that the disease gene in this family is located between markers 1 and 3.

Extensions to Complex Disease

Typically, parametric LOD score analysis (two-point and multipoint) and haplotype analysis are used in Mendelian disease; however, Mendelian transmission of a "complex" trait may be apparent in a subset of families, such as in AD (Post et al., 1997; Pericak-Vance et al., 1991; Schellenberg et al., 1992), ALS (Rosen et al., 1993), and breast cancer (Hall et al., 1992). The study of complex disorders present special, yet not insurmountable, challenges in linkage analysis for gene localization. Elucidation of the genetic defect in a subset of families may provide insights into the pathogenesis of the non-Mendelian form of the disease.

SUMMARY

The study of genes, chromosomes, and patterns of transmission of human traits within families has led to remarkable discoveries that are useful in genetic counseling for recurrence risk, presymptomatic testing, and prenatal diagnosis (see Chapter 4) and in the understanding of the pathogenesis of diseases. The genetic basis of Mendelian disease is relatively straightforward and in many cases is well understood. The situation in common complex disorders is markedly different from the study of Mendelian disease, since more than one gene as well as various nongenetic factors are typically associated with trait phenotype expression. Yet, many of the same principles hold true in complex disease: Mendel's laws regarding the transmission of genes and alleles at loci are as important to the study of resemblance between relatives in genetically complex disease as in Mendelian disease; the same holds true for the extent and result of the differing types of mutation. Because of the completion of the human genome sequence, the advances will accumulate more rapidly; human genomic study can only expand dramatically.

Linkage analysis is a powerful tool for the primary detection of genes leading to human disease. Methods for localizing genes in diseases that are clearly Mendelian have been highly successful. Extreme care and caution must be exercised when planning a linkage analysis, particularly with complex diseases, as the effects of model misspecification can be severe. Although linkage analysis of more complex diseases, in which the mode of inheritance is unclear, is less straightforward, methods are available to address such situations. The potential rewards of localizing and cloning genes causing the complex and most common diseases are abundant, especially with respect to public health and policy.

REFERENCES

Bellus GA, Hefferon TW, Ortiz de Luna RI, Hecht JT, Horton WA, Machado M, Kaitila I, McIntosh I, Francomano CA (1995): Achondroplasia is defined by recurrent G380R mutations of *FGFR3*. 56:368–373.

Beri RK, Marley AE, See CG, Sopwith WF, Aguan K, Carling D, Scott J, Carey F (1994): Molecular cloning, expression and chromosomal localisation of human AMP-activated protein kinase. FEB 356:117–121.

Boehnke M (1986): Estimating the power of a proposed linkage study: A practical computer simulation approach. Am J Hum Genet 39:513–527.

Buetow KH (1991): Influence of aberrant observations on high re-solution linkage analysis and reporting. Am J Hum Genet 49:985–994.

Burke JR, Wingfield MS, Lewis KE, Roses AD, Lee JE, Hulette C, Pericak-Vance MA, Vance JM (1994): The Haw River Syndrome: Dentatorubropallidoluysian atrophy (DRPLA) in an African-American family. Nat Genet 7:521–524.

Clerget-Darpoux F, Bonaiti-Pellie C, Hochez J (1986): Effects of misspecifying genetic parameters in lod score analysis. Biometrics 42:393–399.

Conneally PM, Edwards JH, Kidd KK, Lalouel JM, Morton NE, Ott J, White R (1985): Report of the committee on methods of linkage analysis and reporting. Cytogenet Cell Genet 40:356–359.

Corder EH, Saunders AM, Risch NJ, Strittmatter WJ, Schmechel DE, Gaskell PC Jr, Rimmler JB, Locke PA, Conneally PM, Schmader KE, Small GW, Roses AD, Haines JL, Pericak-Vance MA (1994): Protective effect of apolipoprotein E type 2 allele for late onset Alzheimer disease. Nat Genet 7:180–184.

Corder EH, Saunders AM, Strittmatter WJ, Schmechel D, Gaskell PC, Rimmler JB, Locke PA, Conneally PM, Schmaer KE, Tanzi RE, Gusella J, Small GW, Roses AD, Pericak-Vance MA, Haines JL (1995a): Apolipoprotein E, survival in Alzheimer disease patients and the competing risks of death and Alzheimer disease. Neurology 45:1323–1328.

Corder EH, Saunders AM, Strittmatter WJ, Schmechel DE, Gaskell PCJ, Roses AD, Pericak-Vance MA, Small GW, Haines JL (1995b): The apolipoprotein E E4 allele and sex-specific risk of Alzheimer's disease [letter; comment]. JAMA 273:373–374.

Cottingham RW, Idury RM, Schaffer AA (1993): Faster sequential genetic linkage computations. Am J Hum Genet 53:252–263.

Dausset J, Cann H, Cohen D, Lathrop M, Lalouel JM, White R (1990): Centre d'Etude du Polymorphisme Humain (CEPH): Collaborative genetic mapping of the human genome. Genomics 6:575–577.

DeClue JE, Papageorge AG, Fletcher JA, Diehl SR, Ratner N, Vass WC, Lowry DR (1992): Abnormal regulation of mammalian p21 ras contributes to mailgnant tumor growth in von Recklinghausen (type 1) neurofibromatosis. Cell 69:265–273.

Falk CT (1991): A simple method for ordering loci using data from radiation hybrids. Genomics 9:120–123.

Farrer LA, Cupples LA, Haines JL, Hyman B, Kukull WA, Mayeux R, Myers RH, Pericak-Vance MA, Risch N, Van Duijn CM (1997): Effects of age, sex, and ethnicity on the association between apolipoprotein E genotype and Alzheimer disease. A meta-analysis. APOE and Alzheimer Disease Meta Analysis Consortium. JAMA 278: 1349–1356.

Fu YH, Kuhl DPA, Pizzuti A, Pieretti M, Sutcliffe JS, Richards S, Verkerk AJMH, Holden JJA, Fenwick RG Jr, Warren ST, Oostra BA, Nelson DL, Caskey CT (1991): Variation of the CGG repeat at the fragile X site results in genetic instability: Resolution of the Sherman paradox. Cell 67:1047–1058.

Garrod AE (1902): The incidence of alkaptonuria: A study in chemical individuality. Lancet ii:1616–1620.

Haldane JBS (1919): The combination of linkage values and the calculation of distances between two loci of linked factors. J Genet 8:299–309.

Hall JM, Friedman L, Guenther C, Lee MK, Weber JL, Black DM, King MC (1992): Closing in on a breast cancer gene on chromosome 17q. Am J Hum Genet 50:1235–1242.

Hardy GH (1908): Mendelian proportions in a mixed population. Science 28:41–50.

Juneja T, Dave S, Pericak-Vance MA, Siddique T (1997): Prognosis in familial ALS: Progression and survival in patients with E100G and A4V mutations in Cu, Zn superoxide dismutase. Neurology 48:55–57.

Knowles JA, Vieland VJ, Gilliam TC (1992): Perils of gene mapping with microsatellite markers. Am J Hum Genet 51:905–909.

Koenig M, Monaco AP, Kunkel LM (1988): The complete sequence of dystrophin predicts a rod-shaped cytoskeletal protein. Cell 53:219–228.

Kosambi DD (1944): The estimation of map distances from recombination values. Ann Eugen 12:172–175.

Lander ES, Green P (1987): Construction of multilocus genetic linkage maps in humans. Proc Natl Acad Sci USA 84:2363–2367.

Lange K, Boehnke M (1992): Bayesian methods and optimal experimental design for gene mapping by radiation hybrids. Ann Hum Genet 56:119–144.

Lathrop GM, Lalouel JM, Julier C, Ott J (1984): Strategies for multilocus linkage analysis in humans. Proc Natl Acad Sci USA 81:3443–3446.

Li W, Fann CSJ, Ott J (1998): Low-order polynomial trends of female-to-male map distance ratios along human chromosomes. Hum Hered 48:266–270.

Matise TC, Gitlin JA (1999): MAP-O-MAT: Marker-based linkage mapping on the World Wide Web. Am J Hum Genet 65:A435.

Matise TC, Perlin M, Chakravarti A (1994): Automated construction of genetic linkage maps using an expert system (MultiMap): A human genome linkage map. Nat Genet 6:384–390.

McInnis MG, McMahon FJ, Chase GA, Simpson SG, Ross CA, DePaulo JR Jr (1993): Anticipation in bipolar affective disorder. Am J Hum Genet 53:385–390.

Morton NE (1955): Sequential tests for the detection of linkage. Am J Hum Genet 7:277–318.

O'Connell JR, Weeks DE (1995): The VITESSE algorithm for rapid exact multilocus linkage analysis via genotype set-recoding and fuzzy inheritance. Nat Genet 11:402–408.

Ott J (1989): Computer simulation methods in human linkage analysis. Proc Natl Acad Sci USA 86:4175–4178.

Ott J (1992): Strategies for characterizing highly polymorphic markers in human gene mapping. Am J Hum Genet 51:283–290.

Ott J (1999): Analysis of Human Genetic Linkage, 3rd ed. Baltimore, MD: Johns Hopkins University Press.

Penrose LS (1955): Parental age and mutation. Lancet ii:312–313.

Pericak-Vance MA, Bebout JL, Gaskell PC, Yamaoka LH, Hung WY, Alberts MJ, Walker AP, Bartlett RJ, Haynes CS, Welsh KA, Earl NL, Heyman A, Clark CM, Roses AD (1991): Linkage studies in familial Alzheimer's disease: Evidence for chromosome 19 linkage. Am J Hum Genet 48:1034–1050.

Pericak-Vance MA, Johnson CC, Rimmler JB, Saunders AM, Robinson LC, D'Hondt EG, Jackson CE, Haines JL (1996): Alzheimer's disease and apolipoprotein E-4 allele in an Amish population. Ann Neurol 39:700–704.

Ploughman LM, Boehnke M (1989): Estimating the power of a proposed linkage study for a complex genetic trait. Am J Hum Genet 44:543–551.

Post SG, Whitehouse PJ, Binstock RH, Bird TD, Eckert SK, Farrer LA, Fleck LM, Gaines AD, Juengst ET, Karlinsky H, Miles S, Murray TH, Quaid KA, Relkin NR, Roses AD, George-Hyslop PH, Sachs GA, Steinbock B, Truschke EF, Zinn AB (1997): The clinical introduction of genetic testing for Alzheimer disease. An ethical perspective. JAMA 277:832–836.

Raskind WH, Pericak-Vance MA, Lennon F, Wolf J, Lipe HP, Bird TD (1997): Familial spastic paraparesis: Evaluation of locus heterogeneity, anticipation and haplotype mapping of the SPG4 locus on the short arm of chromosome 2. Am J Med Genet 74:26–36.

Rosen DR, Siddique T, Patterson D, Figlewicz DA, Sapp P, Hentati A, Donaldson D, Goto J, O'Regan JP, Deng HX, Rahmani Z, Krizus A, McKenna-Yasek D, Cayabyab A, Gaston SM, Berger R, Tanzi RE, Halperin JJ, Herzfeldt B, Vanden Bergh R, Hung WY, Bird T, Deng G, Molder DW, Smyth C, Laing NG, Soriamo E, Perickk-Vance MA, Haines JL, Rouleau GA, Gusella JS, Horuitz HR, Brown Jr RH (1993): Mutations in Cu/Zn superoxide dismutase gene asr associated with familial amyotrophic lateral sclerosis. Nature 362:59–62.

Rousseau F, Bonaventure J, Legeai-Mallet L, Pelet A, Rozet J-M, Maroteaux P, Le Merrer M, Munnich A (1994): Mutations in the gene encoding fibroblast growth factor receptor-3 in achondroplasia. Nature 371:252–254.

Saunders AM, Strittmatter WJ, Schmechel D, George-Hyslop PH, Pericak-Vance MA, Joo SH, Rosi BL, Gusella JF, Crapper-MacLachlan DR, Alberts MJ (1993): Association of apolipoprotein E allele epsilon 4 with late-onset familial and sporadic Alzheimer's disease. Neurology 43:1467–1472.

Schaffer AA, Gupta SK, Shriram K, Cottingham RW (1994): Avoiding recomputation in linkage analysis. Hum Hered 44:225–237.

Schellenberg GD, Bird TD, Wijsman EM, Orr HT, Anderson L, Nemens E, White JA, Bonnycastle L, Weber JL, Alonso ME, Potter H, Heston LL, Martin GM (1992): Genetic linkage evidence for a familial Alzheimer's disease locus on chromosome 14. Science 258:668–671.

Scott WK, Gaskell PC, Lennon F, Wolpert C, Menold MM, Aylsworth AS, Warner C, Farrell CD, Boustany RMN, Albright SG, Boyd E, Kingston HM, Cumming WJK, Vance JM, Pericak-Vance MA (1997): Locus heterogeneity, anticipation, and reduction of the chromosome 2p minimal candidate region in autosomal dominant familial spastic paraplegia. Neurogenetics 1:95–102.

Sherman SL, Jacobs PA, Morton NE, Froster-Iskenius U, Howard-Peebles PN, Nielsen KB, Partington MW, Sutherland GR, Turner G, Watson M (1985): Further segregation analysis of the fragile X syndrome with special reference to transmitting males. Hum Genet 69:289–299.

Shiang R, Thompson LM, Zhu Y-Z, Church DM, Fielder TJ, Bocian M, Winokur ST, Wasmuth JJ (1994): Mutations in the transmembrane domain of FGFR3 cause the most common genetic form of dwarfism, achondroplasia. Cell 78:335–342.

Speer MC, Gilchrist JM, Stajich JM, Gaskell PC, Westbrook CA, Horrigan SK, Bartoloni L, Yamaoka LH, Scott WK, Pericak-Vance MA (1998): Evidence for anticipation in autosomal dominant limb-girdle muscular dystrophy. J Med Genet 35:305–308.

Stoll C, Roth M-P, Bigel P (1982): A reexamination of parental age effect on the occurrence of new mutations for achondroplasia. In: Papadatos CJ, Bartscocas CS, eds. Skeletal Dysplasias. New York: Alan R. Liss.

Strachan T, Read AP (1996): Human Molecular Genetics, 1st ed. New York: John Wiley & Sons.

Tawil R, Forrester J, Griggs RC, Mendell J, Kissel J, McDermott M, King W, Weiffenbach B, Figlewicz D (1996): Evidence for anticipation and association of deletion size with severity in facioscapulohumeral muscular dystrophy. Ann Neurol 39:744–748.

Tjio JH, Levan A (1956): The chromosome number of man. Hereditas 42:1–6.

Valero MC, Valasco E, Moreno F, Hernandez-Chico C (1994): Characterization of four mutations in the neurofibromatosis type 1 gene by denaturing gradient gel electrophoresis (DGGE). Hum Mol Genet 3:639–641.

Watson JD (1968): The Double Helix. New York: Atheneum.

Watzke RC, Weingeist TA, Constantine JB (1977): Diagnosis and management of von Hippel-Lindau disease. In: Peyman GA, Apple DJ, Sanders DR, eds. Intraocular Tumors, 1st ed. New York: Appleton\Century\Crofts.

Weeks DE, Lehner T, Squires-Wheeler E, Kaufmann C, Ott J (1990): Measuring the inflation of the lod score due to its maximization over model parameter values in human linkage analysis. Genet Epidemiol 7:237–243.

Weeks DE, Ott J, Lathrop GM (1994): Detection of genetic interference: Simulation studies and mouse data. Genetics 136:1217–1226.

Weeks DE, Sobel E, O'Connell JR, Lange K (1995): Computer programs for multilocus haplotyping of general pedigrees. Am J Hum Genet 56:1506–1507.

Weinberg W (1908): Uber den nachweis der vererbung beim Menschen. Jahreshefte des vereins fur vaterlandische naturkunde in wurttemberg. Wurttemberg 64:368–382.

Wu BL, Austin MA, Schneider GH, Boles RG, Korf BR (1995): Deletion of the entire NF1 gene detected by the FISH: Four deletion patients associated with severe manifestations. Am J Med Genet 59:528–535.

Zatz M, Suely K, Marie M, Passos-Bueno R, Vainzof M, Campiotto S, Cerqueira A, Wijmenga C, Padberg G, Frantz R (1995): High proportion of new mutations and possible anticipation in Brazilian facioscapulohumeral muscular dystrophy families. Am J Hum Genet 56:99–105.

■■■■■■ CHAPTER 2

Defining Disease Phenotypes

ARTHUR S. AYLSWORTH

INTRODUCTION

During the second decade of the twentieth century, Thomas Hunt Morgan and his students made discoveries and developed intellectual concepts that laid the groundwork for the Human Genome Project and our current attempts to identify genes that cause or predispose to human diseases. Working in the famous "fly room" at Columbia University, they discovered that chromosomes are the physical basis of heredity and that genes are specific entities arranged along the lengths of individual chromosomes. An additional important contribution was the delineation of spontaneous and induced *Drosophila* mutant phenotypes. Morgan and his students understood that accurate phenotype definition is an essential requirement for any scientific inquiry into mechanisms of heredity.

The importance of phenotype definition and delineation cannot be overemphasized. For what we refer to now as "gene mapping" is actually *phenotype mapping*, and identification of new gene loci is merely a by-product of our current ignorance. Even now, after most functional loci in the human genome have been mapped and sequenced, there will still remain huge gaps in our knowledge about genotype–phenotype correlations. As long as there is interest in identifying genes associated with human traits and conditions, there will be a continuing need for accurate phenotyping.

Modern-day human genetic research deals with much more complex diseases and genetic mechanisms than imagined by our early twentieth-century forebears. The previous chapter introduced basic concepts of chromosomal structure, Mendelian inheritance, novel types of mutations, and susceptibility as opposed to causation. This chapter continues these themes, with further discussion of some mechanisms that underlie the variations from traditional Mendelian patterns that characterize many clinical traits and disorders. We conclude with a consideration of issues involved in choosing a phenotype for study.

Genetic Analysis of Complex Diseases, Second Edition, Edited by Jonathan L. Haines and Margaret Pericak-Vance

Throughout this chapter, a few specific references are provided, but for descriptions and up-to-date references regarding specific genetic diseases and traits, the reader is encouraged to utilize the online version of *Mendelian Inheritance in Man* (http://www.ncbi.nlm.nih.gov/Omim/). Many topics discussed here include references to pertinent "MIM numbers" (e.g., MIM xxxxxx; MIM xxxxxx.xxxx refers to specific mutant alleles). Some topics have too many relevant OMIM entries to list. These can be searched in OMIM using topical keywords and Boolean operators. The OMIM database is an extremely valuable resource for those engaged in disease gene discovery studies. It is a repository for information about human phenotypes and their correlation with mapping and genotyping data. As such, OMIM serves as a companion resource to other public databases that store, organize, and make available sequence data. Because OMIM is updated daily, its entries will provide the reader with a source of up-to-date information about the human diseases and traits used as examples in this chapter.

EXCEPTIONS TO TRADITIONAL MENDELIAN INHERITANCE PATTERNS

One goal of human genetics is to delineate the causal associations and pathogenetic processes that relate genotype to phenotype. Human pedigree patterns often vary drastically from the classic Mendelian patterns described in Chapter 1. Mendel spent a number of years selecting the particular traits on which he finally collected extensive crossing data. This was necessary because most traits in plants and animals do not "Mendelize"; that is, most phenotypes are not caused by single, causative genes and do not follow the straightforward patterns of unitary inheritance discovered by Mendel. Instead, most pedigrees for common conditions in both children and adults resemble Figure 2.1, where distant relatives are affected and no Mendelian inheritance pattern is obvious.

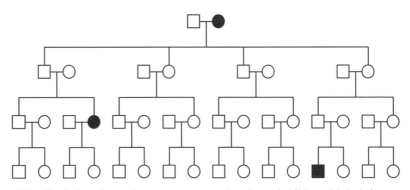

Figure 2.1. Typical pedigree for many *common* disorders of children (birth defects, mental retardation) and adults (heart disease, cancer, diabetes, stroke, hypertension, arthritis, psychiatric disease). Such disorders or traits frequently appear to be familial but have no obvious, straightforward Mendelian pattern of inheritance. Solid symbols = affected.

A number of factors can affect gene expression and cause clinical expression to be different from deterministic, straightforward dominant or recessive patterns. Heterozygous mutant allele expression ranges from complete (the deterministic pattern traditionally called *Mendelian dominant*) to silent (*Mendelian recessive*). In the latter case, a clinical phenotype is expressed only by the mutant homozygote or allelic compound. In between these two extremes lies a range of partial or intermediate expression, where the phenotype produced by an allele may depend on one or more interactions with products of the other homologous allele, alleles at other loci, environmental factors, and/or chance. Such variations from straightforward Mendelian inheritance patterns confound attempts to identify genetic causation. Following are some examples of exceptions to traditional nuclear inheritance patterns that may cause confusion in genetic analysis studies.

Pseudodominant Transmission of a Recessive

Vertical inheritance (expression in subsequent generations) of a genetic disorder, as shown in Figure 2.2*a*, usually implies dominant expression of a mutant allele in the heterozygote. Mendelian autosomal recessive phenotypes may appear in a parent and one or more children, however, thus mimicking a dominant form of the phenotype. For rare recessives, the usual explanation for vertical transmission is consanguinity, where an affected, homozygous parent mates with a relative who is heterozygous by virtue of descent from a common ancestor, as shown in Figure 2.2*b*. On the other hand, for conditions where the mutant allele frequency

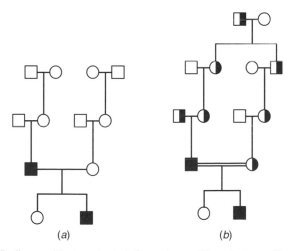

(a)   (b)

Figure 2.2. (*a*) Pedigree with phenotypic information: solid symbols = affected. This appears to be a straightforward dominant, but is it necessarily so? (*b*) Pseudodominant transmission of a recessive trait. The same pedigree as in (*a*) but with genotype data and one more generation added. Solid symbols = homozygous and affected; half-open symbols = heterozygous and unaffected.

is relatively common, such as sickle cell disease and hemochromatosis, an affected individual can easily have a child with the same autosomal recessive condition by mating with an unrelated heterozygote from the general population.

Pseudorecessive Transmission of a Dominant

A typical characteristic of autosomal recessive inheritance is that siblings are frequently affected without any other family history of affected relatives (Fig. 2.3). Other situations may produce such a situation, however, and masquerade as apparent recessive inheritance. Some of these will be discussed in greater detail later. Following is a brief list of some of the more common causes.

Mistaken/Misassigned Paternity. Multiple affected siblings with apparently unaffected parents will be ascertained when the identity of an affected biological father is hidden from the research team and the putative father is unaffected. The term "nonpaternity," commonly misused for this phenomenon, is obviously non-sensical and its usage should be discouraged and replaced by a more correct term such as *mistaken* or *misassigned paternity.*

Incomplete Penetrance and Variable Expression. These are defined and discussed below in more detail. *Incomplete penetrance* causes carriers of a dominantly expressed mutant allele to appear unaffected and be classified incorrectly. Such a "nonpenetrant" parent who carries a dominantly expressed mutant allele may have several affected children. Careful examination of apparently unaffected relatives for mild, early, or subclinical manifestations may be necessary to resolve this uncertainty. Ideally, the phenotype under study should be clearly defined and identified so that assignment of affected and unaffected individuals can be accurate. This may not be possible, especially for disorders that have extremely variable expressivity or complex inheritance. *Genomic imprinting*, also discussed further below, is another mechanism that results in nonpenetrance. Modification of allele

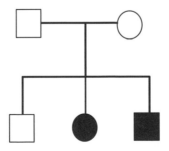

Figure 2.3. Pseudorecessive transmission of a dominant. There are many other mechanisms that need to be considered before concluding that autosomal recessive causation is most likely. Solid symbols = affected.

expression is based on the gender of the transmitting parent. An expanded pedigree analysis may help identify this phenomenon.

Germline Mosaicism. As discussed below, a mutation that occurs after the formation of the zygote results in "mosaicism," where the developing organism contains some cells with the mutant allele and some without. When a mutation occurs in a cell destined only for the germline, then such an unaffected carrier can transmit the mutation with full phenotypic expression to multiple children.

Mosaicism

Mosaicism can be present at some level in all tissues of the body or may be unevenly distributed, predominating in some tissues but not others. We see specific phenotypic effects associated with *somatic mosaicism* and *germline mosaicism*.

Somatic Mosaicism. Irregular distribution of mosaicism for a mutant allele in body tissues and organs can cause a very asymmetric pattern of phenotypic expression. For example, an individual with somatic mosaicism for a mutation in the gene that causes neurofibromatosis type 1 (NF1) (MIM 162200) may have café-au-lait spots and neurofibromas that involve only one segment of the body (Fig. 2.4). If distribution is more even throughout the body, the result can be an unusually mild expression of the phenotype. This is demonstrated by families where parents with phenotypes of Stickler syndrome (MIM 108300) or spondyloepiphyseal dysplasia congenita (MIM 183900) were mosaic for mutations in the type II collagen gene (MIM 120140) and gave birth to children with a more severe condition known as Kniest dysplasia (MIM 156550). By this same mechanism, a parent with a mild form of osteogenesis imperfecta (OI) could have a child with a severe, lethal form of

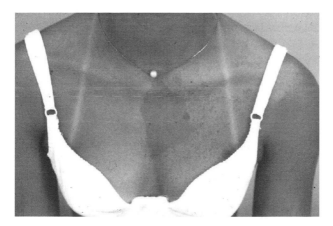

Figure 2.4. Somatic mosaicism for the *NF1* gene. The affected individual has small cutaneous neurofibromas (not visible) and café-au-lait spots on her left (but not her right) arm and hand, shoulder, axilla, back, and chest. Note the large spot that stops at the midline. She has no Lisch nodules and no other features of *NF1* elsewhere.

OI (MIM 166210). Normally, milder forms of OI tend to "breed true" within families, with affected members having approximately similar clinical manifestations. But a parent with mild OI *who is the first one in the family to be affected* may be mildly affected because the mutation occurred after fertilization and is present in only a fraction of the body cells. Therefore, somatic mosaicism for a mutation that usually causes a severe or even lethal phenotype can be associated with a much less severe phenotype.

Another example of this is neurofibromatosis type 2 (NF2) (MIM 101000). Evans et al. (1998) concluded that "somatic mosaicism is likely to be a common cause of classic NF2 and may well account for a low detection rate for mutations in sporadic cases."

Germline Mosaicism. Mentioned above as a cause of multiple affected siblings from unaffected parents, this situation is one in which an individual has mosaicism for a mutation that is limited to gonadal cells. A carrier of such a germline mutation has the potential to transmit a mutant allele to multiple offspring, each of whom would have full phenotypic expression. Osteogenesis imperfecta (see MIM 166210 and several dozen other related entries) is an example of this phenomenon of germline mosaicism. The observed sibling recurrence rate of severe OI (where neither parent is affected) has been reported to be as high as 5–6%. This was originally interpreted as evidence for approximately 25% of cases being autosomal recessive (with a 25% recurrence risk) and 75% being new dominant mutations (with essentially a negligible recurrence risk). Now, however, it is known that most severe cases of OI are caused by heterozygous mutations in type I collagen genes, *COL1A1* (MIM 120150) and *COL1A2* (MIM 120160). This suggests that most of the observed recurrences must be due to germline mosaicism in one of the parents (Cole and Dalgleish, 1995).

Mitochondrial Inheritance

General references: Wallace, 1995; Grossman, 1996; Castro et al., 1998; Rothman, 1999; Chinnery et al., 1999; Disotell, 1999; Beal, 2000; Cottrell et al., 2000.

A zygote receives most of its mitochondria from the egg, not the sperm, although rare exceptions are documented (Schwartz and Vissing, 2002). Expression of mitochondrial gene mutations, therefore, typically follows a pattern of "maternal inheritance," where all of the offspring of an affected woman receive copies of the mutant gene but there is no transmission through affected males. Maternal mitochondria (and, therefore, traits encoded by mitochondrial genes) are transmitted to both sexes. Pedigrees may resemble the idealized one in Figure 2.5, which resembles autosomal dominant inheritance. On closer inspection, however, the key characteristic maternal inheritance is apparent. Variability of expression of mitochondrial mutations within families may be extensive. This is due to *heteroplasmy*, the existence of differing proportions of mutant and wild-type mitochondrial DNA in different relatives. Interesting examples of human conditions with mitochondrial inheritance include several myopathic syndromes (MIM 530000, 540000,

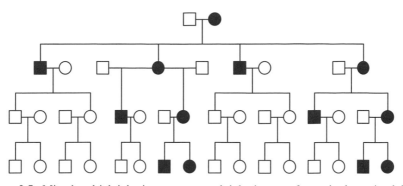

Figure 2.5. Mitochondrial inheritance: maternal inheritance of a trait determined by a mutation in mitochondrial DNA. The key features illustrated are that all offspring of affected women are affected, while none of the offspring of affected males are affected. Note that one woman in the F_2 generation has had an affected son and daughter by two different mates. Solid symbols = affected.

545000), type 2 diabetes and deafness (MIM 520000), chloramphenicol resistance and toxicity (MIM 515000), and aminoglycoside-induced hearing loss (MIM 580000). Note that this last entity, aminoglycoside-induced hearing loss, involves a genetic predisposition and also requires an environmental exposure for expression of the trait. A typical pedigree might look like the one in Figure 2.6. When the female in the last generation presents for evaluation and a family history is taken, it obviously would be very easy to overlook the possibility of a mitochondrial trait. Currently OMIM lists over 60 mitochondrial entries and close to two dozen phenotypes related to mitochondrial DNA mutations.

Investigators studying complex phenotypes should keep in mind that while most of the genes influencing a trait of interest may be nuclear, one or more subgroups

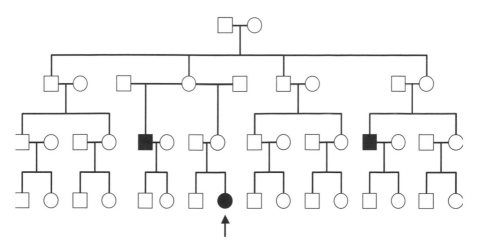

Figure 2.6. Mitochondrial inheritance with expression dependent on environmental exposure [e.g., aminoglycoside-induced hearing loss (MIM 580000)]. Solid symbols = affected.

may have a significant genetic component that is mitochondrial. For example, there is good evidence for autosomal recessive inheritance of the Wolfram syndrome (MIM 222300), also referred to as DIDMOAD (diabetes insipidus, diabetes mellitus, optic atrophy, and deafness). It is caused by mutations in a gene at 4p16.1, *WFS1*, which encodes a putative transmembrane protein that appears to function in survival of islet β cells and neurons, thus explaining the pleiotropic features of the Wolfram phenotype. But evidence also exists for a rarer, mitochondrial form (MIM 598500). Another example is idiopathic dilated cardiomyopathy, a phenotype with over a dozen OMIM entries and many different regions of the nuclear genome implicated. Mitochondrial deletions in an affected mother-and-son pair, however, suggest this mechanism may be responsible for a subset of cases (MIM 510000). Finally, mitochondrial genes have been implicated in the causation of both Alzheimer's disease (MIM 502500) and Parkinson's disease (MIM 556500), complex phenotypes of great interest to the medical community because of prevalence and morbidity.

Incomplete Penetrance and Variable Expressivity

Penetrance refers to the proportion of individuals with a particular mutant genotype that express the mutant phenotype. Penetrance is a proportion that ranges between 0 and 1.0 (or 0 and 100%). When 100% of individuals with alleles mutant individuals express the phenotype, penetrance is complete. If some individuals with mutant alleles do not express the phenotype, penetrance is said to be *incomplete* or *reduced*. Dominant conditions with incomplete penetrance, therefore, are characterized by "skipped" generations with unaffected, obligate gene carriers. In the heterozygote, a recessively expressed allele is, by definition, nonpenetrant.

Penetrance, therefore, is the proportion of heterozygotes that express a dominant phenotype or the proportion of homozygotes (or allelic compounds) that express the recessive phenotype (Fig. 2.7). Conditions known for incomplete penetrance include split hand/foot malformation (MIM 183600) and hemochromatosis (MIM 235200). Note also that penetrance depends on definition of the phenotype. Carriers of the sickle cell gene (MIM 141900) express (i.e., are penetrant for) the trait of in vitro sickling but do not express (i.e., are nonpenetrant for) the disease sickle cell anemia.

Expressivity (expression) refers to the variability in degree of phenotypic expression (i.e., severity) seen in different individuals with the mutant genotype. Expressivity may be extremely variable or fairly consistent. *Intrafamilial* variability of expression may be due to factors such as epistasis, environment, chance, and mosaicism. *Interfamilial* variability of expression may be due to these factors as well but may also be due to allelic or locus genetic heterogeneity. As noted elsewhere, X-linked gene expression is potentially "intermediate" and inherently variable in heterozygous females because of lyonization. Also, mitochondrial mutation expression typically is extremely variable within families because of heteroplasmy.

Because of the possibility of incomplete penetrance and variable expression, phenotypers should *examine* relatives of probands rather than just take a verbal history. Failure to examine relatives carefully for minor expressions of syndromes

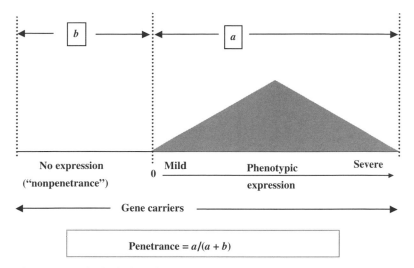

Figure 2.7. Schematic depiction of penetrance and expression. Penetrance is the proportion of individuals with a particular mutant genotype that express the mutant phenotype, while the degree of expression may vary widely from mild to severe.

or early manifestations of late-onset disease can result in significant phenotyping errors.

The term *variable penetrance* is commonly, but *incorrectly*, substituted for *variable expressivity*. Variable penetrance is strictly appropriate only for conditions that have different penetrance figures for different populations that are stratified by parameters such as age, race, or pedigree position. For example, all disorders of postnatal onset have penetrance values that vary with age. But since misapplication of variable penetrance has robbed the term of its usefulness, alternative terms for instances of this phenomenon are used below.

Incomplete or Semidominant Expression. Mendel's original definition of *dominant* said that homozygous expression is the same as heterozygous expression. While some human disorders that we call *autosomal dominant* may actually fit this definition, many may not. Our knowledge of homozygous phenotypes is often limited by the rarity of heterozygote matings for most so-called dominant human conditions. Current evidence suggests that Huntington disease (MIM 143100) is a true dominant, but achondroplasia (MIM 100800), to name just one example, clearly is not. Homozygous achondroplasia is much more severe than heterozygous achondroplasia, with homozygotes usually dying in the neonatal period. Therefore, while the term s*emidominant* is used in mouse and other experimental genetic systems where homozygous phenotypes are known, dominant is commonly used in human medicine for traits that show vertical transmission, regardless of whether the homozygous phenotype is known to be the same as the heterozygous phenotype. Investigators studying large families, especially inbred ones, should try to identify and define homozygous phenotypes and be alert to the possibility of different phenotypic expression in homozygotes.

Gender-Influenced Penetrance and Expression. Expression may be affected by the sex of the individual carrying a particular gene. For example, male pattern baldness (MIM 109200) appears to be expressed as an autosomal dominant trait in males and as an autosomal recessive in females. Susceptibility to breast cancer preferentially affects females, and susceptibility to ovarian cancer, by necessity, is limited to females. Similarly, genetically caused or influenced congenital anomalies of the internal or external genitalia (such as hypospadias or vaginal septum) are, by necessity and definition, limited by gender.

Age-Related Penetrance and Expression (Variable Age at Onset). In all conditions of postnatal onset, the proportion of genetically susceptible individuals that are affected varies with age. Therefore, one must specify age when describing penetrance in all conditions that are not congenital. For some well-studied conditions such as Huntington disease, age-of-onset data are fairly good, but data are incomplete for many others. Disease gene discovery studies should include a thorough search of the literature and the study population to define as well as possible, age-related penetrance values.

Transmission-Related Penetrance and Expression (Dynamic Mutations). General references: O'Donnell and Zoghbi, 1995; Warren and Ashley, Jr., 1995; Monckton and Caskey, 1995; Sutherland and Richards, 1995; Warren, 1996; Lindblad and Schalling, 1996, 1999; Li and el Mallakh, 1997; Koshy and Zoghbi, 1997; Nance, 1997; La Spada, 1997; Mitas, 1997; Tsuji, 1997; Koeppen, 1998; Basu et al., 2000; Schelhaas et al., 2000; Stevanin et al., 2000; Lin et al., 2000; Nussbaum and Auburger, 2000; Morrison, 2000; Cummings and Zoghbi, 2000; Ohara, 2001.

The phenomenon called *true anticipation* causes expression to become more severe and/or have an earlier age of onset as a condition is transmitted from generation to generation. This is due to a region of the gene being "unstable" because it contains a trinucleotide repeat. Normally, a few such repeats will be "stable" while a moderate number of repeats (the critical number varies with the disorder) will produce "instability" of the sequence, leading to a larger number of repeats and disrupted gene function in subsequent generations. Examples include fragile X syndrome (MIM 309550), myotonic dystrophy (MIM 160900), Kennedy disease (MIM 313200), Huntington disease (MIM 143100), dentatorubro-pallidoluysian atrophy (DRPLA) and Haw River syndrome (MIM 125370, 140340), spinocerebellar ataxias (SCA1,2,6,7) (MIM 164400, 183090, 183086, 164500), and SCA3 (Machado-Joseph disease) (MIM 109150).

Note that another explanation for *apparent anticipation* is biased ascertainment. For conditions with a wide range of intrafamilial expressivity, one is much more likely to ascertain families where symptoms are more severe (or started earlier) in later generations than those in which the condition was much more severe in earlier generations. Biased ascertainment must always be considered as a possible explanation for apparent anticipation.

Genomic Imprinting

General references: Hall, 1990; Gold and Pedersen, 1994; Driscoll, 1994; Latham et al., 1995; Feinberg et al., 1995; Ledbetter and Engel, 1995; Barlow, 1995; Sapienza, 1995; Cassidy, 1995; Latham, 1996; Jirtle, 1999; Tilghman, 1999; Horsthemke et al., 1999; Brannan and Bartolomei, 1999; Falls et al., 1999; Hall, 1999b; Sleutels et al., 2000; Pfeifer, 2000; Nicholls, 2000; Paulsen and Ferguson-Smith, 2001; Reik and Walter, 2001; Tycko and Morison, 2002; Sleutels and Barlow, 2002; see also genomic imprinting websites at http://www.geneimprint.com and http://cancer.otago.ac.nz/IGC/Web/home.html.

Genetic contributions from both a mother and father are required for mammalian development. This requirement is related to mechanisms of genetic regulation that cause some normal genes to be differentially expressed, depending on the gender of the parent from whom they are inherited. Imprinting implies suppression of expression. Such parent-of-origin differences in gene expression have been identified in humans largely through the study of genetic disorders that are influenced by this mechanism (Warren and Ashley, Jr., 1995; Reeve, 1996; Rougeulle and Lalande, 1998; Shaffer et al., 1998; Jiang et al., 1998; Mann and Bartolomei, 1999; Greally, 1999; Skuse, 1999; Jirtle, 1999; Lalande et al., 1999). Figure 2.1 was originally used as an example of complex, non-Mendelian inheritance, where distantly related individuals are affected, with no affected relatives in the lines of descent between them. Closer inspection, however, reveals that it is compatible with *maternal imprinting*, where both gene expression and the phenotype are suppressed in maternal transmission. In other words, this pedigree could be produced by a single, dominantly expressed mutation in an autosomal gene that is maternally imprinted (i.e., the maternally transmitted allele is not expressed). Offspring of affected mothers are, therefore, protected from the effects of a deleterious mutant allele, but the mutant phenotype is expressed when inherited from a father, who may or may not be affected depending on the parent from whom he inherited the gene. Figure 2.8 clarifies the situation. In addition to the three affected individuals, there are three others who are obligate gene carriers, because each has both an affected ancestor and an affected descendent. Also, there are another 13 relatives whose carrier status is unknown. Note that layers of complexity can be added to the interpretation of such pedigrees if one considers the possibility of an imprinted repressor of an unimprinted mutant gene, or mutations that affect the imprinting process directly.

Uniparental Disomy. General references: Ledbetter and Engel, 1995; Cassidy, 1995; Lalande, 1996; Mutter, 1997; Kotzot, 1999; Hall, 1999a; Robinson, 2000.

Inheritance of both homologues of a chromosome pair from one parent with corresponding loss of the chromosomal contribution from the other parent is called *uniparental disomy* (UPD), an obvious exception to Mendel's "law" of

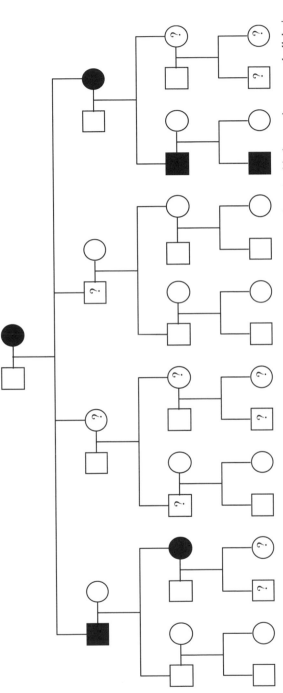

Figure 2.8. Maternal imprinting of a phenotype caused by a dominantly expressed, autosomal mutation. At this locus, the maternal allele is not expressed, and the paternal allele is expressed. Affected individuals, therefore, receive the mutant gene from their fathers. Solid symbols = affected, that is, the phenotype is expressed. Speckled symbols = obligate carriers. Symbols with a ? represent relatives for whom carrier status is unknown.

segregation of alleles. Note that if this corresponding loss of a chromosome from the other parent did not occur, then the zygote would be trisomic. Uniparental disomy may cause developmental abnormalities if the embryo/fetus is homozygous for a recessive, mutant allele, which would be carried by only one parent. If the chromosome involved in UPD contains imprinted loci, then there will be gene product deficiency (two copies of an imprinted gene) or gene product overproduction (two copies of a normally expressed allele at an imprinted locus). Note that uniparental disomy can result in male-to-male transmission of an X-linked trait if a son receives both his X and Y chromosomes from his father.

Phenocopies and Other Environmentally Related Effects

The term *phenocopy* refers to an environmentally caused phenotype that mimics a genetic trait or syndrome. For example, while most cases of DiGeorge syndrome (MIM 188400; 600594) are associated with a chromosome 22q deletion, the phenotypic features can also be produced by prenatal exposure to 13-*cis*-retinoic acid. By the same token, some affected relatives in a family segregating an adult-onset disease such as cancer or diabetes may not share the predisposing gene of interest that is carried by most other affected relatives. Instead, they may have a condition due entirely to factors unrelated to the segregating susceptibility gene of interest, which would cause them to be diagnosed and coded incorrectly in a linkage study.

Some generalize the term phenocopy to include phenomena thought to be a part of the normal aging process, such as age-related macular degeneration, which can mimic autosomal dominant macular dystrophy. This usage would be appropriate if an environmental agent is identified as a major causative factor. But not everyone develops "age-related" macular degeneration, and, in fact, there is great variability in morbidity with aging. It seems likely, therefore, that genetic factors play major roles in determining individual susceptibilities to phenomena considered part of normal aging. Other major genetic causes of these conditions are, then, examples of *genetic heterogeneity*. An oxymoron, "genetic phenocopy" is sometimes used in situations when genetic heterogeneity is meant. The term phenocopy should be reserved for situations where environmental factors closely mimic the effects of mutant gene expression.

Note that environmentally caused phenotypes may also mimic genetic inheritance by being familial. For example, the detrimental effects of prenatal alcohol exposure (so-called fetal alcohol syndrome, or FAS) cause a phenotype usually characterized by short stature and mental retardation, with or without a variety of other congenital anomalies. This phenotype may be seen in several close relatives over several generations, suggesting dominant transmission, if alcoholism occurs in several generations.

While the actual environmental exposure is presumed to be the major causative factor for conditions like FAS, one must also keep in mind the possibility that an underlying *genetic susceptibility* in the embryo, fetus, or adult may affect an individual's response to environmental factors. Chemicals, drugs, maternal disease, infection, and physical agents may disrupt both prenatal morphogenesis and postnatal function. These susceptibilities, which involve gene–environment interaction, fall into the category of "complex" disease. Clearly of great importance to individual and public health, they are receiving increasing research interest and are discussed further in Chapter 14.

HETEROGENEITY

Genetic Heterogeneity

A number of different genetic mutations may produce phenotypes that are identical or similar enough to have been traditionally lumped together under one diagnostic heading. In other words, a particular phenotype may be caused by more than one genotype. For example, the medical conditions of anemia, mental retardation, cancer, and dwarfism are obviously causally heterogeneous. On the other hand, conditions such as tuberous sclerosis (MIM 191092, 191100) and adult polycystic kidney disease (MIM 173900) were originally thought of as genetically homogeneous disorders until the relatively recent discovery of genetic heterogeneity. There are two types of genetic heterogeneity, *allelic heterogeneity* and *locus heterogeneity*.

Allelic Heterogeneity. A phenotype may be caused by more than one allele at a specific gene locus; that is, different mutations at a single locus cause the same phenotype. It is now clear that one of the earliest studied genetic diseases, sickle cell disease (MIM 141900), a disorder caused by one specific base substitution, is the exception rather than the rule. Instead, most disease phenotypes are more like the β-thalassemias, in which a variety of different kinds of mutations throughout the β-globin locus cause the disease phenotype. A few other examples of this common phenomenon are shown in Table 2.1.

Locus Heterogeneity. A phenotype may be caused by mutations at more than one gene locus; that is, mutations at different loci cause the same phenotype or a group of phenotypes that appear so similar that traditionally they have been classified as a single disease, clinical "entity," or diagnostic spectrum. Several examples are listed in Table 2.2.

TABLE 2.1. Allelic Heterogeneity

Different Mutations In	Cause
COL1A1	Osteogenesis imperfecta
CFTR	Cystic fibrosis
β-Globin	β-Thalassemia
FGFR2	Crouzon syndrome
dystrophin	Duchenne dystrophy
NF1	Neurofibromatosis type 1

TABLE 2.2. Locus Heterogeneity

Mutations In	Are Implicated in Causing
Genes on 9q and 16p	Tuberous sclerosis
COL1A1 and *COL1A2*	Osteogenesis imperfecta
BRCA1 and *BRCA2*	Breast cancer
Genes on 17(q21, 25), 14, and 12	MPS III (Sanfilippo syndrome)
Genes on 4 and 16	AD polycystic kidney disease
Genes on 2p, 3q, 8p, 8q, 10q, 11q, 12p, 12q, 14q, 15q, 16q, 19q, and Xq	Familial spastic paraplegia
FGFR1, FGFR2	Pfeiffer syndrome

When a distinctive phenotype is produced by alleles at more than one gene locus, it suggests that the gene products interact in some way or that they are necessary components of a critical developmental or metabolic pathway. An instructive example is the *Sanfilippo syndrome*, an autosomal recessive disorder originally classified as a single phenotype, mucopolysaccharidosis type III. There are now four recognized subtypes, A, B, C, and D (MIM 252900, 252920, 252930, 252940). Each subtype is caused by deficiency of a different lysosomal enzyme, all four of which are required for the stepwise degradation of heparan sulfate. Deficiency of any one of these results in heparan sulfate accumulation in body tissues, causing the clinical manifestations of the Sanfilippo disease phenotype.

Another example is the Mendelian dominant, adult polycystic kidney disease (PKD or APKD) (MIM 173900, 601313, 173910, 600666). Two proteins, polycystin 1 encoded by *PKD1* on chromosome 16p13.3–p13.12 and polycystin 2 encoded by *PKD2* on chromosome 4q21–q23, dimerize at the cell membrane to produce a new ion channel that regulates kidney function and morphology. Neither protein alone is sufficient, and germline mutations in either gene can lead to inactivation of this critical channel activity when a second, somatic mutational "hit" occurs. Furthermore, a third locus is suspected because some families are not linked to either the *PKD1* or *PKD2* loci. Studies of one of these unlinked families, however, resulted in a fascinating discovery (Pei et al., 2001). A large family was excluded from linkage to either the *PKD1* locus or the *PKD2* locus, but inspection of haplotypes led to discovery of a *PKD2* mutation in 12 affected relatives. Then when their phenotypes were recoded as "unknown," significant linkage was found to markers at the *PKD1* locus, strongly suggesting a *PKD1* mutation in 15 other affected relatives. Two relatives had both the *PKD1* haplotype and the *PKD2* mutation, and they were more severely affected than the others. Future studies of locus heterogeneity using candidate gene approaches will be facilitated by studying the protein interactions involved in critical pathways.

Phenotypic Heterogeneity

More than one phenotype may be caused by allelic mutations at a single locus. There are now numerous examples of this phenomenon, a few of which are shown in Table 2.3.

TABLE 2.3. Phenotypic Heterogeneity

Different Mutations In	Cause
COL2A1	Hypochondrogenesis, Kniest dysplasia, spondyloepiphyseal dysplasia congenita, and some cases of Stickler syndrome
CFTR	Cystic fibrosis, congenital bilateral absence of vas deferens
FGFR2	Crouzon, Jackson–Weiss, Pfeiffer, and Apert syndromes
dystrophin	Duchenne and Becker muscular dystrophies
L1CAM	X-linked hydrocephalus, MASA, familial spastic paraplegia
FGFR3	Achondroplasia, hypochondroplasia, thanatophoric dysplasia types I and II, and several other phenotypes

Once again, the β-globin locus (MIM 141900) provides a good example. Different β-globin mutations cause sickle cell disease, methemoglobinemia, numerous thalassemia variants, and hundreds of other hemoglobinopathies.

Discoveries that mutations in single genes can cause different phenotypes have resulted in the creation of phenotype "families." For example, different mutations in the fibroblast growth factor receptor 3 gene (*FGFR3*; MIM 134934) cause a spectrum of bone dysplasias that range from different types of severe, lethal thanatophoric dysplasia (MIM 187600) to classic achondroplasia (MIM 100800) and the milder hypochondroplasia (146000), with several other distinct phenotypes including nonsyndromic craniosynostosis of the coronal suture (MIM 602849). Similarly, the range of type II collagen mutations ranges from severe, frequently lethal dwarfism (achondrogenesis/hypochondrogenesis; MIM 200610) to very mild types of Stickler syndrome (108300) and many other different skeletal dysplasias with intermediate phenotypes. The OMIM entries include a subheading for "allelic variants." These lists clearly illustrate the phenomenon of phenotypic heterogeneity by cross-referencing different phenotype entries that have mutations identified at a single locus.

Other examples of phenotypic heterogeneity include phenotypes that are more dramatically different. Multiple endocrine neoplasia (MEN) type IIA and Hirschsprung disease (HSCR) are caused by different mutations in the *RET* protooncogene (MIM 164761). Specific *RET* mutations in exon 10, however, cause both conditions to cosegregate within families.

The examples cited so far involve rare mutations with strong, causative effects. But by the same token, common variants or polymorphisms may demonstrate phenotypic heterogeneity. Apolipoprotein E (apoE; MIM 107741), encoded by the *APOE* gene on chromosome 19q13.2, is involved in removing fats from the blood stream [chylomicrons and very low density lipoprotein (VLDL)]. As is true for many serum proteins, numerous variants and polymorphisms due to single-base substitutions have been identified by electrophoretic techniques over the years. Variants that decrease the function of apoE result in various degrees of hyperlipoproteinemia with elevated serum cholesterol and triglycerides, predisposing to atherosclerosis with premature coronary artery and peripheral vascular disease. On the other hand, the allele that encodes one of the common isoforms, apoE4, is associated with the common, late-onset, familial and sporadic forms of Alzheimer's disease.

COMPLEX INHERITANCE

The terms *complex inheritance* and *complex disease* imply that a single, causative, completely penetrant gene does not always produce the phenotype in question. Instead, a combination of effects from more than one gene or one or more genes with environmental (nongenetic) factors may produce the phenotype. "Complex" phenotypes are *causally heterogeneous*. This usually means that over an extended population, the causes of a particular phenotype, trait, or disease include both low-frequency, high-penetrance "causative" alleles and common, low-penetrance "susceptibility" alleles interacting with environmental factors. Most of the common disorders of children and adults are complex phenotypes. The common disorders of childhood include birth defects, mental retardation, short stature, and cancer. Common disorders in adults include cancer, diabetes, cardiovascular disease, hypertension, stroke, and the psychoses. Most of these examples are so obviously causally heterogeneous that their inclusion seems trivial. But the same causal heterogeneity is clearly true for conditions once thought to be specific single entities such as Alzheimer's disease, macular degeneration, Parkinson's disease, and spina bifida.

Polygenic and Multifactorial Models

For many, perhaps most, traits, expression of an allele depends at some level on interactions with other genes and/or environmental factors, where the interactions may be genetically programmed or purely random (i.e., stochastic or chance events). Even the phenotypic manifestations of conditions usually considered to be straightforward monogenic disorders may be the result of gene–environment interaction. For example, in the case of phenylketonuria (PKU; MIM 261600), phenotypic expression (i.e., mental retardation) depends not only on genotype but also on exposure to the amino acid phenylalanine in dietary protein. If the amount of phenylalanine in our dietary protein was much less than it is, a deficiency of phenylalanine hydroxylase would be a benign polymorphism in the human population, rather than an inborn error of metabolism that causes mental retardation.

Nonsyndromic birth defects (i.e., single, isolated structural malformations in individuals who do not have any other, causally related abnormalities) and common diseases of adult life are frequently familial, but pedigrees, such as the one in Figure 2.1, usually do not suggest a straightforward pattern of Mendelian inheritance. Note that in such a family, evidence for straightforward Mendelian inheritance may be missed if all relatives are not carefully examined for manifestations of the trait or condition being studied. For example, Potter syndrome or bilateral renal agenesis was thought to be nongenetic because it usually occurs sporadically. Subsequently, careful ultrasonographic studies showed that a significant proportion of first-degree relatives have asymptomatic renal anomalies such as unilateral renal agenesis, suggesting a subtle but definite genetic basis for this malformation.

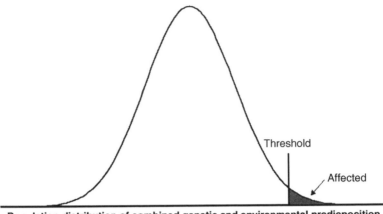

Population distribution of combined genetic and environmental predisposition

Figure 2.9. Multifactorial/threshold model. Continuous distribution in the general population of genetic liability tends to follow a normal distribution. The point within that liability distribution beyond which individuals are affected is called the threshold. Where and how the threshold is set is an important step in defining phenotype(s) for study.

To explain the observed *non-Mendelian* patterns of familial recurrence shown in Figure 2.1, mathematical models were developed (1965, 1969) that assumed a continuous distribution of genetic "liability" to malformation or disease in the general population (Carter, 1967, 1969a,b). *Polygenic* determination refers to the mathematical model in which a number of genes with small, additive effects provide an underlying genetic predisposition to malformation or disease. Some quantitative traits and clinical disorders in humans have been studied and found to be compatible with this mechanism of determination (Holt, 1961; Smith and Aase, 1970; Dragani et al., 1996a,b; Parisi and Kapur, 2000).

Multifactorial describes models in which environmental factors interact with genetic predisposition, which is frequently thought of as polygenic (i.e., the additive effect of a large number of genes, each of which contributes a small effect). In the case of quantitative traits such as blood pressure, weight, and height, a normal curve would represent the distribution of measurements in a population. This model was adapted to account for discontinuous traits by the addition of a threshold, the point within that liability distribution beyond which individuals are affected (Fig. 2.9). A number of observations in human populations and experimental animals are consistent with this model of multifactorial inheritance (Carter, 1969b, 1973, 1976; Fraser, 1976, 1980, 1981; Mitchell and Risch, 1993; Mitchell, 1997; ICRP publication 83, 1999).

Monogenic Predisposition. A proposed alternative to the polygenic/multifactorial models is that underlying genetic liability is determined by a single gene with incomplete penetrance and/or variable expressivity (Biddle and Fraser, 1986; Temple et al., 1989; Moore et al., 1990; Hecht et al., 1991; Marazita et al., 1992).

For many clinical conditions, the existence of less severely affected relatives seems consistent with the hypothesis of a single allele running through the family that *predisposes* carriers to develop the disorder. But penetrance and expressivity are phenomena that may be affected by or depend upon environmental factors, time, parent of origin, and/or the expression of other genes. Since the factors that influence penetrance and expressivity may be a mix of genes and environmental factors, the monogenic model, in practice, may actually be a very complex "mixed" model of major single-gene effects on a multifactorial background, where liability may be determined by more than one gene. Therefore, the term multifactorial would still appear to be an appropriate one if it is not restricted to imply only a polygenic contribution without major single-gene effects.

Digenic Causation. This designates yet another subset in which diseases or traits result from mutations in two genes at different loci. An example is that of two unlinked genes that encode polypeptide subunits of a functional protein complex. Mutations at either or both genes, therefore, cause dysfunction and a clinical phenotype. While this may seem like a simple variation on the theme of monogenic inheritance, it is included under "complex inheritance" because its existence confounds standard linkage analysis. Examples include some forms of retinitis pigmentosa (Goldberg and Molday, 1996; Wang et al., 2001), nonsyndromic hearing loss (Balciuniene et al., 1998; Chen et al., 2000; Borg et al., 2000), junctional epidermolysis bullosa (Floeth and Bruckner-Tuderman, 1999), and the example discussed above of severe autosomal dominant PKD (Pei et al., 2001).

Oligogenic Causation. *Oligogenic* designates a subset of polygenic causation in which diseases or traits result from the effects of relatively few genes, some of which may have rather large effects. Many current research efforts are directed toward identifying a small number of interacting genes that produce clinical phenotypes and diseases, without the necessity of identifying a quantitative, polygenic background of susceptibility (Verge et al., 1998; Holberg et al., 1999; Cookson, 1999; Ginsburg et al., 1999; Rice et al., 2000; Brown et al., 2000; King and Ciclitira, 2000; Yang and Rotter, 2000; Gershon, 2000; Warwick et al., 2000; Holberg et al., 2001).

An excellent example of a successful study is that of Hirschsprung disease (HSCR), a disorder characterized by abnormal bowel peristalsis due to congenital absence of ganglion cells in the wall of the colon. Eight incompletely penetrant genes from three different but interacting signaling pathways are implicated in causing HSCR (Passarge, 2002). Five of these are strongly associated with the rarer, more frequently familial, long-segment type (L-HSCR). As mentioned above, *RET* is the major gene involved, but coding region mutations are found in only half of familial cases. A genomewide search in families with the less common, more often nonfamilial, short-segment type (S-HSCR) has demonstrated oligogenic determination where *RET* interacts with two previously undetected loci (Gabriel et al., 2002).

Role of Environment

The environmental factors that influence both embryonic development and adult disease are currently of great research interest and controversy, especially with regard to the interaction of these environmental factors with an individual's underlying genetic susceptibility. For example, the risk of developing emphysema is greater in individuals with both an environmental exposure such as smoking *and* a genetic predisposition such as α-1-antitrypsin deficiency (MIM 107400) than it is for individuals with only one of these risk factors or none at all. Severe respiratory distress in asthmatics is triggered by exposure to environmental triggers like pollens, tobacco smoke, dust mites, cockroach allergens, acid aerosols, and ozone. On the other hand, nutritional factors may play a protective role in chronic lung diseases (Romieu and Trenga, 2001). Those with a familial susceptibility to skin cancer can alter greatly their risk of developing cancer by modifying their sun exposure. Obesity strongly affects morbidity in individuals with inherited susceptibilities to type II diabetes. There is great interest in studying whether/how subclinical maternal deficiency of folic acid (or other related metabolites), interacting with or due to mutations or polymorphisms in folate metabolism pathway genes, confers to the embryo a susceptibility to defective neural tube closure.

The memorable phrase *"Genetics loads the gun and environment pulls the trigger,"* coined by Judith Stern, seems to best summarize gene–environment interactions. In fact, future approaches to the primary prevention of many diseases will focus on the identification of environmental *cofactors* that lead to clinical disease in individuals with *susceptibility genotypes* (Khoury and Wagener, 1995; Ottman, 1995). Our future ability to prevent or ameliorate many common and chronic diseases will require knowledge about both genetic and environmental contributions to disease phenotypes. The positive aspect of this kind of prevention is that one can alter environmental triggers much more easily than one can manipulate a mutant allele. The translation of new genetic knowledge about susceptibility alleles into effective intervention programs will require extensive population-based epidemiological studies to characterize these cofactors and quantify their effects in genetically susceptible individuals.

Role of Chance in Phenotype Expression

Another factor, traditionally ignored in discussions about the cause of genetic disease, is chance. Kurnit and colleagues (1987) used computer modeling to show that some of the non-Mendelian familial clustering of anomalies usually attributed to concepts such as *reduced penetrance* and *multifactorial inheritance* may be accounted for by simple, random chance. Studies of this stochastic, probabilistic model allow us to conclude that we should think of incomplete penetrance in terms of single genes that predispose the organism to develop an anomaly or disease but do not always result in an abnormal phenotype (Cohen, 1989). Central to this concept is the idea that biological processes (like many other phenomena that we encounter daily) are error prone because they are not unvarying or totally controlled by constants; that is, they are nonlinear systems. The complex processes of embryologic development may be very sensitive to initial, small perturbations (which are, by themselves, "within normal limits") that subsequently, *by chance*, are amplified by the randomness that is inherent in all such systems.

The relative merits of the popular, hypothetical models discussed above (polygenic, multifactorial threshold, and major single gene with incomplete penetrance) have been debated vigorously over the years. In general, it is probably reasonable to assume that complex phenotypes potentially have many different causes, including stochastic combinations of genetic and environmental factors, where the genetic contribution may be determined by one or more genes.

PHENOTYPE DEFINITION

Classification of Disease

The classification systems traditionally used in medicine are somewhat arbitrary. Anatomic systems are commonly used by the surgical specialties, and physicians who see the body as a collection of organ systems frequently use physiological or other functionally oriented classifications. But a classification system based on *causal factors* or *pathogenetic mechanisms* is more appropriate for genetic analysis. Each of the first two approaches is important in other aspects of medical practice, but the causal approach should be utilized for phenotype definition in disease gene discovery studies. In this approach, one attempts to classify phenotypes into categories that have etiological and pathogenetic implications.

The study of congenital anomalies, for example, begins by deciding whether one is dealing with one or more deformations, disruptions, or malformations or an underlying dysplasia (Spranger et al., 1982). This approach, while originally defined for congenital anomalies, can be expanded to apply to genetically caused or influenced functional disorders, including those of postnatal onset. For example, in defining a new or poorly delineated phenotype, it is helpful to consider whether it represents a dysplasia, an inborn error of metabolism, susceptibility to environmental factors, or a degenerative or dystrophic process.

Major clinical features are those that have significant medical, surgical, functional, or cosmetic consequences. On the other hand, minor clinical features are those that do not have important medical, surgical, or cosmetic implications for the patient. One should recognize these minor features, however, because they may be important to include when defining phenotypes. A good example is the significance of Lisch nodules in the diagnosis of NF1 (MIM 162200). These are of no clinical significance to the patient, but they are so specific and common in NF1 that they provide a very important marker for those that carry a mutant allele. Many malformation syndromes are defined by their pattern of minor features, rather than by their major features. Down syndrome is a good example of this. Minor features and familial variants, which may provide clues about the cause or pathogenesis of major problems, are important to assess when identifying probands for study. The finding of minor features segregating with a major medical problem suggests pleiotropy, and these minor features may be of great help in phenotype definition. Therefore, the assignment of individuals to study groups should take into account minor as well as major clinical features.

Nonsyndromic Phenotypes

A condition with only a single, major clinical feature is *nonsyndromic*. Examples include common, isolated birth defects such as spina bifida, cleft lip with or without cleft palate, and congenital heart disease, as well as conditions of postnatal onset (e.g., retinitis pigmentosa, deafness, spastic paraplegia) when they occur as isolated clinical findings.

Pathogenetic Sequences. A *sequence* is a pattern of multiple clinical features that are all part of one pathogenetic sequence of events. A single sequence may have several different potential primary causes. A sequence may be the only obvious clinical finding in an individual or it may be part of a broader pattern of clinical expression as part of a syndrome or pleiotropic phenotype in which several pathogenetic sequences may arise from a single, primary mutation. For example, a baby with spina bifida, hydrocephalus, and clubfoot should not be thought of as having a syndrome with "multiple congenital anomalies." Rather, the hydrocephalus and club foot are secondary manifestations, caused by a single primary malformation, the spinal neural tube defect. Classifying such an infant as having a "hydrocephalus and clubfoot syndrome" would be like classifying a familial cancer syndrome by the sites of metastases, rather than by the primary malignancy.

Similarly, a stroke may be secondary to one or more underlying abnormalities, including hypertension, atherosclerosis, and congenital vascular malformation. Therefore, a study of the genetic causes of strokes would need to subdivide subjects by underlying causative factors. As a general rule, one should attempt to use the most primary causes known as the basis for phenotypic classification.

Causes of Nonsyndromic Disorders. Nonsyndromic disorders usually fall in the category of *complex genetic phenotypes* with causal heterogeneity that includes rare, causative alleles, common, susceptibility alleles involved, and environmental or stochastic factors.

Syndromic Phenotypes

If more than one tissue, organ system, developmental field, or region of the body is dysmorphic, dysplastic, dysfunctional, or dystrophic, the collection of features can be considered *syndromic*. Used in this way, the term does not necessarily imply that the phenotype constitutes a recognizable or previously well-defined syndrome. The word *syndrome* means "a running together," and it is used to refer to a pattern of multiple abnormalities thought to be causally and/or pathogenetically related.

Potential causes of syndromes include single alleles that manifest pleiotropy (e.g., a mutant allele having effects in multiple body sites, organs, or tissues), multigenic (genomic or contiguous gene) mutations, environmental factors, and conditions caused by a combination of genetic and nongenetic factors. For example, birth defects like spina bifida and cleft lip with or without cleft palate may be found in children who also have other primary malformations. Individuals with syndromes usually should be studied separately from those with isolated or nonsyndromic phenotypes. When retinitis pigmentosa (RP) is found in adults who also have severe, congenital sensory deafness, the condition should be classified as syndromic (see MIM 276900 and others) rather than as nonsyndromic RP. This is a good example of how the OMIM database can help one begin studying a trait of interest. An OMIM search for "retinitis pigmentosa" returns over 150 entries, including several dozen in which RP is syndromic, associated with abnormalities of the nervous system, skeleton, kidneys, and other organs.

Associations and Syndromes of Unknown Cause

A recurrent and recognizable pattern of clinical features may be called a "syndrome" even though there is no suggestion of a genetic or environmental cause. Such *unknown genesis syndrome* phenotypes are initially observed to occur sporadically (i.e., only once in each family). At this initial stage of phenotype delineation, one should keep in mind the probability of causal heterogeneity. Sporadic occurrence of a "genetically lethal" phenotype (where affected individuals do not reproduce) may be due to a new, dominantly expressed mutation in a single gene or a small chromosome deletion that is undetectable by routine standard cytogenetic banding techniques. Some phenotypes previously classified as unknown genesis syndromes are now known to be caused by identifiable genetic or environmental factors. Well-known examples of such syndromes include Williams (MIM194050), Prader-Willi (MIM 176270), Angelman (MIM 105830), Rubinstein–Taybi (MIM 180849), and DiGeorge (MIM 188400).

The term *association* refers to the observation that multiple primary clinical features, not known to be related by common etiological or pathogenetic mechanisms, occur together significantly more often than expected by chance. An early example in the childhood malformation literature is the VATER association, where the name is an acronym representing the areas involved in malformation (vertebral, anal, tracheoesophageal fistula, and radial and/or renal anomalies). Associations are a *causally nonspecific category* used to keep track of heterogeneous groups of undelineated syndromes and sequences. Because they are statistical rather than biological entities, associations are not definitive diagnoses. Therefore, one cannot *begin* with a strategy of mapping an "association" because of the assumed underlying causal heterogeneity. Rather, the goal should be to delineate causally specific syndromes and disease complexes from within association categories for eventual mapping studies. For example, there is evidence for Mendelian inheritance of VATER anomalies when these are associated with hydrocephalus (MIM 276950, 314390).

Causes of Syndromes. These include chromosome abnormalities, single-gene mutations (autosomal dominant and recessive, X linked, mitochondrial), and environmental teratogens.

Importance of Chromosomal Rearrangements in Mapping

Contiguous Gene Mutations. Chromosome deletions or partial duplications that involve more than one gene may be cytogenetically visible or submicroscopic, identifiable only by molecular techniques such as fluorescence in situ hybridization (FISH) or Southern blot. Duplications and deletions can be transmitted vertically through families in a dominant pattern, sometimes with large variability in expressivity. It can be very helpful to identify patients that express two or more unrelated, major genetic phenotypes, because of the possibility that genes causing these conditions are closely linked and affected by a genomic rearrangement such as a submicroscopic deletion. If one of these conditions has been mapped, one can look for the other in the same chromosomal region. For example, one patient had three different X-linked disorders—a form of chronic granulomatous disease (MIM 306400), Duchenne muscular dystrophy (MIM 310200), and retinitis pigmentosa (MIM 312610). A subtle interstitial deletion at Xp21 was evidence that the three disorders were caused by closely linked, contiguous genes.

Translocations. Unbalanced translocations may affect several distant relatives who are related through balanced translocation carriers. Figure 2.1 showed the typical appearance of such a pedigree. The solid symbols indicate individuals with unbalanced karyotypes related through balanced translocation carriers. A number of genes have been mapped by finding balanced translocations in individuals who express the trait or condition of interest. Examples include NF1 (MIM162200) and Duchenne dystrophy (MIM 310200). Translocation breakpoints can be used to clone the gene of interest, which is interrupted by the translocation. Therefore, one looks for families that segregate the trait of interest and also have multiple individuals with features suggesting a major chromosome abnormality. Those features usually include growth retardation, mental retardation, one or more structural birth defects, and minor anomalies or dysmorphic features. Individuals carrying balanced translocations often have notable reproductive histories, including neonatal death, stillbirth, and miscarriage.

Qualitative (Discontinuous) and Quantitative (Continuous) Traits

Most medical conditions and "abnormal" traits are recognized by being "significant" deviations from the norm. Signs and symptoms determine whether a disease state is present or absent. Monogenic traits with complete penetrance and easily defined phenotypes are optimal for standard linkage analysis. For example, mapping the achondroplasia locus (MIM 100800) was facilitated by the existence of a readily recognizable phenotype in both children and adults. On the other hand, adult-onset disorders with variable expression and incomplete penetrance will be much more difficult to diagnose and assign to diagnostic groups.

Many traits such as height, blood pressure, head circumference, and IQ can be quantified, with the resulting measurements distributed in a continuous fashion across a population (see Polygenic and Multifactorial Models above). Conversion of continuous traits into discontinuous ones by defining a "threshold" for diagnosis as in Figure 2.9 may be used in linkage studies, but the criteria used for such conversion should be as biologically meaningful as possible to minimize problems of underlying genetic heterogeneity. For the examples above of height, blood pressure, head circumference, and IQ, thresholds are used to define short stature/tall stature, hypertension/hypotension, microcephaly/macrocephaly, and mental retardation/genius. The true state of nature, however, is that quantitative, continuous traits are genetically complex with potentially a large number of loci that determine and affect each trait. This high degree of genetic complexity is assumed also to be true for most of the common "multifactorial," medical conditions in both children and adults.

DEFINING PHENOTYPES FOR ANALYSIS OF COMPLEX GENETIC DISORDERS

To perform a genetic analysis, one must first define the phenotype to be studied. Complex genetic disorders are assumed to be genetically heterogeneous, often the result of polygenic or oligogenic predisposition with a possible environmental component (i.e., multifactorial), occasionally monogenic, and occasionally part of a broader pattern of disease (i.e., syndromic). Following are some steps to take in defining phenotypes of complex disorders for analysis.

Select Most Biologically Meaningful Phenotype

Based on current scientific knowledge, begin with a phenotype that is as well defined and biologically meaningful as possible. For example, instead of studying "diabetes" in general, one would begin by recognizing at least two types, insulin-dependent, or type I, usually with childhood onset, and non-insulin-dependent, or type II, usually with adult onset. Instead of setting out to study facial clefting, one should recognize that cleft lip with or without cleft palate and posterior cleft palate without cleft lip are different biological entities.

Partition Phenotype or Dataset by Cause and Associated Pathology

One must think of syndromic and nonsyndromic phenotypes separately. Stratify the syndromic portion by recognizable syndrome versus unclassified syndromic association. Study the recognizable syndromes as individual entities. Similarly, those phenotypes where the cause is already known should be separated from the study group. As shown in Figure 2.10, this partitioning results in four major categories for consideration: (1) syndromic, cause(s) known or suspected; (2) nonsyndromic,

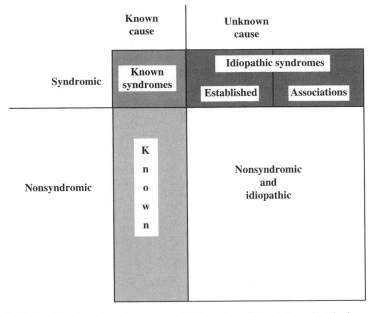

Figure 2.10. Partitioning the phenotype of interest or dataset by what is known about possible causes and associated pathology: (1) syndromic, cause(s) known (upper left quadrant); (2) nonsyndromic, cause(s) known or suspected (lower left quadrant); (3) syndromic, cause unknown (upper right quadrant); (4) nonsyndromic, cause unknown (lower right quadrant) (see text for explanation).

cause(s) known or suspected; (3) syndromic, cause unknown; and (4) nonsyndromic, cause unknown.

Syndromic—Cause(s) Known. Syndromes due to known chromosome abnormalities, environmental teratogens, and single causative genes may include the phenotypic trait of research interest. Monogenic syndromes provide candidate loci for susceptibility alleles. Chromosome abnormalities can be used to identify a candidate region. If a phenotypic trait of interest is coexpressed with a known syndrome that does not usually include that trait, consider the possibility of a chromosomal abnormality such as a microdeletion. Known environmental factors provide a basis for stratifying the cause-unknown groups. Most importantly, cases in this category should be separated from the analysis of those of unknown cause. This may be difficult because of variable expressivity. For example, studies of cleft palate should eliminate cases of Stickler syndrome, which is known to be underdiagnosed in cleft palate clinics. Similarly, undiagnosed patients with Stickler syndrome might be included in separate studies of osteoarthritis or myopia. In studying susceptibility to diabetes, one would want to eliminate patients with Wolfram syndrome, diabetes–deafness syndrome, and so on. Researchers studying deafness must screen their study population for well over a hundred syndromes associated with hearing loss. Investigators studying autism will want to screen out those with fragile X syndrome.

Nonsyndromic—Causes Known or Suspected. Use the monogenic causes as candidate loci for susceptibility alleles. Study the biology of the protein products of these genes to find essential interactions with other proteins, and consider the genes encoding these other proteins as candidates also. Separating these patients from group 4 will require testing everyone with cause unknown who has a phenotype compatible with the condition of known cause. For example, in studying Alzheimer's disease, one would want to eliminate families with causative presenilin mutations. Researchers studying deafness are faced with an extensive list of genes now known to cause nonsyndromic hearing loss.

Syndromic—Cause Unknown. Well-defined syndromes of unknown cause suggest a chromosome abnormality, environmental teratogen, or single-gene causation. Less well-defined syndromic associations may be multifactorial. These syndromic cases should be analyzed separately from nonsyndromic cases of unknown cause. As discussed in item 1, syndromes with subtle phenotypes may be mistakenly included in category 4. Another clefting example with extremely variable expression is the autosomal dominant van der Woude syndrome (MIM 119300) with associated lip pits. These families could easily be included in studies of nonsyndromic clefting if all individuals are not examined carefully by someone who is familiar with the subtle characteristics of this syndrome. If analyzing a syndrome separately results in positive linkage, then that locus can be used as a candidate for the nonsyndromic group. In addition, if the syndrome of interest includes a very rare associated anomaly, then one might consider combining other families with that rare anomaly along with the primary study group. For example, lip pits and clefting also occur in the popliteal pterygium syndrome, and it has been shown that both van der Woude and popliteal pterygium syndromes are caused by mutations in interferon regulatory factor 6 (IRF6) (Kondo et al., 2002).

Nonsyndromic—Cause Unknown. This group contains cases presumably of multifactorial causation. Affected relatives are presumed to have some predisposing genes in common but may actually have different combinations of susceptibility genes, and other unaffected relatives will also carry some of these predisposing alleles. Analysis of the phenotype of interest can be attempted on this group as a whole, but other manipulations such as further stratification of affected patients and their relatives or redefining the phenotype of interest may be useful.

An initial step is to divide this group (4) by familial versus sporadic cases. Search the familial pedigrees for clues about major genetic effects with or without complete penetrance (i.e., AD, AR, XL, Y linked, mitochondrial inheritance patterns), and look for parent-of-origin effects.

Next, *narrowing the phenotype* and *expanding or broadening the phenotype* are complementary techniques that can be used. Phenotypes are narrowed in order to *stratify* a study population, with the hope that major gene effects may be found in different subsets of the population. Phenotypes may be narrowed or expanded to identify measurable, subclinical traits that are biologically significant, related to the complex phenotype of interest, and expressed in both affected individuals and unaffected relatives. Finally, phenotypes may be broadened in order to redefine them to include associated, usually more subtle, features that show familial segregation as a syndrome.

One approach is to develop datasets, large enough for analysis, with carefully characterized and narrowly defined phenotypes. This approach assumes that such a subset of individuals will be genetically more homogeneous and thus more easily characterized than the population of individuals with more broadly or loosely defined phenotypes. Such a priori subsetting is a standard approach in genetic analysis of simple Mendelian and complex disease. It has proven its utility in such diseases as breast cancer (Hall et al., 1990), Parkinson's disease (Kitada et al., 1998), and Alzheimer's disease. In Alzheimer studies, families were stratified based on age of onset to identify initial linkages (Pericak-Vance et al., 1991) and ultimately genes of major effect (Corder et al., 1993).

Although age of onset is a commonly used stratifying variable, other characteristics can be used as well. For example, diabetes was originally stratified by childhood-onset versus adult-onset cases. But a more meaningful subsetting is to begin by dividing the population into those that require insulin (mostly childhood-onset cases) and those who do not (the typical adult-onset type). Stratifying coronary heart disease patients by lipid levels would be necessary to study the more narrowly defined subset of patients with a familial hyperlipidemia (Pajukanta and Porkka, 1999).

One can try to identify biologically significant, measurable traits that reflect the expression of susceptibility genes in both fully affected subjects and their clinically unaffected relatives. Such quantitative or qualitative traits are sometimes called *risk factors* in internal medicine and *endophenotypes* in psychiatric research. Such a trait should be measurable and show deviant values in both affected individuals with full expression of the complex disorder and their unaffected but predisposed relatives who carry some susceptibility genes.

A simple way to understand this is to extrapolate from a disease gene discovery approach for an autosomal recessive inborn error of metabolism. Suppose that homozygous affected individuals have mental retardation and an abnormal elevation of some blood metabolite, the phenotype being caused by complete deficiency of an enzyme. If the underlying enzyme deficiency were unknown, one would have to identify rare families with more than one affected sibling and do linkage analysis using only the rare clinical phenotype under a recessive model. On the other hand, if the underlying enzyme deficiency is known and measurable, then phenotypically normal heterozygous carriers will have the biochemical trait of half-normal levels of enzyme activity. One could then identify many more individuals expressing that trait in relatives of affected individuals and perform a more straightforward linkage analysis for the autosomal dominant trait of half-normal enzyme activity.

A difficult challenge in analyzing complex disorders is to identify phenotypic traits that can be measured efficiently and accurately. Two general approaches are used. One can dissect the complex phenotype into contributing components or the phenotypic definition can be expanded to include other quantitative or qualitative findings that segregate with the complex phenotype of interest.

For example, asthma is characterized by elevated serum immunoglobulin E (IgE) levels and bronchial hyperresponsiveness, both of which have been used by asthma researchers as phenotypes for disease gene discovery studies (Ulbrecht et al., 2000; Sampogna et al., 2000). On the other hand, elevated IgE levels and hypersensitivity to antigens characterize allergic diseases in general, suggesting that broadening the phenotype to include all atopic disease, not just asthma, might help identify loci important in asthma (Ono, 2000).

When studying nonsyndromic congenital malformations, one may try to identify a *forme fruste* among close relatives of probands. The malformation cleft lip with or without cleft palate (CL/P) has been extensively studied without much headway having been made in identifying susceptibility genes. An interesting finding, therefore, among first-degree relatives of patients with CL/P, is that some have occult abnormalities of the upper lip musculature that can be identified by ultrasound (Martin et al., 1993, 2000). The use of this discontinuity of the orbicularis oris muscle, which has been called a *subepithelial cleft*, may be extremely valuable in future mapping studies. Similarly, many anthropometric measurements have been made of the skull and face, and it has been suggested that first-degree relatives of patients with CL/P have wider eye spacing than the general population. One pathogenetic sequence involved in clefting and other "fusion" defects appears to be that two surfaces are capable of undergoing fusion only during a specific window of time during development and that anything that delays the apposition of the two surfaces will result in nonfusion. It seems reasonable to assume, therefore, that quantitative trait loci (QTLs) involved in head and facial growth, size, and shape will prove to be susceptibility loci for CL/P. For analysis, it would be ideal to identify a reasonably easily measured quantitative trait such as interpupillary distance.

Psychiatric disorders present complicated and unique problems to genetic epidemiologists who wish to identify a narrowly defined phenotype. Significant effort has been placed on refining diagnostic criteria and definitions in the *Diagnostic and Statistical Manual of Mental Disorders*, 4th ed. (DSM-IV), and attempts have been made to standardize diagnostic assessments for consistency across studies (Nurnberger et al., 1994). Psychiatric geneticists, however, have begun to doubt that further refinement of the diagnostic instruments will be productive (Berrettini, 1998). Instead, some have emphasized the importance of identifying *endophenotypes* or *alternative phenotypes*, objective measurements that presume to accurately predict a subset of cases (Berrettini, 1998; Craddock and Jones, 1999). The idea is that understanding the genetic basis of susceptibility for even a small subset of cases should be valuable in the early stages of studying the genetic architecture of a complex disorder. For example, hyperintense foci in the white and grey matter of the brain may represent an endophenotype for bipolar disorder (Ahearn et al., 1998). This finding is more frequent in bipolar cases than in age-matched controls

(Dupont et al., 1987, 1990; Swayze et al., 1990; Figiel et al., 1991a,b), and it is not associated with age-related vascular changes (Figiel et al., 1991a,b). In addition, these hyperintense foci appear to segregate with bipolar disorder in at least one large pedigree (Ahearn et al., 1998). Similarly, a discrete trait found in patients with schizophrenia (failure to inhibit the P50 auditory-evoked response to repeated stimuli) has been used in linkage studies and mapped to the α-7-nicotinic acetylcholine receptor on chromosome 15q14 (Freedman et al., 1999, 2000).

Broadening the phenotype of a complex disorder may help define a new phenotype for study. This involves an attempt to identify subtle clinical features that can be identified in relatives of fully affected individuals. For example, a broad autism phenotype (BAP) is being defined by studying and comparing neurodevelopmental characteristics presumably related to autism in relatives of autistic probands and controls. The rationale is that since typical pedigrees of families with autism and other complex disorders usually include only a few affected individuals separated by many unaffected relatives (Fig. 2.1), broadening the phenotype may improve the accuracy of diagnostic coding for these "unaffected" intervening relatives (Piven, 2001). An early twin study of autism found a concordance in monozygotic (MZ) versus dizygotic (DZ) twins of 36 and 0%, respectively. But when the phenotype was *broadened* to include either autism or a milder cognitive disorder, concordance was 82% in MZ compared to 10% in DZ twin pairs (Folstein and Rutter, 1977). Subsequent studies in relatives of autistic probands have showed other neurobehavioral deficits in personality, language, and cognitive abilities (Piven and Palmer, 1997; Piven et al., 1997).

Summary: Approach to Phenotype Definition

1. Begin by searching for the trait or disease phenotype of interest in OMIM using more than one search strategy to identify all possible entries. This approach may uncover a surprising amount of literature that otherwise would not have been considered. In addition, look for other model organisms, especially the mouse, that express a phenotype similar to the one of interest. Browse the multiple Internet resources and databases that are now available.

2. Start with inclusive definitions. Assume that the phenotype of interest is causally heterogeneous and include cases that share major features. Utilize the phenotypic spectrum of literature cases to assess the feasibility of defining a biologically specific trait or syndrome in the study population.

3. Subdivide the phenotype or existing dataset into four categories: (1) syndromic, cause known; (2) nonsyndromic, cause known; (3) syndromic, cause unknown; and (4) nonsyndromic, cause unknown.

4. From the syndromic groups, try to ascertain patients or families that segregate the trait of interest *plus* other major or minor features and look for chromosomal rearrangements to identify candidate regions. To find these patients and families, be on the lookout for those with other major findings that are not usually associated with the trait or syndrome of interest, especially the nonspecific features that are typical of chromosome abnormalities such as

mental retardation, short stature, multiple birth defects, and dysmorphic features. These features represent altered patterns of morphogenesis that may be due to deletion or rearrangement of contiguous genes. Such patients should then be studied carefully for chromosomal rearrangements that might provide candidate regions for further mapping studies. In addition, even just a family history of multiple malformations and/or mental retardation and/or growth problems in relative(s) of subjects with the study trait of interest should be pursued. Such a family history might be an indication that study subjects who have the trait being studied may carry a *balanced* translocation that disrupts a causative or predisposing gene.

5. Consider carefully the biological significance of quantitative traits. In converting quantitative traits to qualitative traits, try to include other biologically significant criteria when deciding threshold definitions. For example, if one wanted to map a gene that causes only mild or borderline short stature, one would try to combine a height threshold with one or more other significant (major or minor) clinical or radiographic features. The more overlap there is between the general population and the affected population, the more important are these associated features in establishing criteria for a diagnostic classification.

6. In families where the cause is known, study nonpenetrant individuals and publish the clinical phenotypes.

7. In groups where the cause is unknown, analyze familial cases for evidence of maternal (i.e., mitochondrial) inheritance and Mendelian inheritance patterns modified by imprinting, anticipation, and gender-related penetrance or expression.

8. Be inclusive in data collection, even though you may have narrower diagnostic criteria. Collect clinical data that might be used to narrow or expand the phenotype, to identify an endophenotype, or to better define the phenotype and natural history once a gene has been identified. Use multiple indices and specific definitions of data items to be collected in order to reduce variability among multiple raters and to ensure consistency over time if only one or a few are collecting clinical data.

9. Every gene-mapping project should also be a phenotype delineation study, since most phenotypes are not well defined in terms of relationships between the genotype and the pathogenesis, expression, and natural history of the disorder or trait being studied. Therefore, it is appropriate and important to include clinical geneticists in research protocols, in addition to the medical consultants with specific expertise in the disorder under study. Teams involved in family studies should use their data to explore questions of clinical and genetic importance such as intrafamilial and interfamilial variability of expression and natural history. Funding agencies should recognize the importance of these clinical aspects of phenotype delineation in human genome research by emphasizing and supporting expertise in phenotyping in the same manner that they require and support expertise in genotyping.

Resources for Information About Clinical Genetics and Phenotype Definition

Another informative and useful online source is GeneClinics (http://www.geneclinics.org/), self-described as "an expert-authored, peer-reviewed clinical genetic information resource consisting of concise descriptions of specific inherited disorders and authoritative, current information on the role of genetic testing in the diagnosis, management, and genetic counseling of patients with these inherited conditions." It provides information on clinical phenotypes and molecular genetic testing for over 100 of the more common genetic disorders, including many disorders with complex inheritance. There are links to a genetics laboratory directory, genetics clinic directory, and educational materials.

Mouse–human homologies are rapidly being delineated, and these should provide the human geneticist with a wealth of additional, useful phenotype and mapping information. The Jackson Laboratory (http://www.jax.org/) provides online services and links to resources, including the Mouse Genome Database (http://www.informatics.jax.org/), which will be integrated with the new Mouse Phenome Project (http://www.jax.org/phenome). The purpose of the Mouse Phenome Project is to establish a collection of baseline phenotypic data on commonly used and genetically diverse inbred mouse strains through a coordinated international effort.

The Davis Human/Mouse Homology Map (http://www3.ncbi.nlm.nih.gov/Omim/Homology/) compares genes in homologous segments of DNA from human and mouse sources, sorted by position in each genome.

Another animal model database is Online Mendelian Inheritance in Animals (OMIA) (http://www.angis.su.oz.au/Databases/BIRX/omia/), accessible from the OMIM home page, providing data and links to data on phenotypes and/or genetic maps for over 130 different animal species.

REFERENCES

Ahearn EP, Steffens DC, Cassidy F, Van M, Provenzale JM, Seldin MF, Weisler RH, Krishnan KR (1998): Familial leukoencephalopathy in bipolar disorder [see comments]. Am J Psychiatry 155:1605–1607.

Balciuniene J, Dahl N, Borg E, Samuelsson E, Koisti MJ, Pettersson U, Jazin EE (1998): Evidence for digenic inheritance of nonsyndromic hereditary hearing loss in a Swedish family. Am J Hum Genet 63:786–793.

Barlow DP (1995): Gametic imprinting in mammals [Review]. Science 270:1610–1613.

Basu P, Chattopadhyay B, Gangopadhaya PK, Mukherjee SC, Sinha KK, Das SK, Roychoudhury S, Majumder PP, Bhattacharyya NP (2000): Analysis of CAG repeats in SCA1, SCA2, SCA3, SCA6, SCA7 and DRPLA loci in spinocerebellar ataxia patients and distribution of CAG repeats at the SCA1, SCA2 and SCA6 loci in nine ethnic populations of eastern India. Hum Genet 106:597–604.

Beal MF (2000): Energetics in the pathogenesis of neurodegenerative diseases [Review]. Trends Neurosci 23:298–304.

Berrettini W (1998): Progress and pitfalls: Bipolar molecular linkage studies [Review]. J Affect Disord 50:287–297.

Biddle FG, Fraser FC (1986): Major gene determination of liability to spontaneous cleft lip in the mouse. J Craniofac Genet Dev Biol 2(Suppl):67–88.

Borg E, Samuelsson E, Dahl N (2000): Audiometric characterization of a family with digenic autosomal, dominant, progressive sensorineural hearing loss. Acta OtoLaryngol 120:51–57.

Brannan CI, Bartolomei MS (1999): Mechanisms of genomic imprinting [Review]. Curr Opin Genet Dev 9:164–170.

Brown MA, Laval SH, Brophy S, Calin A (2000): Recurrence risk modeling of the genetic susceptibility to ankylosing spondylitis. Ann Rheum Dis 59:883–886.

Carter CO (1967): Clinical aspects of genetics, the genetics of common malformations and diseases. Trans Med Soc Lond 83:84–91.

Carter CO (1969a): Genetics in the aetiology of disease. Lancet 1:1014–1016.

Carter CO (1969b): Genetics of common disorders. Br Med Bull 25:52–57.

Carter CO (1973): Multifactorial genetic disease. In: McKusick VA, Claiborne R, eds. Medical Genetics. New York: HP Publishing, pp 199–208.

Carter CO (1976): Genetics of common single malformations. Br Med Bull 32:21–26.

Cassidy SB (1995): Uniparental disomy and genomic imprinting as causes of human genetic disease [Review]. Environ Mol Mutagen 25:13–20.

Castro JA, Picornell A, Ramon M (1998): Mitochondrial DNA: A tool for populational genetics studies [Review]. Int Microbiol 1:327–332.

Chen AH, Fukushima K, McGuirt WT, Smith RJ (2000): DFNB15: Autosomal recessive non-syndromic hearing loss gene chromosome 3q, 19p or digenic recessive inheritance? Adv Oto-Rhino-Laryngol 56:171–175.

Chinnery PF, Howell N, Andrews RM, Turnbull DM (1999): Clinical mitochondrial genetics [Review]. J Med Genet 36:425–436.

Cohen MM Jr (1989): Syndromology: An updated conceptual overview. VI. Molecular and biochemical aspects of dysmorphology. Int J Oral Maxillofac Surg 18:339–346.

Cole WG, Dalgleish R (1995): Perinatal lethal osteogenesis imperfecta. J Med Genet 32:284–289.

Cookson WO (1999): Disease taxonomy—polygenic [Review]. Br Med Bull 55:358–365.

Corder EH, Saunders AM, Strittmatter WJ, Schmechel DE, Gaskell PC, Small GW, Roses AD, Haines JL, Pericak-Vance MA (1993): Gene dose of apolipoprotein E type 4 allele and the risk of Alzheimer's disease in late onset families [see comments]. Science 261:921–923.

Cottrell DA, Blakely EL, Borthwick GM, Johnson MA, Taylor GA, Brierley EJ, Ince PG, Turnbull DM (2000): Role of mitochondrial DNA mutations in disease and aging [Review]. Ann NY Acad Sci 908:199–207.

Craddock N, Jones I (1999): Genetics of bipolar disorder [Review]. J Med Genet 36:585–594.

Cummings CJ, Zoghbi HY (2000): Trinucleotide repeats: Mechanisms and pathophysiology [Review]. Annu Rev Genom Hum Genet 1:281–328.

Disotell TR (1999): Human evolution: Origins of modern humans still look recent [see comments] [Review]. Curr Biol 9:R647–R650.

Dragani TA, Canzian F, Pierotti MA (1996a): A polygenic model of inherited predisposition to cancer [Review]. FASEB J 10:865–870.

Dragani TA, Manenti G, Pierotti MA (1996b): Polygenic inheritance of predisposition to lung cancer [Review]. Annali dell Istituto Superiore di Sanita 32:145–150.

Driscoll DJ (1994): Genomic imprinting in humans [Review]. Mol Genet Med 4:37–77.

Dupont RM, Jernigan TL, Butters N, Delis D, Hesselink JR, Heindel W, Gillin JC (1990): Subcortical abnormalities detected in bipolar affective disorder using magnetic resonance imaging. Clinical and neuropsychological significance [see comments]. Arch Gen Psychiatry 47:55–59.

Dupont RM, Jernigan TL, Gillin JC, Butters N, Delis DC, Hesselink JR (1987): Subcortical signal hyperintensities in bipolar patients detected by MRI. Psychiatry Res 21: 357–358.

Edwards JH (1969): Familial predisposition in man. Br Med Bull 25:58–64.

Evans DG, Wallace AJ, Wu CL, Trueman L, Ramsden RT, Strachan T (1998): Somatic mosaicism: A common cause of classic disease in tumor-prone syndromes? Lessons from type 2 neurofibromatosis. Am J Hum Genet 63:727–736.

Falconer DS (1965): The inheritance of liability to certain diseases, estimated from the incidence among relatives. Ann Hum Genet 29:51–76.

Falls JG, Pulford DJ, Wylie AA, Jirtle RL (1999): Genomic imprinting: Implications for human disease [Review]. Am J Pathol 154:635–647.

Feinberg AP, Rainier S, DeBaun MR (1995): Genomic imprinting, DNA methylation, and cancer [Review]. Monog Nat Cancer Inst 21–26.

Figiel GS, Krishnan KR, Doraiswamy PM, Rao VP, Nemeroff CB, Boyko OB (1991a): Subcortical hyperintensities on brain magnetic resonance imaging: A comparison between late age onset and early onset elderly depressed subjects. Neurobiol Aging 12:245–247.

Figiel GS, Krishnan KR, Rao VP, Doraiswamy M, Ellinwood EHJ, Nemeroff CB, Evans D, Boyko O (1991b): Subcortical hyperintensities on brain magnetic resonance imaging: A comparison of normal and bipolar subjects. J Neuropsychiatry Clin Neurosci 3:18–22.

Floeth M, Bruckner-Tuderman L (1999): Digenic junctional epidermolysis bullosa: Mutations in COL17A1 and LAMB3 genes. Am J Hum Genet 65:1530–1537.

Folstein S, Rutter M (1977): Infantile autism: A genetic study of 21 twin pairs. J Child Psychol Psychiatry Allied Disciplines 18:297–321.

Fraser FC (1976): The multifactorial/threshold concept—Uses and misuses. Teratology 14:267–280.

Fraser FC (1980): Evolution of a palatable multifactorial threshold model. Am J Hum Genet 32:796–813.

Fraser FC (1981): The genetics of common familial disorders—major genes or multifactorial? Can J Genet Cytol 23:1–8.

Freedman R, Adams CE, Adler LE, Bickford PC, Gault J, Harris JG, Nagamoto HT, Olincy A, Ross RG, Stevens KE, Waldo M, Leonard S (2000): Inhibitory neurophysiological deficit as a phenotype for genetic investigation of schizophrenia. Am J Med Genet 97:58–64.

Freedman R, Adler LE, Leonard S (1999): Alternative phenotypes for the complex genetics of schizophrenia [Review]. Biol Psychiatry 45:551–558.

Gabriel SB, Salomon R, Pelet A, Angrist M, Amiel J, Fornage M, Attie-Bitach T, Olson JM, Hofstra R, Buys C, Steffann J, Munnich A, Lyonnet S, Chakravarti A (2002): Segregation at three loci explains familial and population risk in Hirschsprung disease [see comments]. Nat Genet 31:89–93.

Gershon ES (2000): Bipolar illness and schizophrenia as oligogenic diseases: Implications for the future [Review]. Biol Psychiatry 47:240–244.

Ginsburg E, Livshits G, Yakovenko K, Kobyliansky E (1999): Genetics of human body size and shape: Evidence for an oligogenic control of adiposity. Ann Hum Biol 26:79–87.

Gold JD, Pedersen RA (1994): Mechanisms of genomic imprinting in mammals [Review]. Curr Topics Dev Biol 29:227–280.

Goldberg AF, Molday RS (1996): Defective subunit assembly underlies a digenic form of retinitis pigmentosa linked to mutations in peripherin/rds and rom-1. Proc Nat Acad Sci USA 93:13726–13730.

Greally JM (1999): Genomic imprinting and chromatin insulation in Beckwith-Wiedemann syndrome [Review]. Mol Biotechnol 11:159–173.

Grossman LI (1996): Mitochondrial mutations and human disease. Environ Mol Mutagen 25:30–37.

Hall JG (1990): Genomic imprinting: Review and relevance to human diseases. Am J Hum Genet 46:857–873.

Hall JG (1999a): Human diseases and genomic imprinting [Review]. Results and Problems in Cell Differentiation 25:119–132.

Hall JG (1999b): Human diseases and genomic imprinting [Review]. Results and Problems in Cell Differentiation 25:119–132.

Hall JM, Lee MK, Newman B, Morrow JE, Anderson LA, Huey B, King MC (1990): Linkage of early-onset familial breast cancer to chromosome 17q21. Science 250:1684–1689.

Hecht JT, Yang P, Michels VV, Buetow KH (1991): Complex segregation analysis of nonsyndromic cleft lip and palate. Am J Hum Genet 49:674–681.

Holberg CJ, Erickson RP, Bernas MJ, Witte MH, Fultz KE, Andrade M, Witte CL (2001): Segregation analyses and a genome-wide linkage search confirm genetic heterogeneity and suggest oligogenic inheritance in some Milroy congenital primary lymphedema families. Am J Med Genet 98:303–312.

Holberg CJ, Halonen M, Wright AL, Martinez FD (1999): Familial aggregation and segregation analysis of eosinophil levels [see comments]. Am J Respiratory Critical Care Med 160:1604–1610.

Holt SB (1961): Quantitative genetics of finger-print patterns. Br Med Bull 17:247–250.

Horsthemke B, Surani A, James T, Ohlsson R (1999): The mechanisms of genomic imprinting [Review]. Results and Problems in Cell Differentiation 25:91–118.

ICRP publication 83 (1999): Risk estimation for multifactorial diseases. Ann ICRP 29:1–2.

Jiang Y, Tsai TF, Bressler J, Beaudet AL (1998): Imprinting in Angelman and Prader-Willi syndromes [Review]. Curr Opin Genet Dev 8:334–342.

Jirtle RL (1999): Genomic imprinting and cancer [Review]. Exper Cell Res 248:18–24.

Khoury MJ, Wagener DK (1995): Epidemiological evaluation of the use of genetics to improve the predictive value of disease risk factors. Am J Hum Genet 56:835–844.

King AL, Ciclitira PJ (2000): Celiac disease: Strongly heritable, oligogenic, but genetically complex. Mol Genet Metabolism 71:70–75.

Kitada T, Asakawa S, Hattori N, Matsumine H, Yamamura Y, Minoshima S, Yokochi M, Mizuno Y, Shimizu N (1998): Mutations in the parkin gene cause autosomal recessive juvenile parkinsonism [see comments]. Nature 392:605–608.

Koeppen AH (1998): The hereditary ataxias [Review]. J Neuropathol Exper Neurol 57:531–543.

Kondo S, Schutte BC, Richardson RJ, Bjork BC, Knight AS, Watanabe Y, Howard E, Ferreira D, Daack-Hirsch S, Sander A, McDonald-McGinn DM, Zackai EH, Lammer EJ, Aylsworth AS, Ardinger HH, Lidral AC, Pober BR, Moreno L, Arcos-Burgos M, Valencia C, Houdayer C, Bahuau M, Moretti-Ferreira D, Richieri-Costa A, Dixon MJ, Murray JC (2002): Mutations in IRF6 cause Van der Woude and popliteal pterygium syndromes. Nat Genet 32:285–289.

Koshy BT, Zoghbi HY (1997): The CAG/polyglutamine tract diseases: Gene products and molecular pathogenesis [Review]. Brain Pathol 7:927–942.

Kotzot D (1999): Abnormal phenotypes in uniparental disomy (UPD): Fundamental aspects and a critical review with bibliography of UPD other than 15 [Review]. Am J Med Genet 82:265–274.

Kurnit DM, Layton WM, Matthysse S (1987): Genetics, chance, and morphogenesis. Am J Hum Genet 41:979–995.

Lalande M (1996): Parental imprinting and human disease [Review]. Annu Rev Genet 30:173–195.

Lalande M, Minassian BA, DeLorey TM, Olsen RW (1999): Parental imprinting and Angelman syndrome [Review]. Adv Neurol 79:421–429.

La Spada AR (1997): Trinucleotide repeat instability: Genetic features and molecular mechanisms [Review]. Brain Pathol 7:943–963.

Latham KE (1996): X chromosome imprinting and inactivation in the early mammalian embryo [Review]. Trends Genet 12:134–138.

Latham KE, McGrath J, Solter D (1995): Mechanistic and developmental aspects of genetic imprinting in mammals [Review]. Int Rev Cytol 160:53–98.

Ledbetter DH, Engel E (1995): Uniparental disomy in humans: Development of an imprinting map and its implications for prenatal diagnosis [Review]. Hum Mol Genet 4:1757–1764.

Li R, el Mallakh RS (1997): Triplet repeat gene sequences in neuropsychiatric diseases [Review]. Harvard Rev Psychiatry 5:66–74.

Lin X, Antalffy B, Kang D, Orr HT, Zoghbi HY (2000): Polyglutamine expansion down-regulates specific neuronal genes before pathologic changes in SCA1 [see comments]. Nat Neurosci 3:157–163.

Lindblad K, Schalling M (1996): Clinical implications of unstable DNA repeat sequences [Review]. Acta Paediatr 85:265–271.

Lindblad K, Schalling M (1999): Expanded repeat sequences and disease [Review]. Semin Neurol 19:289–299.

Mann MR, Bartolomei MS (1999): Towards a molecular understanding of Prader-Willi and Angelman syndromes [Review]. Hum Mol Genet 8:1867–1873.

Marazita ML, Hu DN, Spence MA, Liu YE, Melnick M (1992): Cleft lip with or without cleft palate in Shanghai, China: Evidence for an autosomal major locus. Am J Hum Genet 51:648–653.

Martin RA, Hunter V, Neufeld-Kaiser W, Flodman P, Spence MA, Furnas D, Martin KA (2000): Ultrasonographic detection of orbicularis oris defects in first degree relatives of isolated cleft lip patients. Am J Med Genet 90:155–161.

Martin RA, Jones KL, Benirschke K (1993): Extension of the cleft lip phenotype: The subepithelial cleft. Am J Med Genet 47:744–747.

Mitas M (1997): Trinucleotide repeats associated with human disease [Review]. Nucl Acids Res 25:2245–2254.

Mitchell LE (1997): Genetic epidemiology of birth defects: Nonsyndromic cleft lip and neural tube defects [Review]. Epidemiol Rev 19:61–68.

Mitchell LE, Risch N (1993): The genetics of infantile hypertrophic pyloric stenosis. A reanalysis [Review]. Am J Dis Child 147:1203–1211.

Monckton DG, Caskey CT (1995): Unstable triplet repeat diseases [Review]. Circulation 91:513–520.

Moore GE, Ivens A, Newton R, Balacs MA, Henderson DJ, Jensson O (1990): X chromosome genes involved in the regulation of facial clefting and spina bifida. Cleft Palate J 27:131–135.

Morrison PJ (2000): The spinocerebellar ataxias: Molecular progress and newly recognized paediatric phenotypes [Review]. Eur J Paediatr Neurol 4:9–15.

Mutter GL (1997): Role of imprinting in abnormal human development [Review]. Mutat Res 396:141–147.

Nance MA (1997): Clinical aspects of CAG repeat diseases [Review]. Brain Pathol 7:881–900.

Nicholls RD (2000): The impact of genomic imprinting for neurobehavioral and developmental disorders [Review]. J Clin Invest 105:413–418.

Nurnberger JII, Blehar MC, Kaufmann CA, York-Cooler C, Simpson SG, Harkavy-Friedman J, Severe JB, Malaspina D, Reich T (1994): Diagnostic interview for genetic studies. Rationale, unique features, and training. NIMH Genetics Initiative. Arch Gen Psychiatry 51:849–859.

Nussbaum R, Auburger G (2000): Neurodegeneration in the polyglutamine diseases: Act 1, Scene 1 [news; comment]. Nat Neurosci 3:103–104.

O'Donnell DM, Zoghbi HY (1995): Trinucleotide repeat disorders in pediatrics. Curr Opin Pediatr 7:715–725.

Ohara K (2001): Anticipation, imprinting, trinucleotide repeat expansions and psychoses [Review]. Prog Neuro-Psychopharmacol Biol Psychiatry 25:167–192.

Ono SJ (2000): Molecular genetics of allergic diseases [Review]. Annu Rev Immunol 18:347–366.

Ottman R (1995): Gene-environment interaction and public health. Am J Hum Genet 56:821–823.

Pajukanta P, Porkka KV (1999): Genetics of familial combined hyperlipidemia. Curr Atherosclerosis Repts 1:79–86.

Parisi MA, Kapur RP (2000): Genetics of Hirschsprung disease. Curr Opin Pediatr 12:610–617.

Passarge E (2002): Dissecting Hirschsprung disease [letter; comment]. Nat Genet 31:11–12.

Paulsen M, Ferguson-Smith AC (2001): DNA methylation in genomic imprinting, development, and disease [Review]. J Pathol 195:97–110.

Pei Y, Paterson AD, Wang KR, He N, Hefferton D, Watnick T, Germino GG, Parfrey P, Somlo S, St. George-Hyslop P (2001): Bilineal disease and trans-heterozygotes in autosomal dominant polycystic kidney disease. Am J Hum Genet 68:355–363.

Pericak-Vance MA, Bebout JL, Gaskell PCJ, Yamaoka LH, Hung WY, Alberts MJ, Walker AP, Bartlett RJ, Haynes CA, Welsh KA, et al (1991): Linkage studies in familial Alzheimer disease: Evidence for chromosome 19 linkage. Am J Hum Genet 48:1034–1050.

Pfeifer K (2000): Mechanisms of genomic imprinting [Review]. Am J Hum Genet 67:777–787.

Piven J (2001): The broad autism phenotype: A complementary strategy for molecular genetic studies of autism. Am J Med Genet 105:34–35.

Piven J, Palmer P (1997): Cognitive deficits in parents from multiple-incidence autism families. J Child Psychol Psychiatry Allied Disciplines 38:1011–1021.

Piven J, Palmer P, Landa R, Santangelo S, Jacobi D, Childress D (1997): Personality and language characteristics in parents from multiple-incidence autism families. Am J Med Genet 74:398–411.

Reeve AE (1996): Role of genomic imprinting in Wilms' tumour and overgrowth disorders [Review]. Med Pediatr Oncol 27:470–475.

Reik W, Walter J (2001): Genomic imprinting: Parental influence on the genome [Review]. Nat Rev Genet 2:21–32.

Rice T, Rankinen T, Province MA, Chagnon YC, Perusse L, Borecki IB, Bouchard C, Rao DC (2000): Genome-wide linkage analysis of systolic and diastolic blood pressure: The Quebec family study. Circulation (Online) 102:1956–1963.

Robinson WP (2000): Mechanisms leading to uniparental disomy and their clinical consequences [Review]. Bioessays 22:452–459.

Romieu I, Trenga C (2001): Diet and obstructive lung diseases [Review]. Epidemiol Rev 23:268–287.

Rothman SM (1999): Mutations of the mitochondrial genome: Clinical overview and possible pathophysiology of cell damage [Review]. Biochem Soc Symp 66:111–122.

Rougeulle C, Lalande M (1998): Angelman syndrome: How many genes to remain silent? [Review]. Neurogenetics 1:229–237.

Sampogna F, Demenais F, Hochez J, Oryszczyn MP, Maccario J, Kauffmann F, Feingold J, Dizier MH (2000): Segregation analysis of IgE levels in 335 French families (EGEA) using different strategies to correct for the ascertainment through a correlated trait (asthma). Genet Epidemiol 18:128–142.

Sapienza C (1995): Genome imprinting: An overview [Review]. Dev Genet 17:185–187.

Schelhaas HJ, Ippel PF, Beemer FA, Hageman G (2000): Similarities and differences in the phenotype, genotype and pathogenesis of different spinocerebellar ataxias [Review]. Eur J Neurol 7:309–314.

Schwartz M, Vissing J (2002): Paternal inheritance of mitochondrial DNA [see comments]. N Engl J Med 347:576–580.

Shaffer LG, McCaskill C, Adkins K, Hassold TJ (1998): Systematic search for uniparental disomy in early fetal losses: The results and a review of the literature [Review]. Am J Med Genet 79:366–372.

Skuse DH (1999): Genomic imprinting of the X chromosome: A novel mechanism for the evolution of sexual dimorphism [Review]. J Lab Clin Med 133:23–32.

Sleutels F, Barlow DP (2002): The origins of genomic imprinting in mammals [Review]. Adv Genet 46:119–163.

Sleutels F, Barlow DP, Lyle R (2000): The uniqueness of the imprinting mechanism [Review]. Curr Opin Genet Dev 10:229–233.

Smith DW, Aase JM (1970): Polygenic inheritance of certain common malformations. Evidence and empiric recurrence risk data. J Pediatr 76:652–659.

Spranger J, Benirschke K, Hall JG, Lenz W, Lowry RB, Opitz JM, Pinsky L, Schwarzacher HG, Smith DW (1982): Errors of morphogenesis: Concepts and terms. Recommendations of an international working group. J ediatr 100:160–165.

Stevanin G, Durr A, Brice A (2000): Clinical and molecular advances in autosomal dominant cerebellar ataxias: From genotype to phenotype and physiopathology [Review]. Eur J Hum Genet 8:4–18.

Sutherland GR, Richards RI (1995): Simple tandem DNA repeats and human genetic disease [Review]. Proc Natl Acad Sci USA 92:3636–3641.

Swayze VW, Andreasen NC, Alliger RJ, Ehrhardt JC, Yuh WT (1990): Structural brain abnormalities in bipolar affective disorder. Ventricular enlargement and focal signal hyperintensities. Arch Gen Psychiatry 47:1054–1059.

Temple K, Calvert M, Plint D, Thompson E, Pembrey M (1989): Dominantly inherited cleft lip and palate in two families. J Med Genet 26:386–389.

Tilghman SM (1999): The sins of the fathers and mothers: Genomic imprinting in mammalian development [Review]. Cell 96:185–193.

Tsuji S (1997): Molecular genetics of triplet repeats: Unstable expansion of triplet repeats as a new mechanism for neurodegenerative diseases [Review]. Internal Med 36:3–8.

Tycko B, Morison IM (2002): Physiological functions of imprinted genes [Review]. J Cell Physiol 192:245–258.

Ulbrecht M, Hergeth MT, Wjst M, Heinrich J, Bickeboller H, Wichmann HE, Weiss EH (2000): Association of beta(2)-adrenoreceptor variants with bronchial hyperresponsiveness. Am J Respir Crit Care Med 161:469–474.

Verge CF, Vardi P, Babu S, Bao F, Erlich HA, Bugawan T, Tiosano D, Yu L, Eisenbarth GS, Fain PR (1998): Evidence for oligogenic inheritance of type 1 diabetes in a large Bedouin Arab family. J Clin Invest 102:1569–1575.

Wallace DW (1995): Mitochondrial DNA variation in human evolution, degenerative disease, and aging. Am J Hum Genet 57:201–223.

Wang Q, Chen Q, Zhao K, Wang L, Wang L, Traboulsi EI (2001): Update on the molecular genetics of retinitis pigmentosa [Review]. Ophthal Paediatr Genet 22:133–154.

Warren ST (1996): The expanding world of trinucleotide repeats. Science 271:1374–1375.

Warren ST, Ashley CT Jr (1995): Triplet repeat expansion mutations: The example of fragile X syndrome [Review]. Annu Rev Neurosci 18:77–99.

Warwick D, Payami H, Nemens EJ, Nochlin D, Bird TD, Schellenberg GD, Wijsman EM (2000): The number of trait loci in late-onset Alzheimer disease. Am J Hum Genet 66:196–204.

Yang H, Rotter JI (2000): The genetic background of inflammatory bowel disease [Review]. Hepato-Gastroenterology 47:5–14.

CHAPTER 3

Determining Genetic Component of a Disease

ALLISON ASHLEY-KOCH

INTRODUCTION

Genetics contributes to the etiology of almost every human disease. Even for infectious diseases that have been traditionally described as environmental, such as tuberculosis and human immunodeficiency virus (HIV), scientists are discovering that genetics is implicated either in the susceptibility to infection or in the severity of the disease (Bellamy, 1998; Bellamy et al., 2000; Gonzalez et al., 2001; Shields and Dell, 2001; Hill, 1999). Understanding the role of genetics in disease etiology will allow development of successful therapies that improve the quality of life for affected individuals and their families. While recent advances in the statistical and molecular methods to identify human disease genes have made achieving this goal increasingly more feasible, genetic analyses remain technically challenging, labor intensive, and financially expensive. Furthermore, the ethical challenges of applying this new knowledge are substantial. Consequently, before embarking upon a large-scale study to identify the genetic factors involved in a particular disease, one must first have clear evidence that genes play an important role in that disease.

For Mendelian disorders, such as sickle cell anemia, cystic fibrosis, and Duchenne muscular dystrophy, establishing a role for genetics is straightforward. These diseases have predictable, recognizable inheritance patterns and their etiologies can be attributed to variations in single genes. The phenotypic expression of the disease is highly correlated with the genotype at the disease locus. For example, all individuals who carry two copies of the hemoglobin (Hb) S variant of the β-globin gene are affected with sickle cell anemia. While there is variation in the severity of the symptoms, all individuals who are homozygous for Hb S will exhibit some of the symptoms of the disorder. In contrast, complex disorders, such as cardiovascular

Genetic Analysis of Complex Diseases, Second Edition, Edited by Jonathan L. Haines and Margaret Pericak-Vance
Copyright © 2006 John Wiley & Sons, Inc.

disease, cancers, and psychiatric disorders, usually do not display distinct inheritance patterns. Such disorders have a significant genetic component but are caused by an intricate web of genetic and environmental interactions. As a result, establishing a role for genetics in the etiology of complex disorders can be more difficult. While the genotype at a single locus may contribute to the susceptibility to develop a disease, it is expected that other loci are involved as well. That is, multiple loci may be interacting with each other or the environment to contribute to disease. Moreover, different loci may be contributing to disease susceptibility in different families. A classic example of a complex disease susceptibility gene is the association between the *APOE* locus and Alzheimer's disease (AD). Corder et al. (1993) demonstrated that individuals who carry the 4 allele of the *APOE* locus have a higher risk and earlier age of onset for late-onset AD when compared with individuals who do not carry the 4 allele. Furthermore, this association is dose dependent. Individuals who have two copies of *APOE-4* are at greater risk for the disease than individuals who carry one copy of *APOE-4*. Some Alzheimer families carry the *APOE-4* genotype and some do not. Thus, the *APOE* gene alone does not explain the etiology of the disease. In addition to the diminished correlation between genotype and phenotype, the high frequency of complex diseases in the general population reduces the certainty that two individuals within the same family have developed the disease as a result of the same genetic liability. For example, the family may exhibit bilineality (family history is present on both the maternal and paternal side). Consequently, each side of the family may contribute different genetic susceptibilities to the disease. Even in the absence of bilineality, there are complications. Some individuals in the family may express the disease as a result of exposure to an environmental insult. Therefore, it can be quite challenging to determine which family members carry the same genetic susceptibilities. For all these reasons it is much more difficult to establish a genetic basis for a complex disease than for a Mendelian disease.

However, even in the presence of such complexities, there are methods available to evaluate whether or not genetics plays an important role in the disease etiology. These methods are the primary topic of this chapter and should be explored prior to embarking on more elaborate analyses such as genomic screens and candidate gene studies. But, before considering any analysis, one must first consider study design.

STUDY DESIGN

The study design process includes all clinical, molecular, bioinformatic, and analytic components of your study. A study population must be defined, as well as the process for selecting individuals from that population for participation (ascertainment, this chapter) and criteria for identifying individuals exhibiting the trait of choice (phenotype definition, Chapter 2). One must determine the molecular technology to be used (e.g., microsatellite versus single-nucleotide polymorphisms, genomic screen versus candidate genes, Chapter 6) as well as establish the electronic process for storing and retrieving the clinical and molecular data (Chapters 5 and 8) and the analysis methods to be used (e.g., linkage versus association, Chapters 10–12). Finally, one must ensure all ethical, legal, and social issues are addressed (Chapter 4). Each component in the study design process should be given careful consideration, as the decisions one makes at each step will impact the conclusions that can be drawn from the results. Furthermore, many of the decisions in the study design process are interrelated. For example, the ascertainment process will impact the analysis methods that can be performed. Since many of the steps in the study design process are covered in detail in other chapters of the book, they will not be discussed further here. Only the process of selecting a study population and an ascertainment scheme will be covered in this chapter.

Selecting a Study Population

In a well-designed study, the study population is clearly defined and individuals from the study population are systematically selected for participation. The two primary types of study population are population based and clinic based. A population-based study is the gold standard for study design, but a clinic-based population tends to be a more realistic design for most genetic studies.

Population Based. Ideally one will select a study population that is an unambiguous subset of a larger underlying population. For example, this could include individuals selected from a newborn state screening registry (e.g., determining the frequency of congenital hypothyroidism in individuals from the California Newborn Screening Program; Waller et al., 2000), children who attend a particular school system (e.g., screening for *FRAXA* and *FRAXE* in a special-needs population; Meadows et al., 1996), and individuals in a local cancer registry (e.g., examination of the involvement of genetics to fallopian tube cancer in patients identified from the Ontario Cancer Registry; Aziz et al., 2001). This epidemiological approach to selecting a study population is much less susceptible to ascertainment bias than a clinic-based study population, and conclusions made from analysis are usually extendable to the larger underlying population from which the study population was drawn. The disadvantage with this approach is that it may be difficult to achieve the desired patient sample size for rare disorders, and even for common disorders, it can be quite laborious to identify eligible participants. For example, for a disease with a frequency of 1 in 1000, one would need to screen 100,000 individuals to identify even 100 cases, a rather small sample size.

Clinic Based. In reality, most genetic studies rely on clinic-based study populations because they provide the investigators with faster access to patients. There are numerous examples of genetic studies relying on this design. Generally, patients are selected for participation from existing patient populations, such as a specialty clinic (e.g., examining the phenotypic heterogeneity of age-related macular degeneration in patients selected from ophthalmology clinics at two academic centers; De La Paz et al., 1997). Less manpower is required for the initial identification of the patients in this approach compared with the population-based approach. However, the disadvantage of a clinic-based study population is that it can be quite challenging to determine the larger underlying population from which a clinic population was drawn. Most large, academic medical centers attract patients not only from the immediate geographic region but often from all over the world. A further complication of clinic-based samples is that patients ascertained in this manner may exhibit a more severe form of the disease, simply because they have sought medical treatment or have been referred for expert diagnosis and care. For example, some individuals with myotonic dystrophy have cataracts as the only phenotypic manifestation. Therefore, these individuals would not seek medical attention on a frequent basis. In contrast, other individuals with myotonic dystrophy are severely debilitated and seek medical treatment repeatedly. Consequently, ascertaining patients with myotonic muscular dystrophy from a neurology clinic would likely oversample individuals with the most severe form of the disease and miss the individuals who carry the less severe form of the disease (cataracts only). Thus, while clinic-based populations provide fast access to desired patients, results obtained in such studies may not be applicable to the general population or even to all individuals with the disease.

Ascertainment

There are three basic designs of ascertainment for a genetic analysis: collection of a single affected family member, relative pairs, and extended families with multiple affected individuals. Examples of these ascertainment schemes are shown in Figure 3.1. As shown below, certain sampling schemes limit the types of analyses that can be performed. However, one's sampling scheme is often dictated by the natural history of the disease under investigation. For example, with late-onset disorders such as Alzheimer's disease, Parkinson's disease, and chronic obstructive pulmonary disease, collection of the parents of an affected individual is often not feasible as the parents are deceased. In such cases, one may be restricted to collection of affected sibling pairs (sibpairs), or a case–control population.

Single Affected Family Member. Collection of a single affected family member includes the traditional epidemiological case–control design as well as the case–parent trio design. The case–parent trio design is primarily used in family-based association analysis and includes collection of a case and both their parents. A further extension is the case–sibling design or discordant sib pair (DSP), where the sibling does not have the trait in question. An alternative to the case–control and trio designs is the case-only design (Khoury and Flanders, 1996). Similar to the trio and DSP designs, the case-only approach arose from concerns regarding selection of appropriate controls for the study of genetic factors in the more traditional case–control approach. The case-only approach has been promoted as a particularly useful approach in the examination of gene–environment interactions.

From the ascertainment perspective, collection of a single affected family member is quite feasible because in complex disorders large families with multiple affected individuals are often difficult to identify. A disadvantage of this approach is that it

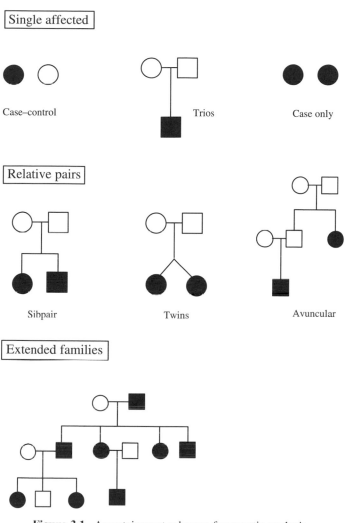

Figure 3.1. Ascertainment schemes for genetic analysis.

limits the statistical genetic analyses to allelic association methods. Traditional linkage analysis cannot be performed on a case–control or trio dataset because multiple affected individuals are needed to determine identical-by-descent (IBD) sharing.

Relative Pairs. The use of relative pairs is a common ascertainment design in the genetic analysis of complex disorders. This approach may include the use of sibpairs that are either concordant for the disorder (affected sibpairs) or discordant. Monozygotic (MZ, identical) and dizygotic (DZ, nonidentical) twins are a special case of sibpairs, and the utility of twins in genetic analysis is described in greater detail later in the chapter. Additionally, there are statistical methods that utilize information from other types of relative pairs, such as parent–child, avuncular pairs (e.g., uncle–niece), cousin pairs, and so on (Weeks and Lange, 1988; Davis et al., 1996; Kruglyak et al., 1996). In this approach, relatives other than the initial pair of ascertained individuals may also be collected so that more detailed linkage analysis can be performed. In the case of MZ twins, only one of the two individuals may be used in the linkage and association analysis because the twins share 100% of their genetic material.

Extended Families. Extended families refer to large families with many affected individuals in several generations. This study design is optimal for traditional linkage analysis but is often a rare occurrence in complex disorders. If such a family is identified, it is possible that the genetic liability in this particular family is due to a single gene, rather than a more complex etiology. Such a family would provide a unique opportunity to localize a single gene that has a large effect on disease risk in that family but may have a more moderate effect on disease etiology in the general population. Association methods may also be used with extended families. However, one must ensure that the association method being used considers the within-family dependence (e.g., the pedigree disequilibrium test; Martin et al., 2000b) or selects only one affected individual from the family to be used in the analysis.

There are also variations on these three ascertainment schemes. For example, in an analysis of breast cancer in Australia, Hopper and colleagues (1999) employed a *case–control–family* design. In this approach, the cases and controls were selected first and subsequently additional family members were recruited based on the family history. If applied correctly, this approach will have the analytic advantages of a family study, and the results can be placed in the context of an epidemiological study. Statistical issues associated with this design have been reviewed by other investigators (Liang and Pulver, 1996; Seybolt et al., 1997) and will not be discussed here.

Many investigators have explored sampling schemes to determine the optimal ascertainment scheme for genetic analysis of complex disorders. McCarthy and colleagues (1998) considered sampling strategies for affected sibpairs and found that the power to detect a disease gene locus is highly dependent on the larger pedigree structure from which the sibpairs were drawn. Furthermore, they concluded that imposing a few restrictions on that pedigree structure (such as the presence of at least one unaffected sibling or parent) can provide a modest increase in power and ascertaining random affected sibpairs (regardless of the larger pedigree structure) tends to be a robust approach under a variety of genetic inheritance models. The advantage of restricting the pedigree structure to one or fewer affected parents is that one can reduce the possibility of bilineality in the pedigree. Terwilliger and Goring (2000) have argued that, even in the case of complex disorders, ascertainment of large pedigrees is a more successful approach for genetic analysis than a case–control approach as the large pedigrees increase the likelihood of genetic homogeneity and, additionally, once ascertaining large pedigrees, one has more flexibility with regard to the types of analyses that may be performed. For example, one can analyze the entire pedigree for linkage analysis and also, by breaking the family structure into smaller units, consider the affected sibpair, affected relative pair, or trio approaches as complementary methods for identifying the disease genes. Badner et al. (1998), however, suggest that there is no benefit to collecting large pedigrees under certain genetic models (a qualitative trait with common alleles under a single locus, additive and multiplicative inheritance models). In spite of the many elegant theoretical considerations of sampling schemes, there does not appear to be any consensus with regard to an optimal sampling scheme for complex disorders (Baron, 1999). The optimal study design for a particular disease is influenced by the underlying genetic model, which is unknown in complex diseases. Consequently, the choice of ascertainment design will be determined primarily by the natural history of the disease under investigation and the available resources (both financial and personnel) rather than theoretical concerns. As mentioned earlier, the type of ascertainment has a major impact on the types of examples that can be employed.

Controls. For some analyses, it is necessary to have control samples to use for comparison with the patient samples. These control samples may include spouses and siblings of affected individuals, classmates, other members of the community, or even untransmitted genetic alleles, as shown in Chapters 12 and 14. Regardless of the relationship of the control sample to the patient sample, one must ensure that the controls are ascertained from the same study population as the patients. Furthermore, the controls must be matched to the patients for confounding factors (any factor that might influence the association between the disease and genotype), such as age, sex, ethnicity, and geographic location. There are two approaches for matching controls to the cases. First, one can select controls such that the overall distribution of cases and controls is comparable with respect to the frequency of the confounders (e.g., for a study of autistic disorder, both cases and controls have a sex ratio of $3:1$ males to females). This is referred to as frequency or category matching. Alternatively, one or more control individuals may be selected to match each case based on the confounding characteristics (e.g., the case and the control are both African-American females, 8 years of age, and reside in Durham County, North Carolina). This approach is called individual matching.

Improper selection of controls can lead to incorrect conclusions. For example, if cases and controls are not appropriately matched on ethnicity and the frequency of alleles for the genetic marker differ by ethnicity, an association study can be doomed. One may falsely conclude an association between a genetic marker and the disease if the "at-risk" marker allele is more prevalent in the predominant ethnicity of the cases versus the controls.

Ascertainment Bias. In genetic studies, research subjects are selected for participation based on the presence or absence of the trait of interest. The family member who comes to the investigator's attention (e.g., through admission to a hospital or solicitation of support groups) is called the proband. Most often, the proband is an individual who exhibits the trait of interest. Ascertainment through an affected individual can lead to a bias in the distribution of the numbers of affected and unaffected family members present in the analysis. Because the ascertainment scheme necessitated that the family have at least one affected individual (proband), families that may be carrying the genetic liability of interest but, by chance, do not contain an affected family member will not be ascertained. This phenomenon is referred to as ascertainment bias and is demonstrated in Figure 3.2. Depending on the analysis, ascertainment bias may greatly influence the outcome of the analyses.

In general, ascertainment bias should not affect the ability to accept or reject linkage in linkage analysis. However, it can affect the estimate of the recombination fraction between the genetic marker and the disease locus (Vieland and Hodge, 1996). Ascertainment bias can also influence familial recurrence risk ratios (λ_R) (Guo, 1998; Cordell and Olson, 2000) and the estimate of the segregation probability of the disease locus in segregation analysis (Stene, 1989; Greenberg, 1986; Ewens and Shute, 1986). Furthermore, it has been argued that in some cases ascertainment bias may be a reasonable explanation for what appears to be genetic anticipation in some pedigrees (Penrose, 1948).

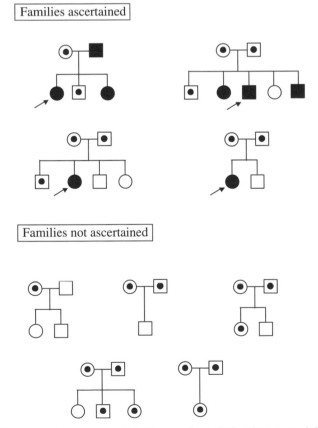

Figure 3.2. Example of ascertainment bias in genetic analysis when ascertaining through an affected individual.

Clearly, ignoring the ascertainment scheme used in an analysis can lead to false conclusions. For example, in the fragile X literature, it had been reported that male offspring of premutation and full-mutation mothers received larger CGG repeat expansions than did female offspring (Rousseau et al., 1994; Loesch et al., 1995). However, once potential sources of ascertainment bias were removed, it was determined that there was no association between sex of the offspring and size of the mutation inherited from the mother (Ashley-Koch et al., 1998). In this example, there were three types of ascertainment bias that were sequentially removed from the analysis. First, in clinically ascertained families, the transmission to the proband was removed. This is a standard ascertainment correction known as the Weinberg correction scheme (Weinberg, 1912). Second, transmissions were excluded where not all individuals within a sibship were tested. That is, in fragile X families, non-symptomatic females are tested more often than nonsymptomatic males because only the females are at risk to have affected offspring. This ascertainment bias increases the number of transmissions to premutation females compared with premutation males. Finally, the transmissions of the CGG repeat to the mothers of the proband were omitted. Because full-mutation females have reduced fitness (Sherman et al., 1984), mothers of probands are more likely to be premutation carriers merely because they have reproduced. Consequently, including transmissions of the CGG repeat from carrier grandmothers to mothers of probands will also increase the number of transmissions to premutation females. Table 3.1 shows how sequentially correcting for the ascertainment bias in fragile X families changes the conclusion regarding the association between offspring sex and the size of the CGG expansion. Notice that the statistical tests (t-test and logistic regression) become insignificant as the correction schemes are applied. Additionally, it illustrates that in some cases standard correction schemes such as the Weinberg method are not sufficient to abolish the bias in the ascertainment scheme. Thus, all potential sources of bias in the dataset must be carefully considered.

TABLE 3.1. Association between Sex of Offspring and Risk of Expansion in Fragile X Syndrome

| Ascertainment Scheme | Number of Cases | Proportion of Offspring with Full Mutation (%) | | *p* Value | |
		Males	Females	*t*-Test	Logistic Regression
Removal of cases associated with ascertainment	434	0.46	0.38	0.06	0.07
Removal of incompletely ascertained sibships	338	0.48	0.43	0.38	0.34
Removal of transmissions to proband's mother	298	0.48	0.50	0.63	0.71

Source: Ashley-Koch et al. (1998).

APPROACHES TO DETERMINING THE GENETIC COMPONENT OF A DISEASE

King et al. (1984) described the steps necessary to define the genetic mechanisms involved in a disease or trait several years ago, but those steps are still valid today. First, the evidence for a familial component to the disease must be established. Next, the cause of familial aggregation must be determined. That is, clustering in a disease may result from common environmental factors, rather than genetic factors, and these two hypotheses must be evaluated. Finally, the specific genetic factors must be identified and the manner in which they interact with each other and with environmental factors to contribute to the disease etiology must be defined.

Below, several approaches are presented to evaluate whether or not genes contribute to the etiology of a disease and to quantitate the contribution of those genes to the disease etiology.

Cosegregation with Chromosomal Abnormalities and Other Genetic Disorders

While complex disorders generally do not exhibit a recognizable inheritance pattern, occasionally in a subset of patients, a complex disorder will segregate with a cyto-genetic abnormality or another known genetic disorder. These associations may provide valuable information regarding the location of at least one locus involved in the disease etiology. For example, individuals with trisomy 21, or Down syndrome, have an increased risk for developing AD. This increased risk is due to amyloid pla-ques resulting from an increased dosage of the amyloid precursor protein (APP) (Rumble et al., 1989). Thus it was not surprising to find that a subset of families with early-onset AD are linked to chromosome 21 and segregate mutations in the *APP* gene (St. George-Hyslop et al., 1987; Goate et al., 1991). Another example of cosegregation of a complex disease and cytogenetic abnormality is the association between autistic disorder and 15q11–q13 abnormalities. There are numerous examples of isolated patients with autistic disorder and duplications or inversions involving 15q11–q13 [see Wolpert et al. (2000) for review]. In many cases the de novo rearrangements are thought to be maternal in origin (Lindgren et al., 1996). In addition, in the absence of these cytogenetic abnormalities, families with two or more individuals exhibiting autistic disorder display evidence for linkage (Philippe et al., 1999; Bass et al., 2000) and linkage disequilibrium (Cook et al., 1998; Martin et al., 2000a; Menold et al., 2000) in this region. Although as yet no gene has been identified, the convergence of cytogenetic, linkage, and association data suggests that a locus involved in susceptibility to autistic disorder is located at 15q11–q13.

Familial Aggregation

One of the characteristics of a genetic disorder is that it aggregates, or clusters, within families. If familial clustering of a disorder is observed, there are several approaches, described below, to determine if this observation is statistically signi-ficant. However, familial clustering may also be due to a common familial environment or simply to chance. So, while statistical evidence of familial clustering may support the involvement of genetics in the disorder under investigation, it will be necessary to identify the underlying genes to confirm this.

Family History Approach. In this approach one ascertains the presence or absence of family history in study participants (with and without disease) and then tests for statistically significant associations between family history and disease. Khoury et al. (1993) defines three variations on this approach, depending on the level of detail obtained on the relatives. The *abbreviated family history approach* involves simply questioning the study participants on the presence or absence of disease in other family members. This method is by far the fastest and cheapest but also the most susceptible to misclassification because no effort is made to confirm the participant responses. That is, the investigator is relying on the accuracy of the informant's memory or perception of the family members, rather than directly observing the disease in the relatives. The *detailed family history approach* requires the investigator to obtain detailed information regarding disease status (including age of onset) and demographic characteristics on each family member. This is the approach routinely utilized in genetic studies. Because the information requested on each family member is more detailed, there is less likelihood of misclassification as with the abbreviated approach. Finally, one can examine and/or interview the relatives directly. This is referred to as the *family study approach*. Often, in this approach, medical records are also reviewed to confirm the presence or absence of disease in the relatives. While time consuming and more expensive, this approach certainly provides the most accurate information regarding the family history of a trait.

Once the necessary information has been collected, there are several methods to test for statistically significant association between family history and disease. If the study design is case–control, the presence or absence of family history may be treated as a "risk factor" for disease, and the standard epidemiological 2 × 2 table may be used (see Table 3.2). From here, an odds ratio may be calculated:

$$\text{Odds ratio} = \frac{ad}{bc} \tag{3.1}$$

The odds ratio derived here is then a measure of the association between family history and disease. In this case, the odds ratio is the odds of having a positive family history in individuals with the disease compared with the odds of having a positive

TABLE 3.2. Standard Epidemiological 2 × 2 Table

Disease in Study Participant	Family History	
	+	−
+	a	b
−	c	d

family history in individuals who are unaffected. This is not to be confused with a risk ratio, which is a prospective measure with respect to an individual's affection status. That is, the risk ratio is the ratio of the incidence of the disease in individuals with a positive family history compared with those who have a negative family history. However, when the incidence of the disease is low, the odds ratio should closely approximate the risk ratio (Rothman and Greenland, 1998).

In keeping with the epidemiological approach, one can also measure the amount of the disease that can be "attributed" to the presence of a positive family history. This in effect provides information regarding what proportion of the disease is due to genetic causes. There are several methods for calculating *attributable fractions*, depending on what information is available. Khoury et al. (1993) review three of these formulas (Kelsey et al., 1986; Levin, 1953; Miettinen, 1974) in their book. The Miettinen formula is:

$$\text{Attributable fraction} = \text{fraction of cases with the risk factor} \times \frac{\text{relative risk} - 1}{\text{relative risk}}$$

(3.2)

Example Calculating Attributable Fraction. In a population-based study examining the contribution of sickle cell disease to the occurrence of developmental disabilities, the Miettinen formula (Eq. 3.2) was applied to demonstrate that 34% of the cases of stroke-related developmental disabilities in black children were due to sickle cell disease (Ashley-Koch et al., 2001). In this example, the fraction of cases with stroke-related disabilities with the risk factor (sickle cell disease) was 0.34 (13 children with sickle cell disease among a total of 38 children with stroke-related developmental disabilities), and the relative risk was 130. Thus, the attributable fraction by the Miettinen formula is 34% [$0.34 \times (129/130)$].

For all the ease with the family history approach, there are several pitfalls that may occur. One should be aware that a positive family history is a function of many factors, including the frequency of the disease in the general population, the size of the family, the natural history of the disease, and the underlying genetic mechanism. For example, if a disease is relatively common in the population, a positive family history may simply result from the presence of phenocopies in the family. Also, in a disease with late onset, young relatives may be misclassified as "disease free." (They may not express the disease simply because they are too young to express it.) For example, if one is studying Alzheimer's disease and finds that a 55-year-old sibling of an affected proband has no signs of dementia, one should reexamine that sibling at regular intervals to determine if he remains asymptomatic over time. It is possible that, at age 55, the sibling is too young to express symptoms. However, if he were reexamined at 60 years of age, he would exhibit symptoms. Thus, if the only physical examination for that sibling was performed at 55 years of age, he would be classified as "normal." But if another physical examination was performed at age 60, then the individual would be classified as "affected." Whenever resources allow, it is best to follow study participants over time for changes in affection status. In addition to variation in the age of onset of a disease, there may be variable expression of the disease. Therefore, family members who have minor manifestations of the disease or are in the early stages of the disease process may not be recognized as expressing the disease. Furthermore, in complex genetic disorders, individuals in the family may not express the disorder because they possess only a fraction of the necessary genetic and environmental factors to express the disease. For all these reasons, one may want to examine the association of family history with several types of relatives to make sure that the same conclusion is reached for all relative types.

Correlation Coefficients. Another simple approach to determining familial aggregation is the use of scatter plots. In the case of a quantitative trait, plotting the trait measurement of one relative against the trait measurement of another relative will provide the correlation of the trait among the pair. The slope of the line that is formed by the data is the square of the correlation coefficient. As shown later in the chapter, the correlation coefficient can also provide information regarding the heritability of a trait. Figure 3.3 examines the correlation of age of onset of Alzheimer's disease among affected siblings. These data were randomly selected from a larger study examining the genetic contribution to the age at onset in Alzheimer's disease and Parkinson's disease (Li et al., 2002). The proband's age of onset was plotted against the sibling's age of onset. The resulting slope of the line was 0.16, corresponding to a Pearson correlation coefficient of 0.40 ($p < 0.0001$, $n = 400$ sibpairs), suggesting that there is a significant correlation of age of onset among siblings affected with Alzheimer's disease. Khoury and colleagues (1985) demonstrate the use of the loglinear model to determine correlation for qualitative traits.

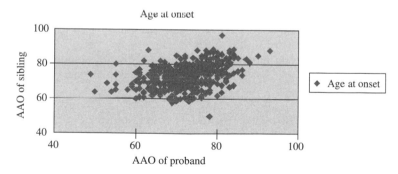

Figure 3.3. Correlation of age of onset among siblings affected with AD.

Twin and Adoption Studies

Twin and adoption studies can be quite useful as they provide an opportunity to tease apart the role of genetics and a common familial environment. The most difficult aspect of these types of studies is obtaining reasonable sample sizes. This is especially true in the United States, where twin, adoption, and disease registries are less common than in European countries.

The premise of twin studies is the comparison of the disease concordance in MZ with DZ twins. Since MZ twins share 100% of their genetic make-up and DZ twins share on average 50% of their genetic make-up, a greater disease concordance in MZ compared with DZ twins is consistent with the involvement of genetics. An advantage of this approach is that it controls for a common familial environment, but this is generally only applicable for exposures during childhood. It may not entirely control for prenatal exposures because the twins, especially DZ, may not have shared placentas, chorions, and amniotic sacs. However, the intrauterine and extrauterine environments are generally more similar for DZ twins than siblings. It is even less likely that adult exposures are similar among the twins, especially if they reside in different geographic locations. Other possible confounding variables controlled for in the twin study approach include age and sex (provided same-sex DZ twins are utilized). However, for a disease with variable age of onset it may be necessary to follow the twins for several years to conclusively determine that a set of twins is concordant or discordant for the disease. But most importantly, prior to beginning any twin study, one must determine the zygosity of the twins, as misclassification can have a devastating effect on the outcome of the analysis (Ellsworth et al., 1999). Often families will know the zygosity of the twins, but it is prudent to determine this experimentally via genotyping (Chapter 6).

Table 3.3 shows examples of concordance rates that might be observed under various disease etiologies. In practice, the results will not be easily interpreted for complex disorders, much like the results obtained in the last row of this table. In general, however, if the frequency in concordance is greater in MZ compared with DZ twins, it is accepted as evidence for at least a minor role of genetics in the disease etiology.

Adoption studies can also be used to examine the evidence for genetic versus common familial environmental factors. In this approach, cases are ascertained and then the frequency of the disease in the biological parents is compared with the frequency of the disease in the adoptive parents. As shown in Table 3.4, a higher disease frequency in biological parents argues for the presence of genetic factors, while a higher disease frequency in the adoptive parents argues for the

TABLE 3.3. Association Between Disease Concordance Rates in Twins and Disease Etiology

Frequency of Disease Concordance in Twins (%)		Possible Etiology
MZ	DZ	
85	85	Common familial environment
100	25	Mendelian recessive genetic factor
100	50	Mendelian dominant genetic factor
63	17	? Genetic and environmental factors

TABLE 3.4. Adoptive Studies and Disease Etiology

Frequency of Disease in Parents (%)		
Biological	Adoptive	Possible Etiology
85	4	Genetic factors
4	85	Common familial environment

presence of a common familial environmental factor. A variation on adoption studies is the adoptive-twin approach where twins who have been reared apart are examined to determine if genetics (high concordance rate) or environment (low concordance rate) plays a role in the disease etiology.

While adoption studies are simple in theory, in reality they are quite difficult to carry out. Achieving the necessary sample size can be challenging because of the need to identify the biological parents. Moreover, children are often placed in demographic environments similar to that of the biological parents, making it difficult to ensure that one can distinguish between genetics and environment. However, when these issues can be overcome, this approach can provide critical information regarding the etiology of complex traits. For example, this approach has been used to demonstrate that familial aggregation of multiple sclerosis is due to genetics and not a shared environment (Ebers et al., 1995) and to differentiate the role of genes and environment on various psychological measures (Bouchard et al., 1990).

Recurrence Risk in Relatives of Affected Individuals

A trio of seminal manuscripts in complex genetic analysis (Risch, 1990a,b,c), evaluated methods for linkage analysis and determining the mode of inheritance of disease loci using affected relative pairs. The basis for these analyses is familial recurrence risk ratios (λ_R):

$$\lambda_R = \frac{\text{risk in relatives of type R}}{\text{risk in the general population}} \qquad (3.3)$$

A greater risk for disease in the relatives (of type R) of an affected individual compared with the risk in the general population is consistent with the involvement of genetics in the disease etiology. Because λ_R is a ratio, usually any value over 1.0 is considered to provide evidence for the involvement of a familial component/genetics in a disease. It is generally accepted that the larger the recurrence risk ratio, the stronger the role of genetics in the disease etiology, but this does not always hold true. Notice that the denominator in Eq. (3.3) is the risk of the disease in the general population. Thus, for some common diseases λ_R is not large, but this is due to the high frequency of the disease in the general population, not to the lack of involvement of genetics. For example, $\lambda_{\text{sibling}} = 4-5$ for Alzheimer's disease (Roses et al., 1995) compared with $\lambda_{\text{sibling}} = 100$ for autistic disorder (Bolton et al., 1994; Bailey et al., 1995). In both of these complex disorders, genetics plays a major role, but the frequency of Alzheimer's disease in the general population is much higher than that of autistic disorder, which contributes to the difference in the $\lambda_{sibling}$ values. In spite of the low $\lambda_{sibling}$ in Alzheimer's disease, four genes that affect risk in Alzheimer's disease have been identified (Farrer et al., 1997) compared with none for autistic disorder. Recurrence risk ratios are also population specific as the disease risk in the general population may vary by ethnicity or geographic location. Keep in mind that a high λ_R could result from a common familial environment rather than from genetics.

Risch (1990a) also demonstrated that the reduction in recurrence risk by degree of familial relation is related to the mode of inheritance. For a single-gene model, $\lambda_R - 1$ is expected to decrease by a factor of 2 with each decreasing degree of relation. For a multiplicative model, $\lambda_R - 1$ will drop off more quickly with each degree of relation. Thus, by comparing the λ_R for various types of relatives, one can determine if the disorder is consistent with a particular inheritance pattern. Brown et al. (2000) found that the recurrence risks for ankylosing spondylitis were 8.2% for first-degree relatives, 1.0% for second-degree relatives, and 0.7% for third-degree relatives. Because this drop-off rate is more rapid than a factor of 2, a single-gene model for genetic susceptibility to ankylosing spondylitis is not likely. Using the information on recurrence risks, the authors determined that the most likely mode of inheritance for ankylosing spondylitis was an oligogenic model with a multiplicative interaction between the loci.

One can also use recurrence risks to make a rough estimate of the number of genes involved in a disorder if one assumes a particular inheritance pattern. For example, $\lambda_{sibling} = 100$ for autistic disorder (Bolton et al., 1994; Bailey et al., 1995). It is clear that autistic disorder does not follow a Mendelian mode of inheritance and, in fact, is more consistent with an epistatic interaction of multiple loci (Jorde et al., 1991). If we assume that there are four genes contributing an equal, additive effect to the genetic susceptibility, $\lambda_{sibling}$ for each locus would be 25 $(25 + 25 + 25 + 25 = 100)$. If there are four genes contributing an equal multiplicative effect, $\lambda_{sibling}$ for each locus would be approximately 3.2 $(3.2 \times 3.2 \times 3.2 \times 3.2 = 105)$. This type of information may be useful in determining the power to detect loci in a dataset. That is, with a sample size of 100 sibpairs, the power to detect linkage at a locus with $\lambda_{sibling} = 3.0$ is about 85%, and the power to detect linkage at a locus with $\lambda_{sibling} = 25$ would approach 100% (Risch, 1990b, Fig. 1).

Heritability

Heritability (h^2) is a measurement of the genetic contribution to a trait, or how much of the trait variability can be attributed to genetics rather than other causes. Heritability is generally calculated for quantitative traits and may be defined in a broad or narrow sense. Heritability in the broad sense is defined as

$$h^2 = \frac{G}{V} = \frac{G}{G + B + E} \tag{3.4}$$

where V is the total phenotypic variance, G is the genetic variance of the phenotype, B is the within-family variance of the phenotype, and E is the random environmental variance of the phenotype. The genetic variance can be further broken down into additive (effects of alleles at a particular locus), dominance (effects of allelic inter-actions at a particular locus), and epistatic (effects of allelic interactions from multiple loci) variance. Heritability in the narrow sense is a measure of the additive genetic variation in a trait. Estimates of heritability will vary between 0 and 1, with $h^2 = 0$ representing traits that are entirely controlled by nongenetic factors and $h^2 = 1$ representing traits that are entirely controlled by genetic factors. A more formal and detailed discussion of heritability with respect to quantitative trait analysis is presented in Chapter 9. Here we will focus on simpler methods of calculating heritability in the context of analyses defined in previous sections of the current chapter.

The correlation coefficient that one obtains from plotting a trait value in one relative versus the trait value in another relative may be used to calculate heritability in the narrow sense, provided that the relatives under consideration are not full siblings:

$$h^2 = \frac{r^2}{\text{fraction of genes shared IBD}} \tag{3.5}$$

For example, if one plots the age of onset for a disease in a proband versus the age of onset in an affected parent, the square of the correlation coefficient (r^2, or the slope of the line) obtained from the plot would be divided by $\frac{1}{2}$, the number of alleles shared IBD for a parent–child pair. When considering full siblings, this particular approach to calculating heritability does not represent heritability in the narrow sense because one has to take into account dominance variance as well. See Hartl and Clark (1997) for a more thorough discussion of using the covariance between two relatives to estimate heritability.

Rice and colleagues (1997) extended this approach to incorporate correlations from multiple relative pairs (r_{sib}, siblings; $r_{\text{p-o}}$, parent–offspring; r_{sp}, spousal):

$$h^2 = \frac{(r_{\text{sib}} + r_{\text{p-o}})(1 + r_{\text{sp}})}{(1 + r_{\text{sp}}) + (2r_{\text{sp}} \times r_{\text{p-o}})} \tag{3.6}$$

Example Using Correlation Coefficients to Calculate Heritability. In a large study that assessed the genetic contribution to normal variation of pulmonary function, Wilk and colleagues (2000) calculated familial correlations for spouses, parent–offspring pairs, and siblings from 455 Caucasian families for forced expira-tory volume (FEV1; a measure of airflow), forced vital capacity (FVC; a measure of lung volume), and the ratio of FEV/FVC and using Eq. (3.6) obtained heritability estimates for the three traits (Table 3.5).

Correlations for a trait among MZ and DZ twins can also be used to calculate heritability:

$$h^2 = 2\,(r_{\text{MZ}} - r_{\text{DZ}})$$

TABLE 3.5. Familial Correlation and Heritability Estimates of Pulmonary Function

Trait	r_{sib}	r_{p-o}	r_{sp}	h^2
FEV1	0.259	0.239	-0.067	0.52
FVC	0.257	0.242	-0.135	0.54
FEV1/FVC ratio	0.27	0.19	0.071	0.45

Source: Wilk et al. (2000, Table II).

Heritability estimates can be quite useful for data mining of a phenotype prior to initiating a genome scan. For example, if several phenotypic measures for a disease have been collected, heritability estimates may provide guidance as to which of those measures will be most useful for genetic analysis. That is, those phenotypes with the highest h^2 values should be prioritized for analysis. A good rule of thumb is that the h^2 for a trait should be at least 0.3 to be useful for analysis.

Segregation Analysis

Segregation analysis is a modeling tool that is used to examine the patterns of disease in families and determine if the patterns are indicative of traditional genetic inheritance models (such as autosomal dominant, autosomal recessive, polygenic) or are more consistent with nongenetic (environmental) models (Elston, 1981; Morton, 1982; Lalouel, 1984). The likelihood of the data to fit a particular inheritance model is computed. By comparing the likelihood of several models, one can determine which model provides the "best fit" to the data. Segregation analysis does not prove that a particular inheritance model is correct but will determine if the data are consistent with that inheritance model.

The advantage of segregation analysis is that it can provide an inheritance model and parameters that may be used in parametric linkage analysis. However, generally segregation analysis can only model one or two loci, which may not be very useful for most complex diseases. That is, if none of the inheritance models examined in the segregation analysis can adequately accommodate the complexities of the underlying inheritance model of the disorder, even the best-fitting model will not provide much information.

Additionally, this approach is extremely sensitive to ascertainment bias. In genetic analysis, families are often collected based on the presence of many affected individuals. Thus, for segregation analysis, there may be a high proportion of families with numerous affected individuals that are used in the analysis, when in reality these types of families only make up a small percentage of the disease population and most cases may be observed in families with only one or two affected individuals. For example, the probability that an affected individual will be ascertained as a proband is π, and when $\pi = 1$, all the individuals in the study population who have the disease have been ascertained. This is referred to as *complete ascertainment*. When the probability that an affected individual is a proband is very low (π approaches zero), each sibship is expected to have only a single proband. This is called *single ascertainment*. With this latter ascertainment approach, the probability that a family will come to the investigator's attention and be included in the study is related to the number of affected individuals in the family. That is, the more affected individuals in the family, the higher the likelihood that this particular family will be ascertained, thus introducing a biased distribution of family types in the analysis.

If one does not correct for the method in which the families were ascertained, the estimate of the segregation probability of the disease allele can be biased, which can affect the best-fitting genetic model. In some cases, the ascertainment bias may be so great that it causes the investigator to incorrectly conclude that the disorder is consistent with a single-gene model (Greenberg, 1986). For a discussion on approaches for correcting ascertainment bias, see Khoury et al. (1993).

There are several analytic approaches used for segregation analysis, each with its strengths and weaknesses. The most commonly used methods include the mixed model (Morton and MacLean, 1974; Lalouel and Morton, 1981; MacLean et al., 1984); the transmission probability model (Elston and Stewart, 1971); the unified model (Lalouel et al., 1983), which draws on the strengths of both the mixed and transmission probability models; and the regressive model (Bonney, 1984). All of these approaches are computationally intensive but are readily available for use in several software packages, such as POINTER (Lalouel and Morton, 1981), SAGE (http://darwin.cwru.edu/octane/sage/sage.php), and PAP (Hasstedt, 1993) to name a few.

Segregation analysis is not widely used in the evaluation of complex diseases because it is susceptible to the presence of genetic heterogeneity, phenocopies, and gene–gene and gene–environment interactions, which are quite difficult to model. Consequently, segregation analysis results that support the involvement of a major gene in a disease are much easier to interpret than results that do not support such an effect. This is especially true if there are other data to support the involvement of genetics, such as twin data or familial clustering. In spite of these difficulties there are examples of the successful application of segregation analysis to complex disorders, even to psychiatric disorders, which tend to have the additional complexity of defining a precise phenotype for investigation. For example, there is substantial evidence for a Mendelian genetic factor contributing to obsessive compulsive disorder (Notarnicola et al., 2000; Nestadt et al., 2000).

SUMMARY

The analyses presented in this chapter will aid in determining whether genes play an important role in a disease or trait. It is important to establish a genetic basis before embarking on more elaborate analyses such as whole-genome scans using linkage and/or association analysis because of the ethical, financial, and labor challenges associated with them. However, the first step in the analytic process should be a thorough examination of the literature. Often, because complex disorders are common and of public health import, many of the analyses discussed in this chapter will have been previously completed by other investigators. Therefore, there may already exist sufficient evidence to support the involvement of genetics in the etiology of the disorder under investigation. However, evaluate the literature critically, keeping in mind how the study population (including age and ethnicity) and phenotype were defined in each of the published analyses. Significant deviations from the definition of one's own study population and phenotype may suggest that prior results do not generalize to the present study and may necessitate repeating of the analyses described in this chapter.

If one determines that the types of analyses presented in this chapter (or any of the other chapters) must be performed, give careful consideration to study design prior to collecting and analyzing the data. Study design will influence the types of biases introduced into the data and can limit the types of analyses that may be performed.

Pay careful attention to the types of bias in the data. Ignoring ascertainment biases and confounding factors (e.g., age, ethnicity) may lead one to the wrong conclusions. If control samples are utilized, one must ensure that they are properly matched to the cases for confounding factors. When historical data are obtained on family members, follow up with examination of medical records or physical examination of those individuals to corroborate the information whenever resources allow. If a late-onset disorder is being studied, reexamine "unaffected" individuals after a period of time to confirm that they are truly unaffected and not misclassified because they were too young to express the disorder.

For complex genetic disorders, there may be a high percentage of the patient population for whom the disorder is sporadic (no other affected family members), and the genetic mechanisms that lead to complex disorders may or may not be the same in familial and sporadic cases. For example, the *BRCA1* gene that has been implicated in familial cases of breast cancer is not frequently mutated in cases of sporadic breast cancer (Futreal et al., 1994; Easton et al., 1993). This mantra is also true when comparing genetic mechanisms for early-onset versus late-onset familial versions of a disorder.

Finally, recognize that statistical associations for familial clustering may not represent a genetic etiology for a disorder but may be due to a common familial environment or simply chance. The definitive evidence that a disorder is genetic is the identification of the specific genetic variations that lead to the disease.

REFERENCES

Ashley-Koch A, Murphy CC, Khoury MJ, Boyle, CA (2001): Contribution of sickle cell. Genet Med 3:181–186.

Ashley-Koch AE, Robinson H, Glicksman AE, Nolin SL, Schwartz CE, Brown WT, Turner G, Sherman SL (1998): Examination of factors associated with instability of the FMR1 CGG repeat. Am J Hum Genet 63:776–785.

Aziz S, Kuperstein G, Rosen B, Cole D, Nedelcu R, McLaughlin J, Narod SA (2001): A genetic epidemiological study of carcinoma of the fallopian tube. Gynecol Oncol 80:341–345.

Badner JA, Gershon ES, Goldin LR (1998): Optimal ascertainment strategies to detect linkage to common disease alleles. Am J Hum Genet 63:880–888.

Bailey A, Le Couteur A, Gottesman I, Bolton P, Simonoff E, Yuzda E, Rutter M (1995): Autism as a strongly genetic disorder: Evidence from a British twin study. Psychol Med 25:63–77.

Baron M (1999): Optimal ascertainment strategies to detect linkage to common disease alleles. Am J Hum Genet 64:1243–1248.

Bass MP, Menold MM, Wolpert CM, Donnelly SL, Ravan SA, Hauser ER, Maddox LO, Vance JM, Abramson RK, Wright HH, Gilbert JR, Cuccaro ML, DeLong GR, Pericak-Vance MA (2000): Genetic studies in autistic disorder and chromosome 15. Neurogenetics 2:219–226.

Bellamy R (1998): Genetic susceptibility to tuberculosis in human populations. Thorax 53:588–593.

Bellamy R, Beyers N, McAdam KPWJ, Ruwende C, Gie RP, Samaai P, Bester D, Meyer M, Corrah T, Collin M, Camidge DR, Wilkinson D, Hoal-van Helden E, Whittle HC, Amos W, van Helden P, Hill AVS (2000): Genetic susceptibility to tuberculosis in Africans: A genome-wide scan. Proc Natl Acad Sci USA 97:8005–8009.

Bolton P, Macdonald H, Pickles A, Rios P, Goode S, Crowson M, Bailey A, Rutter M (1994): A case-control family history study of autism. Child Psychol Psychiatry Allied Disciplines 35:877–900.

Bonney GE (1984): On the statistical determination of major gene mechanisms in continuous human traits: Regressive models. Am J Med Genet 18:731–749.

Bouchard TJ Jr, Lykken DT, McGue M, Segal NL, Tellegen A (1990): Sources of human psychological differences: The Minnesota Study of Twins Reared Apart. Science 250:223–228.

Brown MA, Laval SH, Brophy S, Calin A (2000): Recurrence risk modelling of the genetic susceptibility to ankylosing spondylitis. Ann Rheum Dis 59:883–886.

Cook EH Jr, Courchesne RY, Cox NJ, Lord C, Gonen D, Guter SJ, Lincoln A, Nix K, Haas R, Leventhal BL, Courchesne E (1998): Linkage-disequilibrium mapping of autistic disorder, with 15q11-13 markers. Am J Hum Genet 62:1077–1083.

Cordell HJ, Olson JM (2000): Correcting for ascertainment bias of relative-risk estimates obtained using affected-sib-pair linkage data. Genet Epidemiol 18:307–321.

Corder EH, Saunders AM, Strittmatter WJ, Schmechel DE, Gaskell PC, Small GW, Roses AD, Haines JL, Pericak-Vance MA (1993): Gene dose of apolipoprotein E type 4 allele and the risk of Alzheimer's disease in late onset families. Science 261:921–923.

Davis S, Schroeder M, Goldin LR, Weeks DE (1996): Nonparametric simulation-based statistics for detecting linkage in general pedigrees. Am J Hum Genet 58:867–880.

De La Paz MA, Pericak-Vance MA, Haines JL, Seddon JM (1997): Phenotypic heterogeneity in families with age-related macular degeneration. Am J Ophthalmol 124:331–343.

Easton D, Ford D, Peto J (1993): Inherited susceptibility to breast cancer. Cancer Surv 18:95–113.

Ebers GC, Sadovnick AD, Risch NJ (1995): A genetic basis for familial aggregation in multiple sclerosis. Canadian Collaborative Study Group. Nature 377:150–151.

Ellsworth RE, Ionasescu V, Searby C, Sheffield VC, Braden CC, Kucaba TA, McPherson JD, Marra MA, Green ED (1999): The *CMT2D* locus: Refined genetic position and construction of a bacterial clone-based physical map. Genome Res 9:568–574.

Elston RC (1981): Segregation analysis. Adv Hum Genet 11:63–120.

Elston RC, Stewart J (1971): A general model for the genetic analysis of pedigree data. Hum Hered 21:523–542.

Ewens WJ, Shute NC (1986): A resolution of the ascertainment sampling problem. I. Theory. Theor Popul Biol 30:388–412.

Farrer LA, Cupples LA, Haines JL, Hyman B, Kukull WA, Mayeux R, Myers RH, Pericak-Vance MA, Risch N, Van Duijn CM (1997): Effects of age, sex, and ethnicity on the association between apolipoprotein E genotype and Alzheimer disease. A meta-analysis. APOE and Alzheimer Disease Meta Analysis Consortium. JAMA 278:1349–1356.

Futreal PA, Liu Q, Shattuck-Eidens D, Cochran C, Harshman K, Tavtigian S, Bennett LM, et al (1994): BRCA1 mutations in primary breast and ovarian carcinomas. Science 266:120–126.

Goate A, Chartier-Harlin MC, Mullan M, Brown J, Crawford F, Fidani L, Giuffra L, Haynes A, Irving N, James L, Mant R, Newton P, Rooke K, Roques P, Talbot C, Pericak-Vance MA, Roses A, Williamson R, Rossor M, Owen M, Hardy J (1991): Segregation of a missense mutation in the amyloid precursor protein gene with familial Alzheimer's disease. Nature 349:704–706.

Gonzalez E, Dhanda R, Bamshad M, Mummidi S, Geevarghese R, Catano G, Anderson SA, Walter EA, Stephan KT, Hammer MF, Mangano A, Sen L, Clark RA, Ahuja SS, Dolan MJ, Ahuja SK (2001): Global survey of genetic variation in CCR5, RANTES, and MIP-1alpha: Impact on the epidemiology of the HIV-1 pandemic. Proc Natl Acad Sci USA 98:5199–5204.

Greenberg DA (1986): The effect of proband designation on segregation analysis. Am J Hum Genet 39:329–339.

Guo SW (1998): Inflation of sibling recurrence-risk ratio, due to ascertainment bias and/or overreporting. Am J Hum Genet 63:252–258.

Hartl DL, Clark AG (1997): Principles of Population Genetics. Sunderland, MA: Sinauer Associates.

Hasstedt SJ (1993): Variance components/major locus likelihood approximation for quantitative, polychotomous, and multivariate data. Genet Epidemiol 10:145–158.

Hill AV (1999): Immunogenetics. Defence by diversity. Nature 398:668–669.

Hopper JL, Chenevix-Trench G, Jolley DJ, Dite GS, Jenkins MA, Venter DJ, McCredie MR, Giles GG (1999): Design and analysis issues in a population-based, case-control-family study of the genetic epidemiology of breast cancer and the Co-operative Family Registry for Breast Cancer Studies (CFRBCS). J Natl Cancer Inst Monogr 95–100.

Jorde LB, Hasstedt SJ, Ritvo ER, Mason-Brothers A, Freeman BJ, Pingree C, McMahon WM, Petersen B, Jenon WR, Mo A (1991): Complex segregation analysis of autism. Am J Hum Genet 49:932–938.

Kelsey JL, Thompson WD, Evans AS (1986): Methods in Observational Epidemiology. New York: Oxford University Press.

Khoury MJ, Beaty TH, Cohen BH (1993): Fundamentals of Genetic Epidemiology. New York: Oxford University Press.

Khoury MJ, Beaty TH, Tockman MS, Self SG, Cohen BH (1985): Familial aggregation in chronic obstructive pulmonary disease: Use of the loglinear model to analyze intermediate environmental and genetic risk factors. Genet Epidemiol 2:155–166.

Khoury MJ, Flanders WD (1996): Nontraditional epidemiologic approaches in the analysis of gene-environemnt interaction: Case-control studies with no controls. Am J Epidemiol 144:207–213.

King MC, Lee GM, Spinner NB, Thomson G, Wrensch MR (1984): Genetic epidemiology. Annu Rev Public Health 5:1–52.

Kruglyak L, Daly MJ, Reeve-Daly MP, Lander ES (1996): Parametric and nonparametric linkage analysis: A unified multipoint approach. Am J Hum Genet 58:1347–1363.

Lalouel JM (1984): Segregation analysis: A gene or not a gene. In: Genetic Epidemiology of Coronary Heart Disease: Past, Present, and Future, Alan R. Liss, pp 217–243.

Lalouel JM, Morton NE (1981): Complex segregation analysis with pointers. Hum Hered 31:312–321.

Lalouel JM, Rao DC, Morton NE, Elston RC (1983): A unified model for complex segregation analysis. Am J Hum Genet 35:816–826.

Levin ML (1953): The occurrence of lung cancer in man. Acta Unio Int Contra Cancrum 9:531–541.

Li YJ, Scott WK, Hedges DJ, Zhang F, Gaskell PC, Nance MA, Watts RL, et al (2002): Age at onset in two common neurodegenerative diseases is genetically controlled. Am J Hum Genet 70:985–993.

Liang KY, Pulver AE (1996): Analysis of case-control/family sampling design. Genet Epidemiol 13:253–270.

Lindgren V, Cook EH Jr, Leventhal BL, Courchesne R, Lincoln A, Shulman C, Lord C, Courchesne E (1996): Maternal origin of proximal 15q duplication in autism. Am J Hum Genet 39(Suppl):688.

Loesch DZ, Huggins R, Petrovic V, Slater H (1995): Expansion of the CGG repeat in fragile X in the FMR1 gene depends on the sex of the offspring. Am J Hum Genet 57:1408–1413.

MacLean CJ, Morton NE, Yee S (1984): Combined analysis of genetic segregation and linkage under an oligogenic model. Comput Biomed Res 17:471–4806.

Martin ER, Menold MM, Wolpert CM, Bass MP, Donnelly SL, Ravan SA, Zimmerman A, Gilbert JR, Vance JM, Maddox LO, Wright HH, Abramson RK, DeLong GR, Cuccaro ML, Pericak-Vance MA (2000a): Analysis of linkage disequilibrium in gamma-aminobutyric acid receptor subunit genes in autistic disorder. Am J Med Genet 96:43–48.

Martin ER, Monks SA, Warren LL, Kaplan NL (2000b): A test for linkage and association in general pedigrees: The pedigree disequilibrium test. Am J Hum Genet 67:146–154.

McCarthy MI, Kruglyak L, Lander ES (1998): Sib-pair collection strategies for complex diseases. Genet Epidemiol 15:317–340.

Meadows KL, Pettay D, Newman J, Hersey J, Ashley AE, Sherman SL (1996): Survey of the fragile X syndrome and the fragile X E syndrome in a special education needs population. Am J Med Genet 64:428–433.

Menold MM, Bass MP, Gilbert JR, Wolpert CM, Donnelly SL, Poole CW, Shao YJ, Ravan SA, McClain C, von Wendt L, Zimmerman A, Wright HH, Abramson RK, DeLong GR, Cuccaro ML, Pericak-Vance MA (2000): SNP analysis of GABA$_A$ receptor subunit in autistic disorder. Am J Hum Genet 67(339):1886A.

Miettinen OS (1974): Proportion of disease caused or prevented by a given exposure, trait or intervention. Am J Epidemiol 99:325–332.

Morton NE (1982): Segregation and linkage analysis. In: Human Genetics, Part B: Medical Aspects, Alan R. Liss, pp 3–14.

Morton NE, MacLean CJ (1974): Analysis of family resemblance. 3. Complex segregation of quantitative traits. Am J Hum Genet 26:489–503.

Nestadt G, Lan T, Samuels J, Riddle M, Bienvenu OJ III, Liang KY, Hoehn-Saric R, Cullen B, Grados M, Beaty TH, Shugart YY (2000): Complex segregation analysis provides compelling evidence for a major gene underlying obsessive-compulsive disorder and for heterogeneity by sex. Am J Hum Genet 67:1611–1616.

Notarnicola M, Cavallini A, Cardone R, Pezzolla F, Demma I, Di Leo A (2000): K-ras and p53 mutations in DNA extracted from colonic epithelial cells exfoliated in faeces of patients with colorectal cancer. Digest Liver Dis 32:131–136.

Penrose LS (1948): The problem of anticipation in pedigrees of dystrophia myotonica. Ann Eugen 14:125–132.

Philippe A, Martinez M, Guilloud-Bataille M, Gillberg C, Rastam M, Sponheim E, Coleman M, Zappella M, Aschauer H, Van Maldergem L, Penet C, Feingold J, Brice A, Leboyer M, van Malldergerme L (1999): Genome-wide scan for autism susceptibility genes. Paris Autism Research International Sibpair Study. Hum Mol Genet 8:805–812.

Rice T, Despres JP, Daw EW, Gagnon J, Borecki IB, Perusse L, Leon AS, Skinner JS, Wilmore JH, Rao DC, Bouchard C (1997): Familial resemblance for abdominal visceral fat: The HERITAGE family study. Int J Obes Relat Metab Disord 21:1024–1031.

Risch N (1990a): Linkage strategies for genetically complex traits. I. Multilocus models. Am J Hum Genet 46:222–228.

Risch N (1990b): Linkage strategies for genetically complex traits. II. The power of affected relative pairs. Am J Hum Genet 46:229–241.

Risch N (1990c): Linkage strategies for genetically complex traits. III. The effect of marker polymorphism on analysis of affected relative pairs. Am J Hum Genet 46:242–253.

Roses AD, Devlin B, Conneally PM, Small GW, Saunders AM, Pritchard M, Locke PA, Haines JL, Pericak-Vance MA, Risch N (1995): Measuring the genetic contribution of APOE in late-onset Alzheimer disease (AD). Am J Hum Genet 57:A202.

Rothman KJ, Greenland S (1998): Modern Epidemiology. Philadelphia: Lippincott-Raven.

Rousseau F, Heitz D, Tarleton J, MacPherson J, Malmgren H, Dahl N, Barnicoat A, et al (1994): Higher rate of transition from fragile X premutations into full mutation in males than in females suggest post-conceptional expansion of the CGG repeats. Am J Hum Genet 55:A240.

Rumble B, Retallack R, Hilbich C, Simms G, Multhaup G, Martins R, Hockey A, Montgomery P, Beyreuther K, Masters CL (1989): Amyloid A4 protein and its precursor in Down's syndrome and Alzheimer's disease. N Engl J Med 320:1446–1452.

Seybolt LM, Vachon C, Potter K, Zheng W, Kushi LH, McGovern PG, Sellers TA (1997): Evaluation of potential sources of bias in a genetic epidemiologic study of breast cancer. Genet Epidemiol 14:85–95.

Sherman SL, Morton NE, Jacobs PA, Turner G (1984): The marker (X) syndrome: A cytogenetic and genetic analysis. Ann Hum Genet 48(Pt 1):21–37.

Shields R, Dell H (2001): Genes for HIV susceptibility. Trends Genet 17:19.

St George-Hyslop PH, Tanzi RE, Polinsky RJ, Haines JL, Nee L, Watkins PC, Myers RH, Feldman RG, Pollen D, Drachman D (1987): The genetic defect causing familial Alzheimer's disease maps on chromosome 21. Science 235:885–890.

Stene J (1989): The incomplete, multiple ascertainment model: Assumptions, applications, and alternative models. Genet Epidemiol 6:247–251.

Terwilliger JD, Goring HH (2000): Gene mapping in the 20th and 21st centuries: Statistical methods, data analysis and experimental design. Hum Biol 72:63–132.

Vieland VJ, Hodge SE (1996): The problem of ascertainment for linkage analysis. Am J Hum Genet 58:1072–1084.

Waller DK, Anderson JL, Lorey F, Cunningham GC (2000): Risk factors for congenital hypothyroidism: An investigation of infant's birth weight, ethnicity, and gender in California, 1990–1998. Teratology 62:36–41.

Weeks DE, Lange K (1988): The affected-pedigree member method of linkage analysis. Am J Hum Genet 42:315–326.

Weinberg W (1912): Zur verebung der anlage der bluterkrankheit mit methodologischen ergaenzungen meiner geschwistermethode. Arch Rass Ges Biol 6:694–709.

Wilk JB, Djousse L, Arnett DK, Rich SS, Province MA, Hunt SC, Crapo RO, Higgins M, Myers RH (2000): Evidence for major genes influencing pulmonary function in the NHLBI family heart study. Genet Epidemiol 19:81–94.

Wolpert CM, Menold MM, Bass MP, Qumsiyeh MB, Donnelly SL, Ravan SA, Vance JM, Gilbert JR, Abramson RK, Wright HH, Cuccaro ML, Pericak-Vance MA (2000): Three probands with autistic disorder and isodicentric chromosome 15. Am J Med Genet 96:365–372.

CHAPTER 4

Patient and Family Participation in Genetic Research Studies

CHANTELLE WOLPERT, AMY BAZYK CRUNK, and
SUSAN ESTABROOKS HAHN

INTRODUCTION

The genetic analysis methods described elsewhere in this book are extraordinary tools that have been used to identify many genes associated with different human traits and disorders. As a result of the success of these gene-mapping and positional cloning methods, clinical investigators new to this area of research are doing more and more genetic family studies (also called pedigree or linkage studies). Although such studies offer researchers opportunities for identifying and studying disease genes, they also present specific challenges and ethical quandaries different from those faced by investigators in traditional medical research. Therefore, it is necessary for investigators to understand the particular issues related to family genetic research studies before the start of any such study.

It is first important to distinguish between human genetic research and traditional medical research or clinical trials. The *Merriam-Webster Medical Desk Dictionary*, 1996 edition, defines a clinical trial as "a scientific test of the effectiveness and safety of a therapeutic agent (as a drug or vaccine) using consenting human subjects." Clinical trials involve only the individual testing the therapy, whereas genetic research involves the entire family. Many family members may be asked to provide medical information and to donate samples for DNA analysis. Participants may be asked to provide personal and medical information regarding many other family members. Even genetic studies performed only with individuals can potentially generate genetic information that has implications for the whole family.

In addition, clinical trials typically offer participants more immediate benefits than genetic studies. Individuals are usually motivated to participate in clinical

Genetic Analysis of Complex Diseases, Second Edition, Edited by Jonathan L. Haines and Margaret Pericak-Vance
Copyright © 2006 John Wiley & Sons, Inc.

trials because the treatment or therapy may personally benefit them. Genetic studies, on the other hand, rarely relate to direct clinical care and often offer no immediate or direct benefits to individuals. Families and individuals generally participate in genetic studies for altruistic reasons, hoping to help future generations through the knowledge gained from their participation in the genetic study.

Furthermore, clinical research and genetic studies pose different risks for participants. While clinical research may be associated with a risk of individual physical harm, the risks involved with participating in a genetic study are mainly psychological and social and may also affect other family members. For instance, individuals with or at risk for a genetic disorder may have feelings of low self-esteem, guilt, and anxieties about the future health status of themselves, their children, or other family members.

Such risks are due both to the possible predictive nature of genetic information and to the conceivable use of personal genetic information as a basis for stigmatization and discrimination. Finally, while there are extensive regulations governing clinical research, codes of conduct for performing genetic family studies are still evolving. In this chapter, we present a step-by-step approach for organizing family genetic research studies (hereafter referred to as "family studies"). Practical methods are provided for maximizing family participation and increasing the efficiency of the research. In addition, we present recommendations for providing high-quality research services while minimizing potential problems. Finally, suggested ethical and legal guidelines for conventional practices of genetic research are reviewed.

STEP 1: PREPARING TO INITIATE A FAMILY STUDY

Confidentiality

When conducting a genetic research study, participant confidentiality is of paramount importance. Several steps can be taken to protect a research participant's identity. One measure that health care providers can take is to refrain from documenting participation in a genetic research study in a patient's clinical record. Unlike clinical trials that affect the treatment of the participant, genetic research studies generally do not affect the individual's medical care. Thus, information regarding a genetic research study does not need to be placed into clinical records. Additionally, all data obtained or generated by the research study must be secured to maintain individual confidentiality (MacKay, 1993). Likewise, databases should be secured and access to databases and research files should be restricted to a designated individual(s), such as the family studies director. This person should secure any files or databases containing identifying information such as names, addresses, and so on. Finally, continuing review by the institutional review board (IRB) ensures that study protocols adhere to the strict enforcement of confidentiality as described during the informed-consent process (MacKay, 1993).

Certificate of Confidentiality

Most researchers agree that they have an ethical duty to protect personal genetic information. Although no law exists that expressly guarantees genetic data protection, data are protected from third parties (e.g., insurers and employers) when accompanied by a certificate of confidentiality. This protection, authorized under Section 301(d) of the Public Health Service (PHS), includes protection from subpoenas by third parties and from court orders. However, the certificate does not protect information that is voluntarily provided by research participants to third parties (Early and Strong, 1995) or information that is requested by the Department of Health and Human Services (DHHS) for use in an audit or program evaluation or if it is required by the federal Food, Drug and Cosmetics Act. The certificate is available to researchers regardless of whether the research is federally funded or not, but the research must be IRB approved and the information obtained must a carry a risk of significant harm or damage to the subjects if disclosed. Researchers may apply for a certificate of confidentiality through the National Institutes of Health (NIH). In general, researchers should apply for the certificate through the specific NIH institute funding their research [e.g., NHLBI, Centers for Disease Control and Prevention (CDC)]. Those researchers whose studies are not funded by the NIH should apply for the certificate through the National Institute of Mental Health (NIMH).

Most studies maintain confidentiality and protection of information from third parties. However, despite efforts to uphold confidentiality, such as data coding, securing files, and using certificates of confidentiality, a possibility still exists that information will be inadvertently disclosed to third parties (e.g., indirectly, through a notation in a medical record alluding to the study). Such information could preclude a person's chance for obtaining health or life insurance coverage (Wertz, 1992) (see below, Genetic Discrimination). Therefore, some professionals recommend including a discussion of the risk of future insurability, however small, in the informed-consent process and document (Kass, 1993). Furthermore, researchers should be aware that disclosure of personal information to third parties without a participant's consent might be subject to legal action (Andrews, 1990).

Need for a Family Studies Director

The numerous and diverse responsibilities associated with conducting a family study include the administration of the study, family recruitment, obtaining informed consent from research participants, coordination of sample collection and field trip visits to families, and working with the IRB for study approval. Tasks may also include genetic counseling and other clinical activities. One professional should be designated as the family studies director to handle these diverse but vital responsibilities. (See Box 4.1 for a listing of duties of the family studies director.) Because of the diverse skills required, the family studies director should ideally be a genetic counselor, social worker, psychologist, or other health professional with training in counseling and medical genetics. Some of the duties of the family studies director are discussed in greater detail elsewhere in this chapter.

BOX 4.1. DUTIES OF FAMILY STUDIES DIRECTOR

Administrative

Submit IRB applications and renewals.

Ascertain families for research studies.

Serve as liaison between research team and families, referring professionals, and support organizations.

Monitor and ensure compliance with genetic research regulations.

Draft and implement protocols to ensure quality control measures.

Assist with preparation of grant applications.

Educational

Prepare presentations for professional conferences and disorder support groups.

Prepare yearly research updates (newsletters) for families.

Prepare educational materials as needed for research studies.

Assist with preparation of abstracts, manuscripts, and other publications.

Clinical Activities

Obtain informed consent from research participants.

Elicit family histories.

Elicit medical histories.

Complete study-specific questionnaire, if relevant.

Phlebotomy.*

Directed physical exams.*

Genetic counseling.*

Logistical

Arrange for sample collection.

Arrange field study visits.

Request medical records.

Maintain databases.

*Dependent on family studies director's background.

Working with Families. The family studies director plays an important role in facilitating research while addressing the research participants' needs (McConkie-Rosell and Markel, 1995). Serving as the family's contact person with the research team, the family studies director works to establish rapport with families. In particular, the family studies director helps foster realistic expectations about the study. In addition, individuals may have more questions related to the research study or to the genetic aspects of their condition or they may need to be resampled or reexamined at a later date. By being available to families to answer questions and offer support, the family studies director helps meet the needs of participants, including the need for information, the need to be informed of the study progress, and the need to be understood. The family studies director is essential in providing referrals to families that would benefit from clinical services.

Working with Referring Professionals. In addition to working with families, the family studies director serves as the laboratory's representative to referring professionals and facilities. By being available to answer clinicians' questions about the research study, the study director builds credibility among clinical professionals who may serve as a referral source for families. Also, the director provides current and accurate information on the status of genetic research for a particular disease. For example, the director may provide referring professionals with regular bulletins or notifications of research discoveries with clinical implications. Thus, by maintaining a rapport with outside institutions, the family studies director or research coordinator enhances referrals and family participation.

Researcher as Family Studies Director. Those who do research in genetics are often among the few experts who can provide patients and professionals with accurate and comprehensive information about the disorder they are studying. While providing such information to research participants and referring professionals can be a time-consuming process, it is a vital part of conducting family studies. Many families with genetic disorders, especially rare disorders, may have little, incorrect, or incomplete information about the condition. Additionally, newly diagnosed individuals may need more support and information about the disorder, its mode of inheritance, and genetics in general. Individuals who require formal counseling and genetic testing should be referred for these services when they are not provided as part of the research protocol.

A single investigator with limited funding may not be able to hire a professional for family ascertainment. One option in such instances is to consider collaborating with a research group that has a family studies director who can assist with family ascertainment. If, alternatively, the investigator chooses to serve as the family studies director, he or she should allot time for communicating with the families, recognize the necessity of maintaining ongoing communication with families, and be aware of the families' needs and their reasons for participating in the study. Since the needs of a family may go beyond the expertise of the researcher, it is helpful for all researchers to have a resource list of genetic counselors and other health care professionals as well as disease support groups for family referral.

Working with Human Subjects

> The greatest challenge to securing continuing funding for the Human Genome Project does not originate from concerns about privacy, confidentiality, or coercive genetic testing. It is eugenics, manipulating the human genome to improve or enhance the human species, which is the real source of worry. This is reflected in the content of the futuristic horror scenarios spun by the Project's critics. It is also rooted in the historical reality of social policies based on eugenics that led to the deaths of millions in this century.
>
> —Arthur Caplan (1992a, p. 138)

In light of the history of abuse of genetic science, researchers should proceed with an awareness of the possible harms that can arise from the misuse of genetic information. Perhaps the most notorious example of abuse in the field of human genetics was the implementation of "racial hygiene" programs aimed at eliminating persons considered "genetically defective" by the Nazi regime in Germany (Proctor, 1988; Mueller-Hill, 1992). As in Europe, eugenics proponents in the United States in the early-twentieth century supported compulsory sterilization of those with mental and physical disabilities as well as prison inmates and other "undesirables" (Reilly, 1991). In addition, over half of the states enacted laws barring marriages between whites and other ethnic groups, reinforcing existing stigmas of racism and segregation (Proctor, 1992). Several decades after the disintegration of the eugenics movement, critical accounts of modern genetic technologies continue to caution the scientific community and the public at large about the use and misuse of genetic information (Hubbard, 1986; Holtzman, 1989; King, 1992; Paul, 1994).

In addition to the unique concerns in genetic research, all types of research studies involving human subjects may contain potential risks to participants. During the infamous Tuskegee syphilis study, vulnerable research subjects infected with syphilis were denied antibiotic therapy even though penicillin had become the treatment of choice for this condition (Caplan, 1992b). These subjects were also recruited with misleading promises of free treatment and were enrolled without their informed consent. The public disclosure of these abuses ultimately led to the National Research Act of 1974 [US Office of Protection from Research Risks (OPRR), 1993], which mandates the approval of all federally funded proposed research with human subjects by an IRB.

Institutional Review Board. In accordance with federal regulations, research studies involving human subjects, including noninvasive family studies, that are funded by federal agencies (e.g., NIH, Department of Energy, Food and Drug Administration) must be reviewed by an IRB (US OPRR, 1993). Typically, a facility or medical center IRB reviews the research protocols for all studies that will be conducted there. Research facilities without an existing IRB must either establish an IRB or submit their project proposals for review by an external IRB, which satisfies federal requirements [US Department of Energy (DOE) brochure]. Although the majority of privately funded companies and institutions are connected to federal agencies via funding or facilities, research conducted strictly from private finances is not currently subject to federal regulations or IRB review (US DOE, 10CFR745; US DHHS, 1991). However, independent researchers may choose to contract with an external IRB for protocol review to ensure that participants' interests are protected.

Comprised of a multidisciplinary group of professionals, an IRB is defined as "an administrative body established to protect the rights and welfare of human research subjects to participate in research activities conducted under the auspices of the institution with which it is affiliated" (US OPRR, 1993, p. 1-1). The IRB process is designed to be, as much as possible, beneficial to both the researcher and the study participant. In accordance with protecting the rights and welfare of participating families, IRB members offer additional insight into the study design and strengthen the research by assuring that family studies are conducted in accordance with federal and state regulations.

Human studies protocols are subject to a critical IRB review for scientific and ethical validity before a study begins. There are two types of review processes that family studies may generally be eligible for: full-board review and expedited review. While full-board review involves all or most of the IRB committee members, expedited review is usually performed by a designated member (e.g., the committee chair) or a subset of members (US OPRR, 1993). Typically, the review process involves providing the IRB with a detailed description of the study protocol, documentation of funding, and model consent document for study participants. In addition, the researcher should present a protocol for maintaining identifying information, if any, on study participants and the methods, such as data coding, used to ensure participants' anonymity.

Identifying versus Anonymous Data. The majority of genetic research studies necessitate the use of identifying information, such as pedigree information and documentation of diagnosis through medical records, to establish genetic relationships among family members and to perform genotype–phenotype analyses. In addition, some researchers maintain identifying information so that participants can be recontacted for a variety of reasons as discussed later (see below, Maintaining Contact with Participants). Full-board review is generally required for all studies that maintain identifying information about research subjects. Per the Office of Human Research Protections (OHRP), previously Office for Protection from Research Risks, regulations [Section 46.102(f)], a living individual becomes a human subject when an investigator conducting research obtains data through intervention or interaction with the individual or obtains identifying private information about that individual. This includes any information that can be used to directly or indirectly link an individual with his or her sample (e.g., numerical code, name, address, pedigree information, medical records). These regulations may also apply to "secondary subjects" (individuals on whom information is obtained by the investigator from someone else) who meet the definition of a human subject. This would include nonparticipating family members on whom identifying information is collected during a family history interview.

With regard to the IRB and regulations regarding secondary research subjects, if the information collected in the process of obtaining family history information does not put the secondary subject at reasonable risk, the IRB does not have to consider the data collection in its review. However, the IRB must review the protocol when the proposal includes the collection of family history data deemed to involve more than a minimal risk and determine whether informed consent is necessary from these individuals. The IRB's discussion, particularly the rationale for deciding whether or not informed consent is necessary, must be clearly documented. For further guidance regarding what constitutes minimal risk and when a decision should be made to obtain informed consent on secondary subjects, see the OHRP website, http://www.hhs.gov/ohrp/.

The IRB review is generally not required for studies using preexisting anonymous data (Clayton et al., 1995). Anonymous data cannot be linked to a participant, either directly (e.g., a name) or indirectly (e.g., a numerical code). In contrast, protocols utilizing data that have been "anonymized" through the irreversible removal of identifiers from samples are subject to IRB approval. In both cases, the use of anonymous or anonymized data offers the researcher the advantage of decreasing certain potential risks to individual study participants, such as genetic discrimination (see below, Genetic Discrimination), and social or psychological risks. Even with individually anonymous data, these risks may still exist for groups or populations if ethnic or social group information is retained (Clayton, 1995).

STEP 2: ASCERTAINMENT OF FAMILIES FOR STUDIES

Family Recruitment

As part of the study protocol, family recruitment proposals must receive IRB approval before the start of the study and prior to beginning family ascertainment. As with all voluntary studies, no coercion should be used in ascertainment for family genetic studies. In addition, these recruitment methods must be conducted to ensure that potential participants are not contacted by the researcher until they have given their permission through a referring professional or through a family member. After permission to contact a potential research participant has been obtained, it is often appropriate to write a letter introducing the research team and the study and inviting the person to contact the family studies director for more information. Alternatively, the researcher may call the person directly and explain the study over the telephone.

Families can be ascertained using several methods, each with its own advantages and disadvantages. The difficulty in finding families interested in research studies will vary depending on the incidence of the disorder being investigated and the methods used to publicize the study. Four main sources of family referrals are disease support groups and similar organizations, health care providers, public databases and websites, and medical clinics (Box 4.2).

BOX 4.2. SOURCES OF FAMILY REFERRAL

Support groups

Medical clinics

Advertisements

Collaborators

Inpatient referral

Patient registry/database

Helix directory of clinical and research DNA laboratories

Genetics professional societies

Internet

Support Groups and Organizations. Support groups for specific genetic disorders can serve as an excellent referral source. Focus groups conducted with research participants by Melvin et al. (1999) revealed that research studies endorsed by their advocacy/support group were viewed as more credible. The support group not only lends integrity to the research but also can provide general information about research studies to its membership and refer interested families. In recent years, support groups have begun to take a more active role in research involvement. Some lay advocacy groups have already taken a stance to become equal partners in research. Terry et al. (2000) presented case examples of how lay advocacy groups have worked with researchers as partners. They conclude that some of the most obvious benefits to the researcher of this method are the education of the research participants, the media, funding agencies, and policymakers about the importance of the research. In addition, these groups are optimally positioned to directly collect epidemiological and phenotypic data through recruitment.

Likewise, researchers can refer participants to support groups as a valuable resource for families seeking further information and services. Among the many genetic support groups in (and out of) the United States, there are three umbrella organizations that can facilitate finding a specific disorder support group: the Genetic Alliance, the National Organization of Rare Disorders (NORD), and the Self-Help Clearinghouse (a federally funded program that creates an information office in each U.S. state) (Box 4.3).

Referrals from Health Care Providers. Advertising in medical journals, announcements at medical conferences, and word-of-mouth referrals among professional colleagues are additional methods for reaching families (see Box 4.4). Focus groups conducted with research participants by Melvin et al. (1999) revealed that one of the primary ways individuals would like to be informed about research studies is from their personal physician. For those studies utilizing professional referrals, a study conducted by Melvin et al. (2000) indicates that professionals would like to have the following information before a referral is made: information

BOX 4.3. SUPPORT GROUP ORGANIZATIONS AND RESOURCES

Genetic Alliance
4301 Connecticut Avenue, NW, #404
Washington, DC 20008-2304
(202) 966-5557
(800) 336-GENE
info@geneticalliance.org;
http://www.geneticalliance.org/

**National Organization for Rare
Disorders (NORD)**
P.O Box 8923
New Fairfield, CT 06812-8923
(800) 999-NORD
http://www.rarediseases.org/

American Self-Help Clearinghouse
Saint Clares-Riverside Medical Center
25 Pocono Road
Denville, NJ 07834
201-625-7107
(*The Self-Help Sourcebook: Finding
and Forming Mutual Aid Self-Help
Groups*, 1992)

BOX 4.4. PROFESSIONAL REFERRAL SOURCES

GeneTests™
Directory of Medical Genetics
Laboratories
Children's Hospital and Medical Center
P.O. Box 5371, Mail Stoop CH-94
Seattle, WA 98105-0371
(206) 527-5742
Fax: (206) 527-5743
genetests@genetests.org;
http://www.genetests.org/

**American College of Medical
Genetics**
9650 Rockville Pike
Bethesda, MD 20814-3998
(301) 530-7127
Fax: (301) 571-1895
acmg@faseb.org;   www.faseb.org/
genetics/ashg/ashgmenu.html;
http://www.faseb.org/genetics/
acmg/index.html

**National Society of Genetic
Counselors**
233 Canterbury Drive
Wallingford PA 19086-6617
(610) 872-7608
nsgc@aol.com;
http://www.nsgc.org

**American Society of Human
Genetics**
9650 Rockville Pike
Bethesda, MD 20814-3998
(301) 530-7127
Fax: (301) 530-7079

regarding the disclosure of results, the study protocol, cost, informed consent, and any travel that would be required by the participant. When the health professional makes the referral, the family should be given the researcher's name and contact information or, alternatively, the referring professional should obtain the family's permission to be contacted by the researcher directly. Again, it is very important that families never be contacted directly unless permission to do so has been obtained through a referring professional or support group.

Research Databases and the Internet. Researches can register and list their study descriptions and contact information directly with several databases. Health care professionals seeking out researchers to whom they can refer their patients usually access these databases. With access to the World Wide Web via the Internet now becoming an integral part of communication, it is feasible, and perhaps preferable, to consider ascertaining families through Internet connections. This can be done by posting an announcement online with a disorder support group or by developing a home page where families can contact the researcher (Biesecker and Derenzo, 1995; Scaffe et al., 2000). Researchers should prepare for a high volume of inquiries in response to Internet advertising.

Institution Databases. Many institutions maintain patient databases encompassing a variety of information that may be useful to the researcher. A database may be so large as to encompass all patients served by a particular hospital and the treatments they received or may be generated out of a clinic and contain only disorder- or specialty-specific information. Researchers may be able to utilize such databases to identify eligible study participants and should inquire about their existence at their institution. Due to privacy issues, particularly surrounding the Health Insurance Portability and Accountability Act of 1996 (HIPAA), use of the database will have many restrictions. It is not acceptable for the researcher to contact members of the database regarding the research study without prior permission from the potential participant to be contacted. Most often the physicians whose patients are in the database agree to sort the database based on research eligibility requirements and send letters to their patients announcing the study. The letter should contain a brief explanation of the study and the investigator's contact information should they be interested. This can be an extremely efficient method to introduce the study to a large number of eligible individuals in a short period of time (Estabrooks et al., 1999). If individuals are asked to send something back to indicate their interest in the study, they should be provided with an addressed envelope and not an addressed postcard to ensure that the patient's confidentiality is maintained.

Family Ascertainment through Medical Clinics. Researchers who ascertain families through medical clinics must be especially careful to clearly define the difference between the patient's clinical care and his or her participation in a study. Patients may mistakenly believe that the research study is part of their medical care. Patients may also have the impression that their medical care is provided contingent upon participation in the proposed research study (which contradicts the principle of a voluntary study). In addition, potential study participants may confuse the research study with a clinical service and may anticipate test results or treatment. Therefore, the researcher should clearly state the difference between the clinical and research settings and emphasize their independence (e.g., a research laboratory does not provide clinical services). The researcher should also be sure that the participant fully understands this difference.

In addition to clearly delineating the line between research and clinical services, clinician researchers who also provide medical care to research participants should maintain study files separate from medical records (Earley and Strong, 1995). Furthermore, medical practitioners should avoid referencing the research study in the patient's medical chart. In this way, information gathered by the study is not subject to disclosure requests by insurance companies, except in legal proceedings. As described previously, however, researchers may apply for a certificate of confidentiality that grants research data immunity from such subpoenas.

Recruitment by Family Members. In most genetic studies, especially linkage studies, samples from entire families are necessary for analysis. As has been pointed out, studies that involve individual participants may provide "[genetic] information about relatives who are not in the study and who therefore did not have the opportunity to give or withhold informed consent to their participation" (Frankel and Teich, 1993, p. 31). All family members who participate in a genetic research study must do so voluntarily. This is the most important issue in work with families (Parker and Lidz, 1994). In some instances, people may be tempted to pressure their own family members to join a research study and may in fact do so. A researcher should guard against coercion, albeit well intentioned, by speaking with all family members individually and eliciting their opinions about participating in the research study. Logistically, it may be necessary for one family member to inform all the other family members about the study, but the researcher should ask for permission to communicate individually with each potential participant.

Informed Consent and Family Participation

Participation in a genetic research study involves the comprehension of complex information. This information must be presented to each family member participating in the study in a clear and concise manner (Box 4.5). The process of communicating this information to potential study participants is the basis on which that individual can give his or her informed consent. The researcher must provide the potential participant with an appropriate explanation of the study, for only in this way can informed consent assure voluntary participation and informed decision making on behalf of the individual. When families understand the research protocol, they will be able to set realistic expectations about the benefits and the limitations of participating in a genetic research study.

In response to documented human rights violations in the area of human studies research, regulations have evolved to ensure the protection of research participants.

BOX 4.5. ELEMENTS OF INFORMED CONSENT

Emphasize the voluntary aspect of the study.

Answer all questions thoroughly.

Offer a clear and concise explanation of the study, including information on:

- Risks (e.g., medical, nonmedical/psychosocial, future insurability)
- Benefits
- Type and timing of information disclosure (if any)
- Limits of confidentiality
- Time/financial commitment
- Right to withdraw
- Right not to be recontacted or to know research results
- Ownership of DNA sample
- Secondary uses of samples

The importance of "voluntary consent" in research did not emerge until the end of World War II in accordance with the Nuremberg Code. The required elements of the informed-consent process are set forth by federal regulations (US OPRR, 1993). In addition to federal regulations, state laws and case law precedents may apply to research involving human subjects.

Potential research participants should understand that their participation is optional and that they may withdraw their consent at any time after enrolling in the study without harm or penalty to themselves or to their family members. Participants should be told the purpose of the study as well as the risks, benefits, alternatives to participation, policy on disclosure of results (if any), and financial and time commitment involved (Genetic Alliance brochure; US DOE, brochure; Weir and Horton 1995a,b).

In accordance with the Clinical Laboratories Improvement Amendments (CLIA) of 1988, laboratories that disclose results to participants must obtain approval to do so, as described later (see below, Results, under Guidelines for Releasing Genetic Information) and should set a policy on the results to be disclosed and the circumstances of disclosure. This policy should be clearly explained to potential research participants as part of the informed-consent process. In addition to disclosure of results, genetic family studies raise unique issues with respect to confidentiality and the protection of genetic information (previously discussed), genetic discrimination, psychological risks, and DNA banking and ownership. These issues require further discussion and are addressed later in this chapter.

The informed-consent process should be documented in two forms. This typically is done with a signed consent form (see the Appendix at the end of this chapter) plus an additional documentation, such as recording the date the consent form was signed in the study database. Participants should be given a copy of the consent form for their files. While standard research practice requires a signed consent form before enrolling a research participant in a protocol, the existence of such a document does not guarantee that a participant fully understands all the ramifications of the research. To comply with the IRB approval process, consent forms and procedures must meet the facility's standards as well as federal and state regulations.

Vulnerable Populations. According to federal regulations, individuals belonging to vulnerable populations are afforded special protection with regard to their participation in research protocols. Such populations include children, individuals with a mental disability, prisoners, pregnant women, fetuses, and economically or educationally disadvantaged persons (US OPRR, 1993). In addition to federal laws, studies involving research participants who receive services from state-funded agencies or institutions may require additional approval by those state agencies (e.g., state department of mental retardation). Therefore, as part of the IRB review process, researchers must define the populations to be included and provide a rationale for their inclusion in genetic research protocols.

Minors. By nature of their design, genetic studies, such as linkage analyses, typically involve sampling entire families. While a parent or legal guardian ultimately gives consent for a child's participation in a research protocol, federal regulations suggest that children should receive an explanation of the study at an age-appropriate level and, whenever possible, should give their assent in which they affirm their agreement to participate in the study, especially if the study has no direct medical benefit to the child (US OPRR, 1993). While the validity of assent obtained from children continues to be debated, some researchers feel that children should have the opportunity to refuse to participate in a study (Frankel and Teich, 1993). In addition, according to the specific protocol and the associated risks, limits may be placed on the amount of blood, if any, which may be drawn from a child for a research study (US OPRR, 1993).

The two most likely scenarios involving the sampling of minors for genetic studies are sampling of an affected child for linkage and/or mutation studies and sampling of a child (regardless of affected status) to enable the reconstruction of parental genotypes. However, as with all research results, results should be disclosed only by CLIA-approved laboratories within the context of genetic counseling, as discussed later (see below, Results, under Guidelines for Releasing Genetic Information).

Persons with Cognitive Impairment. Many family studies involve researching the genetic basis of disorders that affect mental and cognitive abilities in a congenital, progressive, or late-onset manner. Persons lacking mental capacity (e.g., persons with mental retardation or dementia) who are thus unable to give informed consent for participation in a study are afforded special consideration. As discussed above with regard to children, in conjunction with obtaining consent from the legal guardian, assent should be sought from the potential participant whenever possible. Researchers should check with their IRB about federal and state guidelines for including persons with mental impairment in their research protocols, since state and local laws vary on who can and cannot provide consent in these situations.

STEP 3: DATA COLLECTION

As required when designing a study protocol, researchers should define their methods for obtaining DNA samples prior to initiating sample collection. The study requirements will determine which samples should be collected. (See Chapter 5.) Depending on the requirements of the study, DNA samples can be collected by the researcher at a medical clinic or obtained via field studies (visits to participants at their homes). Alternatively, sample collection kits can be mailed to research participants. In these instances the researcher or the participant must arrange for the collection of blood or other tissue samples. The researcher should clarify in the protocol whether the participant will be reimbursed for any expenses associated with sample collection, such as blood-drawing fees, parking, and food. In addition, researchers should check with their IRB to ensure that packaging of biological samples sent through the mail or delivery service meets federal Occupational Safety and Health Administration (OSHA) requirements, especially when samples are shipped between countries.

Confirmation of Diagnosis

A good study design requires the confirmation of family history information and medical diagnoses. An accurate pedigree is critical to establishing the genetic relationships among family members necessary for genetic research. Likewise, diagnostic information on affected individuals is essential for proper analysis. Diagnostic confirmation can be obtained in several ways depending on many factors, including the type of disorder being investigated, proximity of the patients to the researcher, research funding for travel, and previous medical workup of research participants. Participants can be examined in a clinic or during a field study (see below, Art of Field Studies). Alternatively, researchers may confirm diagnoses or traits by reviewing medical records.

For some genetic studies, especially linkage studies, it is as important to rule out a diagnosis in an apparently unaffected family member as it is to confirm the diagnosis in an affected individual. This rigorous clinical evaluation may lead to the diagnosis of a genetic disorder in an unsuspecting family member who did not know that he or she was affected. Before participating, research participants should be informed that there is a possibility that they will be found to have the disorder being investigated. The participants should be given the option of deciding whether to be advised if such information develops and should be reminded that their participation in the study is voluntary. If the researcher feels it is appropriate, patients interested in learning whether they have a disorder can be referred to a health care professional for clinical and follow-up care (see below, Need for Additional Medical Services). The research team with the IRB should decide in advance how potential dilemmas should be handled. As with all studies, unanticipated problems arise and should be dealt with on a case-by-case basis in conjunction with the researcher's IRB.

Accurate clinical diagnoses of affected individuals are crucial to performing genetic studies aimed at identifying disease genes. Occasionally, a diagnosis cannot be confirmed unless additional clinical studies are done. In such instances, participants should be informed of the rationale for requesting additional tests and told how the costs of further clinical evaluations will be covered (e.g., by the researcher, by the participant, or by the participant's insurer). Again, people should be reminded of their right to refuse testing. Furthermore, participants should be told that, as with all medical information, results obtained from clinical evaluations or testing done as part of the research will be placed in their medical records. As discussed previously (under Family Ascertainment through Medical Clinics), it is important for the researcher to clearly delineate the distinction between research participation and clinical care to ensure the family's continued informed and voluntary participation in the protocol. Any clinical diagnostic procedure performed specifically as part of the research protocol must be approved by the IRB.

Art of Field Studies

The field study has a long tradition in genetic research. While no formal assessment exists, many researchers attest to the success of field studies as the single most valuable practice to ensure success in family ascertainment for genetic studies. Typically, researchers who do field studies travel to meet with study volunteers in their homes during times convenient for the families. This increases the efficiency of the research study by minimizing the toll on participating family members of arranging travel for physical examinations and tissue sampling.

Time Spent Ascertaining and Collecting Samples. Ascertaining families and performing sample collection for family studies can be extremely time consuming. A family studies director spends a large amount of time making phone calls to patients, family members, and physician's offices, visiting outpatient clinics to recruit and enroll participants, field studies to collect samples and information, and searching medical records to identify potential subjects. There is also time involved in writing letters, reading medical records of participants, database entry, pedigree construction, and organization of field trips.

How a family studies director's time is spent will vary depending on the criteria of the research study and the data being collected. Studies requiring data that are best or often collected off-site may require a larger percentage of the family studies director's time to be spent performing field studies. Other studies may require that the data be collected on-site, resulting in larger amounts of the family studies director's time being spent in outpatient clinics. On average, a family studies director can ascertain and perform sample collection on 14 participants a month (Bazyk et al., 2000). Of course, this will vary depending on the ascertainment criteria and amount of information needed for the particular family study.

Field Study Preparation. A field study takes place only after a family has met the criteria for participation in the study, the diagnosis in the family has been confirmed by reviewing the available medical records of the individuals with the disorder, and all family members who are interested in participating in the research have received documentation of the purpose of the study and other detailed information. Detailed information about what a field study involves should be provided to each individual in the family who may participate. This information includes a meeting location (e.g., the research subject's home or another site, such as a clinical facility) and an estimate of how much time will be required for each individual and for the group.

A research team should consist of a clinician who can perform the directed physical examinations to confirm the diagnosis of the disorder being studied, an experienced phlebotomist, and a researcher who can coordinate the field visit, elicit a family history, explain the purpose of research, answer participants' questions, administer other study-related tests, and obtain informed consent. Typically, a research team consists of a physician or physician assistant and a genetic counselor. For field visits that involve working with a large family, more professionals may be needed to facilitate the examination, sampling, and informed-consent processes.

In conducting a field study, it is customary to introduce the entire research team to the family and to explain what each team member will be doing and to whom questions should be directed. The research team should try to complete the examinations and sample collection in a timely fashion and allow time at the end of the visit for closing questions and comments by family members. Before leaving the family, one member of the research team (usually the family studies director) should define how contact and follow-up will be maintained with the family and research center. Finally, it is courteous to follow up a visit to a family with a thank-you letter. (See Boxes 4.6–4.8.)

Special Issues in Family Studies

Although all studies involving human subjects entail such issues, there is heightened concern that home studies somehow might be viewed as coercive. Therefore, it is especially important to remind research participants prior to and during the visit that their participation is voluntary. To minimize the possibility of family coercion to participate in the study (Parker and Lidz, 1994), it is a good idea whenever possible to obtain informed consent from individual family members before visiting the home. It is vital that the informed-consent process include an explanation of what should and should not be expected during the field visit. For instance, while a directed physical examination may be performed to confirm the diagnosis, clinical services, like medical treatment, are usually not provided by researchers in the field. In addition, individual decisions about whether to participate in the study should not be discussed with other family members unless permission has been given to do so. A written summary of these discussions needs to be sent to all the members of the family who will participate. Interacting with families is qualitatively different from interacting with individuals and

---

BOX 4.6. SUGGESTED GUIDELINES FOR FIELD STUDIES

Do:

Introduce the field team to everyone in the family.

Explain the purpose of the research to each family member (see Box 6.3).

Travel with two researchers.

Have complete supplies available (see Box 6.6).

Tell individuals approximately how long a visit will take.

Be respectful about being in someone's home.

Refer family to a physician or genetic counselor, when indicated.

Do not:

Put a pedigree on a table for everyone to see.

Reveal information about one family member to another family member.

Offer diagnostic information (unless part of an IRB-approved clinical
  protocol).

BOX 4.7. FIELD STUDY PREPARATION

Pre–Field Trip Checklist

Obtain medical records.

Review pedigree and medical records.

Approve field visit to family.

Contact all relevant family members and confirm attendance.

Send confirmation letter to all relevant family members.

Confirm travel arrangements.

Notify laboratory of the field trip date and expected number of samples.

Pack field trip bag (see Box 6.8).

Post–Field Trip Checklist

Update pedigree (including affected status changes, age of onset, physical
  exam findings, dates of birth).

Match clinical evaluation forms with blood samples and consent forms.

Take blood samples to laboratory.

Write post–field trip note synopsis of what took place, significant clinical
  findings, and follow-up plan (e.g., if other family members need to be
  seen in the future).

Send letter of appreciation to family.

**BOX 4.8. FIELD STUDY SUPPLY CHECKLIST FOR
FIELD TRIPS**

Clinical Supplies

8.5-mL yellow stopper tubes
   (containing ACD solution)*
Other tubes for special studies
   (e.g., chromosomes, creatine
   phosphokinase)
$21 \times \frac{3}{4}$, 129 Butterfly tubing
   (purple)
$23 \times \frac{3}{4}$, 129 Butterfly tubing
   (orange)
Alcohol preps
Gauze pads
Band-Aids
Luer adapters
Vac holders
10-mL Syringes
Tourniquets (latex-free
   tourniquets)
Dirty needle box
Marker pens and two regular pens

Wire test tube for stopper tubes
Buccal swabs
Reflex hammer
Tuning fork
Other medical equipment
   (as needed)
Gloves (latex free)
4 Mailing kits

Paperwork

Summary of the study*
Consent forms*
Pedigree forms*
Clinical evaluation forms
Study-specific questionnaire
Medical record release forms*
Post–field trip checklist
   (see Box 6.5)
Educational materials (e.g.,
   fact sheet, support
   group information)*

*These materials as well as packaging material for return shipment with instructions should be
included in all kits mailed to families.

presents unique circumstances for both the families and the researcher. Visiting families in their homes allows for a more detailed discussion of family history and the family folklore surrounding a disorder. However, while some families are very open and accepting of their disease, others view it more as a stigma and treat the disorder as a taboo within the family. Thus, in addition to observing normal social conventions and being courteous when in someone's home, it is important for the researcher to anticipate and be sensitive to different attitudes of family members toward the disease in question.

STEP 4: FAMILY FOLLOW-UP

Need for Additional Medical Services

When appropriate, the family studies director should refer individuals to other health care professionals or to clinics for additional medical services. Research participants with a family history of a genetic disorder may have specific questions about their personal reproductive risk and should be referred for genetic counseling. In addition, families with an affected child may seek community and federal services such as early intervention programs and financial compensation for medical care. The family studies director can refer these families to a local genetic counselor, social worker, or support group to serve as a family advocate by exploring available programs. Finally, family members needing supportive counseling or additional medical or psychological care should be referred to the appropriate clinician when indicated.

Duty to Recontact Research Participants

In addition to the hope of learning research results (see below, Results, under Guidelines for Releasing Genetic Information), many families participate in genetic studies because they want to contribute to the research effort and are interested in following the progress of the research. These families often express a need to be given information about the study following their participation and may periodically contact the laboratory to inquire about the status of the project.

While the release of individual research results is regulated by federal legislation (Andrews et al., 1994), the legal duty of a research laboratory to recontact study participants to prevent potential harm—the "duty to warn"—is ambiguous and controversial. According to one report, "The fact that the health care provider or the laboratory had only brief contact with the person providing the DNA or that much time has passed since the DNA was originally deposited *does not provide a defense for not recontacting the patient*" (Andrews, 1990, p. 225, italics added). Nevertheless, unlike the case of clinical practice (*Tarasoff v. Regents of University of California*, 1976; Macklin, 1992; Wertz et al., 1995), the researcher's duty to warn participants of potential reproductive or other medical risks based on their genetic status or to provide them with general information (e.g., availability of commercial DNA testing) has not been clearly decided by the courts because there have been too few court cases on the subject. Of interest, there are two very similar court cases in two different states regarding duty to warn third parties of genetic information in the clinical setting (*Safer v. Pack*, 1996; *Pate v. Threlkel*, 1995). Despite their striking similarity, they had opposite rulings. Therefore, there is no clear legal precedent in these types of cases. In the absence of a law or a court precedent creating a duty to recontact participants in the research setting, the question of whether researchers have a moral duty to recontact participants is being debated among geneticists, ethicists, lawyers, public health professionals, and consumers. Some professionals believe that genetic researchers have an obligation to inform participants of the progress of the research study (MacKay, 1993), especially if that information may impact participants' treatment options and reproductive and medical decision making. Furthermore, it has been suggested that families be sent an abstract or reprint of any published studies that resulted from their participation (de Leon and Lustenberger, 1990) and be given a summary of the data gathered during a study (Kodish et al., 1994).

Regardless of the research team's view of the ethical duty to recontact, participants should be told, as part of the informed-consent process, whether they will receive any information about the status of the research following their participation in the study. Facilities that do maintain contact with study participants should clarify how the information will be relayed, as discussed below.

Maintaining Contact with Participants

Maintaining contact with study participants provides both the researchers and the participants with several advantages. The participant is informed of pertinent information relating to a particular genetic disorder, including any breakthroughs that have resulted from the research study. The researcher maintains contact with the family, providing the potential for recontact, reexamination, and/or resampling in the future. Ideally, given adequate funding and staffing, the family studies director provides participants with yearly updates on the progress of the study as well as bulletins when research breakthroughs occur.

Contact can be maintained through a variety of ways: a follow-up letter to the families, periodic updates on the status of the research (e.g., a newsletter), and telephone calls. It is helpful to designate a contact person in each family and to direct all correspondence to that person. In addition, each person should receive the telephone number of the researcher (e.g., family studies director) in the event that future questions arise. It is beneficial to inform families who wish to be recontacted about their responsibility to notify the research team of any change in address or phone number so that they may receive timely reports on research progress and breakthroughs. In case contact is lost with a family, mailing a certified or registered (return receipt requested) letter to the family at their last known address is suggested.

Guidelines for Releasing Genetic Information

> If researchers are aware of identities of the individuals whose samples they have received, do they have a moral or legal obligation to notify the participants of information that may save, or at least change, [the participants'] lives?
> —Researcher of Huntington disease, quoted in Frankel and Teich (1993, p. 16)

The nature of genetic information is unique in comparison to other types of personal information for three key reasons as delineated by Annas (1995, p. 1196): "It can predict an individual's likely medical future; it divulges personal information about one's parents, siblings, and children; and it has a history of being used to stigmatize and victimize individuals." Thus the disclosure of genetic information must occur in accordance with federal laws and ethical guidelines.

CLIA Regulations: Separation of Research and Clinical Laboratories. The

Clinical Laboratories Improvement Amendments of 1988 (CLIA88) are federal legislation enacted to regulate the quality assurance of testing and reporting of results by clinical laboratories. In accordance with CLIA88 and other federal regulations, genetic testing protocols differ between research and clinical laboratories (Andrews et al., 1994). Under CLIA88, laboratories that disclose test results to patients for the purposes of the diagnosis, prevention, or treatment of a disease are subject to regulation. Research laboratories that release results to families for clinical decision making (e.g., laboratories that study rare diseases for which commercial testing is not available) are included under this provision. In addition to legal considerations, study participants should understand this separation of research and clinical laboratories as part of their decision to participate in the research (Miller, in preparation).

It is essential that research laboratories define the extent of information that will be disclosed as part of the informed-consent process to determine whether the CLIA88 regulations are applicable and to develop a policy about releasing results to review with research participants. Many research laboratories follow a protocol of complete nondisclosure of information to families (i.e., they do not provide research results to individuals). However, some research laboratories may desire authorization to release results under CLIA regulations. Laboratories that disclose results or other information to participants must consider additional issues as discussed below.

Although most research laboratories do not release genetic results, researchers should draft their study protocol in anticipation of scenarios that could necessitate the unforeseen disclosure of genetic information to participants, such as incidental findings with medical implications. Reilly has suggested that researchers define three categories of disclosure: "(1) findings that are of such potential importance to the subject that they *must* be disclosed immediately; (2) [findings] that are of importance to subjects, but about which [the researcher] should exercise judgment about the decision to disclose; (3) [findings] that do *not* require special disclosure" (Reilly, 1980, pp. 5, 12, original italics).

Incidental Findings and Special Disclosure. Incidental findings, such as nonpaternity or adoption status, reveal information other than that sought by the researchers. Such information, while important in establishing genetic relationships, is both potentially harmful when revealed to unsuspecting family members (Mac Kay, 1993) and potentially helpful when used in conjunction with a clinical protocol. For example, if nonpaternity were uncovered in a family in which the father had Huntington disease, such information could at once be devastating to and welcomed by the children, who were assumed to carry a 50% risk for developing this neuro-degenerative disease. In addition, a genetic disorder other than the one being studied may be revealed in an individual's sample. Cases that are not clearly defined under the study protocol or involve serious or immediate medical implications should be referred to an ethics committee or an IRB on an individual basis to determine appropriate action.

Results. When results are to be provided to research participants for the purpose of obtaining a diagnosis, susceptibility risk, and/or other medical purposes, the results must come from a CLIA-approved laboratory. If the research laboratory itself is not CLIA approved, then a mechanism needs to be in place to offer testing in a CLIA-approved laboratory. This testing must be offered in a manner sensitive to participants' desire, or lack thereof, to receive it. While participants may volunteer for family studies understanding the potential for identification of causative genes, they may not have fully anticipated the consequences and concerns that arise from the reality of the availability of testing, especially presymptomatic or predictive testing. Also, researchers disclosing results must define a policy detailing which results will be disclosed and the method of disclosure. Researchers who include nongenetic clinical testing, such as cholesterol testing, as part of their study protocol will also need to decide whether they will release these results. There may be instances where nongenetic data, but not genetic data, will be released to participants. For those studies releasing genetic results, a distinction needs to be made regarding linkage and mutation analysis results. All information pertaining to the type and release of results should be covered fully in the informed-consent process. In addition, laboratories should provide participants with a genetic counseling service (US OPRR, 1993), either directly or, more typically, through referral to an appropriate center. The problematic issues associated with disclosing genetic information, such as the potential personal and social ramifications of learning one's genetic status, are addressed in the genetic counseling process (MacKay, 1993; Earley and Strong, 1995). In this way, test results are disclosed to the family and interpreted in a meaningful manner.

Psychological Risks. The potential harm to an individual from learning his or her genetic status—for example, by eliciting feelings of guilt, low self-esteem, or similar responses—is well recognized (US OPRR, 1993; Clayton et al., 1995). Such psychological and social risks should be discussed with participants as part of the informed-consent process *and* prior to disclosing genetic results. This discussion should enable each individual to reach an informed, independent decision about whether to receive test results. In this manner, an individual's "right not to know" is regarded equally with the right to know his or her genetic status.

Genetic Testing of Children

Genetic testing of an apparently unaffected child in the absence of any direct medical benefit is a controversial practice (Wertz et al., 1994) and should be done only within the context of genetic counseling and according to specific guidelines (ASHG/ACMG Report, 1995; Council on Ethical and Judicial Affairs, 1995). Concern arises over the potential risk of stigmatization of a child known by family members, future employers, and insurers to carry a genetic mutation (US OPRR, 1993). In addition, questions about the child's autonomy and his or her right not to know personal genetic information may conflict with parental autonomy. Parents wishing to have their children tested for genetic disorders should be referred to a genetic counselor to discuss the potential risks and benefits of testing.

Genetic Discrimination

The potentially stigmatizing nature of genetic information and the history of its abuse necessitate special provisions for protection against discrimination based on genetic information (Holtzman and Rothstein, 1992a). Although a basic right to privacy has been established by the U.S. Supreme Court (*Roe v. Wade*, 1973), the nonspecific nature of this privacy protection fails to adequately guard against unauthorized disclosure of personal genetic information. Reports show that employers and insurers wishing to avoid potential medical expenses for individuals with a family history of a genetic disease, including asymptomatic individuals, practice genetic discrimination (Gostin, 1991; Billings et al., 1992; Natowicz et al., 1992a; Wertz, 1992; Billings, 1993).

The Rehabilitation Act of 1973, revised in 1992, was the first law to prohibit employment discrimination by federally funded agencies and institutions based on physical disability. This legislation defines an individual with a disability as any person who "is regarded as having [a physical or mental] impairment" [1992 Reauthorization of the Rehabilitation Act of 1973, Section 7(8-B)]. Such a definition could include asymptomatic gene carriers perceived as being "sick" by their employer. The 1990 Americans with Disabilities Act (ADA) significantly broadened the scope of the 1973 legislation by prohibiting discrimination against disabled individuals in most areas of employment and public transportation. While the ADA's definition of disability excludes persons with a "characteristic predisposition to illness or disease," some professionals believe that the legislation allots sufficient protection to persons against genetic discrimination in the workplace (cited in Natowicz et al., 1992a). Although there is disagreement about whether persons with a genetic predisposition to disease are adequately protected under the ADA (Holtzman and Rothstein 1992a,b; Natowicz et al., 1992b), recent rulings by the Equal Employment Opportunity Commission (EEOC) have interpreted the ADA as affording protection to individuals from employer discrimination based on genetic test results (Leary, 1995). Nevertheless, the judicial system may ultimately be called upon to define the extent of protection from genetic discrimination that the ADA legislation provides.

Although the ADA prohibits employer discrimination, this law, along with the majority of state laws, does not adequately protect those with genetic disorders—both symptomatic and asymptomatic—against discrimination by insurers (Natowicz et al., 1992a). Thus, one could argue that employment opportunities may be limited, especially by smaller employers, who may refuse to offer employment because of a desire to avoid paying the higher premiums demanded by insurers. The Health Insurance Portability and Accountability Act of 1996 (HIPAA) protects individuals with a genetic disorder from being denied access to group health insurance. However, this legislation offers no protection for those with individual health insurance policies, such as those often provided by smaller employers or purchased by the self-employed. Equally disconcerting is the potential for employers to institute genetic screening programs prior to offering employment (Gostin, 1991). Given the legal latitude afforded to insurance companies, genetic information could become part of the standard application process for insurance policies (McEwen et al., 1992). Family medical history is already a part of the application process for most life insurance policies, and insurers may interpret this information to indicate an increased risk for a genetic disorder, regardless of the results of any formal genetic evaluations or testing.

The HIPAA applies to the transmission and storage of all medical information (Holder and Sugarman, 2002). Since medical information is defined as any identifiable information collected by a health care provider or other parties such as researchers, the HIPAA will have an enormous impact on how genetic research is conducted. Researchers should work with their respective IRBs to ensure that genetic research projects are being conducted according to HIPAA regulations.

There have been reports of denial of insurance coverage based on participation in genetic studies (Earley and Strong, 1995; Hudson et al., 1995). Therefore, it is essential that all information collected and generated as part of the family study be kept confidential. Because of recognition of the potential harm of disclosure of a person's genetic status to a third party, the need for federal legislation guaranteeing a right to genetic privacy is gaining support (Reilly, 1992; Annas, 1995). In addition, working groups have been formed to evaluate the impact of genetic information on individual insurability and on the insurance industry (US NIH-DOE Working Group on Ethical, Legal, and Social Implications of Human Genome Project, 1993; ASHG Ad Hoc Committee on Genetic Testing/Insurance Issues, 1995).

DNA Banking

Genetic research laboratories are inherently DNA banking facilities. Therefore, research laboratories should develop policies that address banking issues, including length of DNA storage and accessibility to banked samples. These policies should be incorporated into the informed-consent process (Hall et al., 1991; Weir and Horton, 1995b).

The use of previously collected DNA samples for a new research study presents a problematic scenario in terms of informed consent. Suggestions for developing consent forms that address the issues of DNA banking and informed decision making have been presented (ASHG Ad Hoc Committee on DNA Technology, 1988; Knoppers and Laberge, 1989; Gold et al., 1993; ACMG Storage of Genetics Materials Committee, 1995; Weir and Horton, 1995b). Research facilities that bank DNA samples as a service to families in order to maintain a sample for future diagnostic purposes should clarify the differences between the banking and use of research versus clinical DNA samples.

Some practical guidelines for informing the participants about DNA banking options include the following:

1. *Incorporation of Research Center's Policy on DNA Banking into Informed-Consent Process.* This document should state clearly whether DNA samples collected for research purposes will be available to the research participants or their surviving family members.

2. *Advising Research Participants about availability of Commercial DNA Banking Facilities.* Commercial DNA banking of a DNA sample enables an individual, or his or her family members as designated by the individual, to access a DNA sample and pursue diagnostic genetic testing when it becomes available. This advice may be particularly relevant for affected elderly family members or when a sole individual in a family has an unknown or terminal disorder.

It is advantageous for both the family and the research team to have interested family members deposit DNA samples with a commercial DNA bank. This allows individuals, not researchers, to designate which family members may or may not have access to the banked DNA sample.

DNA Ownership. In a landmark decision in 1980, the U.S. Supreme Court ruled that biological materials could be patented, opening the door for the patenting of genes and gene products (*Diamond v. Chakrabarty*, cited in Annas and Elias, 1992, p. 235). Ten years later, the California Supreme Court ruled that a person has no ownership rights to his or her cells once they have been removed from his or her body (*Moore v. Regents of University of California*, 1990). Therefore, although the legal status of DNA ownership has yet to be completely resolved, one could argue that the DNA derived from discarded tissue or from research participants is the property of the researcher.

The right of a research participant to control the fate of his or her sample remains a sensitive and ambiguous issue. Several reports suggest that in addition to DNA storage and accessibility issues, participants should be offered a choice in determining the fate of their samples in terms of research and commercial uses (Clayton et al., 1995). Guidelines for DNA banking as well as legislation establishing an individual's ownership rights to his or her DNA have been proposed (Annas, 1993, 1995). Although the dispute over DNA ownership is an important one, it appears that unless a sample proves to have commercial value (*Moore v. Regents of University of California*, 1990), research participants are generally more concerned that the information derived from genetic studies be secure than about the fate of their individual sample material. Still, it may be helpful to include a clause in the informed-consent form specifying that the individual's DNA becomes the property of the institution at the time of the blood draw.

FUTURE CONSIDERATIONS

As mentioned at the beginning of this chapter, many of the issues in genetic research studies discussed here are not completely resolved, and the procedures currently in use are constantly evolving. Some of the strongest debates center around questions such as what constitutes sufficient informed consent (and for which age groups), what is the status of DNA ownership, and how DNA banking should be handled. Two other issues still being hotly debated are the status of samples collected under earlier (and generally less detailed) informed-consent procedures and laboratory legal liabilities and responsibilities to the participants and their families. The resolution of these debates will have a major and lasting impact on the research community.

APPENDIX

You are being asked to take part in a research study at the [list institution]. The purpose of the study is to identify genetic factors that contribute to or cause [list disease being studied]. Once identified, these factors can be studied further to identify how they contribute to or cause this disease. Ultimately, this type of research may result in improved diagnosis, improved treatment, and possibly prevention. The sponsor of this study is [list funding agency, institution performing research, and the researcher's name].

You are being asked to consent to those items checked below:

Initials

[ ] [ ] (a) A telephone or direct (in-person) interview to review family, medical, and/or environmental risk factor history.

[ ] [ ] (b) The removal of up to 24 cc (2 tablespoons) of blood by vein puncture. There is a small risk of fainting or bruising and a 1 in 1000 risk of infection at the site where the blood is drawn.

[ ] [ ] (c) The use of a portion of your blood sample for chromosome studies/cytogenetic studies, including testing to identify rearrangements in the chromosomes that may help to locate a gene or genes for (list disease). If cytogenetic studies are performed, these results are available to your health care provider for interpretation following written request by you.

[ ] [ ] (d) The removal of cells from the inside of your mouth with a cheek brush (soft bristle swab). This involves twisting the brush on the inside of each cheek for 30 s. It is not painful and does not involve needles or puncture to the skin.

[ ] [ ] (e) A physical examination specific to [list disease being studied].

[ ] [ ] (f) Videotaping or audio taping of the interview(s).

[ ] [ ] (g) A photograph of your face.

[ ] [ ] (h) Review of your medical records.

Your genetic material (DNA) will be removed from cells and used in the research studies.

The following basic principles relate to your participation in the genetic studies:

(a) *Participation in Study.* Taking part in this study is entirely voluntary. Individuals participating in the study either have [list disease being studied] or are related to a person or persons known to have [list disease being studied]. The studies described are for research purposes only. Therefore, you will receive no results from this study. Participation may include repeated contacts/interviews to obtain current or additional medical information.

It is not the purpose of this study to look for or provide you with any medical information or diagnosis relating to your present condition or to any other disorder or condition. Your participation in this study is not a substitute for your regular medical care or check-ups.

(b) *Study Recruitment and Duration.* Over the course of this research study [XXXX] individuals will be recruited for study. This research study will continue until the gene or genes related to [list disease being studied] are identified and characterized, which will take several years.

(c) *Ownership of Samples.* Cells, tissue, tumor tissue, or blood removed from you during the course of this study or the genetic material (DNA) removed from such specimens may be valuable for scientific, research, or teaching purposes or for the development of a new medical product. By agreeing to participate in this research, you authorize [list research institution] and members of its staff to use your cells, blood, or other specimens for these purposes. In the event this research project results in a product, which could be sold commercially, [list institution] and its collaborators will assert the exclusive right to any revenue from the sale of such a product.

(d) *Use of Specimens.* The cells, tissue, blood, genetic material (DNA), or other routine biological samples removed from you during the course of this study will be maintained indefinitely by [list research institution] for research purposes or until the samples are gone. These samples are not available for clinical (diagnostic) purposes. Therefore, any future diagnostic testing as a result of this or other research must be performed using a new sample.

Your samples may be shared with colleagues at other institutions who are working with us on identifying genes for [list disease]. In such cases, the samples are coded such that our colleague has no way to identify you or your family. Under no conditions will we release information that identifies you to our colleague(s) without your written permission.

(e) *Secondary Uses of Specimens.* Your cells, tissue, blood, or genetic material (DNA) may be shared anonymously with other investigators for study of disorders unrelated to the one(s) in your family. Such usage will be strictly anonymous, in that no identifying information about you is provided to the researcher. Such usage will in no way compromise the study of the disorder(s) in your family. You have the option to permit or deny the secondary use of your samples, by placing your initials in one of the following boxes:

[   ]   Yes, I agree to share my samples.

[   ]   No, I do *not* agree to share my samples.

(f) *Research Results.* This study is for research purposes only. In general, the results from laboratory studies utilizing data collected as part of this research study will be preliminary. The implications of any laboratory findings may not be understood for years and may not be useful to any given individual. Therefore, individual results from the genetic studies will not be shared with your health care provider.

During your participation in this study, suspicious physical examination findings may be identified. If you so request, such physical findings will be discussed with you. You will be encouraged to consult with your physician to clarify any such findings. Costs associated with a medical evaluation that is not part of the research study are the responsibility of the research subject.

A written summary of the clinical evaluation finding can be provided to you if you request. You should be aware that this summary is not a comprehensive psychological evaluation of you.

[   ]   Yes, I would like a written summary of my clinical evaluation.

[   ]   No, I would *not* like a written summary of my clinical evaluation.

(g) *Incidental Findings.* It is possible that this study will identify information about you that was previously unknown, such as disease status risk. Such incidental findings, if any, will not be shared with you or anyone related to you unless the incidental finding regards an inherited risk for a disease known at the time of testing to be likely to cause premature death if untreated. Should such life-threatening results be uncovered through these genetic research studies and if they are directly applicable to you or to your minor children, you will be notified via certified mail to contact the staff at the [list research institution]. Notification will be sent to the last address you provided to the [list research institution]. The [list research institution] staff will not release these specific research findings over the telephone or in the mail. The staff at the [list research institution] will arrange for you to meet with a physician and/or a genetic counselor or other appropriate health care provider at either [list research institution] or another medical institution near your residence to review the research information.

(h) *Confidentiality*

Confidentiality and access to information. Study records that identify you will be kept confidential as required by law. Federal privacy regulations provide safeguards for privacy, security, and authorized access. Except when required by law, you will not be identified by name, social security number, address, telephone number, or any other direct personal identifier in the study records disclosed outside of [list research institution]. All information collected during genetic research studies, including information you provide about your family history, clinical information collected during a medical history or physical examination, standard laboratory tests used for research purposes, and analysis of data derived from the study of DNA, will not be released to anyone, including other family members, without your written consent or that of your legal representative. Any written request is verified by telephone interview with the requester.

Confidentiality among research Staff. Access to information about you, your family, or your cells, tissue, blood, or genetic material (DNA) is restricted to staff involved in enrolling research subjects at the [list research institution]. Information including your cells, tissue, blood, or genetic material (DNA) is coded using unique identifiers.

Confidentiality in professional settings. When research results from this study are reported in a professional setting, such as a medical journal or a scientific meeting, the identity of research subjects taking part in this study is withheld.

Limits of confidentiality. There is a risk for discrimination against individuals who are at risk for a medical disorder or have a medical disorder/condition in their family. Discrimination may include barriers to obtaining health, life, or long-term care insurance or employment. Extensive efforts are made to protect all research subjects from prejudice, discrimination, or uses of this information that will adversely affect them. Specifically, clinical and research information with respect to this study is maintained in a research file separate from hospital medical records and will not be placed in the official [list research institution] medical record by research staff. Research data from which you may be identified will not be disclosed to third parties except with your permission or as may be required by law.

Certificate of confidentiality protection of privacy. To further protect your privacy, the investigators have obtained a confidentiality certificate from the Department of Health and Human Services. With this certificate, the investigators cannot be forced (for example, by court subpoena) to disclose information that may identify you in any federal, state, or local civil, criminal, administrative, legislative, or other proceedings. Disclosure will be necessary, however, upon request of the Department of Health and Human Services for audit or program evaluation purposes. You should understand that a confidentiality certificate does not prevent you from voluntarily releasing information about yourself or your involvement in this research. Also, it does not prevent the investigators from voluntarily releasing information if you give your permission in writing for them to do so. Finally, if we learn that keeping specific information private would immediately put you in danger or put in danger someone else we know about, we may discuss it with you, if possible, or seek help from others to protect you or another person.

(i) *Benefit.* Though you may not gain immediate personal benefit from taking part in this study, knowledge will be gained that may ultimately benefit your family, others, or yourself.

(j) *Costs.* There will be no charge to you for participating in this research. Routine medical care for your condition can neither be paid for nor provided by the [list research institution]. Nor can the [list research institution] cover expenses associated with genetic counseling, genetic testing for clinical purposes, or investigation of findings outside the scope of this research study.

(k) *Withdrawal from Study.* If you agree to be in the study, you are free to change your mind. At any time you may withdraw your consent to be in this study and for us to use your data. If you withdraw from the study, you will continue to have access to health care at [list research institution]. If you do decide to withdraw, we ask that you contact the researcher [list name] in writing and inform him/her that you are withdrawing from the study. At that time we will ask your permission to continue using all information about you that has already been collected as part of the study prior to your withdrawal.

(l) *Maintaining Contact with Researchers.* It is your responsibility to notify the researchers at the [list research institution] at [area code, list telephone number or if there is a change in your mailing address and/or telephone number]. It may be necessary for the researchers to contact you to update information needed for the research study, to send current newsletters, and to notify you of research breakthroughs or availability of testing.

(m) *More Information.* For more information about the research study, you may contact [list name], study coordinator, at [insert telephone number]. Information is also available at our website: [list URL].

The purpose of this study, procedures to be followed, risks, and benefits have been explained to me. <u>I have been allowed to ask questions, and my questions have been answered to my satisfaction.</u> I have read the above and have been given the opportunity to discuss it and to ask questions. I have been informed that I may contact the lead researcher to answer any questions I may have during the investigation and that I may contact the Office of Risk Management at [list telephone number] for any question concerning my rights as a research subject. I agree to participate as a subject with the understanding that I may withdraw at any time without interfering with my regular care. I have been told that I will be given a signed copy of this consent form.

_____   _____

Signature of Subject   **Date**

_____   _____

Signature of Person Obtaining Consent   **Date**

We are committed to maintaining contact with individuals and families who participate in these genetic research studies. As the research continues, we will update you through an annual newsletter sent to the address below.

Please <u>initial</u> here _____ if you would prefer not to receive this mailing. If you move, please contact us with your new address. Thank you.

SUBJECT: **Name (print):** _____

              Home Address: _____

                            _____

                            _____

              Home Phone: (____)_____

                     E-Mail: _____

REFERENCES

ACMG Storage of Genetics Materials Committee (1995): ACMG statement: Statement on storage and use of genetic materials. Am J Hum Genet 57:1499–1500.

Andrews LB (1990): DNA testing, banking, and individual rights. In: Knoppers BM, Laberge CM, eds. Genetic Screening: From Newborns to DNA Typing. London: Elsevier Science Publishers.

Andrews LB, Fullarton JE, Holtzman NA, Motulsky AG (1994): Assessing Genetic Risks: Implications for Health and Social Policy. Washington, DC: National Academy Press, Institute of Medicine.

Annas GJ (1993): Privacy rules for DNA databanks: Protecting coded "future diaries." JAMA 270:2346–2350.

Annas GJ (1995): Editorial: Genetic prophecy and genetic privacy—Can we prevent the dream from becoming a nightmare? Am J Public Health 85:1196–1197.

Annas GJ, Elias S (1992): Gene Mapping: Using Law and Ethics as Guides. New York: Oxford University Press.

ASHG/ACMG Report (1995): Points to consider: Ethical, legal, and psychosocial implications of genetic testing in children and adolescents. Am J Hum Genet 57: 1233–1241.

ASHG Ad Hoc Committee on DNA Technology (1988): DNA banking and DNA analysis: Points to consider. Am J Hum Genet 42:781–783.

ASHG Ad Hoc Committee on Genetic Testing/Insurance Issues (1995): Background statement: Genetic testing and insurance. Am J Hum Genet 56:327–331.

Bazyk AE, Neumeister E, Lynch B, Watson-Clevenger P, McFarland L, Haines JL (2000): Ascertainment effort in genetic research studies for common disease. Am J Hum Genet 67(4, Suppl):208.

Biesecker LG, Collins FS, DeRenzo EG, Grady C, MacKay CR (1995): Case: Responding to a request for genetic testing that is still in the lab. Cambridge Q Healthcare Ethics 4:387–400.

Biesecker LG, DeRenzo EG (1995): Internet solicitation of research subjects for genetic studies. Am J Hum Genet 57:1255–1256.

Billings PR (1993): Genetic discrimination. Healthcare Forum J Sept–Oct:35–37.

Billings PR, Kohn MA, de Cuevas M, Beckwith J, Alper JS, Natowicz MR (1992): Discrimination as a consequence of genetic testing. Am J Hum Genet 50:472–482.

Caplan AL (1992a): If gene therapy is the cure, what is the disease? In: Annas GJ, Elias S, eds. Gene Mapping: Using Law and Ethics as Guides. New York: Oxford University Press, pp 128–141.

Caplan AL (1992b): Twenty years after: The legacy of the Tuskegee syphilis study. Hastings Center Report 6:29–32.

Clayton EW (1995): Panel comment: Why the use of anonymous samples for research matters. J Law Med Ethics 23:375–377.

Clayton EW, Steinberg KK, Khoury MJ, Thomson E, Andrews L, Ellis Kahn MJ, Kopelman LM, Weiss JO (1995): Informed consent for genetic research on stored tissue samples. JAMA 274:1786–1792.

Council on Ethical and Judicial Affairs (1995): Testing children for genetic status. Code of Medical Ethics Reports, American Medical Association 6:47–58.

de Leon D, Lustenberger A (1990): Issues raised in gene linkage studies. Birth Defects (original article series) 26:1391–1394.

Earley CL, Strong LC (1995): Certificates of confidentiality: A valuable tool for protecting genetic data. Am J Hum Genet 57:727–731.

Estabrooks S, Dowdy E, Granger C, Pericak-Vance M, Kraus M, Hauser E (1999): Family ascertainment strategies in common complex genetic disorders: An example in early onset coronary artery disease. J Genet Counseling 8(5, Abstract):400.

Frankel MS, Teich AH (1993): Ethical and Legal Issues in Pedigree Research. Washington, DC: Directorate for Science and Policy Programs, American Association for the Advancement of Science.

Genetic Alliance (brochure): Informed consent: Participation in genetic research studies. Available by calling 1-800-336-GENE.

Gold RL, Lebel RR, Mearns EA, Dworkin RB, Hadro T, Burns JK (1993): Model consent forms for DNA linkage analysis and storage. Am J Med Genet 47:1223–1224.

Gostin L (1991): Genetic discrimination: The use of genetically based diagnostic and prognostic tests by employers and insurers. Am J Law Mcd 17:109–144.

Hall J, Hamerton J, Hoar D, Korneluk R, Ray P, Rosenblatt D, Wood S (1991): Policy statement concerning DNA banking and molecular genetic diagnosis. Clin Invest Med 14: 363–365.

Holtzman NA (1989): Proceed with Caution: Predicting Genetic Risks in the Recombinant DNA Era. Baltimore: Johns Hopkins University Press.

Holtzman NA, Rothstein MA (1992a): Invited editorial: Eugenics and genetic discrimination. Am J Hum Genet 50:457–459.

Holtzman NA, Rothstein MA (1992b): Reply to Natowicz et al. Am J Hum Genet 51:897.

Hubbard R (1986): Eugenics and prenatal testing. Int J Health Serv 6:227–242.

Hudson KL, Rothenberg KH, Andrews LB, Ellis Kahn MJ, Collins FS (1995): Genetic discrimination and health insurance: An urgent need for reform. Science 270: 391–393.

Jordan E (1992): Invited editorial: The Human Genome Project: Where did it come from, where is it going? Am J Hum Genet 51:1–6.

Kass NE (1993): Participation in pedigree studies and the risk of impeded access to health insurance. IRB 15(September–October):7–10.

King PA (1992): The past as prologue: Race, class, and gene discrimination. In: Annas GJ, Elias S, eds. Gene Mapping: Using Law and Ethics as Guides. New York: Oxford University Press, pp 94–111.

Knoppers BM, Laberge C (1989): DNA sampling and informed consent. Can Med Assoc J 140:1023–1028.

Kodish E, Murray TH, Shurin S (1994): Cancer risk research: What should we tell subjects? Clin Res 42:396–402.

Leary WE (1995): Using gene tests to deny jobs is ruled illegal. New York Times, April 8.

MacKay CR (1993): Discussion points to consider in research related to the human genome. Hum Gen Ther 4:477–495.

Macklin R (1992): Privacy and control of genetic information. In: Annas GJ, Elias S, eds. Gene Mapping: Using Law and Ethics as Guides. New York: Oxford University Press, pp 157–172.

McConkie-Rosell A, Markel D (1995): Facilitating research: The many roles of the genetic counselor. Perspectives in Genetic Counseling (newsletter, National Society of Genetic Counselors) 17:1, 4.

McEwen JE, McCarty K, Reilly PR (1992): A survey of state insurance commissioners concerning genetic testing and life insurance. Am J Hum Genet 51:785–792.

Melvin E, Estabrooks S, Rotblatt M, Wolpert C, Pericak-Vance M, Speer M (1999): Report of focus groups to identify the attitudes, motivations, and preferences of families participating in genetic research. J Genet Counsel 8(6, Abstract):419.

Melvin E, Estabrooks S, Wolpert C, Pericak-Vance M, Speer M (2000): Releasing research results for gene identification studies. J Genet Counsel 9(6, Abstract):524.

Merriam-Webster's Medical Desk Dictionary (1996) Springfield, MA: Merriam-Webster.

Miller JL (in preparation): Needs and rights of participants in human genetic research: Perspective of the subjects. Evanston, IL: Northwestern University, Department of Obstetrics and Gynecology.

Moore v. Regents of University of California, 793 P2d 479 Sup. Ct. CA (1990).

Mueller-Hill B (1992): Eugenics: The science and religion of the Nazis. In: Caplan AL, ed. When Medicine Went Mad: Bioethics and the Holocaust. Totowa, NJ: Humana, pp 43–52.

Natowicz MR, Alper JK, Alper JS (1992a): Genetic discrimination and the law. Am J Hum Genet 50:465–475.

Natowicz MR, Alper JK, Alper JS (1992b): Genetic discrimination and the Americans with Disabilities Act. Am J Hum Genet 51:895–897.

Parker LS, Lidz CW (1994): Familial coercion to participate in genetic family studies: Is there cause for IRB intervention? IRB 16(January–April):6–12.

Pate v. Threlkel, 661 So.2d 278 (1995).

Paul DB (1994): Is human genetics disguised eugenics? In: Weir RF, Lawrence SC, Fales E, eds. Genes and Human Self-Knowledge: Historical and Philosophical Reflections on Modern Genetics. Iowa City: University of Iowa Press, pp 67–83.

Pelias MZ (1991): Duty to disclose in medical genetics: A legal perspective. Am J Med Genet 39:347–354.

Proctor R (1988): Racial Hygiene: Medicine under the Nazis. Cambridge, MA: Harvard University Press.

Proctor RN (1992): Genomics and eugenics: How fair is the comparison? In: Annas GJ, Elias S, eds. Gene Mapping: Using Law and Ethics as Guides. New York: Oxford University Press, pp 57–93.

Public Health Service Act § 301(d); *42 United States Code 241(d)*.

Reilly P (1980): When should an investigator share raw data with the subjects? IRB 2 (November):4–5, 12.

Reilly PR (1991): The Surgical Solution: A History of Involuntary Sterilization in the United States. Baltimore: Johns Hopkins University Press.

Reilly PR (1992): ASHG statement on genetics and privacy: Testimony to United States Congress. Am J Hum Genet 50:640–642.

Roe v. Wade, 410 U.S. Sup. Ct. 113 (1973).

Safer v. Pack, 291 N.J. Sup. 619, 677 A.2d 1188 (1996).

Scaffe A, Pollin T, Xu J, Kittner S (2000): An evaluation of different recruitment strategies used in a family study. J Genet Counsel 9(6, Abstract):530.

Shore D, Berg K, Wynne D, Folstein MF (1993): Legal and ethical issues in psychiatric genetic research. Am J Med Genet 48:17–21.

Tarasoff v. Regents of University of California, 551 P2d 334 Sup. Ct. CA (1976).

Terry S, Whittemore V, Terry P, Cody J, Davidson M (2000): Partners in research. Am J Hum Genet 67(4, Suppl):243.

US Department of Energy (DOE) Title 10, Code of Federal Regulations, Part 745 (10CFR745): Federal Policy for the Protection of Human Subjects—Notices and Rules.

US Department of Energy (DOE) (brochure): Protecting Human Research Subjects within the U.S. Department of Energy. Brochure available by calling 301-903-5037.

US Department of Health and Human Services (DHHS) (1991): Title 45, Code of Federal Regulations, Part 46 (45CFR46), Protection of Human Subjects.

US NIH-DOE Working Group on Ethical, Legal, and Social Implications of Human Genome Research (1993): Genetic information and health insurance: Report of the Task Force on Genetic Information and Insurance. NIH publication 93-3686. Bethesda, MD: National Institutes of Health.

US Office for Protection from Research Risks (OPRR) (1993): Protecting Human Research Subjects: Institutional Review Board Guidebook. Washington, DC: Department of Health and Human Services, National Institutes of Health.

Weir RF, Horton JR (1995a): DNA banking and informed consent—Part 1. IRB 17(July–August):1–4.

Weir RF, Horton JR (1995b): DNA banking and informed consent—Part 2. IRB 17(September–December):1–8.

Wertz DC (1992): Ethical and legal implications of the new genetics: Issues for discussion. Soc Sci Med 35:495–505.

Wertz DC, Fanos JH, Reilly PR (1994): Genetic testing for children and adolescents: Who decides? JAMA 272:875–881.

Wertz DC, Fletcher JC, Berg K (1995): Guidelines on ethical issues in medical genetics and the provision of genetic services, WHO/HDP/GL/ETH/95.1. Geneva: World Health Organization, Hereditary Diseases Programme.

CHAPTER 5

Collection of Biological Samples for DNA Analysis

JEFFERY M. VANCE

This chapter provides practical knowledge on sample collection in genetic studies, so that an individual can make informed decisions concerning the options available for sample collection at various levels of cost and expertise.

ESTABLISHING GOALS OF COLLECTION

It is crucial to the success of any family study for the investigator to clearly define the goal of the investigation. If the intent of the study is to perform genomic screening and gene localization, then a relatively large and consistent amount of DNA is required; usually such a supply is obtained through the collection of whole blood. If only a few specific tests are planned, the smaller amount of DNA obtainable from a buccal sample may be appropriate. Mutational analysis in unknown disorders is greatly enhanced by the availability of messenger RNA (mRNA) (Noguchi et al., 1995); thus lymphoblasts or tissue samples from selected affecteds may be desired in these instances. The geographic location of families and patients and the technical background of those collecting the samples are also important items to consider. For example, an individual in a remote location, where venipuncture would not be practical, may easily be included in a study if a mailed buccal sample is used instead.

TYPES OF DNA SAMPLE COLLECTION

Venipuncture (Blood)

Venipuncture should preferably be done using a 21-gauge or larger needle. While a 23-gauge needle can be used, in our experience the risk of hemolysis is greater.

Genetic Analysis of Complex Diseases, Second Edition, Edited by Jonathan L. Haines and Margaret Pericak-Vance
Copyright © 2006 John Wiley & Sons, Inc.

Because field studies often occur in less than optimum conditions, we use butterfly needles, which are easier to handle than Vacutainer needles, to collect our samples. In addition, if the collected blood tubes are to be frozen (immediately or at the final storage facility), the use of plastic rather than glass blood collection tubes is highly recommended. At freezer temperatures from -20 to $-80°C$, normal glass vials are more likely to break when thawed. The S-Monovette Blood Collection System manufactured by Sarstedt is ideal when samples are to be frozen for long periods of time.

For DNA extraction, the S-Monovettes containing potassium ethylenediaminetetraacetic acid (EDTA) (No. 01.1605.100) are recommended. For collection, these tubes employ an integral pull plunger to withdraw the blood sample and form a vacuum. Upon completion, the plunger is simply broken off.

Gustafson et al. (1987) demonstrated that DNA is most stable at room temperature when acid citrate dextrose or EDTA is used as the anticoagulant. The anticoagulant of choice for all routine hematological investigations is EDTA. Usually, the final concentration is approximately $1.2–2$ mg EDTA/mL blood after collection. Maximum dilution caused by EDTA is 1%. In heparin, DNA is not stable and residual heparin has been reported to interfere with Taq polymerase (Kirby, 1990). Therefore, use of this anticoagulant should be avoided.

The volume of blood required for sampling depends on the purpose of the study. There are many different methods for DNA extraction, differing primarily in the volume of blood they can process easily, the toxicity of the components, and the number of steps or pieces of equipment required. DNA is obtained only from the lymphocytes (red cells are without a cell nucleus); therefore the amount of DNA in a blood sample can vary from individual to individual. This may contribute to the wide variation in estimates of the amount of DNA in a milliliter of whole blood, ranging in the majority of cases from 25 to 40 μg DNA/mL whole blood (Kirby, 1990).

Immediately after collection, the collection tube should be gently inverted $5–10$ times to ensure proper mixing of the anticoagulant. This is critical: Omission of this step is a common mistake that leads to clotted samples. If slated for DNA extraction, the blood should be stored at $4°C$ as soon as possible. This slows the metabolism of the living cells. However, samples to be used for cell culture need to be held at room temperature, since chilling these samples will reduce the transformation success. In this situation, the less expensive, standard glass blood tubes containing ACD are recommended. Caution must be used when shipping samples during extreme weather conditions. The higher temperatures during summer and lower temperatures during winter can have adverse affects upon the sample. Adequate insulation when packing the samples is advised rather than using standard mail carriers. Also, if the samples are extremely valuable, overnight shipment should be considered a priority.

Receiving samples by mail also requires planning. Blood samples are best sent by overnight delivery. This is absolutely true for samples destined for transformation, which need to be received and processed as soon as possible, certainly within $2–3$ days of collection. However, it may be cheaper to send samples intended for DNA extraction by postal service, where time is important but not as critical. Again, be aware of current weather and temperature conditions at shipping and receiving points. There are specific regulations for sending blood products via mail both within the United States and internationally. The investigator should check with the service selected to determine what specific rules apply.

Buccal Samples

In the last few years buccal brushes have become increasingly popular for collecting DNA specimens (Richards et al., 1994). These noninvasive devices do not require the technical skills needed for venipuncture and eliminate the chance of an accidental needle stick during the collection of the sample. Prior to extraction, the DNA on the brushes degrades slowly. Buccal brushes are particularly good for young children, obese patients, and those situations requiring a smaller amount of DNA. There are some disadvantages to buccal brushes. The amount of DNA can vary significantly from one individual to another; typically, even under the best collection conditions, it is significantly smaller than the quantity collected by venipuncture. However, this may not be a problem unless a large genomic screening project is planned. The DNA from buccal samples is usually isolated crudely using NaOH. Thus, the presence of proteases and nucleases makes the sample much less stable than DNA purified from blood, so it must be aliquoted upon extraction and kept frozen until thawed for single use. Since only DNA consisting of small fragment sizes will be available using this method, it is not suitable for techniques requiring large fragments, such as pulsed-field gel electrophoresis (PFGE). Recently, a modification of the Puregene DNA Isolation Kit (Gentra Systems) has allowed a less aggressive approach to isolating DNA from buccal brushes. Cell lysis detergents in combination with proteinase K can yield from 0.2 to 2 µg DNA per brush.

Before a sample is collected, the individual washes his or her mouth out gently with water to remove large foreign particles. The brush is then twirled as it is moved firmly over the inner surface of one cheek for at least 30 s. A second brush is then used against the other cheek for another 30 s. These steps are critical, since many failures in buccal sampling appear to occur at this stage. One is more likely to obtain sufficient DNA from buccal collection when the samples are collected under the supervision of an experienced worker than when samples are collected by the subjects themselves without supervision. It may be that many unsupervised subjects perform just a cursory brushing which will not collect enough cells for a successful polymerase chain reaction (PCR) analysis. To alleviate this problem, we use longer brushing times as specified above.

We have kept buccal brushes after collection of a sample in their container tube for over a month with no detrimental effect on PCR results, but in our experience the DNA will start to degrade if brushes are stored over several months. They are good for field studies (where the samples cannot be sent to the laboratory within several days), for extended field trips, and for collecting samples from patients in isolated locales.

More recently, cheek cells are being collected using a liquid wash rather than a brush. The patients themselves can collect their buccal cell samples very easily and mail their samples to the DNA laboratory for isolation and analysis. Here, the patient uses approximately 10 mL of Original Mint Scope Mouthwash to swish orally 10 times before spitting back into a 50-mL disposable centrifuge tube. In this procedure, for best results, the patient should wait as least 1 h after eating or drinking to collect buccal cells. These cells are stable at room temperature for at least 7 days in the mouthwash solution. The expected yield of DNA is typically much larger than with brushes, ranging from 4 to 40 μg, with the mean equal to 16 μg. Laboratory processing time is about 3 h total.

Dried Blood

Dried blood samples can also be used as a source of DNA when a relatively small number of PCRs are required (Guthrie and Susi, 1963; McCabe, 1991). Matsubara et al. (1992) have suggested that dried blood can be a source of mRNA as well. We use these samples as a method to check for sample mix-up, so that if problems arise later in the genotyping analysis, we will have a second sample to test. From some individuals—children, for example—one can also procure samples via fingerpricks. At the time of venipuncture, a small amount of blood is placed on relatively inexpensive "Guthrie cards" (Guthrie and Susi, 1963). These are then dried and stored in photo albums at room temperature. Since these samples are obtained at the time of collection, we can be assured that they do indeed represent the DNA of the intended participant. They also provide backup DNA on an individual.

Tissue

Both fresh (frozen) and fixed tissues can provide samples for study. Fresh tissue from biopsies or autopsies, when flash frozen in liquid nitrogen, can be used in mutational analysis for molecular studies. It is useful to observe and plan for those opportunities where these samples may be obtained. Fixed tissue may provide samples on affected or deceased individuals for genotyping or association studies. In most cases, paraffin-embedded tissue (PET) will be the source of DNA for genetic study (Kosel and Graeber, 1994). Such samples, used routinely for analysis of surgical biopsies and subsequently archived, represent a resource for retrospective study. Recently, however, issues about informed consent have been raised concerning the use of some of these samples (Marshall, 1996; see also Chapter 4).

The central problem of fixed tissue is that the fixatives themselves degrade DNA. Thus, the size of the products of PCR obtainable from fixed tissue is usually limited. Therefore, obtaining results from fixed tissue can be difficult and should be reserved for individuals where no other DNA source is available. Greer et al. (1991) and Smith et al. (1987) studied the variation in the size of PCR products that can be obtained using different fixatives and found that 95% ethanol routinely allowed the largest PCR fragments to be obtained; OmniFix and acetone are next in order of preference. Buffered neutral formalin produced the smallest PCR fragments.

DNA EXTRACTION AND PROCESSING

Blood

There are numerous methods to extract DNA. These methods usually differ with respect to (1) the volume of blood that can be easily and readily processed; (2) time scale, the number of steps and therefore the number of samples that can be effectively processed; (3) toxicity; (4) equipment needed; and (5) purity required (a crude preparation for PCR or a more traditional extraction). Since each author believes his or her method to be "simpler" than earlier approaches, the reader should choose the method that best fits present needs and laboratory facilities. Usually a pellet is isolated which consists mainly of lymphocytes, the only component of blood that contains DNA. Centrifugation in the presence of a medium that lyses the red blood cells but leaves the lymphocytes intact can accomplish this isolation. An alternative to this approach is to spin the blood in a proprietary medium whose specific gravity causes the blood to separate into visible fractions from which lymphocytes can be harvested (buffy coat), for example, ficoll gradients (Sigma, ICN) or prepared gel tubes (Becton-Dickinson).

In either case, the enriched pellet is then treated by one of several methods to lyse the remaining lymphocytes and isolate the DNA. A primary method for many years has been a mixture of sodium dodecyl sulfate (SDS) and proteinase K to rupture cell walls and degrade protein (Kirby, 1990). To remove other proteins and peptides from the DNA, the sample is treated to a standard phenol-chloroform extraction and, finally, is ethanol precipitated in the presence of a simple salt (Kirby, 1990). Several other techniques, seeking to avoid the toxicity of phenol and chloroform, either substitute compounds for these chemicals (Johns and Paulus-Thomas, 1989; Planelles et al., 1996) or use a different approach, such as phase separation using guanidine thiocyanate (Chomczynski, 1993) or isolation employing a high salt concentration (Miller et al., 1988). The PCR methods for small samples also include proteinase K digestion (Kawasaki, 1990). Several commercial products are available, including those based on guanidine thiocyanate (DNAzol, Molecular Research Center) and the popular Puregene kits sold by Gentra Systems, Minneapolis, MN.

The quality of the DNA may be critical, depending on the study. For restricted fragment length polymorphism (RFLP) or enzyme restriction analysis, the quality of the DNA should be very high to ensure complete digestion. However, since the bulk of genetic analyses is now done by PCR, small amounts of DNA can be easily obtained without significant purification. Such is the case with buccal brushes. Here, the DNA must be absolutely free of contaminants that would otherwise interfere with Taq polymerase and the ability to produce a clean or correct product. However, the highest quality DNA possible is the most suitable for genotyping studies and also provides the most stable sample for long-term storage.

Quantitation

The most commonly used technique for measuring nucleic acid concentration is the determination of absorbance at 260 nm (A_{260}). Using a standard ultraviolet (UV)

TABLE 5.1. Manufacturers

Gentra Systems
13355 10th Avenue North, Suite 120
Minneapolis, MN 55441
Phone: 612-543-0678; fax 612-543-0699
Toll Free: 888-476-5283
www.gentra.com

Molecular Probes, Inc.
PO Box 22010
Eugene, OR 97402-0469
Phone: 541-465-8300; fax 541-344-6504
www.probes.com

Amersham Pharmacia Biotech, Inc.
800 Centennial Avenue
Piscataway, NJ 08855-1327
1-800-526-3593
www.apbiotech.com

spectrophotometer, one of the major disadvantages of this method is the large relative contribution of nucleotides and single-stranded nucleic acids to the signal (e.g., all nucleic acid species are detected). This is good if one needs to quantify these species, but the presence of RNA can falsely inflate DNA yield. Other disadvantages are the interference caused by contaminants commonly found in nucleic acid preparations (such as proteins), the inability to distinguish between DNA and RNA, and the relative insensitivity of the assay. An optical density (OD) of 0.1 corresponds to a 5-μg/mL double-stranded (ds) DNA solution and an OD of 1.0 corresponds to 40 mg/mL for single-stranded (ss) DNA or RNA. Therefore, the concentration for dsDNA (in milligrams per milliliters) is OD260 $\times$ 50 $\times$ dilution factor.

The purity of the DNA sample is usually determined using the ratio of the absorbance at OD260/OD280. While several factors can affect this ratio, DNA without significant contamination usually has a ratio of approximately 1.8. Ratios significantly less than 1.7 suggest contamination by protein, phenol, or other contaminants. Manchester (1995) suggested using a commercially pure DNA sample as a standard for DNA measurements. This excellent suggestion allowed the correction of inaccuracies of the individual spectrophotometer, which can be significant enough to lead to errors in concentration estimates and purity ratios.

There are some advantages to using the traditional method of UV spectroscopy, however. Generally the technique is simple and easily run as no special dyes or stains are required and only serial dilutions are made. The preparation of standards does not require much time or effort. And, finally, the overall cost of a spectrophotometer is considerably less than that of the new molecular imaging instrumentation.

The fluorometer or fluoroimager is much more sensitive than UV spectrophotometry and can detect DNA in very minute amounts (0.01-μg/mL range).

Herein lies its main advantage over UV absorption. This is also the preferred technique when quantitating very small samples that have little DNA initially. However, a fluorescent dye such as PicoGreen is required to assay the DNA species and ssDNAs will not be detected, which could be a potential problem if this species has to be quantitated. Also, in the past RNA could not be quantitated using a fluorometer, but recently RNA detection dyes have become available through Molecular Probes.

PicoGreen, an ultrasensitive fluorescent nucleic acid stain, can provide more consistent, more accurate, and more sensitive quantitations when used in conjunction with a fluoroimager (BioTechniques 20, 676, 1996). Molecular Probes (Eugene, OR) supplies a PicoGreen dsDNA quantitation reagent that allows measurements as low as 25 pg/mL of dsDNA with a standard spectrofluorometer and fluorescein excitation and emission wavelengths. This sensitivity exceeds that of the Hoechst 33258-based/UV-spectrophotometric assay by 400-fold. Moreover, the linear detection range of the PicoGreen assay extends over more than four orders of magnitude in DNA concentration, from 25 pg/mL to 1000 ng/mL, with a single dye concentration. A variety of other nucleic acid stains are also available, such as SYBR Gold (Molecular Probes Bulletin, No. MP-11494).

The increased sensitivity of using fluorescent dyes in conjunction with flourometry has a few disadvantages: costs, light sensitivity and storage of the dye, and as mentioned earlier, inability to detect certain nucleic species. In addition, the process is very labor intensive when performed manually. If an automated dispensing system or robotics unit is available, this is an excellent choice. Also, this methodology lends itself to large numbers of samples, not just a few.

In choosing one technique over another, consider what your requirements and budget will allow. Whatever technique chosen can be maximized to give accurate and consistent results.

Tissue Culture

Historically the primary reason for immortalizing lymphocytes (lymphoblasts) was to provide a long-term source of DNA even if a subject was unavailable for additional sampling. When the primary method used for genetic analysis was RFLP, this step was very useful, since up to 10 μg of DNA was used for one restriction digest. However, since a single genotype obtained by PCR requires only 30 ng or less of DNA in our laboratory (Ben Othmane et al., 1992), a standard draw of 10–30 mL of blood provides a sufficient amount of DNA for most analyses. Therefore, to transform all samples wastes effort and resources. Finally, even in the most experienced laboratories, transformation will occasionally fail, especially if the blood samples are more than a few days old. In this case, the entire sample committed to the transformation is lost to the study. The prudent investigator may elect to maintain lymphocytes for the purpose of ensuring a supply of DNA for certain key individuals in a study. Such individuals would include one or two affected for later mutational analysis in the study, elderly family members, individuals with critical recombination events, or individuals defining a haplotype in a family.

Another use of lymphoblasts or lymphocytes is to provide high-molecular-weight fragments (≥ 2 Mb) of DNA for use in PFGE. Standard extraction shears DNA into pieces that are too small (<50 kb) for this analysis. This damage can be avoided by embedding whole lymphoblast cells (106–108 cells) in agarose prior to purification (Smith et al., 1988). The protein is then extracted, leaving the DNA intact and preventing any shearing.

Finally, lymphocytes or lymphoblasts may be an easily obtainable source of RNA for studies of mutations in a gene. While point mutations can be identified in ordinary extracted DNA, the process requires knowledge of exon–intron boundaries whose delineation can be very time consuming. However, if the gene under study is expressed in lymphocytes, PCR can be used to directly amplify the mRNA to study the gene product (reverse-transcriptase PCR; Foley et al., 1993).

One deterrent to the routine transformation of lymphocytes into lymphoblasts is the high cost of tissue culture. However, one alternative does exist that can significantly reduce the cost of transformation: freezing lymphocytes [cell-culture-grade dimethyl sulfoxide (DMSO) dissolved in a special freezing medium containing 40–50% serum] in nitrogen prior to transformation (Louie and King, 1991; Pressman and Rotter, 1991). A word of caution when using this technique: Introduce the freeze medium dropwise if DMSO is already mixed or add the DMSO dropwise (final concentration 10%) after resuspension in freeze medium. Intact cells can be kept in long-term storage until the investigator decides whether that sample will require transformation. If the sample is not needed for transformation, it can then be extracted for DNA. One disadvantage of this approach is that freezing before transformation will kill some cells and therefore provide fewer cells for the transformation. Thus, the failure rate for delayed transformation is higher than for direct processing. This is critical, as it could mean the loss of the entire sample from individuals who are no longer are available for repeat sampling. Thus, if transformation of a specific individual is known to be needed for an analysis, it may be better to do it initially, for if failure occurs, the individual can be quickly sampled again. The techniques for tissue culture and lymphocyte transformation, which are beyond the space limitations of this chapter, can be found in many manuals (e.g., Doyle, 1990).

Buccal Brushes

Buccal brushes can be stored after collection at room temperature for weeks or at $-80°C$ for an extended time, prior to cell lysis. However, once the brushes have been processed, they are potentially unstable owing to protease and nuclease contamination and should be frozen for long-term storage.

The primary question always asked concerning buccal samples is "How much DNA will I get?" This is, unfortunately, difficult to answer. We routinely use 1–2 mL of a 660-mL total sample in a 10-mL PCR reaction. This volume limits the use of buccals for genomic screening to secondary samples. However, we have performed experiments to test the potential usefulness of buccal samples as primary samples as well. All samples were collected by experienced physician assistants on patients aged 20–30. While some samples could be diluted 50-fold and still provide an excellent PCR product, samples from other individuals had to be used at full strength for successful results. Therefore, the variability of DNA concentration, even when collected under controlled conditions, as well as the potential for degradation leads us at present to recommend the use of buccal samples only for secondary samples when large genomic studies are performed. However, they remain excellent samples for isolated patients, patients in whom venipuncture is difficult, candidate gene analysis, genotype confirmation, and smaller studies and as a supplementary source of DNA.

Dried Blood Cards

Many techniques are now available for DNA extraction from dried blood on Guthrie cards. Clearly, there is enough DNA on the cards to provide many PCR reactions (Del Rio et al., 1996). The problem is processing the sample in a way that allows Taq polymerase to efficiently access the DNA, which requires more manipulations for the analyses and limits the practical number of PCR reactions that can be performed using a card. Many techniques have been published employing boiling, heat cycles, sonication, and autoclaving to denature the sample prior to the PCR (Carducci et al., 1992; Raskin et al., 1993; Fishbein et al., 1995). In addition, re-PCR of an initial reaction can be used to bolster the amount of PCR product, although aberrant bands are also increased. We have used the method of Del Rio et al. (1996) with excellent results.

Genomic DNA can be extracted from blood cards in sufficient yield and purity for use in PCR using a variety of commercially available kits. The addition of glycogen greatly enhances the yield [Science Tools from Pharmacia Biotech 1,1 (1996)].

Fixed Tissue

There are many published protocols for the extraction of DNA from fixed tissue. The different approaches reflect the difficulty of isolating DNA from these tissues compared to other tissue sources. In our laboratory we use a modified technique in which the paraffin is removed with xylene, the residual tissue washed with ethanol, and a 1-cm^2 piece treated with proteinase K. After digestion, the resulting lysate is used without further purification in a PCR (De Souza et al., 1995).

Whole-Genome Amplification

Often only a small amount of DNA will be available from a sample, with no opportunity to obtain more (e.g., the participant is now deceased). In the past this sample would be of little use as it would only support a small number of individual assays. However, in recent years several approaches have been developed to amplify the amount of native DNA, generating substantial amounts of product available for large numbers of assays. primer extension preamplification (PEP; Zhang et al., 1992) and degenerate oligonucleotide primed PCR (DOP; Telenius, 1992) have been used successfully on DNA from single cells but do not amplify all regions of the genome equally and generate relatively small genomic fragments. Multiple-displacement amplification (MDA; Dean, 2002) uses the ϕ29 DNA polymerase to amplify large fragments (functionally up to 10 kb) with better representation of the genome. Only DNA very near the centromere and telomere are underrepresented.

SAMPLE MANAGEMENT

Management of samples can be critical to the success of any project, especially when the number of samples collected reaches thousands per year. A proven and certified sample sign-in, tracking, and location program becomes essential. In both the public and private sectors, such management software is known as a Laboratory Information Management System (LIMS). All LIMSs are based around sample tracking and laboratory workflow, incorporating features such as instrument integration. The development or purchase of such a database system should include the following elements: sample sign-in or receiving (in conjunction with barcoding, if applicable), verification of individual samples, sample tracking, quantitation, automatic data transfer and result entry, location of exact positions in workflows and storage, and meeting the constraints of the family/patient environment for providing privacy and security.

In our laboratory samples are initially signed in using a genotype form. This allows data to be entered for the sample and collected for quality control issues (volume, presence of hemolysis, etc.). At this time, the sample is assigned a sample number, which is used for labeling all subsequent analyses. The samples for extraction can be held at 4°C before freezing at −80°C. If desired, an aliquot of plasma can be taken prior to freezing. This freezing of whole blood provides a "holding buffer" that allows the sample to be stored until extraction can occur. We have successfully extracted whole-blood samples that were frozen for over one year. A flowchart of sample management is shown in Figure 5.1.

Several factors affect the success of DNA extraction. The volume of the sample and its effect on the chosen extraction method should be considered. After 3 days at room temperature, the amount of DNA that can be successfully extracted from an ACD sample begins to diminish, decreasing rapidly after 5 days (Gustafson et al., 1987). Therefore, it is important to place the sample at 4°C and extract the DNA as soon as possible. When energy available for the living cells begins to be depleted, the cells will die and the DNA will become degraded. If the sample cannot be extracted quickly, it should be stored at −80°C. In addition, light will degrade the DNA, and the samples should be kept in the dark (Kirby, 1990). We have also found that hemolysis is correlated with decreasing DNA yield. It is not clear whether the hemolysis itself is affecting the extraction. Rather, it seems more likely that the factors causing the red blood cell hemolysis have also affected the viability of the white blood cells. Clotting of tubes occurs occasionally, usually as a result of improper mixing. This can affect DNA yield as well, depending on the extraction method chosen.

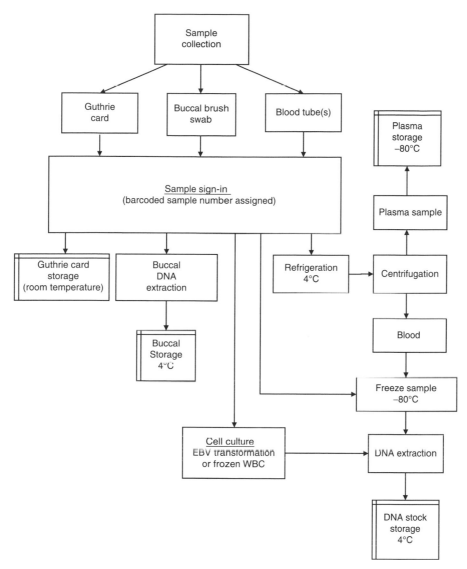

Figure 5.1. Flowchart of sample management.

Once the DNA is extracted, we store the samples in TE (10 mM Tris, 1.0 mM EDTA, sterile filtered) at 4°C with an added drop (20–40 μL per tube) of chloroform (Kirby, 1990). We have stored DNA for over 10 years in this manner without any obvious degradation. The chloroform sterilizes the sample and then evaporates with time due to its inherent volatility. If this technique is used, it is important to avoid vials with any component (e.g., a rubber gasket) that will dissolve in the presence of organic solvents, contaminating the DNA. Less than 2% of our 50,000+

extracted samples have displayed significant evaporation over time, and almost all that did evaporate could be reconstituted by adding sterile water. Choose storage tubes that are manufactured for cold storage and have "full-turn" twists (vs. one-quarter turn) to ensure an airtight seal.

It is debatable whether freezing DNA at $-80°C$ for long-term storage incurs any advantage. Freezing and thawing do cause degradation, especially of higher molecular weight fragments. However, this diminution of quality probably has little effect on today's PCR-based technology, which utilizes smaller fragments. In addition, depending on the number of samples, cost and freezer space can quickly become a problem in storage.

INFORMED CONSENT/SECURITY

DNA samples are not like other types of stored tissue, and their potential for abuse is a concern. A sample of DNA is intensely personal and should be treated as a confidential source of information, similar to a medical record. The investigator must assess the consent forms used for DNA storage and sampling to ensure that participants understand the purpose and limitations of that DNA bank. It is the investigator's responsibility to the participants in each study to ensure that appropriate security is maintained, both for the physical samples and for the data accompanying them. Detailed discussions of informed-consent issues in DNA banking have been the topic of several recent reviews (Yates et al., 1989; Clayton et al., 1995; see also Chapter 6).

REFERENCES

Ben Othmane K, Ben Hamida M, Pericak-Vance MA, Ben Hamida C, Blel S, Carter SC, Bowcock AM, Petruhkin K, Gilliam TC, Roses AD, Hentati F, Vance JM (1992): Linkage of Tunisian autosomal recessive Duchenne-like muscular dystrophy to the pericentromeric region of chromosome 13q. Nat Genet 2:315–317.

Carducci C, Ellul L, Antonozzi I, Pontecorvi A (1992): DNA elution and amplification by polymerase chain reaction from dried blood spots. BioTechniques 13:735–737.

Chomczynski P (1993): A reagent for the single-step simultaneous isolation of RNA, DNA and proteins from cell and tissue samples. BioTechniques 15:532–536.

Clayton EW, Steinberg KK, Khoury MJ, Thomson E, Andrews L, Kahn MJ, Kopelman LM, Weiss JO (1995): Informed consent for genetic research on stored tissue samples. JAMA 274:1786–1792.

Dean FB, et al (2002): Comprehensive human genome amplification using multiple displacement amplification Proc Natl Acad Sci USA 99:5261–5266.

Del Rio SA, Marino MA, Belgrader P (1996): Reusing the same bloodstained punch for sequential DNA amplifications and typing. BioTechniques 20:970–974.

De Souza AT, Hankins GR, Washington MK, Fine RL, Orton TC, Jirtle RL (1995): Frequent loss of heterozygosity on 6q at the mannose 6-phosphate/insulin-like growth factor II receptor locus in human hepatocellular tumors. Oncogene 10:1725–1729.

Doyle A (1990): Establishment of lymphoblastoid cell lines. In: Pollard JW, Walker JM, eds. Animal Cell Culture, Methods in Molecular Biology. Totowa, NJ: Humana.

Fishbein WN, Foellmer JW, Davis JI, Kirsch IR (1995): Detection of very-rare-copy DNA in 0.2 mL dried human blood blots. Biochem Mol Med 56:152–157.

Foley KP, Leonard MW, Engel JD (1993): Quantitation of RNA using the polymerase chain reaction. Trends Genet 9:380–385.

Greer CE, Lund JK, Manos MM (1991): PCR amplification from paraffin-embedded tissues: Recommendations on fixatives for long-term storage and prospective studies. PCR Methods Appl 1:46–50.

Gustafson S, Proper J, Bowie EJW, Sommer SS (1987): Parameters affecting the yield of DNA from human blood. Anal Chem 165:294–299.

Guthrie R, Susi A (1963): A simple phenylalanine method for detecting phenylketonuria in large populations of newborn infants. Pediatrics 32:338–343.

Johns MB, Paulus-Thomas JE (1989): Purification of human genomic DNA from whole blood using sodium perchlorate in place of phenol. Anal Biochem 180:276–278.

Kawasaki ES (1990): Sample preparation from blood, cells, and other fluids. In: Innis MA, Gelfand DH, Sninsky JJ, White TJ, eds. PCR Protocols: A Guide to Methods and Applications. San Diego, CA: Academic, pp 146–152.

Kirby LT (1990): DNA Fingerprinting: An Introduction. New York: Stockton, pp 51–74.

Kosel S, Graeber MB (1994): Use of neuropathological tissue for molecular genetic studies: Parameters affecting DNA extraction and polymerase chain reaction. Acta Neuropathol (Berlin) 88:19–25.

Louie LG, King M-C (1991): A novel approach to establishing permanent lymphoblastoid cell lines: Epstein–Barr virus transformation of cryopreserved lymphocytes. Am J Hum Genet 48:637–638.

Manchester KL (1995): Value of A260/A280 ratios for measurement of purity of nucleic acids. BioTechniques 19:208–209.

Marshall E (1996): Policy on DNA research troubles tissue bankers. Science 271:440.

Matsubara Y, Ikeda H, Endo H, Narisawa K (1992): Dried blood spot on filter paper as a source of mRNA. Nucleic Acids Res 20:1998.

McCabe ERB (1991): Utility of PCR for DNA analysis from dried blood spots on filter paper blotters. PCR Methods Appl 1:99–106.

Miller SA, Dykes DD, Polesky HF (1988): A simple salting out procedure for extracting DNA from human nucleated cells. Nucleic Acids Res 16:1215.

Noguchi S, McNally EM, Ben Othmane K, Hagiwara Y, Mizuno Y, Yoshida M, Yamamoto H, Carsten G, Bönnemann CG, Gussoni E, Denton PH, Kyriakides T, Middleton L, Hentati F, Ben Hamida M, Nonaka I, Vance JM, Kunkel LM, Ozawa E (1995): Mutations in the dystrophin-associated protein (g-sarcoglycan in chromosome 13 muscular dystrophy. Science 270:819–822.

Planelles D, Llopis F, Puig N, Montoro JA (1996): A new, fast and simple DNA extraction method for HLA and VNTR genotyping by PCR amplification. J Clin Lab Anal 10:125–128.

Pressman S, Rotter JI (1991): Epstein–Barr virus transformation of cryopreserved lymphocytes: Prolonged experience with technique. Am J Hum Genet 49:467.

Raskin S, Phillips JA 3rd, Krishnamani MR, Vnencak-Jones C, Parker RA, Rozov T, Cardieri JM, Marostica P, Abreu F, Giugliani R (1993): DNA analysis of cystic fibrosis in Brazil by direct PCR amplification from Guthrie cards. Am J Med Genet 46:665–669.

Richards B, Skoletsky J, Shuber AP, Balfour R, Stern RC, Dorkin HL, Parad RB, Witt D, Klinger KW (1994): Multiplex PCR amplification from the CFTR gene using DNA prepared from buccal brushes/swabs. Hum Mol Genet 2:159–163.

Smith CL, Klco SR, Cantor CR (1988): Pulsed-field gel electrophoresis and the technology of large DNA molecules. In: Davies K, ed. Genome Analysis: A Practical Approach. Oxford, England: IRL.

Smith LJR, Braylan RC, Nutkis JE, Edmundson KB, Downing JR, Wakeland EK (1987): Extraction of cellular DNA from human cells and tissues fixed in ethanol. Anal Biochem 160:135–138.

Telenius H, et al (1992): Degenerate oligonucleotide-primed PCR: General amplification of target DNA by a single degenerate primer. Genomics 13:718–725.

Yates JRW, Malcolm S, Read AP (1989): Guidelines for DNA banking. Report of the Clinical Genetics Society working party on DNA banking. J Med Genet 26:245–250.

Zhang L, et al (1992): Whole genome amplification from a single cell: Implication for genetic analysis. Proc Natl Acad Sci USA 89:5847–5851.

CHAPTER 6

Methods of Genotyping

JEFFERY M. VANCE

One of the critical components of any disease gene mapping study is obtaining the genotypes for the genetic markers used to test for linkage or association. This chapter provides the reader with practical knowledge of different methods and techniques used in genotyping, including options, advantages, and disadvantages.

BRIEF HISTORICAL REVIEW OF MARKERS USED FOR GENOTYPING

Restriction Fragment Length Polymorphisms

Restriction fragment length polymorphisms (RFLPs) were introduced in 1978 (Kan and Dozy, 1978; Botstein et al., 1980) and in 1982 became the first modern genotyping markers to be used in a successful linkage (Huntington disease; Gusella et al., 1983). They are a subset of the single-nucleotide polymorphisms (SNPs) described below. The RFLPs are based on a single-base-pair change that creates or obliterates a cleavage site for a specific DNA restriction enzyme. The resulting variation between individuals can be detected by using that restriction enzyme to digest the DNA. The DNA fragments are electrophoresed in an agarose gel and subsequently transferred to a nylon membrane (Southern blot; Southern, 1975). A labeled (usually ^{32}P) DNA probe that overlaps the restriction fragments of interest is hybridized to the membrane. The resulting variability in fragment size (alleles) is then detected using X-ray film (Vance et al., 1989). Major drawbacks of this technique include the large amount of DNA that is needed (2–10 µg, often requiring the establishment of cell lines for the studied individuals), the low heterozygosity (usually <0.4) of the markers, and the large amount of labor required (Botstein et al., 1980). In addition, since the technique depends on the availability of the hybridized probe, this limits the ability to share or obtain the resources to actually perform the genotyping.

Genetic Analysis of Complex Diseases, Second Edition, Edited by Jonathan L. Haines and Margaret Pericak-Vance

Thus RFLPs traditionally have a sluggish genotyping throughput (often requiring 1–2 weeks before results are obtained). However, with the continuing increase in the availability of public genomic sequence, polymerase chain reaction (PCR; Mullis et al., 1986) can now provide a suitable probe for hybridization. In addition, if the sequence surrounding the RFLP site is known, a more efficient method is to perform PCR across the RFLP site, then digest the PCR product and run these fragments on an agarose or polyacrylamide gel for typing.

Variable Number of Tandem Repeat Markers

The low heterozygosity characteristic of RFLPs was greatly improved by the identification of the variable number of tandem repeats (VNTRs) in 1985 (Nakamura et al., 1987). This new class of probes, also known as minisatellites, is made up of specific sets of consensus repetitive sequences that vary between 14 and 100 bp in length. Although some can be converted for use with PCR, most require probe hybridization like RFLPs. The VNTRs are remarkably polymorphic, with a high heterozygosity rate in the population, and have been commonly used for paternity testing. However, this high heterozygosity rate, coupled with their relatively large size, can make comparisons in allele sizes difficult.

Short Tandem Repeats or Microsatellites

The use of PCR for the analysis of microsatellite polymorphisms has dramatically improved the speed and efficiency of genotyping. Initially described by Weber and May (1989) and Litt and Luty (1989), short tandem repeats (STRs) are widely and evenly distributed in the genome and are relatively easy to score. While the number of repeated motifs in microsatellites varies, the most useful microsatellites consist of a repeated sequence motif of two (dinucleotide), three (trinucleotide), or four (tetranucleotide), bases (for example CA, ATG, GATA). To detect the variability that exists between individuals at these STRs, two unique sequences (primers) are determined on each side of the STR. These short sequences (approximately 20 bp) are then used to synthesize DNA primers, which are subsequently employed in a PCR to amplify the DNA that lies between them (including the repeat). The variable number of repeats in the STR produces PCR fragments of different sizes, which can be easily detected by means of denaturing sequencing gels. This new approach revolutionized genotyping, as genotypic data are produced not in weeks but in hours. An additional advantage is that STRs require only very small (nanogram) amounts of DNA and can easily provide heterozygosities greater than 0.70. Perhaps most importantly, the technique eliminates the need for dependence on physical resources such as probes. Now markers can be obtained electronically, speeding up analysis.

Single-Nucleotide Polymorphisms

Single-nucleotide polymorphisms, which reflect a polymorphism at a single-base-pair location, occur very frequently in the human genome, approximately 1 every

1000 bases. As previously mentioned, when these changes affect a restriction enzyme cutting site, the SNP is known as an RFLP. The high frequency of SNPs makes them very useful for association studies, where the power of association and linkage disequilibrium is strongest over small distances. One of their drawbacks is their low heterozygosity (the same as RFLPs) relative to microsatellites. However, in some cases they can be clustered into haplotypes, which can significantly increase their information content. The biggest practical problem with SNPs is that detection of alleles can be difficult and relatively costly. Unlike microsatellites, detection often requires expensive fluorescent tags. In addition, in the majority of techniques a two-step process is needed: PCR to provide the DNA template and then one of several techniques to actually genotype the SNP allele. This makes SNP detection more costly than microsatellites, particularly as the lower heterozygosity requires more SNPs to be genotyped for a given project.

However, along with the very large number of SNPs throughout the genome, there are other advantages to the use of SNPs. The presence of only two alleles, which gives the low heterozygosity, also provides the potential for automation, which is difficult in a multiallele polymorphism. Multiple platforms are now available that are highly automated and require miniscule (1–2 ng) amounts of DNA per SNP. Thus speed may make up for cost in the use of SNPs, especially as technology progresses.

SOURCES OF MARKERS

Restriction Fragment Length Polymorphisms

It can be difficult to identify RFLPs through databases, with the Genome Database probably the most useful. Certainly various DNA analysis programs can identify restriction sites on known sequence. Probes can often be obtained from the American Type Culture Collection (ATCC).

Microsatellites

Information on several thousand microsatellite markers is electronically accessible to investigators through databases such as the National Center for Biotechnology Information's (NCBI) Locuslink, the Genome Database, the Whitehead Institute, the Marshfield Institute, Applied Biosystems, Généthon, and DeCODE (Table 6.1). Many genetic maps now exist that enable the investigator to use these highly polymorphic markers for linkage analysis. However, it is important to remember that all genetic maps are not constructed alike or with equal readability.

Dinucleotide Repeats. The dinucleotide CA (GT) is the most common repeat, with a highly polymorphic form of the repeat initially estimated to occur on average one every 0.4 cM (Weber, 1990). However, in our experience they are much more frequent. Their polymorphic status usually depends on two factors: the size of the

TABLE 6.1. Sources of Information on Microsatellite Markers

Name	URL
National Center for Biotehcnology Information's Locuslink	http://www.locuslink.com
GDB Human Genome Database	http://gdbwww.gdb.org/
Whitehead Institute for Biomedical Research	http://www.wi.mit.edu/
Marshfield Clinic's Center for Medical Genetics	http://research.marshfieldclinic.org/ genetics/
Généthon	http://www.genethon.fr/php/index.php
DeCODE Genetics	http://www.decode.com/
Ensembl Genome Browser	http://www.ensembl.org/
University of California at Santa Cruz Genome Browser	http://genome.ucsc.edu/
National Center for Biotechnology Information	http://www.ncbi.nlm.nih.gov/
Primer 3	http://frodo.wi.mit.edu/cgi-bin/ primer3/primer3_www.cgi

repeat and whether a perfect (no interruptions) or imperfect repeat is present. The majority of markers with 15 or more perfect repeats are polymorphic (Weber, 1990). While powerful for linkage analysis, dinucleotide repeats can have several technical drawbacks. Since the alleles are only 2 bp apart, it can be difficult to distinguish one allele from another on a gel. More importantly, many dinucleotide repeats have a "stutter" that produces a background ladder of bands 2 bp apart. This feature can make scoring difficult, particularly in differentiating homozygotes from heterozygotes.

Trinucleotide Repeats. Although certain trinucleotide repeats that can cause disease (LaSpada et al., 1994) have received much attention, the majority of these markers are quite stable, have heterozygosities similar to dinucleotide repeats, and have been incorporated into recent maps. They also generally have fewer stutter bands than dinucleotide repeats. Fewer trinucleotide than dinucleotide repeats have been described, however.

Tetranucleotide Repeats. Although less numerous than dinucleotide repeats, tetranucleotide repeats are quite polymorphic (Edwards et al., 1991). These larger repeat motifs typically generate unique PCR bands without a "laddering" artifact, greatly facilitating the identification of alleles. The clarity of the results makes the application of computer algorithms to determine molecular weights much more feasible. Accurate allele identification is particularly important in the study of complex disorders, where the lack of significant family structure decreases the likelihood of detecting genotyping errors by inconsistencies of segregation. Thus, tetranucleotide repeats are the markers of choice in these studies.

Single-Nucleotide Polymorphisms

These markers are now quite numerous, thanks to the efforts of the SNP consortium, the hapmap project, and the human genome sequencing effort. Currently, according to the NCBI, more than 6 million unique SNPs are known in the genome. Maps of SNPs can be best visualized using the public software Ensembl, the "golden path" at the University of California at Santa Cruz (UCSC), and NCBI websites (Table 6.1).

PCR AND GENOTYPING

The optimization of the PCR reaction is one of the most critical factors in efficient genotyping. Reduction of background bands and maximum production of product provides clarity in identifying alleles. Therefore, investigators need to optimize laboratory conditions for genotyping so that they will provide a flexible environment for as many different PCR primers as possible.

Laboratory and Methodology Optimization

In PCR the amplification of the desired target continually competes with non-specific priming. Therefore, it is useful to screen the sequence of a potential primer for redundancy before synthesizing it. In addition, the primer dimerization reaction, in which one of the primers is substituted for the target DNA and extended, can be a problem. Following basic rules of avoiding complementary end sequences can avoid this problem. To help eliminate nonspecific binding problems, the PCR cycle times should be reduced as far as possible without significantly reducing the amount of product. In addition, extended incubations at high temperatures should be avoided, since the half-life ($t_{1/2}$) of Taq polymerase rapidly decreases above 93°C. Therefore, shortening the denaturing time increases the amount of available Taq for subsequent cycles and reducing the annealing and extension times increases specificity and decreases background bands. Beyond 35 cycles, the PCR is usually nearing the plateau phase, and additional cycles tend to increase the background by preferentially amplifying spurious products.

Standard Taq polymerase is minimally active at room temperature. It is therefore recommended to start the thermal cycling immediately after the addition of reagents to the reaction to minimize spurious products. Alternatively, variations of the hot-start method (Chou et al., 1992) can be employed. One form has the individual withhold one of the essential reagents from the reaction mixture until the temperature of the reaction exceeds the annealing temperature of the primers. The missing reagent can be added individually to each reaction in the thermocycler block once the desired temperature has been reached. However, this approach is not practical in dealing with a large number of samples and is prone to cross-contamination. An early solution to this problem consisted in using a solid wax barrier to separate the retained reagent from the bulk of the reaction (hot start). The first cycling step melts the wax, allowing the Taq polymerase to mix with the templates (Bassam and Caetano-Anollés, 1993). This was followed by the engineering of a form of Taq (Taq gold, Perkins-Elmer), which does not have significant activity at room temperature. Subsequently, a form of Taq coated with an antibody (Platinum Taq, Life Technologies) became available. With this product, denaturing of the reaction releases the antibody, allowing the enzyme to become active only once the cycling reaction has begun. TaqStart (Clonetech) is a Taq polymerase antibody that can be bought separately. Use of these products reduces nonspecific activity and thus increases specificity without any additional reagents.

Perhaps the greatest concern in optimizing a laboratory is controlling PCR contamination, which can significantly reduce the consistency and reliability of genotyping reactions. While false positives can be contained by the use of UV irradiation (Sarkar and Sommer, 1991; Corless et al., 2000), uracil DNA glycosylase (Hengen, 1997), and physical barriers (Neiderhauser et al., 1994), false negatives and inhibition of the PCR reaction by previously amplified material are more difficult problems for genotyping. Neiderhauser et al. (1994) demonstrated that intact and partially degraded amplified products (amplicons) and primer artifacts from previous reactions are able to inhibit subsequent reactions, probably by primer competition and DNA blocking. In fact, microsatellite reactions generate more false-negative problems than other types of PCR contamination. Although the PCR primers used in genotyping may be different, the repeat motifs are usually highly similar, allowing the core repeat to block the template and inhibit the reaction. If your laboratory has experienced a cyclic failure of genotyping which affects several individuals, suspect contamination.

Several basic precautions may help minimize the contamination problem. These include pipeting reagents into small aliquots for single use, frequently changing gloves, and avoiding splashes (Kwok and Higuchi, 1989). However, the single most important guideline in preventing contamination is complete physical separation between the location where the reagents are stored and set up and the location where the thermocycling is done and the post-PCR product is handled and stored. Either separate rooms should be employed for pre- and postthermocycler reactions (best) or, minimally, reagent storage and setup performed under a biosafety hood equipped with a UV germicidal lamp. If UV decontamination is used, it is important not to expose the mineral oil to the UV irradiation, since it has been postulated that this will cause the PCR to be inhibited by induced radicals interfering with Taq polymerase (Gilgen et al., 1995).

Optimization of Reagents

In general, one strives to obtain the most economical and flexible set of conditions in which most primers will perform adequately. We perform genotyping PCRs on 30 ng of DNA in a final volume of 10 μL (Ben Othmane et al., 1992). A standard PCR buffer is composed of 20 mM Tris–HCl and 50 mM KCl. A combined deoxynucleotide triphosphate (dNTP) final concentration of 0.6 mM and an Mg^{2+} concentration of 3 mM are usually optimal for a wide range of primers for multiplexing. A surplus of Mg^{2+} reduces the PCR specificity and increases the laddering artifacts, while a shortage of magnesium reduces the reaction efficiency. Since there is a mutual titration between Mg^{2+} and dNTP, it is important always to remember that a significant change in the Mg^{2+} concentration usually requires a similar modification in the amount of dNTP. A final concentration of 0.5 mM of each primer is optimal for most markers. Excess primers will result in the generation of artifacts.

If the amount of PCR product is a problem, increasing the dNTP concentration and adjusting the Mg^{2+} concentration accordingly may obtain an increase. Optimization of the PCR conditions for that primer pair should also increase specificity of the reaction and therefore the amount of desired product as well. Tetramethylammonium is a reagent that has been postulated to increase the amount of specific PCR product (Chevet et al., 1995). Since, however, its effect is to increase the thermal stability of AT base pairs, its utility varies from primer to primer and it should be tested using a battery of concentrations, similar to dimethyl sulfoxide (DMSO; see Other Tools, below).

"I Can't Read a Marker, What Should I Do?"

Despite the best general conditions, some primers provide poor results. What to do to improve them and how hard to try are common questions. Optimizing the PCR conditions for a microsatellite can be a time-consuming task and therefore should be attempted only when genotypes from the marker in question are essential and irreplaceable. There are many valid approaches; some suggestions are offered in the sections that follow.

Get Another Marker. It is widely recognized that the most crucial factor in determining the success of a PCR is the design of primer sequences. Poorly designed primers lead to nonspecific PCR with background, ghost, and stuttering bands. In our laboratory redesigning a primer is one of the most useful optimization steps, and as time goes along, we tend to redesign a primer sooner rather than later, especially as the cost of oligomer synthesis continues to decrease. If one is simply performing a genomic screen, obtaining a better marker in the same chromosomal region or redesigning an existing primer is usually the best way to proceed. Again, now that sequence is much more publically available, this is relatively straightforward. Several programs, such as Primer 3 (see Table 6.1), are available to aid in primer selection. However, efficient primers can often be redesigned if simple criteria are followed. The primers should be as close to the repeat as possible: The smaller the product, the easier it is to read in standard denaturing gels. An optimal primer length is about 20 bp, and while longer primers may improve the PCR stability, they are rarely necessary. The G/C content should be kept near 50%, and the base distribution should be random to minimize secondary structure and self-complementation. The last 3–7 bp in the 3' end of the primer are crucial to the success of the annealing and extension steps of the PCR. Complementarity of the primers to each other, particularly in the 3' end, needs to be avoided, since it may induce primer dimerization (Saiki, 1990). Again, any primer sequence should be screened through BLAST to identify the degree of potential nonspecificity. However, if it is necessary to use the original marker, several options are open.

Optimize Temperature, Mg^{2+}, and pH. Temperature and pH generally are the most useful factors to initially optimize for an individual set of primers, although other factors can be occasionally important as well (Innis and Gelfand, 1990). In our experience, Mg^{2+} concentration, although potentially important, is less useful in optimizing a specific primer set. To optimize temperature, touchdown or step-down PCR techniques are particularly useful (Hecker and Roux, 1996). These techniques begin the PCR at a very high annealing temperature, greatly reducing the amount of product made initially, but making it highly specific. As the PCR continues to cycle, the temperature is slowly allowed to ramp down with each subsequent cycle, eventually reaching normal cycling temperatures. This approach provides the correct template in the form of a short product, early in the reaction. As short products quickly dominate the PCR, nonspecific products generated from the original template are greatly reduced.

It is tempting to use the touchdown method for all coamplified multiplexing. However, the touchdown method can easily emphasize one primer set over another. Thus, while it may be valuable in some PCR multiplex combinations, it is not wise to uniformly apply the touchdown method to all multiplex situations.

Optimizing pH and Mg^{2+} is probably best done by using one of several commercially available kits. A typical optimization scheme may include sequential testing of different annealing temperatures, usually between 50 and 65°C, Mg^{2+} concentrations between 1 and 3.5 mM, and pH conditions between 8.5 and 10. Optimally, one would wish to test these factors simultaneously. In practice, however, one condition usually stands out as the key factor affecting a primer set's performance.

Other Tools. Dimethyl sulfoxide has been used to enhance PCR for some primer pairs, increasing the amount of product and primer specificity. The actual mechanism of this improvement is unknown (Filichkin and Gelvin, 1992; Baskaran et al., 1996). The primer set can be run against a battery of 1, 2.5, 5.0, 7.5, and 10% DMSO to see whether an improvement of product specificity can be obtained. Dimethyl sulfoxide reduces the activity of Taq, so concentrations over 10% are generally not useful.

Betaine (*N,N,N*-trimethylglycine) has been suggested by many reports to help multiplex PCR. Studies have suggested that high GC content and very high melting domains within a PCR target can cause Taq polymerase to pause and have poor efficiency in amplification (Henke et al., 1997; McDowell et al., 1998; Shammas et al., 2001). Several authors have shown that addition of betaine may help overcome these amplification problems (Henke et al., 1997; McDowell et al., 1998). Betaine is believed to destabilize GC base pairing, effectively targeting these regions (McDowell et al., 1998).

MARKER SEPARATION

Manual or Nonsequencer Genotyping

While some authors have used nondenaturing conditions to visualize polymorphisms (Buzas and Varga, 1995), on the whole most researchers select a denaturing (sequencing) setup for its greater sensitivity (Ben Othmane et al., 1992). Various points to be considered in marker separation are presented next.

Apparatus. The selection of the type of apparatus is important because uniform heat distribution is as critical to genotyping as it is to sequencing. Units also can differ in terms of width, which affects both number of samples and detection schemes. We have found units from C.B.S. Corp. (San Diego, CA) work well, but most units on the market are suitable.

Acrylamide. It is very important to have a high-quality matrix for clean, sharp bands. We use freshly made acrylamide solutions because, in our experience, some premade solutions do not always provide the desired consistency. Acrylamide substitutes and variations are also available but are usually more costly. If wrapped to prevent evaporation, gels can be made up several days ahead of time and stored at 4°C.

One of the most common problems in pouring gels is keeping the plates clean. The routine use of cerium oxide (Millard and de Couet, 1995) is highly recommended. Also recommended are tools to help pour gels, like the *Gel Slider* from C.B.S.

Gel Formation. A key factor in this regard is maintaining a standard thickness of the gel during polymerization. Our laboratory uses clamps (Hoefer, SE 6003) on three sides instead of tape, since a millimeter of difference in gel width is enough to prevent straight lanes.

Fixing. The fixing often used in manual sequencing is not necessary for genotyping gels.

Loading. We prefer to use slot gel combs for reading microsatellites. Shark-tooth combs can also be used, depending on the system, although we find shark-tooth combs too difficult to read when large volumes of data are at hand. Currently we use 110 wells per gel. We also apply a bind silane solution [1 mL 100% EtOH, 5 μL Bind Silane (Sigma M-6514), 5 μL 10% acetic acid] in a 1-cm band across the top of the gel. The bind silane interacts with the glass plate and acrylamide so that after polymerization the gel remains tightly bound to the glass plate, enabling the formation of wells around the gel comb.

There are many variations in loading PCR samples on a gel. While loading can be accomplished with a single pipet, it is much faster with a multichannel pipet such as that available from the Hamilton Company.

Loading Variants

The sections that follow describe some of the variations to standard loading of samples.

Multiloading. This is the simple act of loading different aliquots of the same marker on the same gel in multiple loads, separated by 15–30 min, providing many layers of the same marker for analysis on one gel. Multiloading works well for markers with few aberrant bands and can greatly increase the number of genotypes generated for one marker.

Multiplexing. This approach has several variants. In one, many different primer sets (2–50) are amplified individually by PCR, then mixed, concentrated, and run in a single lane of a gel. These overlapping fragments are then Southern blotted to a nylon membrane. This is subsequently hybridized with a single labeled primer of each set (Litt et al., 1991), which provides the specificity for the detection.

The second technique is similar to the first, except the primers are fluorescently labeled. Multiple primers are then loaded in a single lane, and detection depends on the different emissions of the fluorescent moiety (Ziegle et al., 1992). This approach requires a laser for detection (Fig. 6.1).

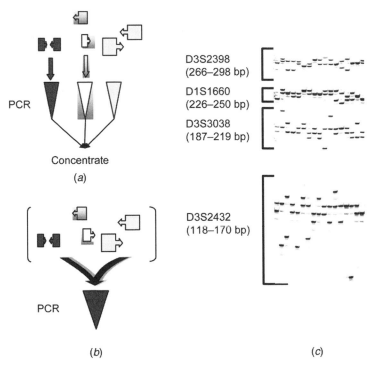

Figure 6.1.

The third technique is employed regularly in our laboratory. Instead of separate PCRs and subsequent pooling of the PCR products, a single PCR amplification is performed with two to five primer sets amplified together in one tube (Fig. 6.1). Primers to amplify together are chosen based on the molecular weight of their products and their PCR compatibility. They are then loaded into a single lane on the gel and separated according to molecular weight. This approach can be combined with fluorescence to provide multiple multiplex sets that can be run in the same lane, a combination of the second and third techniques is given here.

This single-tube multiplex technique can use the identical conditions employed for single-primer PCR, especially when radioactivity is used for detection. The high sensitivity of this isotope allows satisfactory results for less than optimum PCR conditions for many of the primers. However, other detection schemes (e.g., silver stain or fluorescence) may be less forgiving. In these situations, multiplexes may work more optimally when two to three times the dNTP concentration is used for single-primer-pair reactions and a 2–3 mM concentration of Mg^{2+} is employed.

In setting up a single-tube multiplex set, the primers chosen should be run on a small number of test samples as a single reaction as well as in combinations. This allows one to identify interferences due to spurious bands that may fall within the same molecular weight as the alleles from other primers in the set. It also demonstrates which primer sets produce weak or exceptionally strong products when multiplexed. Typically one marker in the multiplex set will have a significantly higher PCR proficiency, relative to the rest of the markers. This will cause it to dominate the reaction early in the amplification. Thus, the amount of this primer set added to the multiplex mixture may need to be adjusted accordingly.

The advantages of single-tube multiplexing are the increase in genotyping output and the reduction in cost that can be accomplished without high-priced equipment, as three to five markers can be analyzed at once. It does, however, require advance work and therefore is really most appropriate for large laboratories with extensive genotyping. Jim Weber has made available his genomic multiplex screening sets (see Marshfield website, Table 6.1), which also use this single-tube multiplex technique.

DNA Pooling and Homozygosity Mapping

The DNA samples from all affecteds may be pooled together and processed as a unique template in a single PCR, a practice that greatly reduces the number of working samples and significantly accelerates the genotyping process (Sheffield et al., 1994). An equal number of unaffected members are pooled and used as a control. This efficient strategy is particularly recommended as an initial screening tool for autosomal recessive conditions especially with inbred families. It has been applied to dominant conditions but is not nearly as useful (Damji et al., 1996). When this is done within a single family or ancestral mutation, the individuals can be expected to share the identical homozygous genotype near the disease gene. This is referred to as *homozygosity mapping*. Here one concentrates on all markers that are homozygous in all affecteds, relative to nonaffected relatives. If one takes different families, with potentially different mutations, then a shift in allele frequencies toward a single homozygous allele in the affected DNA pool, relative to controls, may indicate linkage (the intensity of each allele band will be directly proportional to its relative frequency in the pooled population). This is termed *DNA pooling*. It has been estimated to reduce by as much as 75% the number of markers that will require subsequent pedigree linkage analysis (J. Haines, personal communication).

DETECTION METHODS

The methods to detect microsatellites are numerous. Factors to consider in choosing a method include the number of samples that will be analyzed, cost, and ability to use radioactivity.

Radioactive Methods (^{32}P or ^{33}P)

The use of radioactivity provides the highest sensitivity of any of the detection methods. However, availability of ^{32}P can be a problem in some locales. The safety requirements and waste management protocols that necessarily accompany the use of this method can be additional drawbacks. Two primary methods exist. In one, radioactively labeled dNTP is incorporated into the PCR (Ben Othmane et al., 1992); in the second, hybridization with a labeled primer occurs after the cold microsatellite product has been separated by electrophoresis and Southern blotted (Southern, 1975). To improve the sharpness of the bands, Zagursky et al. (1992) used ^{33}P instead of ^{32}P; the reduced emission energy that results has safety merits as well. However, the high cost of the larger isotope reduces its practicality for most investigators, particularly those in the United States.

The incorporation method is fast, requires no special equipment, and when coupled with a multiplex method has a higher output capability than other radioactive methods. It also requires minimal manipulation. It has been one of the most common and popular methods. The hybridization method is less straightforward, since its sensitivity requires the additional step of efficient hybridization of the labeled primers. Many primers can be incorporated into a single lane, however, meaning that fewer gels are required, and since only the primer is labeled, this method in general produces less background than the incorporation method.

Silver Stain

Silver staining is a method of detection well known to protein chemists. The application of silver staining of nucleic acids in polyacrylamide gels was introduced in 1981 (Merril et al., 1981; Bassam et al., 1991; Von Deimling et al., 1993; Hudson et al., 1995). It is available commercially for DNA, and the process is relatively straightforward. After marker separation, the gel is subjected to several staining steps, with the eventual DNA–silver product developed chemically.

Silver staining is a sensitive method that is a low-cost alternative to radioactivity. It is relatively fast (gels can be developed in approximately 1 h). However, its reactions are time dependent and it requires more hands-on technician time than radioactivity. Gel plates may have to be committed to the silver stain method, since a gel adhesive is applied to the plates to stabilize the gel through the washing steps. In addition, the gels are "developed" at a specific point and fixed. Therefore, light or dark bands can be a problem if wide variation appears in the gel. Gels can be stored, but permanent images are readily made using X-ray duplicating film or photography. Silver staining is an excellent alternative to radioactivity-based methods.

Fluorescence

Fluorescence is the fastest genotyping detection technique, is the most amenable to automation, and has the highest throughput. Two basic formats are available: *static* and *real-time* scanning. Real-time scanning has two additional formats, plate or capillary electrophoresis. All call for intensive capital investment because lasers are required. Usually a DNA sequencer is applied to this approach.

Fluorescent Allele Static Scanning Technique. The application of fluorescent allele static scanning (FASST) technology to genotyping was developed in our laboratory (Vance et al., 1996). The key concept in this approach is to split the separation of alleles from their detection. This eliminates the need for a DNA sequencer to be used for detection of the products. Using coamplified multiplex microsatellite sets, multiple gels can be run at once, in different locations, and then scanned separately on a fluorescent laser scanner (Hitachi FMBIO II), with a scanning time of only a few minutes per gel. Analysis of the data can take place at one of many networked stations, concurrent with scanning. Primers can be prelabeled with a fluorescent moiety, or the PCR product can be stained following electrophoresis for a few minutes using Sybrgold (Molecular Probes) and then scanned, allowing unlabeled primers to be used. The image is downloaded to a networked computer system for image analysis and allele determination (BioImage). This technique has produced 1.25 million genotypes this year in our laboratory, supported by only two FMBIO II laser units (Fig. 6.2).

Real-Time Scanning. In this approach the electrophoresis is done directly on the detector, which is a DNA sequencer, and the data are then analyzed by incorporated software. The equipment is usually provided as a single system. All primers must be prelabeled. Several commercial packages are available.

Advantages and Disadvantages. Fluorescence is the fastest microsatellite detection method available. In addition, because it can be automated, it has the highest potential throughput. The main disadvantage is the large capital outlay required, which makes it prohibitively expensive for smaller laboratories.

There are distinct differences between the formats available. Real-time scanning has the fastest output for a small number of samples, but there are clear disadvantages in some cases. Because electrophoresis must be performed on the machine and only

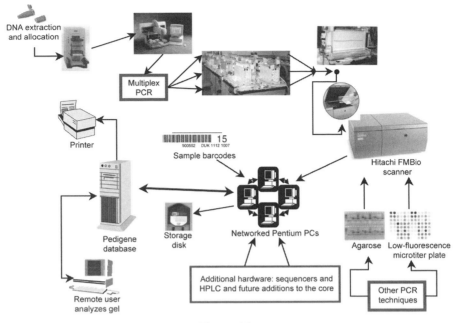

Figure 6.2.

one project can be electrophoresed at once, the machine itself becomes a bottleneck for laboratories that run many projects or if the project is very large. Thus one of the advantages of FASST is that because detection is uncoupled from electrophoresis, a single machine can detect for a large number of projects. Also, much genotyping involves "re-dos" and inconsistency correcting, which can be done easily on static equipment without the interruptions in flow of traffic that occur in real-time scanning. In addition, real-time scanners require prelabeled primers, so previously synthesized primers are not useful. This is not a limitation of static scanners, since a relatively inexpensive fluorescent dye such as SybrGreen or Sybrgold can be used following electrophoresis. Currently some real-time scanners (Applied Biosystems) allow more isofluors per lane than static scanners and hence can analyze more markers per lane. This greater capacity allows a standard to be placed in each lane as well. In our experience, however, in-lane standards are not needed for accurate molecular weight determination, especially with tetranucleotides.

Other Scanning Technologies. Currently, 96 well capillary electrophoresis units are available to perform high-throughput genotyping. They have the advantage of higher automation potential because it is not necessary to hand-load the gels. However the high cost of these machines is prohibitive for most laboratories.

SNP DETECTION

No dominant technology has yet to emerge for SNP detection. More than for any other polymorphic marker, the choice of technique relies heavily on the laboratory's workflow and capital resources. The section below will address some of the major approaches used for SNP genotyping, which are numerous.

DNA Array or "Chip"

DNA chips were one of the initial methodologies proposed for SNP detection. These high-density microarrays are created when oligonucleotides are attached to a solid silicon surface in a known, ordered array. Labeled dNTPs are incorporated when performing PCR of the SNP in the individual being genotyped. This product is then hybridized to the array, the SNP PCR product that perfectly matches the correct allele hybridizing more efficiently than mismatches (Shi, 2001). The advantage of the method is the large number of different SNPs that can be assayed at once. However, for research laboratories, the cost of both constructing the chips and obtaining a "reader" can be substantial. The technique by itself is not as flexible as some others. It is probably best suited to highly repetitive assays, such as would be used in genomic screening or carrier detection, where the cost per reaction of the process and the labor required for optimizing oligonucleotides can be minimized.

Different genotyping platforms fulfill different genotyping needs. On the high-throughput scale, the Illumina BeadArray has proven to be one of the techniques of choice. Essentially a pool of up to 1536 fluorescently labeled, SNP-specific probes are hybridized to target DNA, elongated, and then amplified. These labeled, amplified products are then hybridized to complementary single-stranded DNA sequences that have been immobilized onto the surface of beads, which sit within preetched holes at the ends of an optical fibre. Laser scanning through the fiber then detects the fluorescent signal from the SNP assay which is annealed with its complementary sequence on the fibre. The technique has proven to be very accurate and the software is very easy to use relative to most techniques. While the per-genotype cost is very competitive, the current high level of multiplexing means that the total cost of a single experiment will be very high. Thus this technique is best suited for core facilities or very large studies.

Advances in SNP genotyping technology are allowing ever increasing numbers of SNPs to be genotyped in single experiments. The two most common platforms are made by Affymetrix (http://www.affymetrix.com) and Illumina (http://www.illumina.com). Using the results of the Hapmap project (Altshuler et al., 2005), arrays have been developed with 10,000, 100,000, 250,000, and 500,000 SNPs. Larger arrays will undoubtedly be developed in the future. The goal of these large arrays is to interrogate a substantial portion (estimated at 50–70% for the 250,000 and 500,000 SNP arrays) of the common variation in the genome (e.g. Whole Genome Association). Although they use different proprietary technologies, both systems generate SNP genotypes with very high fidelity.

Oligonucleotide Ligation Assay

Oligonucleotide ligation assay (OLA) utilizes the fact that DNA ligase requires a perfect base-pair match to exist before it will ligate two oligonucleotides immediately adjacent to each other on a PCR template. One member of this

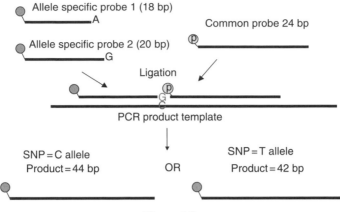

Figure 6.3.

"ligation" pair is composed of one of two allele-specific probes. These probes are designed so that the oligonucleotide ends on the SNP, with one of the two corresponding SNP's complementary base pairs. When the allele probe ending in the correct matching base pair for the SNP hybridizes to the template, it will be ligated to the "anchor" or "detection" allele. This process is shown in Figure 6.3. The ligation can be detected using different approaches. The standard technique has been as a plate-based method (Samiotaki et al., 1996; Grossman et al., 1994). However, Eggerding et al. (1995) have published a gel-based method for genotyping. It has also been adapted to capillary electrophoresis systems (e.g. SNPlex; ABI Foster City, CA).

The OLA technique is sensitive and accurate and can have a reasonably high output. Optimization is usually minimal, although it is good practice to sequence initial genotypes to assure the correct genotypes are being identified. The advantage of OLA is that it is one of the few methods having the potential of multiplex PCR without minimal optimization. It is very cost effective for large datasets but can be expensive for smaller studies.

Fluorescent Polarization

In fluorescent polarization (FP) polarized light becomes refracted when a large moiety, such as a fluorescent tag, is rotating freely in solution. If the tag is taken out of solution and stabilized, the polarized light is not affected. By itself FP is not a SNP detection method, but rather is coupled to one of many techniques (Latif et al., 2001) to detect allele differences (Fig. 6.4).

Taqman

Taqman identifies the correct SNP allele during PCR amplification. It is based on the exonuclease properties of Taq polymerase. A labeled allele-specific primer is hybridized to the short template of the PCR amplification as the PCR amplification

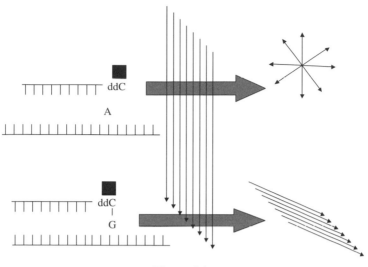

Figure 6.4.

proceeds. This label is prevented from excitation by the presence of a "quencher" on the other side of the allele-specific primer. The quencher represses the detection dye through a process known as *fluorescence resonance energy transfer* (FRET; Chen et al., 1997; Stryer and Haugland, 1967). This is a distance-dependent interaction between the donor molecule and the acceptor or "quencher." When the donor dye is excited, it transfers its excitation to the quencher without emission of a photon.

During amplification, the correct allele-specific primer will hybridize to the PCR template in front of the amplifying Taq polymerase. When the enzyme reaches the hybridized primer, it is "chewed up" as the PCR amplification occurs (like its namesake, the video game Pac-man). This removes the quencher from the fluorescent label, allowing the donor now to emit fluorescence and be detected when excited by a laser. This process is repeated each PCR cycle (Fig. 6.5).

Advantages of this method are its potential speed and simplicity. Once optimized, it has proven to be a robust technique. It can have a very high throughput, and commercially available machines such as the ABI 7900 can perform up to 30,000 genotypes per day. Its main disadvantages are cost (two dyes) and the need to optimize the allele-specific primers. However, it has recently been coupled with FP, which is reported to reduce the need for the quencher (Latif et al., 2001). Currently it cannot be multiplexed, although it has been pooled successfully (Breen et al., 2000).

Single-Base-Pair Extension

Sequencing is an obvious method for genotyping SNPs. However, the only sequence that is needed for genotyping is the one base pair that defines the polymorphism. In single-base-pair extension (SBE) a primer is constructed that is adjacent to the

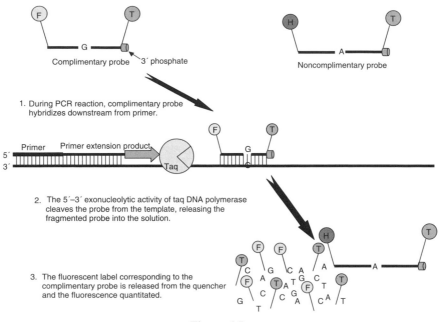

Figure 6.5.

polymorphic SNP base so that the allele base lies immediately 3′ to the primer (Syvanen et al., 1993). A single- or double-base sequencing reaction is then performed, with only dideoxynucleotides that match the SNP base-pair possibilities. This can then be detected directly on a sequencer or, more commonly, by a second method such as polarization.

Pyrosequencing

This is a sequencing reaction in which the incorporation of allele-specific nucleotides is detected by measuring the release of pyrophosphate during the incorporation process. The pyrophosphate is converted to adenosine triphosphate (ATP), which stimulates luciferase, emitting light which is then captured. The ATP and the unincorporated dNPTs are degraded by apyrase, and the next NTP is released. Specific instrumentation for the process is sold by Pyrosequencing AB (Uppsala, Sweden).

Matrix-Assisted Laser Desorption/Ionization Time-of-Flight Spectrometry

The technique matrix-assisted laser desorption/ionization time-of-flight spectrometry (MALDI-TOF) utilizes the high sensitivity of mass spectrometry to determine each SNP. Current systems such as the MassARRAY system from Sequenom have potentially high throughput, but instrument costs are very high for a single laboratory (Fig. 6.6). Continued improvement of this approach should lower the cost in the future.

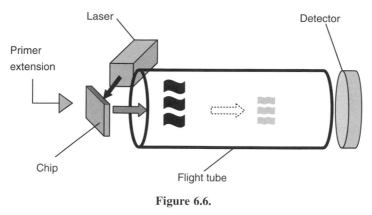

Figure 6.6.

Invader and PCR-Invader Assays

Invader is a discrimination method that relies on the property of a flap endonuclease (Cleavase) to identify single-base-pair differences. As shown in Figure 6.7, the basic procedure consists of two steps. In the first step, two probes are constructed for each allele: (1) the Invader probe that includes the SNP and the sequence 5′ to the polymorphism and (2) the primary probe, which contains the SNP and complementary sequence downstream from the SNP and a universal 5′ tail sequence that matches a signal cassette, used for fluorescent detection in the second step. When both probes are complementary to the template containing the SNP, a three-dimensional structure is formed by the two overlapping oligonucleotides. This structure is recognized by the flap endonuclease, which cleaves the 5′ "signaling" tail of the primary

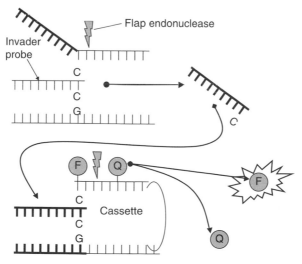

Figure 6.7.

probe. This tail contains complementary sequence for the signal probe and subsequently becomes a second invader probe, this time for the signaling cassette. Again, the enzyme cleaves the three-dimensional structure formed in the signal cassette, releasing a fluorescent dye, separating from its coupled quencher. Each allele is done separately (Fig. 6.7).

The initial Invader method was performed on genomic DNA directly, eliminating the need for PCR, potentially a great advantage. However, the technique has become more popular using a PCR product as the template. Recently, additional modifications have been reported. Hsu et al. (2001) streamlined the PCR-Invader by assaying both alleles in a single reaction, using different labeled single probes for each allele and then detecting the fluorescent molecule by fluorescent polarization. This eliminates the need for a quencher, reducing the cost of the process. Ohnishi et al. (2001) have further reduced the cost and the need for a large amount of DNA to perform large genotyping screens. They performed a multiplex PCR to produce the SNP templates. From this single reaction, 1 μL is placed in each well of specially designed 384-well "cards," which are sealed with plastic tops. These cards are then run through the Invader assay, again with multiple fluorescent probes in the same reaction. Ohnishi et al., estimate they can carry out 300,000–400,000 genotypes per day with this system.

Single-Strand Conformational Polymorphism

This well-known, single-strand conformational polymorphism (SSCP) technique is easy to use and relatively inexpensive (Orita et al., 1989). The PCR product is denatured into single strands. These are then separated on a nondenaturing gel. Under these conditions, the fragment's migration is dependent upon its three-dimensional confirmation, which is partly dependent upon the fragment's sequence. While SSCP is relatively easy to use, it is not as sensitive as other techniques. It is sensitive to a whole host of additional factors besides sequence, including temperature, fragment size, and gel matrix. It is suitable for small-scale SNP detection but not for large genotyping projects.

Denaturing High-Pressure Liquid Chromatography

With denaturing high-pressure liquid chromatography technique (DHPLC; Oefner and Underhill, 1995) the SNP PCR product is denatured and then allowed to reanneal in all of its possible double-stranded combinations (heteroduplex assay; Peeters and Kotze, 1994). If a heterozygote situation is present, one-half of the reannealled products will hold a base-pair mismatch. This mismatch destabilizes the double-stranded molecule and affects its melting temperature. Utilizing the known sequence surrounding the SNP, a specific temperature is chosen in which the mismatch is released from the column before the homozygous strands. The technique is very sensitive for heterozygotes (Gross et al., 2000; Choy et al., 1999). It can be pooled successfully (Wolford et al., 2000).

DATA MANAGEMENT

Data management is one of the true keys to success in genotyping, especially for complex disorders where the number of genotypes required for the analyses will be in the hundreds of thousands. There are a number key factors, which are discussed below.

Objectivity

The importance of objectivity in reading genotypes cannot be overemphasized. No matter how sophisticated a system, there is always data that are not as clear as one would like, and prior knowledge of how the genotype should be (or how one wishes it to be) has swayed many an "objective" decision. Objectivity is enhanced by identifying samples by sample number, not an individual's name or family ID number. This convention is also important for maintaining the confidentiality of results for family members. Genotypes should be read independently and without the use of pedigrees.

Genotype Integrity

Using a gel key, a grouping of samples that always places them in the same order, can enhance sample integrity in genotyping. This is very useful in repeated screening of the same family data and eliminates individual technician placements that can lead to sample switching. Bar code systems to enter genotyping numbers can be employed to reduce coding errors as well. Data management systems such as PEDIGENE (Haynes et al., 1988, 1995) are invaluable in maintaining both clinical and genotypic data.

Scoring

The keys to accurately reading genotypes are many, although "garbage in, garbage out" is probably applicable here in terms of the time spent in optimizing PCR products. Pattern recognition is the key, especially in dinucleotide primers. In reading gels, the best rule of thumb is "If in doubt, don't read it, redo it." This can create a conflict if the individual reading a gel is also responsible for running it.

Standards

Since population allele frequencies are needed for most analyses, alleles should be read by their molecular weight. Two types of standards are needed for genotyping. One standard is an "in-gel" standard that allows the molecular weight of the product to be determined, providing a genotype. The second standard is a "between-gel" standard, to assure that the same individual always produces the same size genotype for a specific marker. This precaution is necessary because there is always variability between gels in actual running conditions and between individuals in interpretation of results. In addition, PCR itself can often vary in a "run" by the addition of one base (Smith et al., 1996), actually changing the molecular weight for that group.

For the first standard we place a sequencing ladder such as that generated using M13, a commercially available ladder, or any individualized set of known PCR products, in at least three lanes distributed across the gel. If the molecular weights of alleles are determined manually, then a sequencing ladder is most useful. More automated allele determination, with the use of software algorithms, will allow ladders with greater spacing between bands to be utilized.

For the gel-to-gel standard, two to three individuals (e.g. CEPH 1331-01, 1331-02, or 1347-02) are amplified with each run and checked to ensure that the same results are obtained each time. Cell lines or DNA on these individuals are available from the Coriell Institute for Medical Research. This has become a general standard approach for most genotyping laboratories. Internal allele-sizing methods may not be consistent from laboratory to laboratory with different standards and genotyping techniques, and this amplification not only allows the same family or group of data to be run on different gels within a single laboratory but also allows data from several laboratories to be efficiently combined for analysis.

Quality Control

One of the increasingly important elements of success in genomics and human genetics is the ability to merge good data management with good molecular biology. This need is no better shown than in genotyping of complex disorders. In these disorders family structure is usually minimal, reducing the ability to use family structure to identify genotyping errors. In addition, the large number of genotypes required increases the chances of error in the process. One must approach the process realizing that errors will be made. The key is to develop a structure that will catch most of these errors, as introduced errors will quickly mask real changes in genetic analyses, reducing years of work to a meaningless mess.

One approach to this problem is to incorporate a quality control (QC) structure into the genotyping process itself. To do this we have developed a two-level system of quality control. The first level consists of duplicating six samples from each 96-well plate that are designated as "QC samples." These samples are randomly spread throughout all plates in a genomic screen, blinded to the technicians using them. This system is very good at identifying laboratory-based errors such as misloading of lanes, skipping lanes, inconsistent reading of alleles, and poor quality of products. It also can serve as an excellent teaching tool for new technicians. By varying the "accepted" level of matches in a study, the researchers can have some flexibility in the level of error they will allow. In large genotyping screens the management of these QCs becomes an arduous process. For this reason we have developed automatic computer checking of QCs.

Finally, when genotyping SNPs, the markers are often close enough together that double or triple crossovers would be unexpected. Therefore, identifying these recombinations likely will lead to genotyping or data management errors.

The importance of following a good, objective data management protocol cannot be overestimated in performing genotyping in human disorders. While it does increase both the cost and time of an analysis, it is well worth it. Once performed, the data then provide the analyst and researchers building further work on these data great confidence that the considerable efforts of the clinical collection and genotyping have true meaning.

REFERENCES

Altshuler D, Brooks LD, Chakravarti A, Collins FS, Daly MJ, Donnelly P (2005): International HapMap Consortium. A haplotype map of the human genome. Nature 437(7063):1241–1242.

Baskaran N, Kandpal RP, Bhargava AK, Glynn MW, Bale AE, Weissman SM (1996): Uniform amplification of a mixture of deoxyribonucleic acids with varying GC content. Genome Res 6:633–638.

Bassam BJ, Caetano-Anollés G (1993): Automated "hot start" PCR using mineral oil and paraffin wax. BioTechniques 14:30–34.

Bassam BJ, Caetano-Anollés G, Gresshoff PM (1991): Fast and sensitive silver staining of DNA in polyacrylamide gels. Anal Biochem 196:80–83.

Ben Othmane K, Ben Hamida M, Pericak-Vance MA, Ben Hamida C, Blel S, Carter SC, Bowcock AM, et al (1992): Linkage of Tunisian autosomal recessive Duchenne-like muscular dystrophy to the pericentromeric region of chromosome 13q. Nat Genet 2:315–317.

Botstein D, White RL, Skolnick M, Davis RW (1980): Construction of a genetic linkage map in man using restriction fragment length polymorphisms. Am J Hum Genet 32:314–331.

Breen G, Harold D, Ralson S, Shaw D, St. Clair D (2000): Determining SNP allele frequencies in DNA pools. BioTechniques 28:464–470.

Buzas Z, Varga L (1995): Rapid method for separation of microsatellite alleles by the Phastsystem. PCR Methods Appl 4:380–381.

Chee M, Yang R, Hubbell E, Berno A, Huang XC, Stern D, Winkler J, Lockhart DJ, Morris MS, Fodor SP (1996): Accessing genetic information with high-density DNA arrays. Science 274:610–614.

Chen X, Zehnbauer B, Gnirke A, Kwok PY (1997): Fluorescence energy transfer detection as a homogeneous DNA diagnostic method. Proc Natl Acad Sci USA 94:10756–10761.

Chevet E, Lemaitre G, Katinka MD (1995): Low concentrations of tetramethylammonium chloride increase yield and specificity of PCR. Nucleic Acids Res 23:3343–3344.

Chou Q, Russell M, Birch DE, Raymond J, Bloch W (1992): Prevention of pre-PCR mispriming and primer dimerization improves low-copy-number amplifications. Nucleic Acids Res 20:1717–1723.

Choy YS, Dabora SL, Hall F, Ramesh V, Niida Y, Franz D, Kasprzyk-Obara J, Reeve MP, Kwiatkowski DJ (1999): Superiority of denaturing high performance liquid chromatography over single-stranded conformation and conformation-sensitive gel electrophoresis for mutation detection in TSC2. Ann Hum Genet 63(Pt 5):383–391.

Corless CE, Guiver M, Borrow R, Edwards-Jones V, Kaczmarski EB, Fox AJ (2000): Contamination and sensitivity issues with a real-time universal 16S rRNA PCR. J Clin Microbiol 38:1747–1752.

Damji KF, Gallione CJ, Allingham RR, Slotterbeck B, Guttmacher AE, Pasyk KA, Pericak-Vance MA, Speer MC, Marchuk DA (1996): Quantitative DNA pooling to increase the efficiency of linkage analysis in autosomal dominant disease. Am J Hum Genet 59:A215.

Edwards A, Civitello A, Hammond HA, Caskey CT (1991): DNA typing and genetic mapping with trimeric and tetrameric tandem repeats. Am J Hum Genet 49:746–756.

Eggerding FA, Iovannisci DM, Brinson E, Grossman P, Winn-Deen ES (1995): Fluorescence-based oligonucleotide ligation assay for analysis of cystic fibrosis transmembrane conductance regulator gene mutations. Hum Mutat 5:153–165.

Filichkin SA, Gelvin SB (1992): Effect of dimethyl sulfoxide concentration on specificity of primer matching in PCR. BioTechniques 12:828–830.

Gilgen M, Hofelein C, Luthy J, Hubner P (1995): Hydroxyquinoline overcomes PCR inhibition by UV-damaged mineral oil. Nucleic Acids Res 23:4001–4002.

Gross E, Arnold N, Pfeifer K, Bandick K, Christian-Albrechts A (2000): Identification of specific BRCA1 and BRCA2 variants by DHPLC. Hum Mutat 16:345–353.

Grossman PD, Bloch W, Brinson E, Chang CC, Eggerding FA, Fung S, Iovannisci DM, Woo S, Winn-Deen ES, Iovannisci DA (1994): High-density multiplex detection of nucleic acid sequences: Oligonucleotide ligation assay and sequence-coded separation. Nucleic Acids Res 22:4527–4534.

Gusella JF, Wexler NS, Conneally PM, Naylor SL, Anderson MA, Tanzi RE, Watkins PC, et al (1983): A polymorphic DNA marker genetically linked to Huntington's disease. Nature 306:234–238.

Hacia JG, Brody LC, Chee MS, Fodor SPA, Collins FS (1996): Detection of heterozygous mutations in BRCA1 using high density oligonucleotide arrays and two-color fluorescence analysis. Nat Genet 14:441–447.

Haynes CS, Pericak-Vance MA, Hung W-Y, Deutsch DB, Roses AD (1988): PEDIGENE—A computerized data collection and analysis system for genetic laboratories [Abstract]. Am J Hum Genet 43:A146.

Haynes CS, Speer MC, Peedin M, Roses AD, Haines JL, Vance JM, Pericak-Vance MA (1995): PEDIGENE: A comprehensive data management system to facilitate efficient and rapid disease gene mapping. Am J Hum Genet 57:A193.

Hecker KH, Roux KH (1996): High and low annealing temperatures increase both specificity and yield in touchdown and stepdown PCR. BioTechniques 20:478–485.

Hengen PN (1997): Optimizing multiplex and LA-PCR with betaine. Trends Biochem Sci 22:225–226.

Henke W, Herdel K, Jung K, Schnorr D, Loening SA (1997): Betaine improves the PCR amplification of GC-rich DNA sequences. Nucleic Acids Res 25:3957–3958.

Hsu TM, Law SM, Duan S, Neri BP, Kwok PY (2001): Genotyping single-nucleotide polymorphisms by the invader assay with dual-color fluorescence polarization detection. Clin Chem 47:1373–1377.

Hudson TJ, et al (1997): PCR methods of genotyping. In: Dracopoli N, ed. Current Protocols in Human Genetics, Supplement 12 ed. New York: John Wiley & Sons.

Innis MA, Gelfand DH (1990): Optimization of PCRs. In: Innis MA, Gelfand DH, Sninsky JJ, White TJ, eds. PCR Protocols: A Guide to Methods and Applications. San Diego, CA: Academic, pp 3–12.

Kan YW, Dozy AM (1978): Polymorphism of DNA sequence adjacent to human beta-globin structural gene: Relationship to sickle mutation. Proc Natl Acad Sci USA 75(11):5631–5635.

Kwok S, Higuchi R (1989): Avoiding false positives with PCR. Nature 339:237–238.

Lander ES, Botstein D (1987): Homozygosity mapping: A way to map human recessive traits with the DNA of inbred children. Science 236:1567–1570.

La Spada AR, Paulson HL, Fischbeck KH (1994): Trinucleotide repeat expansion in neurological disease [Review]. Ann Neurol 36:814–822.

Latif S, Bauer-Sardina I, Ranade K, Livak KJ, Kwok PY (2001): Fluorescence polarization in homogeneous nucleic acid analysis II: 5′- nuclease assay. Genome Res 11:436–440.

Litt M, Luty JA (1989): A hypervariable microsatellite revealed by in vitro amplification of a dinucleotide repeat within the cardiac muscle actin gene. Am J Hum Genet 44:397–401.

Litt M, McPherson MJ, Quirke P, Taylor GR, eds (1991): PCR: A Practical Approach. New York: IRL.

McDowell DG, Burns NA, Parkes HC (1998): Localized sequence regions possessing high melting temperatures prevent the amplification of a DNA mimic in competitive PCR. Nucleic Acids Res 26:3340–3347.

Merril CR, Goldman D, Sedman SA, Ebert MII (1981): Ultra-sensitive stain for proteins in polyacrylamide gels shows regional variation in cerebrospinal fluid proteins. Science 211:1437–1438.

Millard D, de Couet HG (1995): Preparation of glass plates with cerium oxide for DNA sequencing. BioTechniques 19:576–576.

Mullis K, Faloona F, Scharf S, Saiki R, Horn G, Erlich H (1986): Specific enzymatic amplification of DNA in vitro: The polymerase chain reaction. Cold Spring Harbor Symp Quant Biol L1:263–273.

Nakamura Y, Leppert M, O'Connell P, Wolff R, Holm T, Culver M, Martin C, et al (1987): Variable number of tandem repeat (VNTR) markers for human gene mapping. Science 235:1616–1622.

Neiderhauser C, Hofelein C, Wegmuller B, Luthy J, Candrian U (1994): Reliability of PCR decontamination systems. PCR Methods Appl 4:117–123.

Oefner PJ, Underhill PA (1995): Comparative DNA sequencing by denaturing high-performance liquid chromatography (DHPLC). Am J Hum Genet 57:A266.

Ohnishi Y, Tanaka T, Ozaki K, Yamada R, Suzuki H, Nakamura Y (2001): A high-throughput SNP typing system for genome-wide association studies. J Hum Genet 46:471–477.

Orita M, Suzuki Y, Sekiya T, Hayashi K (1989): Rapid and sensitive detection of point mutations and DNA polymorphisms using the polymerase chain reaction. Genomics 5:874–879.

Peeters AV, Kotze MJ (1994): Improved heteroduplex detection of single-base substitutions in PCR-amplified DNA. PCR Methods Appl 4:188–190.

Saiki RK (1990): Amplification of genomic DNA. In: Innis MA, Gelfand DH, Sninsky JJ, White TJ, eds. PCR Protocols: A Guide to Methods and Applications. San Diego, CA: Academic, pp 13–20.

Samiotaki M, Kwiatkowski M, Landegren U (1996): OLA. Dual-color oligonucleotide ligation assay. In: Landegren U, ed. Laboratory Protocols for Mutation Detection. New York: Oxford University Press, pp 96–100.

Sarkar G, Sommer SS (1991): Parameters affecting susceptibility of PCR contamination to UV inactivation. BioTechniques 10:590–594.

Shammas FV, Heikkila R, Osland A (2001): Fluorescence-based method for measuring and determining the mechanisms of recombination in quantitative PCR. Clin Chim Acta 304:19–28.

Sheffield VC, Carmi R, Kwitek-Black AE, Rokhlina T, Nishimura D, Duyk GM, Elbedour K, et al (1994): Identification of a Bardet–Biedl syndrome locus on chromosome 3 and evaluation of an efficient approach to homozygosity mapping. Hum Mol Genet 3:1331–1335.

Shi MM (2001): Enabling large-scale pharmacogenetic studies by high-throughput mutation detection and genotyping technologies. Clin Chem 47:164–172.

Smith JR, Carpten JD, Brownstein MJ, Ghosh S, Magnuson VL, Gilbert DA, Trent JM, Collins FS (1996): Approach to genotyping errors caused by nontemplated nucleotide addition by Taq DNA polymerase. Genome Res 5:312–317.

Southern EM (1975): Detection of specific sequences among DNA fragments separated by gel electrophoresis. J Mol Biol 98:503–517.

Stryer L, Haugland RP (1967): Energy transfer: A spectroscopic ruler. Proc Natl Acad Sci USA 58:719–726.

Syvanen AC, Sajantila A, Lukka M (1993): Identification of individuals by analysis of biallelic DNA markers, using PCR and solid-phase minisequencing. Am J Hum Genet 52:46–59.

Vance JM, Nicholson GA, Yamaoka LH, Stajich J, Stewart CS, Speer MC, Hung W-Y, et al (1989): Linkage of Charcot–Marie–Tooth neuropathy type 1a to chromosome 17. Exp Neurol 104:186–189.

Vance JM, Slotterbeck B, Yamaoka L, Haynes C, Roses AD, Pericak-Vance MA (1996): A fluorescent genotyping system for multiple users, flexibility and high output. Paper persented at the American Society of Human Genetics 46th Annual Meeting, San Francisco, November, p A239.

Von Deimling A, Bender B, Louis DN, Wiestler OD (1993): A rapid and non-radioactive PCR based assay for the detection of allelic loss in human gliomas. Neuropathol Appl Neurobiol 19:524–529.

Weber JL (1990): Informativeness of human (dC-dA)n–(dG-dT)n polymorphisms. Genomics 7:524–530.

Weber JL, May PE (1989): Abundant class of human DNA polymorphisms which can be typed using the polymerase chain reaction. Am J Hum Genet 44:388–396.

Wolford JK, Blunt D, Ballecer C, Prochazka M (2000): High-throughput SNP detection by using DNA pooling and denaturing high performance liquid chromatography (DHPLC). Hum Genet 107:483–487.

Zagursky RJ, Conway PS, Kashdan MA (1992): Use of 33P for Sanger DNA sequencing. BioTechniques 11:36–38.

Ziegle JS, Su Y, Corcoran KP, Nie L, Mayrand E, Hof LB, McBride LJ, et al (1992): Application of automated DNA sizing technology for genotyping microsatellite loci. Genomics 14:1026–1031.

CHAPTER 7

Data Analysis Issues in Expression Profiling

SIMON LIN and MICHAEL HAUSER

INTRODUCTION

Although virtually every cell in the body contains the full complement of genes, individual cells and tissues are distinguished from one another by the subset of RNA or protein products of those genes they express. Expression profiling enables identification of this subset of transcribed genes and the levels at which they are expressed, thereby providing a valuable tool for increasing our understanding of the regulatory and functional complexities of the genome. Expression profiling has been used to characterize signaling and regulatory pathways, identifying genes responsive to p53 induction (Zhao et al., 2000; Madden et al., 1997) and genes expressed in yeast during sporulation (Chu et al., 1998). Much research has focused on using gene expression information to distinguish between different classes of malignant tissue: Diffuse large B-cell lymphomas that respond well to chemotherapy can be distinguished from those that do not on the basis of expression profiling (Alizadeh et al., 2000). Similar approaches have been taken to analyze breast cancer (Perou et al., 2000) and colon cancer (Alon et al., 1999; Zhang et al., 1997). This use of expression analysis to segregate samples into distinct phenotypic groups, called classification analysis, has the potential to greatly improve clinical management of cancer.

Another powerful use of gene expression profiling is the identification of candidate susceptibility genes for complex diseases. Linkage analysis in such disorders frequently identifies large genomic regions that may contain hundreds of genes. Sorting through so many genes to find the few that influence disease susceptibility can be a daunting task. Expression profiling reveals genes whose expression levels are up or down regulated as a function of the disease process or in response to acute disease-related stimuli. Genes with such a pattern of expression that also map to

Genetic Analysis of Complex Diseases, Second Edition, Edited by Jonathan L. Haines and Margaret Pericak-Vance
Copyright © 2006 John Wiley & Sons, Inc.

regions of linkage represent excellent candidates for further molecular analysis. This strategy has been used in conjunction with rat model systems to generate a series of excellent candidate susceptibility genes for mania and psychosis (Niculescu et al., 2000).

Many different experimental methods have been used for expression profiling. All methods begin with the purification of transcribed RNA from the tissue sample(s) of interest. The classic techniques of subtractive hybridization (Lee et al., 1991) and differential display (Liang and Pardee, 1992) use physical methods to isolate and clone messages whose expression levels differ greatly between two samples. In this chapter, we will focus on the two techniques that have emerged as the most powerful and flexible experimental approaches to expression analysis: serial analysis of gene expression (SAGE) (Velculescu et al., 1995) and microarray analysis. These two techniques have the great advantage that they generate large databases of expression information, rather than just a few clones for immediate analysis. A single experiment can provide information about the expression levels of thousands of genes in a given sample. These expression data are now being stored in public databases, providing a tremendously rich source of information that can be mined repeatedly by many different investigators.

In this chapter, we will first describe the construction and analysis of SAGE libraries. Then, after a brief description of the different types of DNA microarrays, we will introduce some more advanced approaches to the statistical analysis of complex expression datasets obtained through SAGE or microarrays. Finally, we will discuss some biological applications of these expression profiling techniques.

SERIAL ANALYSIS OF GENE EXPRESSION

Developed in 1995 by Vogelstein and Kinzler (Velculescu et al., 1995), briefly, SAGE isolates a 14-bp sequence immediately adjacent to the 3'-most NlaIII site within each transcript. These short sequences, called tags, are then cloned in long tandem arrays for sequencing. The approach is conceptually similar to sequencing large numbers of cDNA clones from a tissue but is much more efficient as more than 20 tags can be identified by sequencing a single template. The genes to which the SAGE tags correspond are then identified by using the "tag to gene mapping" database prepared by Alex Lash, available on the National Center for Biotechnology Information (NCBI) website (http://www.ncbi.nlm.nih.gov/SAGE/SAGEtag.cgi). Our current knowledge of the human genome allows the majority of SAGE tags to be mapped to a single UniGene set. The UniGene sets are clusters of mRNA and expressed sequence tags (ESTs) that are believed to represent individual genes. A subset of SAGE tags map to multiple UniGene sets, which may reflect uncertainty in transcript clustering in the UniGene sets, actual sequence similarity causing different genes to have the same SAGE tag, or complexities of transcript processing, including alternative splicing or polyadenylation sites. The expression patterns of many transcripts can be evaluated by sequencing several thousand clones (40,000–50,000 tags), comparing the abundance of individual tags, and identifying the corresponding genes.

The detailed SAGE protocol is freely available for academic use and can be obtained from http://www.sagenet.org/sage.htm. This technique has several important advantages as compared to other techniques for the detection of tissue-specific expression. First, SAGE does not simply isolate a small number of clones, but rather creates a permanent, quantitative record of the entire set of sequences transcribed in a given tissue or cell population. Thus, a SAGE database can be used long after it was originally constructed to ascertain the expression levels of newly identified genes. Second, SAGE can detect small changes in expression levels, even among messages that are expressed at very low levels over-all. Third, transcripts that are over- or underexpressed can be detected equally well. SAGE is an "open-platform" technology: It does not require any preexisting biological or sequence information and can be applied to any species. In contrast, micro-array experiments cannot detect a transcript unless the corresponding gene is already known and has been included in the array. Other open-platform systems for analysis of gene expression such as GeneCalling, TOGA, and READS have been recently reviewed (Green et al., 2001).

Since the original SAGE protocol was published in 1995, many technical improvements have been introduced. Using the I-SAGE kit marketed by Invitrogen (Carlsbad, CA), SAGE can now be performed starting with as little as 2 μg total RNA. Additional modifications allow SAGE libraries to be constructed from as few as 40,000 cells (Datson et al., 1999; Virlon et al., 1999; Peters et al., 1999) or, with the addition of a polymerase chain reaction (PCR) amplification step, from a single oocyte (Neilson et al., 2000). The sequencing of several thousand clones from each SAGE library can now be performed commercially on a fee-for-service basis (Genome Therapeutics Co.). There have also been improvements in the software used for extraction of SAGE tags from individual sequence reads. For example, eSAGE (Margulies and Innis, 2000) allows the user to establish a minimum PHRED quality score for each extracted SAGE tag. This allows quantitative calculations of the frequency of sequencing errors and resulting misspecification of genes.

Analysis of SAGE Libraries

The SAGE libraries are most commonly analyzed by performing a direct comparison of tag abundance between two individual libraries or two groups of libraries using software such as the xProfiler (http://www3.ncbi.nlm.nih.gov/SAGE/sagexpsetup.cgi). The statistical significance of observed differences in tag counts between SAGE libraries can be evaluated in several different ways. Monte Carlo simulations can be used to estimate the likelihood that the experimentally observed differences could arise by chance alone (Zhang et al., 1997). An alternate Bayesian approach has been suggested (Audic and Claverie, 1997), and the necessary software has been made available (http://igs-server.cnrs.mrs.fr). Analysis of SAGE data using simple hypothesis-testing methods such as χ^2 or the Fisher exact test has been reviewed (Man et al., 2000).

The amount of SAGE data in public databases is increasing rapidly through the efforts of groups such as the Cancer Genome Anatomy Project (CGAP, http:// cgap.nci.nih.gov/). Well over 300 libraries have now been constructed from normal and malignant tissue (Lal et al., 1999). As more libraries are constructed from different tissue types, these databases can be used for surveys of tissue-specific expression of individual genes, sometimes called digital northern analysis (http:// www.ncbi.nlm.nih.gov/SAGE/sagevn.cgi). The availability of large amounts of SAGE data will also allow the application of the more sophisticated dimensional reduction and clustering technologies described below.

Microarray Analysis

In contrast to SAGE, microarray technologies are all based on mRNA hybridization, so only genes with available sequences or clones can be detected. First, polyA mRNA is isolated from samples and labeled with fluorescent dye or a radioactive isotope. It is then hybridized to probe sequences arrayed at high density on a solid support (glass, silicon, or nylon membrane). Finally, a scanner measures the fluorescent intensities of each spot on the hybridization array, from which the initial concentrations of the corresponding transcripts are inferred. There are two main types of microarrays: commercial high-density oligonucleotide arrays, called "gene chips," and custom-spotted glass slides.

Affymetrix Corporation (Santa Clara, CA) manufactures high-density gene chips using photolithographic technology. In general, the specific gene probes found on these arrays are standardized and cannot be customized for individual users. Each probe set consists of over 20 different oligonucleotides, some matching the transcript's sequence exactly and some containing intentional mismatches as background hybridization controls. Proprietary software algorithms provided by the manufacturer use the level of hybridization to all of these oligonucleotides to calculate the level of gene expression in a sample.

Custom-spotted glass slide microarrays can be prepared by individual users or core laboratories. Large collections of transcripts of individual genes (PCR products or long oligonucleotides) are available from commercial sources such as Invitrogen Corporation (Carlsbad, CA). Individual transcripts are selected from these probe sets and a small volume of each corresponding DNA solution is spotted onto a glass slide, where it adheres tightly. These glass slides are then hybridized and scanned to quantitate RNA abundance.

Oligonucleotide gene chips are manufactured with a consistent amount of hybridization target at each spot on each chip. For this reason, a chip can be probed with a single RNA sample, and the resulting gene expression levels can be compared with other independent chip experiments after appropriate normalization. To account for variations in the amount of material spotted on each array, custom-spotted glass slide arrays are hybridized with two different RNA samples simultaneously, one experimental and one reference sample. The two samples are labeled with different fluorescent dyes, and the expression data obtained are a ratio of the two samples.

Microarray data analysis consists of three steps: (i) data preparation, in which data are adjusted for the downstream algorithms; (ii) algorithm selection for data analysis; and (iii) interpretation, in which the results from the algorithms are explained in a biological context.

Data Preparation

Data preparation is a critical step in microarray data analysis. Each data preprocessing method can accentuate, create, or destroy findings from downstream analytic algorithms. It has been argued that data preprocessing has had a stronger influence on the final results than the choices of subsequence statistical analysis methods (Hoffmann et al., 2002). Data quality control, transformation, and normalization are usually addressed in this step.

Data Quality Control. Expression profiling measurements usually contain noise and erroneous data points, due to scratches on the hybridization surface, white noise during image scanning, irregular spot morphology, and numerous other stochastic experimental artifacts. The first step in data analysis is quality control to identify those erroneous data points. Flagging such points and eliminating them from further analysis can improve the overall quality of the analysis. More sophisticated algorithms have been developed to quantitatively measure spot qualities (Wang et al., 2001). Li and Wong (2001) have proposed a statistical model to detect outliers and data irregularities for the Affymetrix platform. It has been shown that quality control of the raw data helps normalization and downstream analysis (Yang et al., 2001; Raffelsberger et al., 2002).

Data Transformation. Many statistical tests such as the *t*-test and analysis of variance (ANOVA) assume independence between the mean and the variance of data. However, raw measurements of microarrays usually demonstrate a reasonably strong correlation between the mean expression level of a given gene and its standard deviation (Durbin et al., 2002). Thus, a transformation is necessary to stabilize the variance, such as logarithm transformations (Dudoit et al., 2002) and cubic-root transformations (Tusher et al., 2001). However, log transformations inflate the variance of observations at low expression levels. Rocke (Durbin et al., 2002) and Huber (2002) independently proposed similar models for microarray data transformation to stabilize the variance over the full range of measurements.

Data Normalization. Before any numerical inference about microarray data can be made, it must be clear that those measurements reflect true biological differences, not experimental artifacts such as dye-labeling effects, array-printing pin effects, or overall intensity differences. Those systematic differences caused by experimental artifacts can be eliminated by a normalization procedure. The assumptions of normalization are usually based on constant expression levels of internal housekeeping genes or external markers, overall intensity levels, or constant expression of the majority of genes (Yang et al., 2002; Kepler et al., 2002). Huber et al. (2002) consolidated both data normalization and variance into one variance stabilization and normalization (VSN) model.

Expression Data Matrix

Before we discuss data analysis in more detail, we must first define several concepts used in multivariate analysis and pattern recognition. An *object* is a sample possessing quantitative or qualitative properties that can be described by *features*, also known as attributes or variables. A series of observations made on a set of objects can be organized as a multivariate data matrix—a rectangular array of numbers. Here we use the convention that lets the rows of the matrix represent objects and the columns represent features. The general data matrix $\mathbf{X}$ with m objects and n features can be written as follows:

$$
\mathbf{X} = \begin{bmatrix}
x_{11} & \cdots & x_{1j} & \cdots & x_{1n} \\
\vdots & \ddots & \vdots & \ddots & \vdots \\
x_{i1} & \cdots & x_{ij} & \cdots & x_{in} \\
\vdots & \ddots & \vdots & \ddots & \vdots \\
x_{m1} & \cdots & x_{mj} & \cdots & x_{mn}
\end{bmatrix}
$$

The *element* in row i and column j of matrix $\mathbf{X}$ is denoted x_{ij}. It represents the value of an object i on feature j. A *vector* is a special matrix with a single column or a single row of elements. Thus, rows or columns in a matrix can be viewed as vectors.

In the *gene expression data matrix*, by convention the objects (rows) in the matrix are genes and the features (columns) in the matrix are experimental conditions. Usually, m, the number of genes, is much larger than n, the number of experimental conditions. In this orientation, a complete row of features is called a gene's *profile*. It is expressed as a vector in the n-dimensional feature space. Similarly, a column represents an experiment's profile by a vector in the m-dimensional gene space.

This orientation of the expression data matrix is commonly used for clustering genes into functional groups in response to experimental stimuli. If necessary, the data matrix can be transposed so that the columns represent genes while the rows represent experiments (or patients). Such an arrangement is often used in supervised learning to classify the gene expression profiles of patients.

In addition to feature measurements, there is also *a priori* knowledge of the objects. This external knowledge is called a *label*. For example, if we measure gene expression features of each patient by microarrays, the known diagnostic category to which a patient belongs is the label. Note that in this case a label is acquired by the physician, independently from the features acquired by microarrays. In supervised machine learning, labels are required in the training session, so that unknown samples can subsequently be classified into predefined groups. Unsupervised machine learning (clustering) does not require labels.

Dimension Reduction of Features

To reduce the complexity of the dataset, the objects in the n-dimensional space can be mapped into a lower dimensional space. This procedure is always applied before feeding data into a machine-learning algorithm. Although it may seem counter-intuitive to present fewer features to the machine-learning algorithm, dimension reduction is necessary for two reasons. First, the "curse of dimensionality" (see below) prohibits higher dimensional data from being fed into the machine-learning process. Second, objects can only be visualized when the dimension is less than 3. Dimension reduction is achievable by selecting existing features based on expert knowledge, statistical algorithms (Fowlkes et al., 1987), or by creating new features using dimension reduction algorithms such as principal-component analysis or multidimensional scaling.

Curse of Dimensionality. In studies of disease classification based on micro-array data, there tend to be too many molecular features for each patient, given the small sample size of patients. There is a temptation to throw everything into the machine-learning algorithm, but extra features that are unrelated to the classi-fication tend to dilute the analysis and cause the algorithm to run astray. This is called the "curse of dimensionality." In machine learning, the size of the search space increases exponentially with the dimension of features to model. Thus, the reduction of the features is crucial to circumvent the curse of dimensionality. Vari-able selection prior to clustering is discussed by Fowlkes et al. (1987). In super-vised machine learning, it is recommended that the number of the training samples per class be at least 5–10 times the number of features (Jain and Chandrasekaran, 1982).

Principal-Component Analysis. Principal-component analysis (PCA) reduces the number of features by forming new features to describe objects in lower dimen-sions (Jolliffe, 1986). These new features are labeled principal components, which are linear combinations of the original features. The same idea is also known as Karhunen–Loeve transformation in signal processing and singular-value deposition (SVD) in matrix computation. The principal components have two properties. First, the new components are orthogonal to each other. This indicates that the correlation between the original features has been removed. Second, the principal components have decreasing ability to explain variance in the dataset. The first principal com-ponent explains the largest variance of the objects. As the cardinality of the principal components increases, the explained variance in the dataset decreases. It is often the case that a small number of the early components are sufficient to represent most of the variations in the data.

The orthogonal characteristic of the principal components was exploited by Alter et al. (2000) in the hope of categorizing genes into orthogonal signal transduction pathways. The decreasing importance of the principal components makes it possible to ignore low-impact components with very little loss of information. Principal-component analysis has been used as a means to filter out the noise and to remove redundancy in the dataset (Hilsenbeck et al., 1999).

Multidimensional Scaling. Unlike PCA, multidimensional scaling (MDS) uses the distance between objects as a starting point for analysis, rather than the direct observation of objects in the n-dimensional space. It attempts to represent a multidimensional dataset in two or three dimensions such that the distance between objects in the original n-dimensional space is preserved as faithfully as possible in the projected space. A simple example is to construct a map of Virginia using only road distances between the towns. Multidimensional scaling can be achieved by the principal coordinate algorithm (Gower, 1966) to minimize the variance or by the spring-embedding algorithm (Kruskal, 1964) to minimize the stress. It is the process of visualizing relationships between objects so that human experts can interpret the data. If the data are projected in one dimension, then the ordering of objects is seriated according to their similarities. The application of MDS to microarray data will be discussed in the visualization section below.

Measures of Similarity between Objects

To study the degree of resemblance in the objects, many different mathematical functions are suggested as similarity measures. The most common ones, Euclidean distance and Pearson correlation, are described here. A different choice of similarity measure can lead to a very different interpretation of expression data (Getz et al., 2000).

Euclidean Distance. If objects are geometrically represented as points in an n-dimensional feature space, their proximities can be measured as the distance between pairs of points. Euclidean distance is the square root of the sum of the squared difference between two objects across n features. The Euclidean distance D_{ij} between object i and object j is represented by

$$D_{ij} = \sqrt{\sum_{k=1}^{n}(x_{ik} - x_{jk})^2}$$

where x_{ik} denotes the measurement of object i on feature k, x_{jk} denotes the measurement of object j on feature k, and n is the dimension.

In microarray analysis, Euclidean distance is used to contrast the difference in the absolute expression level of two objects. For example, the absolute expression level of a group of genes can be significantly higher in malignant tumors than in benign tumors, even though their relative shapes as measured by the Pearson correlation coefficient are the same. In this case, Euclidean distance is more appropriate to measure the differences than the Pearson correlation coefficient.

Pearson Coefficient. The Pearson product-moment correlation, or Pearson correlation, is the dot product of two normalized vectors. It ranges from $+1$ to -1. A correlation of $+1$ conveys a perfect positive linear relationship; a correlation of -1 conveys a perfect negative linear relationship; a correlation of zero indicated that there is no linear relationship. The Pearson coefficient Q_{ij} between object i and object j is represented by

$$Q_{ij} = \frac{\sum_{k=1}^{n} (x_{ik} - \overline{x_i})(x_{jk} - \overline{x_j})}{\sqrt{\sum_{k=1}^{n} (x_{ik} - \overline{x_i})^2}\sqrt{\sum_{k=1}^{n} (x_{jk} - \overline{x_j})^2}}$$

where $\overline{x_i}$ and $\overline{x_j}$ are the vector profile means of objects i and j, respectively. The denominator terms represent the scatter. Pearson correlation stresses the similarity between the "shape" of two expression profile vectors and ignores the differences in their absolute magnitudes. This corresponds to the biological notion of two coexpressed genes (Eisen et al., 1998).

The difference between the Euclidian distance and the Pearson correlation also depends on the rescaling of the vectors. When the Euclidean distance is calculated on vectors normalized by zero mean and unit variance transformation, discrimination of elevation and scatter is lost, making the distance closely related to the Pearson correlation (Skinner, 1978).

Unsupervised Machine Learning: Clustering

Clustering uses a collection of unsupervised machine-learning algorithms to group objects into subsets based on their similarities. By revealing these natural groupings in a large dataset, cluster analysis provides an intuitive way to reduce data complexity from m objects to k groups. Clustering does not rely on any previous knowledge of the dataset (no label information is required before clustering); neither does it require a training process. In this sense, it is "unsupervised" learning. Cluster analysis is widely used in biological sciences such as numerical taxonomy (Sneath and Sokal, 1973) and evolution analysis (Fitch and Margoliash, 1967). It is also an appealing approach to microarray data analysis because it promises to find patterns without *a priori* knowledge of the data.

What to Cluster. Clustering methods can be used to group genes that have similar profiles across a range of experimental conditions. Alternatively, experimental conditions in the m-dimensional gene space can be clustered according to their gene expression profiles. Two-way clustering groups both the row and column vectors; it can be done independently by using a two-stage approach (Perou et al., 2000) or by considering both row and column vectors at the same time during clustering (Getz et al., 2000).

Hierarchical Clustering. The hierarchical algorithm results in a treelike representation of the data, often called a dendrogram. It enables an analyst to observe how objects are being merged or split into groups. The process of forming the hierarchy can be "bottom up" or "top down." *Agglomerative clustering* starts with each object in its own cluster. The objects then merge into more and more coarse-grained clusters until all objects are in a single cluster. *Divisive clustering* reverses this process. It starts with all objects in one cluster and subdivides them into many fine-grained clusters.

Most clustering algorithms work agglomeratively. They iteratively join the objects into a tree structure. First, all of the objects are in their own cluster. Then, a heuristic rule is applied to find the "best" pair for merging. The merged clusters then replace the original ones. This process is repeated until only one cluster remains. Algorithms differ in the heuristics used to define the best pair of clusters to merge. Single-linkage, average-linkage, and complete-linkage algorithms all use the minimum-distance criterion. Single linkage defines the distance as the single shortest link between clusters. Complete linkage takes the distance between the most distant members, and average linkage uses the average distance between the two cluster centers. An alternative to the distance-based rule, Ward's method uses the sum-of-squares criterion. This criterion chooses the cluster to be merged based on the smallest increase in the within-group sum of squares.

Given the heuristic algorithm rules described above, users choose an algorithm for a specific purpose. Single linkage often finds large undesirable serpentine clusters, where the chained objects at the opposite ends of the same cluster may be dissimilar. Thus, single linkage is incapable of delineating poorly separated clusters. On the opposite end of the spectrum, complete linkage tends to find excessively small and compact clusters. Average linkage, sometimes used as a compromise between single and complete linkage, is sensitive to the transformation of the distance metrics. In other words, the dendrogram resulting from the average-linkage method might be different if a different transformation is applied to the raw measurements. Different algorithms can reveal different facets of the dataset and may have the potential to complement each other.

Partitional Clustering. Hierarchical clustering, as described above, results in a nested structure of groupings. In contrast, *partitional clustering*, such as k-means (MacQueen, 1967), simply divides the objects into k groups without giving any details of subgrouping within each group. It is especially appropriate when the desired number of groups, k, is known. The algorithm achieves the partition iteratively. Initially, all the objects are assigned at random to the k groups. Then the centroid of each group is computed, the distance of each object to the centroids is recalculated, and each object is assigned to one of the nearest cluster centroids. These two steps are alternated until a convergence criterion is met. Computationally, k-means clustering is less expensive in terms of time and memory consumption than hierarchical clustering. However, in hierarchical clustering, there is no need to specify the number of classes k. One should be aware that due to the iterative nature of k-means algorithm, the initial order of the objects influences the results; the results will be slightly different with each run. Examples of k-means clustering can be found in Ishida et al. (2002) and Brar et al. (2001).

Self-Organizing Map. Kohonen's self-organizing map (SOM) algorithm (Kohonen, 1995) clusters data by taking advantage of the robustness of neural network techniques and has been applied to messy datasets with outliers. The

SOM maps unordered objects to nodes in a one- or two-dimensional grid, such that similar objects are assigned to the same node and similar nodes are topologically close to each other. Chu et al. (1998; online supplement) used a one-dimensional SOM grid to order genes during sporulation, revealing a pattern: The top fourth of the genes are repressed at early time points and then released, while the bottom fourth of the genes are either repressed or reduced at early time points and highly induced at later time points.

Fuzzy Clustering. All the clustering methods described above are exclusive partitions, where no objects belong to more than one subset or cluster. Fuzzy clustering, which uses the degree of membership of an object to describe whether it belongs to a cluster, allows a nonexclusive classification of objects. The degree of membership is based on the fuzzy-set theory developed by Zadeh (1965). This theory conceptually modeled touching or overlapping clusters that do not have well-defined boundaries. A fuzzy-clustering algorithm is fuzzy c-means (Cannon et al., 1986). Gasch and Eisen (2002) demonstrated the relevance of overlapping cluster assignments to the conditional coregulation of yeast genes.

Cluster Validity. The results of many clustering algorithms vary with the choice of parameters. For example, when a hierarchical clustering procedure is applied, where is the best cutting level? How does one determine k in k-means? Should these clustering results be believed? Those are the determinants of cluster validity. Although a visual inspection can help to validate the clustering results, a numeric measure of cluster validity is required in many cases.

Internal Validity of Clusters. Separability can be used as an internal validation measurement of clustering results. Davies and Bouldin (1979) developed an index to indicate the compactness and separation of clusters. A minimum within-cluster scatter and a maximum between-class separation will yield a lower number on the Davies–Bouldin (DB) index, which indicates superior clustering. For k-means clustering, a plot of DB against k will help to identify the best k for clustering.

External Validity. The goal of cluster analysis is to classify objects into meaningful subsets in a biological context. External validation of clusters utilizes information from labels to determine this. Information from labels is an independent source of the features used in clustering. In Spellman et al. (1998), the existence of the MCB element is used as a label to validate the G1 cluster. For the genes in the G1 cluster, 58% (vs. 6% in control) have a copy of the perfect MCB element. This element is bound by MBF, whose activity depends on cyclin activation. A formal way to assess the support of a priori labeling of a certain partition is to use Hubert's Γ statistics (Hubert and Arabie, 1985).

Stability. Resampling techniques can be used to check the stability of clustering. Robust clusters are less likely to be the result of a sample artifact or fluctuation. Felsenstein (1985) used bootstrapping (Efron and Tibshirani, 1993) to estimate the confidence of a dendrogram, which was followed up by Efron et al. (1996). A discussion of the use of bootstrapping to assess the stability of hierarchical clustering of microarray data can be found in Zhang and Zhao (2000).

Conceptualization from Clusters. In Spellman et al. (1998), expression information regarding 800 genes is reduced to a smaller number of exemplar groups by clustering. The authors further conceptualized the clusters into an MCM cluster, MET cluster, histone cluster, etc. These concepts extracted from clusters are easy to understand and remember. They greatly facilitate communication and understanding of the biology. Text data mining strategies can also help the conceptualization process. Inpharmix (Greenwood, IN) software searches the MEDLINE literature database to find conceptual schema in a list of genes identified in the same cluster.

Implementation Strategies. There are a large number of clustering methods reported in the literature. Readers can always get freeware executables from the individual authors (Eisen et al., 1998; Tamayo et al., 1999) or they can code based on the published algorithm. Several commercial data analysis packages provide a collection of documented and tested programs for clustering: SAS (RTP, NC), S-plus (Seattle, WA), and Clustan (Edinburgh, UK). There are also software tools that are specifically designed for analyzing microarray data, such as GeneSpring (Redwood City, CA), BioDiscovery (Los Angeles, CA), Partek (St. Charles, MI), and SpotFire (Cambridge, MA). The final choice of analysis methods is affected by the investigation goal, the availability of the software, and the computational complexity of the algorithm.

Supervised Machine Learning

The problem of supervised machine learning can be stated as follows: Given a training set of samples with known classifications (labels), build a machine that can classify objects without label information. Supervised learning performs a training process in which the system is supervised to learn the previously classified cases by using both features and category labels. In contrast, clustering is called unsupervised learning because neither a training session nor a priori category labels are used in the partition process. Thus, the "unconstrained" nature of clustering makes it appropriate for pattern discovery, such as finding new subclass of diseases, while supervised machine learning is more powerful to classify patients according to predefined diagnostic categories. The current challenge is to classify patients according to their clinical, morphological, and chemical data along with their molecular expression data.

What to Classify. In supervised learning of microarray analysis, the objects for classification are usually patient microarray experiments or biological samples under varying conditions. For example, *m* patients may be described by their gene expression profile as a vector with n dimensions, each dimension a gene. The goal is to classify the patients in the n-dimensional space.

Linear Discriminant Analysis. Mathematically, a classifier is a mapping of m objects in a set X to a much smaller set C of k classes. Thus, a classifier is a mapping function

$$f: X \longrightarrow C$$

where the input to the function f is the pattern of an object and the output is a decision on its classification. The simplest mapping can be achieved by a linear discriminant function that takes a linear combination of the features from the input and correlates them with the output categories. This is called linear discriminant analysis (LDA). The discriminant function L can be written as

$$L = b_1 x_1 + b_2 x_2 + b_3 x_3 + \cdots + b_n x_n$$

where L is the LDA score, $x_1, x_2, \ldots, x_n$ are the features of object x, and $b_1, b_2,$ b_n are the weights. The discriminant function L defines a hyperplane that separates the objects into classes. In a simple case of classifying x into one of the two categories, a cutoff point c can be used to make the classification decision. If $L \geq c$, then object x belongs to category 1; otherwise, it belongs to category 2. In cases where the input is not linearly separable, LDA can fail.

k-Nearest Neighbor. Because it uses the distribution of the probability density functions as an assumption, LDA is based on parametric statistics. In contrast, k-nearest neighbor (k-NN) is a nonparametric classifier. It does not depend on the distribution assumption of the probability density function. Nearest neighbor is one of the simplest approaches to classifying an unknown pattern by matching it to the closest known patterns in the training samples. This learn-by-example approach is suggested by physicians who diagnose patients by matching the symptoms of the present patient to past patients with correct diagnoses. A nearest-neighbor classifier looks at the nearest neighbor, or k nearest neighbors, to a given input and classifies it according to how the majority of its neighbors have been classified. This approach has been used to classify AML from ALL (Zhao et al., 2000).

Artificial Neural Networks. Artificial neural networks (Hertz et al., 1991) were developed rapidly in the 1980s and are now used widely in classification problems. These networks use several layers of interconnected "neurons" to achieve the mapping from X to C. Knowledge of classification is stored in the connection (weights) between the neurons. The main characteristic of neural networks is their ability to learn complex nonlinear input–output relationships. Ellis et al. (2002) used artificial neural networks to classify cancer and noncancer breast biopsies by microarray profiling. The drawback is that the reasoning of neural networks is a "black box."

This black box can work well on a classification job, but the classification process is hard for humans to interpret.

Decision Trees. A decision tree classifies an object by traversing a tree-shaped flow chart of questions. The cascade of yes or no answers at each node will eventually direct the object to the appropriate class assignment. Compared to artificial neural networks, the explicit if–then rules of decision trees make it easier for humans to interpret the results. Software implementation of decision trees can be found in Classification and Regression Trees (CART; Breiman, 1984) and in Quinlan's development of ID3 (Quinlan, 1996), C4.5, and C5.0. Dubitzky et al. (2000) compared decision trees and neural networks in classifying leukemias.

Support Vector Machines. Support vector machines (SVMs) are learning algorithms developed in the 1990s (Cristianini and Shawe-Taylor, 2000). They map the input feature space into a higher dimensional space where the objects are linearly separable. A separating hyperplane is then constructed with the maximum margin to avoid overfitting. To reduce computational complexity, SVMs utilize a kernel method (Burges et al., 1999) to calculate the hyperplane without explicitly carrying the mapping into a higher dimensional space. Support vector machines have been applied in pattern recognition domains, including handwriting recognition, speaker identification, and text categorization (Burges, 1998). For microarray studies, SVM has been recently utilized successfully to classify cancer subtypes (Furey et al., 2000; Valentini, 2002).

Evaluation of a Classifier. To evaluate a classifier, the available samples are divided into a training set and a test set. The classifier is optimized using the training set and evaluated on the samples from the test set. There are three aspects to consider when evaluating a classifier: accuracy, complexity, and interpretability. To measure the accuracy, *training error rate* and *generalization error rate* are usually estimated. Training error rate is the predicted error rate on the training set, while generalization error rate is the predicted error rate on the test set. Other more computationally expensive evaluations of accuracy include *m*-fold cross-validation (Breiman, 1984) and bootstrapping (Efron, 1983). Leave-one-out is a special case of *m*-fold cross-validation. In addition to accuracy, the complexity of the classifier should be optimized. Usually, a bigger decision tree or a larger neural network will result in a better training error rate. Unfortunately, beyond a certain complexity, the classifier is overtrained on the training set and loses its generality. Thus, its generalization error rate increases. The third evaluation criterion of a classifier is interpretability, because the greater motivation for building classifiers in a biomedical domain is to understand the biology behind the logic of the classification. This interpretation of the classifier can help a biologist gain insight into a complex system. Decision trees are generally more interpretable since we can easily identify rules for human experts to understand. Neural networks, on the other hard, are largely black boxes. It is well documented in medical diagnostics that physicians prefer interpretable classifiers rather than classifiers with the best performance measure.

Data Visualization

Visualization is especially useful for obtaining an understanding of the data during the exploratory stage of analysis. It exploits human intuition to help recognize patterns. Although sophisticated computer programs can use shape, color, and motion to represent more dimensions, the gift of human pattern recognition is better exploited if the data are represented in two- or three-dimensional space. Therefore, visualizations in two or three dimensions are more common.

Principal-Component Analysis and Multidimensional Scaling. The relationship among objects can be visually inspected to find their clustering trends if they can be plotted in a two- or three-dimensional space. Thus, the dimensional reduction techniques of PCA and MDS can also be used as means of data visualization. Bittner et al. (2000) used MDS visualization in three dimensions to examine the possibilities of recognizing subtypes of cutaneous malignant melanoma, while Misra et al. (2002) illustrated the application of PCA in pattern exploration of expression data.

Dendrograms. The results of hierarchical clustering can be rendered as dendrograms. A dendrogram is a special type of tree structure that consists of layers of nodes each representing a certain level of granularity. Lines connect nodes of clusters that are nested into one another. The biologist can inspect the grouping visually and try to cut the dendrogram at different horizontal levels to create meaningful partitions (Fig. 7.1). The dendrogram should be interpreted with caution because it is often an unrooted tree. One should not be visually misled by the arrangement of

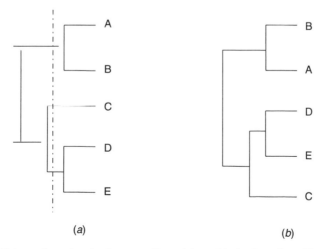

(a)   (b)

Figure 7.1. Interpreting the dendrogram from hierarchical clustering. The dashed line indicates the cut used to yield three clusters. Panels (a) and (b) are equivalent, although the apparent gene orders are different.

the objects in the dendrogram, since the branches at each node can be flipped. In Figure 7.1, dendrograms (*a*) and (*b*) are exactly equivalent, although their orders are completely different.

Data Matrix Visualization. The expression data matrix can be visualized as a color map (each value encoded with a certain color). Eisen et al. (1998) combined the color map with the dendrogram to help biologists rapidly assimilate and interpret microarray data. Our discussion has focused on machine-learning approaches to identify patterns. Here we briefly survey two other approaches to analysis of gene expression data and provide examples of applications to biological problems.

Other Types of Gene Expression Data Analysis

Pattern Recognition. This type of analysis is usually applied to large collections of patient samples (Perou et al., 2000) or to datasets collected by measuring expression changes during a time course (Spellman et al., 1998) under numerous perturbations (Hughes et al., 2000). Machine-learning methodologies, either supervised or unsupervised, are used to recognize patterns in these large datasets.

Detecting Differentially Expressed Genes. The goal of this type of analysis is to statistically detect differentially expressed genes under predefined conditions or treatments. Many criteria have been proposed for identifying significant changes. Early rule-of-thumb methods only consider the expression differences by picking genes with a change of more than twofold. A simple statistical model using Student's *t*-test or ANOVA can take the observed variations into account. However, even when performed in triplicate, microarray experiments usually lack adequate degrees of freedom to reliably estimate the true variance. Thus, modified *t*-tests have been proposed either by using pooled sample variance estimations (Li et al., 2001) or by regularization (Tusher et al., 2001). Bayesian approaches have also been proposed (Baldi and Long, 2001; Lonnstedt and Speed, 2002). Generally, these statistical methods use a gene-by-gene modeling approach, which necessitates corrections for testing multiple comparisons. The utilization of standard Bonferroni correction and Westfall–Young step-down *p*-value adjustments to control family-wise error rate has been investigated (Dudoit et al., 2002), and others have argued that a less conservative method to control false discovery rate is more appropriate in the context of microarray analysis (Tusher et al., 2001).

Understanding Network Regulations. A third and more challenging approach is to analyze expression data in the context of gene regulation networks. A major characteristic of living organisms is that the expression levels of different genes are interconnected with one another through complex regulatory networks and pathways. Microarray data have been reexamined in the context of protein–protein interaction networks (Ge et al., 2001), protein–DNA interaction networks (Aubrey et al., 2001), and literature data-mining networks (Sluka, 2001). An even greater challenge is to use gene expression data to elucidate the underlying blueprint of biological regulatory networks (Greller and Somogyi, 2002). Formal mathematical tools such as Bayesian networks (Friedman et al., 2000), Bayesian decomposition (Bidaut et al., 2001), and SVD (Yeung et al., 2002) have been applied to this challenge, but successful application of this approach will require larger expression datasets than are currently available.

Biological Applications of Expression Profiling

Microarray and SAGE-based expression profiling has been applied to a number of biological problems.

Gene Annotation. The annotation of the human genomic sequence requires the identification of all coding and noncoding exons. These exons can be computationally predicted but must also be experimentally confirmed and assigned to their respective genes. Using the idea that all exons of a given gene will be expressed coordinately, Shoemaker et al. (2001) have placed all known or predicted exons from chromosome 22 q on microarrays and conducted a series of hybridizations to those chips. Genes with similar biological functions also tend to be coexpressed under a variety of experimental conditions. This "guilt-by-association" approach can be used to assign putative functions to uncharacterized genes that do not share any sequence or structural similarity with known proteins. An unsupervised learning approach to clustering was used to infer the function of uncharacterized *Saccharomyces* and human genes in this way (Eisen et al., 1998). Support vector machines have also been utilized to functionally classify genes in a supervised-learning manner (Brown et al., 2000).

Elucidation of Transcriptional Control Pathways. SAGE analysis after induction of p53 by heat shock (Madden et al., 1997) and microarray analysis after zinc-induced p53 expression (Zhao et al., 2000) both identified hundreds of genes whose expression levels are regulated by p53. These genes fell into categories of apoptosis and growth arrest, cytoskeletal function, growth factors, extracellular matrix, and adhesion proteins. These relatively simple experiments were able to rapidly outline the entire regulatory pathway for this important protein. Similar experiments demonstrated that *BRCA1* stimulates the GADD45 and JNK/SAPK apoptosis pathways (Harkin et al., 1999). Interestingly, these experiments have shown that p53 and *BRCA1* work together in a coordinated network. Methods have been proposed to reverse engineer such transcription networks (D'Haeseleer et al., 2000).

Identification of Transcription Factor Binding Sites. Sequences upstream of yeast genes with similar expression profiles have been searched with pattern discovery algorithms to identify conserved gene regulatory elements (Brazma et al., 1998). This approach identified several known yeast transcription factor binding sites. Novel targets for the yeast transcription factors SBF and MBF were identified using an extension of this approach in which protein was chemically crosslinked to DNA in vivo, purified by immunoprecipitation, PCR amplified, and then used to probe microarrays (Iyer et al., 2001). This powerful approach identified over 200 novel gene targets for these transcription factors.

Cancer Research. Molecular classification of malignant tissue was discussed in the introduction to this chapter. Other work has demonstrated that gene expression patterns can be used to identify the tissue of origin of most tumors (Ross et al., 2000). The CGAP has databased more than 100 SAGE libraries constructed from normal and malignant tissue. Analysis of these data has identified hundreds of genes involved in cancer pathogenesis, ranging from angiogenesis factors to cell cycle regulators to transcription factors (Lal et al., 1999). The molecular pharmacology of cancer has been addressed by profiling gene expression of 60 cancer cell lines before and after drug treatments. This work has led to methods for predicting drug sensitivity to chemotherapeutic agents on the basis of gene expression (Scherf et al., 2000). In a related study, chemoinformatics (high-throughput screening informed by crystal structures and bioinformatics) was used to develop kinase inhibitors (Gray et al., 1998). The effects of these inhibitors were then characterized by expression analysis before and after drug treatment.

Identifying and Prioritizing Candidate Genes for Complex Genetic Disorders. The techniques of gene expression profiling can assist in the search for and analysis of candidate genes for complex genetic disease. Linkage analysis (described in detail in Chapters 9–11) can identify genomic regions that harbor susceptibility loci for a complex disease; however, the regions of linkage are frequently large (20 cM or greater) and may contain hundreds of genes. It is impractical or impossible to evaluate all of these positional candidates for possible disease-related polymorphisms, and prioritization of candidates solely on the basis of known or inferred biological activity could miss many relevant genes.

Identifying and demonstrating the effects of polymorphisms in susceptibility genes in complex disease can be challenging. While there exist rare premature stop-codon mutations such as those in the TIGR/myocilin gene that lead directly to glaucoma (Allingham et al., 1998; Suzuki et al., 1997), the more common situation is likely to be the identification of polymorphisms—defined as having >1% frequency in the population—that increase risk of developing disease but may not by themselves directly result in disease. This is the very nature of complex disease: Multiple susceptibility genes will exist, with predisposing polymorphisms in any one gene showing reduced penetrance as well as potential interactions with environmental factors. Thus some individuals will have a given predisposing polymorphism yet will not exhibit disease, while at the same time, other individuals that do exhibit disease will lack that specific predisposing polymorphism. These difficulties in the interpretation of polymorphisms or sequence variants reflect our evolving understanding of the genetic etiology of complex diseases in general.

Precisely because of these difficulties, the experimental approaches with the greatest likelihood of success will combine multiple different kinds of analysis, including family-based linkage and association analysis, gene expression analysis, and functional studies of normal and mutant proteins. This combination of strategies takes advantage of the strengths of each method. Gene expression studies can benefit the search for susceptibility loci in several ways.

Expression Profiling to Prioritize Candidate Genes. Linkage intervals harboring candidate genes can be quite large and encompass hundreds of genes. These genes can be prioritized for mutation and polymorphism detection by using microarray or SAGE analysis on the relevant tissue. For example, the Udall Parkinson Disease Center of Excellence at the Duke University Medical Center has conducted SAGE and microarray analysis of substantia nigra tissue from Parkinson disease (PD) patients and age-matched controls. The substantia nigra is a pivotal tissue in PD—patients exhibit a dramatic loss of dopaminergic neurons in this tissue as disease progresses. Genes whose expression levels are increased or decreased in PD patients as compared to controls are good candidates for PD susceptibility genes, as are genes that are preferentially expressed in the substantia nigra as compared to other neural tissues. Hundreds of genes will be differentially expressed in this way, but only a subset of these differentially expressed genes are located within linkage intervals. These selected genes represent excellent candidates because two independent experimental procedures (linkage mapping and expression profiling) have identified them as potential susceptibility genes. In this way, expression analysis can dramatically reduce the number of candidate genes that must be evaluated following linkage analysis. This combined strategy has also been used to identify a number of high-probability candidate genes for mania and psychosis (Niculescu et al., 2000).

Annotation of Genes within Linkage Intervals. The SAGE approach discussed above will be especially powerful when combined with the recently developed "long SAGE" protocol (V. Velculescu, personal communication). This modified SAGE protocol uses a different tagging enzyme to generate 20-bp SAGE tags rather than the standard 14-bp tags. While this increases sequencing costs, it greatly reduces redundancy in tag-to-gene mapping. Also, primers designed from long SAGE tags can be used in conjunction with oligo dT primers to directly amplify the 3' ends of novel genes from total RNA. Subsequent 5' or 3' rapid amplification of cDNA ends (RACE) allows the isolation of the full-length transcript. Because SAGE is an open-platform technology (it does not rely on prior knowledge of genes), this strategy will allow the identification of entirely novel genes within linkage intervals.

Genes within a linkage interval can also be annotated by constructing a microarray of all known or predicted exons within the interval. This is a large undertaking but is becoming increasingly feasible as both genomic sequence quality and microarray spotting technologies improve. Such an array could then be probed with RNA from the relevant tissue to determine the genes and exons expressed in that tissue. This approach has been used to experimentally annotate the genes expressed on 22q (Shoemaker et al., 2001). Further, if RNA from an affected individual were used to probe such an array, patient-specific defects in gene expression or transcript splicing patterns could be detected directly. Such changes are often very difficult to detect by searching for sequence variations in the genomic DNA of affected individuals.

Gene expression profiling is an extraordinarily powerful research tool. We have concentrated here on two techniques: SAGE and microarray hybridization. These approaches have been used to characterize the regulation of transcription in many organisms and systems, but their application to complex disease is still in its early stages. We have described a few ways in which these techniques might be applied to this area of research, and undoubtedly many more applications will follow as the full potential of gene expression profiling is realized.

REFERENCES

Alizadeh AA, Eisen MB, Davis RE, Ma C, Lossos IS, Rosenwald A, Boldrick JC, et al (2000): Distinct types of diffuse large B-cell lymphoma identified by gene expression profiling. Nature 403:503–511.

Allingham RR, Wiggs JLdlPMA, Vollrath D, Tallett DA, Broomer R, Jones KH, Del Bono EA, Kern J, Patterson K, Haines JL, and Pericak-Vance MA (1998): Gln368STOP myocilin mutation in families with late-onset primary open-angle glaucoma. Invest Ophthalmol Vis Sci 39:2288–2295.

Alon U, Barkai N, Notterman DA, Gish K, Ybarra S, Mack D, Levine AJ (1999): Broad patterns of gene expression revealed by clustering analysis of tumor and normal colon tissues probed by oligonucleotide arrays. Proc Natl Acad Sci USA 96:6745–6750.

Alter O, Brown PO, Botstein D (2000): Singular value decomposition for genome-wide expression data processing and modeling. Proc Natl Acad Sci USA 97:10101–10106.

Aubrey N, Devaux C, di Luccio E, Goyffon M, Rochat H, Billiald P (2001): Androctonus a, biotin, immunoconjugate, single-chain a, fragment, and strep t: A recombinant scFv/ streptavidin-binding peptide fusion protein for the quantitative determination of the scorpion venom neuà. Biol Chem 382:1621–1628.

Audic S, Claverie J-M (1997): The significance of digital gene expression profiles. Genome Res 7:986–995.

Baldi P, Long AD (2001): A Bayesian framework for the analysis of microarray expression data: Regularized t-test and statistical inferences of gene changes. Bioinformatics 17:509–519.

Bidaut G, Moloshok TD, Grant JD, Manion FJ, Ochs MF (2001): Bayesian decomposition analysis of gene expression in yeast deletion mutants. In: Lin SM, Johnson KF, eds. Methods of Microarray Data Analysis, Vol. II. Boston, MA: Kluwer Acadmeic Publishers.

Bittner M, Meltzer P, Chen Y, Jiang Y, Seftor E, Hendrix M, Radmacher M, et al (2000): Molecular classification of cutaneous malignant melanoma by gene expression profiling. Nature 406:536–540.

Brar AK, Handwerger S, Kessler CA, Aronow BJ (2001): Gene induction and categorical reprogramming during in vitro human endometrial fibroblast decidualization. Physiol Genom 7:135–148.

Brazma A, Jonassen I, Vilo J, Ukkonen E (1998): Predicting gene regulatory elements in silico on a genomic scale. Genome Res 8:1202–1215.

Breiman L (1984): Wadsworth Statistics/Probability Series. Belmont: Wadsworth International Group.

Brown MP, Grundy WN, Lin D, Cristianini N, Sugnet CW, Furey TS, Ares M Jr, Haussler D (2000): Knowledge-based analysis of microarray gene expression data by using support vector machines. Proc Natl Acad Sci USA 97:262–267.

Burges CJC (1998): A tutorial on support vector machines for pattern recognition. Data Mining and Knowledge Discovery 2:121–167.

Burges CJC, Burges CJC, Smola AJ (1999): Advances in Kernel Methods: Support Vector Learning. Cambridge: MIT Press.

Cannon RL, Dave JV, Bezdek JC (1986): Efficient implement of the fuzzy C-means clustering algorithms. Trans Pattern Anal Machine Intell 8:248–255.

Chu S, DeRisi J, Eisen M, Mulholland J, Botstein D, Brown PO, Herskowitz I (1998): The transcriptional program of sporulation in budding yeast. Science 282:699–705.

Cristianini N, Shawe-Taylor J (2000): An Introduction to Support Vector Machines and Other Kernel-Based Learning Methods. New York: Cambridge University Press.

D'Haeseleer P, Liang S, Somogyi R (2000): Genetic network inference: From co-expression clustering to reverse engineering. Bioinformatics 16:707–726.

Datson NA, van der Perk-de Jong J, van den Berg MP, de Kloet ER, Vreugdenhil E (1999): MicroSAGE: A modified procedure for serial analysis of gene expression in limited amounts of tissue. Nucleic Acids Res 27:1300–1307.

Davies DL, Bouldin DW (1979): Cluster separation measure. IEEE Trans Pattern Anal Machine Intell 1:224–227.

Dubitzky W, Granzow M, Berrar D (2000): Comparing symbolic and subsymbolic machine learning approaches to classification of cancer and gene identification. In: Lin SM, Johnson KF, eds. Methods of Microarray Data Analysis, Vol. I. Boston, MA: Kluwer Academic.

Dudoit S, Yang YH, Callow MJ, Speed TP (2002): Statistical methods for identifying differentially expressed genes in replicated cDNA microarray experiments. Statistica Sinica 12:111–139.

Durbin BP, Hardin JS, Hawkins DM, Rocke DM (2002): A variance-stabilizing transformation for gene-expression microarray data. Bioinformatics 18(Suppl 1):S105–S110.

Efron B (1983): Estimating the error rate of prediction rule—Improvement on cross-validation. J Am Statist Assoc 78:316–331.

Efron B, Halloran E, Holmes S (1996): Bootstrap confidence levels for phylogenetic trees. Proc Natl Acad Sci USA 93:7085–7090.

Efron B, Tibshirani R (1993): An Introduction to the Bootstrap, 57th ed. New York: Chapman & Hall.

Eisen MB, Spellman PT, Brown PO, Botstein D (1998): Cluster analysis and display of genome-wide expression patterns. Proc Natl Acad Sci USA 95:14863–14868.

Ellis M, Davis N, Coop A, Liu M, Schumaker L, Lee RY, Srikanchana R, Russell CG, Singh B, Miller WR, Stearns V, Pennanen M, Tsangaris T, Gallagher A, Liu A, Zwart A, Hayes DF, Lippman ME, Wang Y, Clarke R (2002): Development and validation of a method for using breast core needle biopsies for gene expression microarray analyses. Clin Cancer Res 8:1155–1166.

Felsenstein J (1985): Confidence-limits on phylogenies—An approach using the bootstrap. Evolution 39:783–791.

Fitch WM, Margoliash E (1967): Construction of phylogenetic trees. Science 155:279–284.

Fowlkes EB, Gnanadesikan R, Kettenring JR (1987): Variable selection in clustering and other contexts. In: Daniel C, Mallows CL, eds. Design, Data, and Analysis, 380th ed. New York: John Wiley & Sons.

Friedman N, Linial M, Nachman I, Pe'er D (2000): Using Bayesian networks to analyze expression data. J Comput Biol 7:601–620.

Furey TS, Cristianini N, Duffy N, Bednarski DW, Schummer M, Haussler D (2000): Support vector machine classification and validation of cancer tissue samples using microarray expression data. Bioinformatics 16:906–914.

Gasch AP, Eisen MB (2002): Exploring the conditional coregulation of yeast gene expression through fuzzy k-means clustering. Genome Biol 3:RESEARCH0059.

Ge H, Liu Z, Church GM, Vidal M (2001): Correlation between transcriptome and interactome mapping data from *Saccharomyces cerevisiae*. Nat Genet 29:482–486.

Getz G, Levine E, Domany E (2000): Coupled two-way clustering analysis of gene microarray data. Proc Natl Acad Sci USA 97:12079–12084.

Gower JC (1966): Some distance properties of latent root and vector methods used in multivariate analysis. Biometrika 53:325–338.

Gray NS, Wodicka L, Thunnissen AM, Norman TC, Kwon S, Espinoza FH, Morgan DO, Barnes G, LeClerc S, Meijer L, Kim SH, Lockhart DJ, Schultz PG (1998): Exploiting chemical libraries, structure, and genomics in the search for kinase inhibitors. Science 281:533–538.

Green CD, Simons JF, Taillon BE, Lewin DA (2001): Open systems: panoramic views of gene expression. J Immunol Methods 250:67–79.

Greller LD, Somogyi R (2002): Reverse engineers map the molecular switching yards. Trends Biotechnol 20:445–447.

Harkin DP, Bean JM, Miklos D, Song YH, Truong VB, Englert C, Christians FC, Ellisen LW, Maheswaran S, Oliner JD, Haber DA (1999): Induction of GADD45 and JNK/SAPK-dependent apoptosis following inducible expression of BRCA1. Cell 97:575–586.

Hertz J, Krogh A, Palmer RG (1991): Introduction to the Theory of Neural Computation. Santa Fe Institute Studies in the Sciences of Complexity. Lecture Notes, Vol. 1. Redwood City: Addison-Wesley.

Hilsenbeck SG, Friedrichs WE, Schiff R, O'Connell P, Hansen RK, Osborne CK, Fuqua SA (1999): Statistical analysis of array expression data as applied to the problem of tamoxifen resistance. J Nat Cancer Inst 91:453–459.

Hoffmann R, Seidl T, Dugas M (2002): Profound effect of normalization on detection of differentially expressed genes in oligonucleotide microarray data analysis. Genome Biol 3:RESEARCH0033.

Huber W, Von Heydebreck A, Sultmann H, Poustka A, Vingron M (2002): Variance stabilization applied to microarray data calibration and to the quantification of differential expression. Bioinformatics 18(Suppl 1):S96–S104.

Hubert L, Arabie P (1985): Comparing partitions. J Classification 2:193–218.

Hughes TR, Marton MJ, Jones AR, Roberts CJ, Stoughton R, Armour CD, Bennett HA, Coffey E, Dai H, He YD, Kidd MJ, King AM, Meyer MR, Slade D, Lum PY, Stepaniants

SB, Shoemaker DD, Gachotte D, Chakraburtty K, Simon J, Bard M, Friend SH (2000): Functional discovery via a compendium of expression profiles. Cell 102:109–126.

Ishida N, Hayashi K, Hoshijima M, Ogawa T, Koga S, Miyatake Y, Kumegawa M, Kimura T, Takeya T (2002): Large scale gene expression analysis of osteoclastogenesis in vitro and elucidation of NFAT2 as a key regulator. J Biol Chem 277:41147–41156.

Iyer VR, Horak CE, Scafe CS, Botstein D, Snyder M, Brown PO (2001): Genomic binding sites of the yeast cell-cycle transcription factors SBF and MBF. Nature 409:533–538.

Jain AK, Chandrasekaran B (1982): In: Krishnaiah PR, Kanal LN, eds. Handbook of Statistics, Vol. 2. Amsterdam: North-Holland.

Jolliffe IT (1986): Principal Components Analysis. New York: Springer-Verlag.

Kepler TB, Crosby L, Morgan KT (2002): Normalization and analysis of DNA microarray data by self-consistency and local regression. Genome Biol 3:RESEARCH0037.

Kohonen T (1995): Self-Organizing Maps. New York: Springer-Berlin.

Kruskal J (1964): Multidimensional scaling by optimizing goodness to fit to nonmetric hypothesis. Psychometrika 29:1–27.

Lal A, Lash AE, Altschul SF, Celculescu V, Zhang L, McLendon RE, Marra MA, Prange C, Morin PJ, Polyak K, Papadopoulos N, Vogelstein B, Kinzler KW, Strausberg RL, Riggins GJ (1999): A Public database for gene expression in human cancers. Cancer Res 59:5403–5407.

Lee SW, Tomasetto C, Sager R (1991): Positive selection of candidate tumor-suppressor genes by subtractive hybridization. Proc Nat Acad Sci USA 88:2825–2829.

Li YJ, Zhang L, Speer MC, Martin ER (2001): Evaluation of current methods of testing differential gene expression and beyond. In: Lin SM, Johnson KF, eds. Methods of Microarray Data Analysis, Vol. II. Boston, MA: Kluwer Academic.

Liang P, Pardee AB (1992): Differential display of eukaryotic messenger RNA by means of the polymerase chain reaction. Science 257:967–971.

Lonnstedt I, Speed T (2002): Replicated microarray data. Statistica Sinica 12:31–46.

MacQueen J (1967): Some methods for classification and analysis of multivariate observations. In: Cam LL, Neyman J, eds. Proceedings of the Fifth Berkeley Symposium on Mathematical Statistics and Probability, Vol. 1. University of California Press.

Madden SL, Galella EA, Zhu J, Bertelsen AH, Beaudry GA (1997): SAGE transcript profiles for p53-dependent growth regulation. Oncogene 15:1079–1085.

Man MZ, Wang X, Wang Y (2000): POWER_SAGE: Comparing statistical tests for SAGE experiments. Bioinformatics 16:953–959.

Margulies EH, Innis JW (2000): eSAGE: Managing and analysing data generated with serial analysis of gene expression (SAGE). Bioinformatics 16:650–651.

Misra J, Schmitt W, Hwang D, Hsiao LL, Gullans S, Stephanopoulos G, Stephanopoulos G (2002): Interactive exploration of microarray gene expression patterns in a reduced dimensional space. Genome Res 12:1112–1120.

Neilson L, Andalibi A, Kang D, Coutifaris C, Strauss JF III, Stanton JL, Green DPL (2000): Molecular phenotype of the human oocyte by PCR-SAGE. Genomics 63:13–24.

Niculescu AB III, Segal DS, Kuczenski R, Barrett T, Hauger RL, Kelsoe JR (2000): Identifying a series of candidate genes for mania and psychosis: A convergent functional genomics approach. Physiol Genomics 4:83–91.

Perou CM, Sorlie T, Eisen MB, van de RM, Jeffrey SS, Rees CA, Pollack JR, Ross DT, Johnsen H, Akslen LA, Fluge O, Pergamenschikov A, Williams C, Zhu SX, Lonning PE,

Borresen-Dale AL, Brown PO, Botstein D (2000): Molecular portraits of human breast tumours. Nature 406:747–752.

Peters DG, Kassam AB, Yonas H, O'Hare EH, Ferrell RE, Brufsky AM (1999): Comprehensive transcript analysis in small quantities of mRNA by SAGE-lite. Nucleic Acids Res 27:e39.

Quinlan JR (1996): Learning decision tree classifiers. ACM Comput Surv 28:71–72.

Raffelsberger W, Dembele D, Neubauer MG, Gottardis MM, Gronemeyer H (2002): Quality indicators increase the reliability of microarray data. Genomics 80:385–394.

Ross DT, Scherf U, Eisen MB, Perou CM, Rees C, Spellman P, Iyer V, Jeffrey SS, van de Rijn M, Waltham M, Pergamenschikov A, Lee JC, Lashkari D, Shalon D, Myers TG, Weinstein JN, Botstein D, Brown PO (2000): Systematic variation in gene expression patterns in human cancer cell lines [see comments]. Nat Genet 24:227–235.

Scherf U, Ross DT, Waltham M, Smith LH, Lee JK, Tanabe L, Kohn KW, Reinhold WC, Myers TG, Andrews DT, Scudiero DA, Eisen MB, Sausville EA, Pommier Y, Botstein D, Brown PO, Weinstein JN (2000): A gene expression database for the molecular pharmacology of cancer. Nat Genet 24:236–244.

Schroeder SR, Oster-Granite ML, Berkson G, Bodfish JW, Breese GR, Cataldo MF, Cook EH, et al (2001): Self-injurious behavior: Gene-brain-behavior relationships. Ment Retard Dev Disabil Res Rev 7:3–12.

Shoemaker DD, Schadt EE, Armour CD, He YD, Garrett-Engele P, McDonagh PD, Loerch PM, et al (2001): Experimental annotation of the human genome using microarray technology. Nature 409:922–927.

Skinner HA (1978): Differentiating contribution of elevation, scatter and shape in profile similarity. Ed Psychol Measur 38:297–308.

Sluka JP (2001): Extracting knowledge from genomic experiments by incorporating the biomedical literature. In: Lin SM, Johnson KF, eds. Methods of Microarray Data Analysis, Vol. II. Boston, MA: Kluwer Academic.

Sneath PHA, Sokal RR (1973): Numerical Taxonomy; the Principles and Practice of Numerical Classification. San Francisco: W. H. Freeman.

Spellman PT, Sherlock G, Zhang MQ, Iyer VR, Anders K, Eisen MB, Brown PO, Botstein D, Futcher B (1998): Comprehensive identification of cell cycle-regulated genes of the yeast *Saccharomyces cerevisiae* by microarray hybridization. Mol Biol Cell 9:3273–3297.

Suzuki Y, Shirato S, Taniguchi F, Ohara K, Nishimaki K, Ohta S (1997): Mutations in the TIGR gene in familial primary open-angle glaucoma in Japan. Am J Hum Genet 61:1202–1204.

Tamayo P, Slonim D, Mesirov J, Zhu Q, Kitareewan S, Dmitrovsky E, Lander ES, Golub TR (1999): Interpreting patterns of gene expression with self-organizing maps: Methods and application to hematopoietic differentiation. Proc Natl Acad Sci USA 96:2907–2912.

Tusher VG, Tibshirani R, Chu G (2001): Significance analysis of microarrays applied to the ionizing radiation response. Proc Natl Acad Sci USA 98:5116–5121.

Valentini G (2002): Gene expression data analysis of human lymphoma using support vector machines and output coding ensembles. Artif Intell Med 26:281–304.

Velculescu VE, Zhang L, Vogelstein B, Kinzler KW (1995): Serial analysis of gene expression. Science 270:484–487.

Virlon B, Cheval L, Buhler JM, Billon E, Doucet A, Elalouf JM (1999): Serial microanalysis of renal transcriptomes. Proc Natl Acad Sci USA 96:15286–15291.

Wang X, Ghosh S, Guo SW (2001): Quantitative quality control in microarray image processing and data acquisition. Nucleic Acids Res 29:e75.

Yang MC, Ruan QG, Yang JJ, Eckenrode S, Wu S, McIndoe RA, She JX (2001): A statistical method for flagging weak spots improves normalization and ratio estimates in microarrays. Physiol Genomics 7:45–53.

Yang YH, Dudoit S, Luu P, Lin DM, Peng V, Ngai J, Speed TP (2002): Normalization for cDNA microarray data: A robust composite method addressing single and multiple slide systematic variation. Nucleic Acids Res 30:e15.

Yeung MK, Tegner J, Collins JJ (2002): Reverse engineering gene networks using singular value decomposition and robust regression. Proc Natl Acad Sci USA 99:6163–6168.

Zadeh LA (1965): Fuzzy sets. Information and Control 8:338–353.

Zhang K, Zhao H (2000): Assessing reliability of gene clusters from gene expression data. Functional & Integrative Genomics 1:156–173.

Zhang L, Zhou W, Velculescu VE, Kern SE, Hruban RH, Hamilton ST, Vogelstein B, Kinzler KW (1997): Gene expression profiles in normal and cancer cells. Science 276:1268–1272.

Zhao R, Gish K, Murphy M, Yin Y, Notterman D, Hoffman WH, Tom E, Mack DH, Levine AJ (2000): Analysis of p53-regulated gene expression patterns using oligonucleotide arrays. Genes Dev 14:981–993.

 **CHAPTER 8**

Information Management

CAROL HAYNES and COLETTE BLACH

The success of a genetic analysis project, or any research project, is significantly influenced by both the quality and quantity of data available for study. The likelihood of finding data to answer a pressing research question or to verify a hypothesis depends on the ability to store and retrieve the necessary data in a timely fashion. Clearly this requires advance planning for the selection of important data points and choosing the software and hardware that will handle the data storage and retrieval requirements. Functionality and usability must be designed into the database system. This chapter details the steps required to produce a functional database and to choose the information tools for a genetic analysis project.

All genetics information management projects follow the decision path below:

1. Develop the experimental design (information planning).
2. Establish the information flow.
3. Create a model for information storage.
4. Determine the hardware/software requirements. Consider the amount of information, cost, performance, security, data integrity, and scalability needed.
5. Define the database structure with a focus on maintaining data integrity (database implementation).
6. Choose the user interface for the selected database. Develop tools to provide the functions required, if necessary.
7. Determine information security requirements and protocols.
8. Select the genetics-specific software tools that will integrate with the database.

This chapter is a primer for information development for a genetic analysis project and will discuss each stage of genetics data development.

Genetic Analysis of Complex Diseases, Second Edition, Edited by Jonathan L. Haines and Margaret Pericak-Vance
Copyright © 2006 John Wiley & Sons, Inc.

219

INFORMATION PLANNING

Needs Assessment

All software development follows a similar process from the initial stage of realization that information processing is required to the project's final deployment. The process begins with the analysis of the data to be handled, establishing how specific users and contributors will utilize the software. The software must be designed with the information's specific functions in mind. This step is often called *needs assessment*.

Genetic information can be divided into three groups: (1) clinical information, (2) blood/tissue sample data, and (3) laboratory findings. Clinical information will include the data necessary to determine a specific phenotype for each individual in the study. Among these are demographic descriptors, physical examination findings, and longitudinal patient status information. Different types of blood and tissue may be collected for examination, and information about informed consent, quantity of collection, and amount of DNA extracted are crucial data. Laboratory findings include the genetic variation results for each individual sample. Personnel performing laboratory tests must be blinded to the origin of the samples in order to avoid biasing the results and to protect the confidentiality of patient data.

The obvious advantage to collecting both laboratory and clinical information is that it enables the investigator to search for patterns of clinical symptoms versus genetic variation. The most successful way to specify information needs is to initially interview the users to establish their requirements. The information that is gathered determines the analysis questions that can be answered at the end of the study. Therefore, knowing what types of questions need to be answered is the key determinant in data collection.

Information Requirements: Clinical Information

Contact Information/Family Ascertainment. Genetic research may require the ascertainment of related individuals (family based) or may use a case/control-type collection or sometimes a combination of both approaches is used. Identifying families that meet study protocol criteria requires a different procedure, and therefore different software, than identifying cases and controls. Another common variable is whether the analyses will be concentrating on a single disease phenotype, multiple phenotypes, or subsets of phenotypes. If the disease phenotype is not sufficiently described before ascertainment begins, important variables that contribute to that phenotype or subtype may not be well defined. Studies can be retrospective or prospective, and the type of contact information required for these may differ. If the study will be longitudinal, the data structures needed to store repeat values are different than those that expect a single value per individual. Also, contact information for families that you expect to follow over time will be more extensive than that gathered for a retrospective study. Many studies utilize periodic newsletters to keep families informed of their research; thus contact information should be collected even for retrospective studies so that each household receives a single copy of the newsletter. For many studies, the subjects are children or are mentally handicapped; contact information for these individuals may include unrelated caregivers.

Individual Information. Collecting related individuals often means that analyses will be performed on specific types of family relationships, that is, affected sibling pairs, discordant sibling pairs, or affected relative pairs. This means that relationship data must be collected and databased in such a way as to enable queries for the requested types of relationships. If matched controls are needed, data must be stored so that matched individuals can be extracted on demand. Individuals may exhibit more than one phenotype, and the database must be able to determine all the different variables that make up any particular phenotype.

Information Requirements: Blood/Tissue Sample Data. Because DNA is an extremely precious resource for genetic research, extra care should be taken with the collection and storage of the data describing blood and tissue sources and the resulting DNA extraction. Taking advantage of an established DNA banking facility would be a wise choice, as up-to-date inventories of the amount of DNA available from each research subject is crucial to successful genotyping. Institutional review board (IRB) concerns make it necessary to destroy all blood and DNA if a subject wishes to withdraw from the study and requests that samples be destroyed. This means that all aliquots of extracted DNA, all frozen blood tubes, and all DNA shared with collaborators must be located and destroyed, which in turn means that such location information must be databased and easily accessible. Other required database variables would be collection date and documentation of informed consent.

Information Requirements: Laboratory Findings. There are two aspects to databasing genotyping results. One is the actual genetic variation, which is linked to a particular individual. The other concerns quality control issues on the genotyping itself.

Genotyping Results—Individuals. The database must be able to keep track of all the genetic variation results for an individual, regardless of how many blood or tissue samples were collected and tested for that individual. Summaries of markers typed for individuals, sets of controls, or families should be easy to query. Aggregate values such as allele frequencies should be produced from the generated genotypes.

Genotyping Results—Quality Control. For quality control purposes, information regarding each genotyping experiment should be databased, recording the laboratory technician performing the experiment and the experimental conditions. The database can also be set up to allow the monitoring of the productivity of individual laboratory technicians. Software can be used to detect possible laboratory errors before genotyping results are allowed into the database.

Bioinformatics. Information regarding primers, including sequence and location, should be databased along with the genotypes. This will facilitate the subsequent statistical analyses of the genetic variation.

INFORMATION FLOW

Information flow refers to tracking the path of the data. For example, magnetic resonance imaging (MRI) may be ordered. When the MRI is received by the data technician, the technician logs the receipt of the image and places the image in the doctor's box to be reviewed. The doctor reads the image and records his or her notes in the patient database. The MRI is stored in the patient file room. The receipt of this MRI must be included in the weekly report summarizing the number of MRIs received.

In this example, the MRI enters this system, is handled by the data technician and doctor, and is finally stored in the file room. Other information is spawned by the MRI's receipt. The image receipt and doctor's comments on the MRI are entered in the database with user input screens. The receipt of the image is incorporated into a summary report. Each of these is an example of information flow.

Information flow is an essential part of any process. For the information to be available as required, the process must be clearly defined (Fig. 8.1):

1. *Define Users and their Roles.* In the example above, the technician and doctor are the users.
2. *Describe the Information Flow.* Information flow describes how information is created, processed, and transformed. Identify what information each user accesses, how it is stored, and with whom it is shared. The MRI is logged, read, and stored. Information about the MRI is logged via user input computer screens and stored in the database.
3. *Identify Information Content.* What data collection tools will be used and what reports need to be produced? The doctor enters text for notes and the technician inputs arrival dates of the image. Summary reports are output weekly.
4. *List Interfaces to the Information.* For example, what software tools will be used and what data formats will be required? In this example, information is entered via computer screen forms.

Object-oriented analysis is an effective way to identify functions and information. First, write the description of the process in sentences. Underline the nouns and noun phrases in the text. Next, using the underlined nouns, verbs, and modifiers, list the information objects (nouns), attributes (modifiers), and corresponding operations (verbs) (Table 8.1). Detail each component of the information flow, such as clinical information, laboratory information, and genotyping results, using top-down methodology. Further, test a variety of situations to confirm that these can be represented in the data flow. To illustrate this process, a genetic interview process is described and used in examples throughout the chapter. This process creates lists of information with its flow and required functions. This description of information flow and functionality requirements can be refined to provide more details.

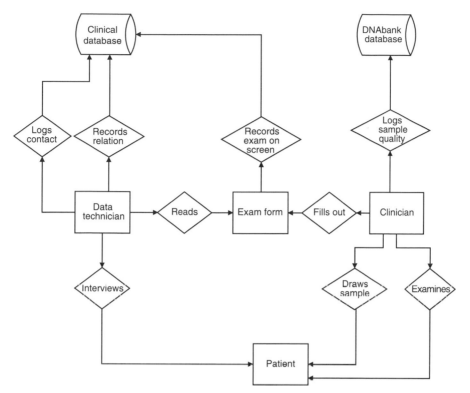

Figure 8.1. Data technicians contact prospective patients with telephone interviews. Those patients who meet criteria are examined by medical staff who fill out clinical exam information on an exam form and draw blood samples. Data technicians enter the exam information via computer screens. The exam records are stored in the clinical database. Sample qualities are recorded in the DNAbank database. The data technicians draw an electronic pedigree and store family relation information into the relations database.

PLAN LOGICAL DATABASE MODEL

Once the requirements are established, they must be converted into a conceptual model. Conceptual models are independent of hardware and software. Instead this model establishes the information to be collected and how the data are related.

TABLE 8.1.

Object	Attribute	Corresponding Operation
Interview	Prospective patient	Contact
Exam form	Clinical	Fill out
Computer screen	Examination form	Enter
Sample qualities	DNAbank database	Record
Pedigree	Electronic	Draw
Relation Information	Relations database	Store

Each object is drawn as a box, and corresponding relations, also described as entities, are drawn as a diamond. Objects are categorized as strong or weak. Strong objects can exist on their own, but weak objects require the presence of another object. For example, an examination would not exist if there was no patient. Each object can have attributes, which are listed next to the object box.

In the example given in Figure 8.2, information about the family relations is stored in the top box. Each family has many individuals. The comment 1 : n under Relation_Patient specifies that each family can have more than one patient but each individual can only belong to a single family. Likewise, each patient can have more than one sample, examination, and contact. For example, there is no restriction that a patient can only be sampled once, but each individual sample can only belong to a single patient.

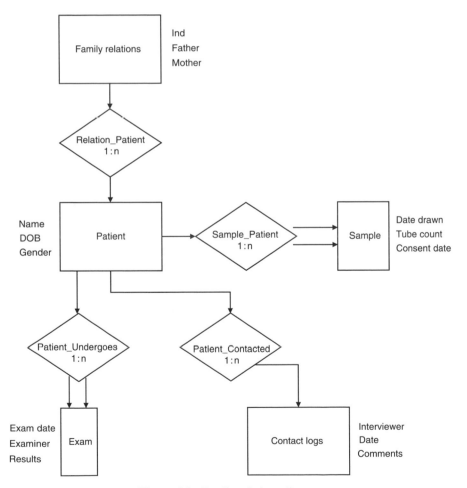

Figure 8.2. Family relations diagram.

Additionally, this conceptual model begins to specify what precise information to collect. For example, for each sample, the date drawn, tube count, and consent date were added outside the object box as attributes to be collected.

Test the entity/relation diagrams with data from real situations to confirm that the diagram handles all contingencies. Consider if additional attributes should be added. For example, for the contact log, should the type of contact, such as telephone, letter, or clinic visit, be recorded? Modify and fine tune the diagram until it satisfies all requirements defined.

Also note that this type of diagram has nothing to do with the database software chosen. The diagram works equally well with commercially available or freeware databases.

HARDWARE AND SOFTWARE REQUIREMENTS

Before the actual database design can take place, the platforms must be selected. The hardware and software together are known as the program's platform. Careful selection is required to achieve an adequate platform for the project's requirements. Selection of hardware, operating system, and software is a puzzle that only fits together in a limited number of ways. Only a few operating systems will run on each hardware platform, and only a few databases will run on each operating system. The obvious challenge, therefore, is to select compatible hardware, operating systems, and programs that will suit the user's needs (Table 8.2).

The computer that will store the database must be fast enough to respond to the user's database questions (called queries) in a reasonable amount of time. The hard drive must be large enough to store the database and allow for additional growth. The computer must have enough memory for the program to

TABLE 8.2. Operating System and Database Characteristics

	Small Database	Large Database
Users	Few users	Many users
Operating system	Windows, Windows NT, MacOS	Windows NT, UNIX
Hardware	Personal computer	Workstation or database server
Database	Spreadsheets, small databases (MS Access)	Sybase, Informix, Oracle, SQLServer
Security, including auditing	Minimal	Extensive
Multiuser facilities	Limited	Extensive
Database administration	Minimal	Required
System administration	Minimal	Required
Transactional processing	Sometimes	Always
Distributed processing	No	Available

respond in a timely fashion. Since the hardware will eventually fail, the computer must have a method to store the data in a protected place. This backup may be on diskettes, CD-ROMs, or zip drives for small amounts of data or on tapes for larger amounts of data. Cost is a primary consideration for selection of a computer. Most computers vary in cost according to size of storage, speed of processor, amount of memory, and inclusion of additional devices such as monitors and tape backups.

Software Selection

A database is defined as the storage of similar information. With this definition, a small database can be stored on a Microsoft Excel spreadsheet or a large database can reside on a multiprocessor Sun workstation running a Sybase or Oracle database. However, the development process involved is independent of size.

Software includes the operating system and programs. The operating system is the platform on which the program will run. At present, workstations will run under UNIX/Linux, Windows NT and other Windows products, and MAC and OS/2 operating systems. Only certain operating systems run on certain hardware platforms. Operating systems differ by degree of security, ability for multiple users and multiple tasks, database programs available, performance, and requirements for system administration.

Microsoft Excel and Lotus Spreadsheet are two common spreadsheets. These have facilities for sorting data, restricting data, and listing a limited number of fields. With a small program, called a macro, two tables can be joined together. For example, a table with the patient's diagnosis and a table with the patient's marker information can be joined together to display both on the same line.

A database, by comparison, accommodates more tables. Databases make possible security measures such as only showing patient names to certain users. Databases also provide protection to maintain data integrity. When two users attempt to modify patient data, the database will ensure that the users do not alter the information in contrary ways.

System Administration

A system administrator sets login passwords, establishes who can log on, assigns storage space to users, and assigns who has access to certain directories and who can read and write to certain files. The system administrator may also handle backing up files and retrieving lost data. Multiuser operating systems, such as UNIX and Windows NT, require a system administrator.

Database Administration

The database administrator is responsible for maintaining data engines and the data in the database. He or she installs database patches as required, maintains database security, and oversees database operations.

DATABASE IMPLEMENTATION

After the program requirements are specified and the hardware and database engine are selected, the database conceptual model can be converted to a schema. The schema is the database definition that is specific to that hardware/software. The conceptual model must be converted to database tables; some design tools perform this automatically. Then the tables are modified for performance and data integrity. Prototype data can be added to test the database's functionality.

Conversion

Although database design is as much an art as a science, a process does exist to create the initial design of tables. By applying several rules, the entity relation diagram can be converted into database tables:

1. *Objects Map to Base Tables.* In our example, the object boxes (family relations, patient, contacts, and samples) will each convert to a table. Use the attributes that were created in the conceptual models as the fields in the table. For each table record, there must be one field or a combination of fields which are always unique. These unique fields are called the primary key and are necessary for data integrity.

2. *Many-to-Many Relations Require the Creation of Additional Tables.* An example of this would be if a model had one object for patients and a second object for doctors. Each patient sees multiple doctors and each doctor treats multiple patients. To relate these two objects successfully, a doctor, a doctor–patient, and a patient table would need to be created (see Tables 8.3–8.5).

With the addition of the doctor–patient table, the fact that Dr. Jones saw both patients but Dr. Xi only saw one can easily be represented.

3. *One-to-Many Relations (1 : n) Require a Foreign Key Next to the Object that Cannot Exist Alone.* Patients and samples is an example of this relationship. The sample should not exist if there were no patient, but the patient can exist without a sample. A foreign key ensures that a record will not exist without an associated record on the referenced table. A foreign key prevents a patient record from deletion while a sample row still exists for that patient or prevents the addition of a sample without the patient.

TABLE 8.3. Doctor Table

Doctor ID	Doctor	Specialty
1	Dr. Jones	Family Medicine
2	Dr. Xi	Gynecology

TABLE 8.4. Doctor–Patient Table

Doctor ID	Patient ID	Date of Visit
1	200	1/2/2001
2	201	4/34/2001
1	200	11/12/2001
1	201	12/23/2001

TABLE 8.5. Patient Table

Patient ID	Patient Name
200	Jack Smith
201	Jane Doe

Tables 8.6–8.10 illustrate the example given in our entity relation diagram (Fig. 8.2). Primary keys must be defined for each table. These force unique values across all records on one field or a set of fields.

Performance Tuning

Many databases do not perform well after the initial design has been implemented. The database may need performance tuning, which consists of adjustments to the database to improve response time. Most often, this requires adding indexes on frequently searched fields. An index is the storing of pointers to the location of the records. Other times a redesign of several tables is required to improve performance. Moving tables to separate physical disks is another technique to use when queries perform poorly. Many improvements can also be gained by rewriting database queries. Depending on the database, different optimization tools are available.

Data Integrity

A key ingredient in any genetic analysis study is the integrity of the data generated and analyzed. The core function of informatics is to maintain an electronic audit trail of all data transactions. Additions, modifications, and deletions to the database are

TABLE 8.6. Relations

Field	Data Type	Data Integrity
Family	Int	Primary key: Family + Ind
Individual	Int	
Father	Int	
Mother	Int	

TABLE 8.7. Patient

Field	Data Type	Data Integrity
Family	Int	Primary key: Family + Ind
Ind	Int	
DOB	DateTime	DOB > 1/1/1800 and DOB < = today
Address	Text (50)	
City	Text (50)	
State_Prov	Text (2)	
Zip	Char (10)	
Telephone	Text (20)	
Gender	Char (1)	Rule: Value 'M' or 'F'

TABLE 8.8. Examination

Field	Data Type	Data Integrity
Family	Int	Primary key: Family + Ind + Exam_No
Ind	Int	Foreign key: Family + Ind
Exam_No	Int	
ExamDate	DateTime	ExamDate>"1/1/2000" and ExamDate<=today
examiner	Text (50)	
Test	Text (50)	
Results	Text (50)	

TABLE 8.9. Contact Log

Field	Data Type	Data Integrity
Family	Int	Primary key: Family + Ind + Log_No
Ind	Int	
Log_No	Int	
Interviewer	Text (50)	
Date	DateTime	ExamDate>"1/1/2000" and ExamDate<=today
Comments	Text (255)	

TABLE 8.10. Sample

Field	Data Type	Data Integrity
Family	Int	Foreign key: Family + Ind
Ind	Int	
SampleNumber	Int	Primary key: SampleNumber
DateDrawn	DateTime	ExamDate>"1/1/2000" and ExamDate<=today
TubeCount	int	TubeCount < 40
ConsentDate	DateTime	ExamDate>"1/1/2000" and ExamDate<=today

all electronically captured, including the date and author of each audit item. The advantages to this are clear: While data entry clerks and laboratory technicians, for example, come and go during the life of a project, the geneticist has the computer audit trail to augment or enhance the records of laboratory and data entry personnel whenever a data point is in question.

Data integrity is not automatic; it must be designed into each table. For example, to retain an audit trail of each transaction, separate tables must be defined to audit the changes. Other tools are used to ensure data integrity. Rules can be defined specifying valid values for each field or valid value ranges. For example, for gender, a rule can specify that the value may be only 'M' or 'F'. Foreign keys prevent invalid changes or deletion of records. This will prevent orphan records, such as examinations that exist without a patient. Primary keys and unique indexes ensure values that should remain unique. In our example, for instance, an individual number in a family cannot be entered twice. Database tools check database storage, catalogs, indexes, and tables for inconsistencies. Check your database user's manual for these features and run them at regularly scheduled intervals.

Disaster Recovery Is Part of Maintaining Data Integrity. A primary component of database implementation is setting up a suitable backup strategy. Large database systems (e.g., Sybase and Oracle) provide "rollback" capability, which allows the user to return the database to its status at a particular point in time. This is particularly useful when the database becomes corrupted for some reason, perhaps during a hardware crash. Spreadsheets and MS Access databases do not have this type of capability; the user will need to reenter all the information entered since the last backup of the database. All types of databases must periodically be copied to off-line media; the following guidelines should be observed:

1. Perform backups whenever enough data have been added (or changes made) to make reentry unpleasant in the event of a disk failure or system crash. Generally, daily backups are the minimum. Preferably, backups should be set up to run automatically.

2. Keep multiple backups. Have a separate set of disks or tapes for each day of the week, so that if a problem with the database occurs on Tuesday but is not noticed until Wednesday, Monday's copy will still be available. Large systems will want to keep several weeks' worth of backups at all times. Then, keep one full backup of the database each month as an archive file. It can be quite useful to look back 6 months to find a record that was deleted or a table before it was modified.

3. Maintain off-site backups. Keep a backup copy of the database in an off-site location in case of a major disaster, such as fire or flood.

4. Periodically test the backups. Tapes and diskettes can become worn out, so test a different tape or diskette from your daily rotation periodically to be sure that the database can be restored from that tape or diskette.

5. Keep the tapes in a secure location. The backup files have the same patient information as the online database and therefore have the same security needs as the online database. Be sure that unauthorized users do not have access to the backup information.

USER INTERFACES

The user interfaces to a database often are described as the *front-end* or graphical user interface (GUI). These are the tools that the user will view. For database user interfaces, both custom-written tools and third-party tools can be used, depending on the functions required and user skill levels. For custom applications, compiled applications, web pages, and electronic forms are commonly used. However, many users prefer to use common tools to access their databases. The database can be queried through Structured Query Language (SQL), which is well documented both on the Internet and in books available at any bookstore. Additionally, a variety of report-writing programs are on the market. Data can also be exported to other formats, such as spreadsheets or statistical tools. With some programming skill, the database can be queried with shell languages or the C language. For those with less programming language experience, most databases can be accessed through MS Access and thereby take advantage of Access forms, for example.

At this step, the functionality specification is used to produce user applications and tools that will provide the functions needed. The most successful user interfaces have been prototyped, tested by someone other than the programmer, and refined in several iterations. Maintain a list of user interface "bugs" and requested features to guide future application design.

SECURITY

The widespread use of the Internet has changed the security parameters of every research protocol. Computers are linked to each other and to every other system in the world via the Internet. The advent of broadband technology means that even home computer systems are vulnerable to attack. It is tempting to think that because there are millions of Internet Procotol (IP) addresses floating around in cyberspace any particular system would be an unlikely target for a hacker. This has proven to be a costly mistake for too many users. Having a secure system requires a knowledgeable computer staff who methodically monitor all aspects of the computer network, systems, and software and diligently apply security patches as they are available. However, careless actions by users can thwart even the most secure system and leave it vulnerable to attack.

Transmission Security

Every database, if it includes any information that could reasonably be expected to identify a study subject (see HIPAA regulations), should at the very minimum be password protected. For a single-user database (e.g., MS Access) that can only be accessed by a user sitting at that particular computer, password protection is probably sufficient. However, as soon as the database resides on a shared file system—one that can be accessed from a computer that is not physically connected to the one on which the database resides—a more stringent approach is needed. In this case, the password must travel across the network and therefore is vulnerable to "sniffing." Any packets of data that leave one computer and travel to another are subject to capture by any other computer on that network, which may well be any other computer in the world. One way to avoid this is by using a firewall. A firewall is a hardware solution that acts as a gatekeeper for a local network. The firewall monitors packets that come in to the local network and rejects those that do not meet the criteria set up by the system administrator. For many university research departments, however, firewalls are impractical either because of the cost and system support needed or because of the topography of the general university network. An even more effective way to deter sniffing a password from the data stream is to encrypt the password before it leaves one computer and travels to the next. That way, even if the password is intercepted, it cannot be decrypted and therefore is useless information and could not be used to break into the database. Some database engines (Sybase and Oracle are examples) include password encryption options. Password encryption can also be set through the operating system.

Passwords are not the only kind of information that needs to be encrypted. All data that pass through the data stream are vulnerable to attack. If personal patient information such as names and addresses are stored in your database and the person entering these data is connecting to the database across the network, then that personal information is subject to capture by outside computers. Software can be installed on your network that creates a secure "tunnel" between the host and client computers. All data traveling through the tunnel are securely transmitted. This software is effective both for local networks and for applications that can be accessed from remote sites.

For data entry or data querying via a web application, using secure socket layer (SSL) encryption for transmission assures complete encryption of the data. One can generally identify SSL encryption by the "http://" web address. An SSL server requires a digital certificate, which is issued by a certification authority (CA). A CA is a trusted third-party company that rigorously verifies the identity of those applying for a digital certificate. In practice, this means that the visitor to an SSL site can be sure that the pages being viewed verifiably belong to the company listed in the digital certificate. In both the Internet Explorer (IE) and Netscape browsers, a small lock icon at the bottom of the screen indicates that a digital certificate has been issued for the page and can be clicked to view the certificate. It is important to note the name of the CA on the certificate, as some CAs are more "trusted" than others.

A digital certificate uses public key cryptography to assure that data transmitted over the Internet cannot be decrypted by anyone except the intended recipient, even if intercepted. It makes use of an extremely clever algorithm to mathematically combine a key (random alphanumeric characters) with the data in such a way as to make it possible to use one key to encrypt a document but requires a completely different key to decrypt it. This means that the encryption key—the "public" key—can be published globally as part of the digital certificate. The owner of the digital certificate keeps the private (decryption) key securely protected locally.

System Security

While the network design is important to the overall security of the system, it is equally important to have secure operating systems in place. Whether your network uses Unix, Linux, Windows, MacOS, or some combination of them, it is important that all the security features of the operating systems be installed and monitored and the operating system updates be installed regularly. One of the greatest security holes, however, occurs at the user level. Every research laboratory should have written procedures to explain how to create security-conscious passwords and have the software in place to implement that requirement. Employees must be aware of the penalty for sharing passwords and user accounts. Even the most secure network can be attacked if users leave their computer screens unlocked and office doors open. Backup tapes, CDs, and diskettes are particularly vulnerable; they should be stored off-site (for disaster recovery) and in locked cabinets, not thrown in a desk drawer. There must be rules and protocols for creating new user accounts, including the types of file and database privileges allowed, and also for deleting user accounts at employee termination. Users have an enormous impact on system security and should be reminded often of their role in maintaining system integrity.

Patient Confidentiality

All information in a genetic database system should be available to individual researchers on a "need-to-know" basis only. The structure of the database must be such that information can be compartmentalized so that the permissions granted to a particular user precisely fit the research needs of that user. For example, a clinician, who by definition would be examining patients, would naturally need access to the full set of identifying information for that patient, including name and address. However, the clinician would be denied any and all access to the research genetic variation information, effectively precluding them from divulging this information to the patient. The laboratory technicians, on the other hand, generate the genetic variation information but should have no access to any information about research subjects except the sample identification number used to report the laboratory results. This type of security can be implemented using both password protection on particular tables and fields through the database engine and password protection on the user interface software. Biometric checks, such as thumbprint or palm print devices, will become more popular for checking the identity of database users as the cost of such devices falls.

PEDIGREE PLOTTING AND DATA MANIPULATION SOFTWARE

Throughout this book are references to numerous genetic analysis programs available on the Internet that perform calculations geneticists could only dream of just a few short years ago. With all that processing power, however, comes the realization that each of those analysis programs requires a different input format and each produces dozens of output files—and then the process must be repeated for hundreds or thousands of genetic markers. Fortunately, there are also many data manipulation programs available to aid in the creation of input files, streamline the batch processing of multiple marker sets, and translate between the different program input formats. Among the more widely known programs are the following:

1. Genetic Analysis Package (GAP): "a comprehensive package for the management and analysis of pedigree data. It offers: (1) Powerful database management tools, specifically designed for family data; (2) Automatic pedigree drawing; and (3) Segregation and linkage analysis." (http://icarus2.hsc.usc.edu/epicenter/gap.html)
2. Linkage Auxillary Package (LAP): includes the extensively utilized MAKEPED and PREPLINK programs for creating the input files to the LINKAGE suite of programs. (ftp://linkage.rockefeller.edu/software/linkage)
3. MEGA2: "a data handling program for facilitating genetic linkage and association analysis." (http://watson.hgen.pitt.edu/register/soft_doc.html)
4. PEDSYS: "a full-scale database system developed as a specialized tool for management of genetic, pedigree, and demographic data." (http://www.sfbr.org/sfbr/public/software/pedsys/pedsys.html)

Note that some of these programs have a license fee.

It is also extremely useful to be able to graphically represent family data in the form of a pedigree plot. Often, the user would like to view genotype information on the pedigree to visualize cross-overs or to determine haplotypes. The plots show affected individuals and other clinical information. There are several commercial and noncommercial plotting programs available. The most well-known commercial products are Cyrillic (http://www.cyrillicsoftware.com) and Progeny (http://www.Progeny2000.com), but these can be costly. Other plotting programs are available either free or for a small fee:

(a) Collaborative Pedigree drawing Environment (COPE)
    (http://www.infobiogen.fr/services/CoPE)
(b) Pedigree/Draw
    (http://www.sfbr.org/sfbr/public/software/pedraw/peddrw.html)
(c) PANGAEA
    (http://www.stat.washington.edu/thompson/Genepi/pangaea.shtml)
(d) PEDPLOT
    (http://wwwchg.duhs.duke.edu/software/pedplot.html)

SUMMARY

Regardless of the size of your study or the sophistication of your computer capabilities, there are certain requirements for genetic information management. Performing needs assessment guarantees that all members of your team know at the start of the study exactly what type of information is expected to be collected over the course of the study. Determining the information flow of the data described in the needs assessment means that all users know their role in the acquisition of information and what tools they will use to supply the needed information. Choosing the right combination of hardware and software is critical to the success of the data collection and also to the subsequent analysis of that data. Training users to correctly employ these tools assures consistent information gathering, which leads to better data analyses. Having a disaster recovery plan, doing periodic data integrity checks, and maintaining a secure computer system are essential to ensure that the study proceeds smoothly to completion.

CHAPTER 9

Quantitative Trait Linkage Analysis

JASON H. MOORE

INTRODUCTION TO QUANTITATIVE TRAITS

The goal of many gene discovery studies is to identify genes that confer an increased susceptibility to a particular disease. In these studies, the clinical endpoint of interest is the presence or absence of disease. Clinical endpoints that have only several possible outcomes (i.e., those that are discretely distributed) are called qualitative traits. In contrast, clinical endpoints whose outcomes can take on a wide range of different values (i.e., those that are continuously distributed) are called quantitative traits (Moore, 2002). Examples of quantitative traits include height, weight, blood pressure, and plasma enzyme levels. Quantitative traits are inherently more difficult to study than discrete traits and thus require different statistical methods.

Why study quantitative traits? The answer to this question is obvious when the clinical endpoint of interest is itself quantitative in nature. For example, clinical pharmacologists are interested in identifying genetic variations that are associated with interindividual variation in the metabolism or pharmacokinetics of specific drugs. Here, the clinical endpoint might be the activity of cardiovascular drug-metabolizing enzymes such as those from the cytochrome P450 family (Nakagawa and Ishizaki, 2000). There are, however, compelling reasons to study quantitative traits even when the clinical endpoint is qualitative. Factors such as the uncertainty of diagnosis due to measurement error and lack of standard objective criteria for diagnosis can make it difficult to discern who, in fact, is affected by disease. This is evident for common chronic diseases such as psychiatric disorders (Rice, 1993; Tsuang et al., 1993) and asthma (Britton, 1998). A strategy to facilitate identification of susceptibility genes for these diseases is to use intermediate-risk factors as the trait of interest. Intermediate-risk factors or endophenotypes (Ott, 1995) may be closer to the underlying biological etiology of the disease and thus may make it easier to identify the underlying genes. For example, Williams et al. (1999) demonstrated that joint consideration of alcoholism diagnosis and quantitative event-related

Genetic Analysis of Complex Diseases, Second Edition, Edited by Jonathan L. Haines and Margaret Pericak-Vance
Copyright © 2006 John Wiley & Sons, Inc.

potentials increases the evidence for linkage of markers on chromosome 4 to an alcoholism susceptibility gene. Thus, for some diseases, quantitative traits may improve the power to identify susceptibility genes.

The purpose of this chapter is to introduce the fundamental concepts and methods necessary for carrying out a quantitative trait linkage analysis. The two primary methods presented are Haseman–Elston regression and variance component linkage analysis. We begin with a review of genetic architecture and various study designs.

GENETIC ARCHITECTURE

Genetic architecture has been defined as (1) the number of genes that influence a trait, (2) the number of alleles each gene has, (3) the frequencies of those alleles in the population, and (4) the influence of each gene and its alleles on the trait (Weiss, 1993). While the first three points fall to some degree under the umbrella of population genetics, the last point lies at the heart of the genetics of quantitative traits. Although there are many different ways that genes can influence quantitative traits, most studies have focused on the influence of particular polymorphisms on the mean value of a trait. In fact, much of the basic theory underlying the methods presented in this chapter focuses on genotypic means. We review this fundamental theory first and then present examples of other ways in which genes can influence quantitative traits.

Let us assume we have a candidate gene polymorphism with two alleles (B and b) and three genotypes (BB, Bb, and bb). In the population, we represent the overall trait mean by μ and each genotype-specific mean by μ_{BB}, μ_{Bb}, and μ_{bb}. Differences between the genotype-specific means can be due to the additive effects of the alleles, represented by a, and the dominant effect of one of the alleles, represented by d. If we define a constant, c, as $(\mu_{BB} + \mu_{bb})/2$, then each genotype-specific mean can be written as $\mu_{BB} = c - a$, $\mu_{Bb} = c + d$, and $\mu_{bb} = c + a$. If we assume $d = 0$, then the allelic effects are purely additive and the trait mean for the Bb genotype is midway between the two homozygotes. However, if $d = -a$, then the B allele has a purely dominant effect. If we assume Hardy–Weinberg equilibrium, the overall quantitative trait mean in the population can be expressed as a function of each genotype-specific mean multiplied by each genotype frequency. Thus, $\mu = p^2(c + a) + 2pq(c + d) + q^2(c - a)$.

An important result of the above formulas for describing genotype-specific means is the idea of heritability. Chapter 3 introduces heritability as a general statistic that is sometimes useful for determining whether a particular trait has a genetic component. The heritability of a quantitative trait that is due to a particular quantitative trait locus (QTL) can be estimated from the ratio of the variability in the trait that is due to variation in the QTL to the total variability of the trait. The amount of variability in the trait that is due to variation in the QTL can be estimated from the heterozygosity of the polymorphism ($2pq$) and the additive effects of the alleles (a) by $2pqa^2$. This is referred to as the additive genetic variance and is denoted by σ_a^2. It is important to note that σ_a^2 reflects the degree to which individual phenotypic values differ from the overall mean, μ, because of additive allelic effects. This is not to be confused with genotype-specific trait variances that represent the degree to which individual phenotypic values differ from the genotype-specific means.

We have briefly reviewed several ways in which genes can influence quantitative trait levels ("level genes"). However, it is important to keep in mind that variances, covariances, and dynamics of quantitative traits may also be genotype specific. One of the earliest studies demonstrating that variability in a quantitative trait can be genotype specific was carried out by Berg et al. (1981). This study, and a confirmatory study by Berg (1988), showed that variability of serum cholesterol is dependent on Kidd blood group locus genotype. More recent studies by Reilly et al. (1991), Haviland et al. (1995), and Nelson et al. (1999) have reinforced the idea that variability genes may play an important role in the genetic architecture of quantitative traits. These studies demonstrated that variability in multiple lipid and apolipoprotein traits were *apolipoprotein E* genotype specific. Why are variability genes important? What are the etiological inferences? Reilly et al. (1991) suggest that intragenotypic variability may be due to gene–gene and gene–environment interactions. Thus, the magnitude of the effects of a particular environmental or genetic factor may differ across genotypes. Reilly et al. (1994), Haviland et al. (1995), and Nelson et al. (1999) also demonstrated that pairwise correlations of various lipids and apolipoproteins were *apolipoprotein E* genotype specific, suggesting that genes can influence trait covariability. An exploratory data analysis by Moore et al. (2002b) has led to the working hypothesis that the correlation of tissue plasminogen activator (t-PA) and plasminogen activator inhibitor 1 (PAI-1) is dependent on genotypes at the *angiotensin-converting enzyme* (*ACE*) and *PAI-1* genes.

A final example of types of gene effects is dynamics. Most phenotypes in an individual are in a constant state of change over time. For example, it is well known that human blood pressure follows a circadian, or 24-h, rhythm. Moore (1999) demonstrated that interindividual variations in genes from the renin–angiotensin system are associated with variations in both linear and nonlinear dynamic features of systolic and diastolic blood pressure. Further, Moore (2001) demonstrated through simulation studies that the use of measures of quantitative trait dynamics can greatly improve the power of a linkage study to identify dynamics genes. Although not an exhaustive list, level, variability, covariability, and dynamics are examples of the types of gene effects that should be kept in mind when investigating the genetic architecture of quantitative traits. Unfortunately, most linkage studies only focus on identifying genes that influence quantitative trait levels.

In addition to considering different types of gene effects, it is also important to realize that each type of gene effect may be dependent on a particular genetic or environmental context. Thus, the influence of a particular polymorphism on quantitative trait levels, variability, covariability, or dynamics may depend on one or more other polymorphisms (in epistasis or gene–gene interaction) and/or one or more environmental factors (e.g., plastic reaction norm or gene–environment interaction). In fact, epistasis may be much more common than previously believed (Templeton, 2000). For example, Nelson et al. (2001) demonstrated strong evidence for epistasis for most of the lipid-related quantitative traits examined using an exploratory data analysis approach called the combinatorial partitioning method (CPM). These epistasis effects were detected in the absence of any independent main effects of the candidate gene polymorphisms. Studies by Moore et al. (2002a,b) have also documented evidence that levels of PAI-1 and covariability of t-PA and PAI-1 are dependent on interactions among polymorphisms in the *ACE* and *PAI-1* genes.

Although context-dependent gene effects play an important role in genetic architecture, our ability to detect and characterize context-dependent effects is limited in several important ways. First, most parametric statistical methods suffer from the curse of dimensionality (Bellman, 1961). For example, most linkage and association methods for quantitative traits use linear regression. As each additional genetic locus is added to the linear regression model, the number of gene–gene interaction terms grows exponentially. Few datasets are large enough to provide the degree of freedom necessary to estimate the number of parameters required to model the epistasis effects. Second, as the total number of markers grows, it quickly becomes computationally infeasible to evaluate all possible combinations. One solution is to use stepwise linear regression to first condition on those markers that might have an independent main effect. However, gene effects that are purely interactive will be missed. The data dimensionality problem and the computability problem will need to be addressed before we can fully characterize the genetic architecture of quantitative traits.

STUDY DESIGN

There are several general approaches to determining the nature of the genetic component of a quantitative trait. The first approach starts with the trait of interest and attempts to draw inferences about the underlying genetics from looking at the degree of trait resemblance among relatives. This approach is sometimes referred to as a top-down or unmeasured genotype strategy because the inheritance pattern of the trait is the focus and no genetic polymorphisms are actually measured. The top-down approach is often the first step taken to determine whether there is evidence for a genetic component for a given trait. Chapter 3 introduces heritability and segregation analysis as two top-down approaches.

With the bottom-up or measured genotype approach, either candidate gene polymorphisms or anonymous genomic markers are measured and then used to draw inferences about which genes might play a role in determining interindividual variability in a quantitative trait. The focus of this chapter is the use of a bottom-up approach to identify QTLs using genetic linkage analysis.

HASEMAN–ELSTON REGRESSION

In 1972, Haseman and Elston published a relatively simple linear regression method for identifying QTLs using full sibling pairs. This approach is based on the idea that if a particular marker locus is linked to a QTL, the squared difference in the trait values between sibling pairs should decrease as the proportion of alleles the siblings share identical by descent (IBD) increases. In a randomly mating population, the mean proportion of alleles shared IBD is expected to be $\frac{1}{2}$. Thus, siblings are expected to share greater than one-half of their alleles IBD if the marker is linked to the QTL. Using linear regression, the dependent variable (Y) in the model is the squared difference in trait values and the independent or predictor variable (π) is the proportion of marker alleles shared IBD ($Y = \alpha + \beta\pi$). When the marker and the QTL are the same, the slope, or β, is equivalent to the additive genetic variance multiplied by -2. When the marker and the QTL are not the same and are separated by some distance, β is equivalent to the additive genetic variance multiplied by $-2(1 - 2\theta)^2$, where θ is the recombination fraction. Assuming linkage equilibrium between alleles at the marker locus and alleles at the QTL, the null hypothesis that there is no relationship between the squared difference in trait values and the proportion of marker alleles shared IBD can be tested using a one-sided t-test where the test statistic is the estimate of the slope divided by the standard error of the slope.

The Haseman–Elston approach is appealing because of its simplicity. However, it is limited in several important ways. First, for diallelic markers the power is fairly low because there are few informative matings for estimating allele sharing (Robertson, 1973). This can be avoided by using markers with four or more alleles (Amos et al., 1989). Second, the distance of the QTL from the marker locus is confounded with the additive genetic variance component such that, when a statistically significant regression is observed, it is difficult to know whether there is a QTL with a large effect far away from the marker locus or a QTL with a small effect close to the marker locus. Fulker and Cardon (1994) addressed this issue by developing an interval mapping extension of Haseman–Elston regression.

With the interval mapping approach (Fulker and Cardon, 1994), two flanking markers separated by a known map distance are used to locate the QTL and estimate the size of its effect. This is carried out using the proportion of alleles shared IBD at each of the marker loci to estimate the proportion of alleles shared IBD at the QTL, assuming a range of recombination values in the marker interval. The independent variable in this regression is the proportion of alleles shared IBD at the QTL. As with the original Haseman–Elston regression, the null hypothesis is tested using a one-sided t-test where the test statistic is the estimate of the slope divided by the standard error of the slope. It should be noted that multipoint extensions of this approach may provide more power and better estimates of QTL location when a dense marker map is available or when marker information is variable (Fulker et al., 1995; Kruglyak and Lander, 1995). However, multipoint methods assume that there is no linkage interference and that the distances between markers are accurately estimated.

Recently, Elston et al. (2000) described a revision of the popular regression approach described above. Here, the dependent variable (Y) becomes the mean corrected product of the sibling trait values ($[x_{1j} - \mu][x_{2j} - \mu]$) instead of the squared differences. Using simulation studies, this change was found to improve both the type I and type II error rates over those associated with the original Haseman–Elston regression approach described above. In addition, Xu et al. (2000) provide a modification of this revised approach that provides greater power when the sibling trait values are correlated. A study by Allison et al. (2000) shows that this new

Haseman–Elston approach is robust to deviations from trait normality. Even so, power may be improved by winsorization (i.e., adjusting extreme values) of nonnormal distributions (Fernandez et al., 2002). A detailed comparison of these recent developments with Haseman–Elston regression and others is presented by Feingold (2002).

An example application of Haseman–Elston regression is a study by Saccone et al. (2000). The goal of this study was to carry out a genome scan to identify QTLs for an intermediate trait or endophenotype for alcoholism. Here, the trait of interest was the maximum number of alcoholic drinks consumed in a 24-h period. A total of 1105 full sibpairs were analyzed using Haseman–Elston regression. Saccone et al. found very good evidence [logarithm of the odds (LOD) = 3.5] for a region on chromosome 4 that might harbor an alcoholism QTL. This region is very close to the alcohol dehydrogenase gene cluster on chromosome 4 which may prove to include several candidate-level genes. The authors conclude that studying a quantitative trait for alcoholism improved their power to identify the candidate region and that similar measures of substance consumption may benefit studies of other substance-abuse disorders.

What are the advantages of the Haseman–Elston regression approach? The most important advantage is simplicity. These methods are relatively easy to understand and easy to interpret. Another key advantage is that this approach is model free, that is, no particular genetic model is assumed. In general, model-free methods have good statistical properties (i.e., they are robust), are easier to extend to analyzing multiple loci simultaneously, and have more power than model-based approaches when the inheritance model has not been correctly specified. An important disadvantage is that this approach has been limited to the analysis of siblings. However, a recent study by Sham et al. (2002) illustrates a multivariate regression approach to Haseman–Elston for the analysis of extended pedigrees. This development should greatly improve the usefulness of regression-based approaches for quantitative trait linkage analysis.

The SAGE and MAPMAKER/SIBS software packages both provide Haseman–Elston regression programs. Information on obtaining SAGE is available online at http://darwin.cwru.edu/octane/sage/sage.php while information about MAPMAKER/SIBS is available online at http://linkage.rockefeller.edu/soft/.

MULTIPOINT IBD METHOD

The multipoint IBD method (MIM) was developed as a multipoint alternative to the single-marker version of Haseman–Elston regression (Goldgar, 1990). With MIM, the proportion of alleles shared IBD in a chromosomal region is of interest instead of the sharing at a single locus. Thus, MIM does not require estimation of the effects of every individual locus since it is able to summarize the impact of a chromosomal region on interindividual variation in the quantitative trait. These IBD estimates are used to determine the proportion of the total genetic variance that is due to QTLs on specific chromosomes or chromosomal regions. This is carried out using all nonindependent sibpairs and assumes that the QTL or QTLs have additive effects. Thus, the trait variance (σ_T^2) can be partitioned into the additive effects of one or more loci within the test chromosomal region (σ_c^2), the additive effects of all other loci (σ_a^2), and the random environmental effects (σ_e^2) such that $\sigma_T^2 = \sigma_c^2 + \sigma_a^2 + \sigma_e^2$. Likelihood ratio testing is used to test the null hypothesis that $\sigma_c^2 = 0$. Using simulation studies, Goldgar (1990) demonstrated that the MIM is more powerful than the single-locus version of Haseman–Elston regression. Further simulation studies by Goldgar and Oniki (1992) demonstrated that MIM has similar power to parametric multipoint linkage analysis when parental data are known, the effect of the major QTL is small and there are additional genetic effects, or the parameters of the major QTL model are misspecified.

Schork (1993) provides a number of extensions of MIM that allow the method to incorporate a wider range of factors and effects. Many of these extensions allow the effects of concomitant variables such as age and body mass index to be considered, thus possibly reducing noise that can confound the genetic effects. Further, these extensions allow gene–gene and gene–environment interactions to be tested. The conclusion of this study is that the MIM can be made more powerful by using closely spaced markers, considering concomitant variables, using informative markers flanking the chromosomal region of interest, using multivariate traits, and using larger sibships (Schork, 1993).

Shugart and Goldgar (1997, 1999) have used simulation studies to compare the MIM with the multipoint Haseman–Elston method as implemented by Kruglyak and Lander (1995). They simulated a quantitative trait with interindividual variability due to two QTLs and compared the power and false-positive rates of both methods for varying marker densities (5–20 cM) and sibship sizes (2–5). The results suggest that MIM has better power and better false-positive rates than multipoint Haseman–Elston, especially as sibship sizes increase. However, it should be noted that the powers of MIM and the recently extended Haseman–Elston approach (Elston et al., 2000) have not been directly compared. This study and others (Goldgar, 1990; Goldgar and Oniki, 1992) suggest that multipoint methods such as MIM have reasonable power to detect a QTL that explains about 25–30% of the interindividual variation in the quantitative trait. It is clear from Goldgar (1990) and Schork (1993) that methods that partition the variance of a quantitative trait into various genetic and nongenetic components (i.e., variance component methods) offer a powerful and flexible framework for the identification of major-effect QTLs. Variance component methods are the subject of the next section.

The MIM software package provides programs for carrying out the multipoint IBD method of Goldgar (1990). Information on obtaining these packages is available online at http://linkage.rockefeller.edu/soft/.

VARIANCE COMPONENT LINKAGE ANALYSIS

Amos (1994) developed a variance component linkage analysis that is different from the MIM method of Goldgar (1990) in two important ways. First, the Amos variance component approach estimates the effects of a single major QTL instead of the polygenic effects in a particular chromosomal region. Further, the Amos approach can be used for relatives other than siblings. Table 1 of Amos (1994) gives the additive major-gene variance components for different relative pairs. For example, the additive major-gene variance component for full siblings is $\left[\frac{1}{2} + (1 - 2\theta)^2 (\pi_{ij} - \frac{1}{2})\right]\sigma_a^2$, where θ is the recombination fraction between the marker locus and the QTL, π_{ij} is the IBD sharing for the marker locus for the ith and jth relatives, and σ_a^2 is the proportion of variance due to the additive effects of the alleles at the major gene. Amos (1994) also outlines how the variance components for polygenes or unmeasured QTLs (σ_G^2) and unexplained environmental or chance factors (σ_e^2) can also be modeled. Maximum-likelihood and estimating equation methods can both be used to estimate the various parameters in the model. Here, σ_a^2 is the primary parameter of interest for hypothesis testing. Under the null hypothesis that a particular QTL is not associated with a quantitative trait, $\sigma_a^2 = 0$. This null hypothesis can be tested using chi-square statistics that are formed from -2 multiplied times the log-likelihood ratio comparing the unrestricted model in which σ_a^2, σ_G^2, σ_e^2, and θ are all estimated to the restricted model where σ_a^2 is set to zero.

Amos (1994) carried out a number of simulation studies to validate the power and robustness of the variance component approach. The simulations were based on a major QTL influencing a normally distributed quantitative trait. These studies confirm that this quantitative trait linkage approach has reasonable power and good statistical properties for identifying QTLs that have a relatively large, independent main effect. This is further confirmed by Amos et al. (1996), who performed additional simulation studies and developed an application of the method to identifying QTLs for several intermediate-risk factor traits for cardiovascular disease. Recent extensions to the variance component approach include the ability to model longitudinal data in pedigrees (de Andrade et al., 2002) and the ability to test for linkage in the presence of imprinting (Shete and Amos, 2002).

Barnholtz et al. (1999) have applied the variance component approach to identifying QTLs associated with monoamine oxidase B, an intermediate-risk factor trait for alcoholism, using families and markers from the Collaborative Study on the Genetics of Alcoholism (COGA). The authors carried out single-point and multi-point linkage analysis with covariates in the model and identified 24 markers on four different chromosomes that showed evidence for linkage with major-gene QTLs. However, when three monoamine oxidase B outliers were removed and the analysis was repeated, no evidence for linkage of any markers to a QTL was found. This study illustrates the importance of exploratory data analysis prior to conducting a genetic study and the potential impact of violations of trait distribution assumptions on variance component models.

The ACT software package carries out the variance component linkage analysis approach described by Amos (1994). Information on obtaining these packages is available online at http://www.epigenetic.org/Linkage/act.html.

The variance component linkage analysis methods of Goldgar (1990) and Amos (1994) are more powerful and more flexible than the Haseman–Elston approaches but are limited to relative pairs. Blangero and Almasy (1997) introduced an extension of the variance component approach that considers the complete multivariate phenotypic vector of a pedigree, thus allowing application of the method to full-pedigree data. Blangero and Almasy (1997) suggest that for a simple model in which some number of QTLs and residual polygenes play a role in determining quantitative trait levels, the jth individual's phenotype (y_j) can be assumed to be a function of the sum of $i = 1, \ldots, n$ QTL effects (q_{ji}), an additive polygenic effect (g_j), a covariate effect (x_j), and a random environmental deviation (e_j). With these assumptions, the jth individual's phenotype can be modeled by the linear function

$$y_j = \mu + \sum q_{ji} + \beta x_j + g_j + e_j$$

where μ is the overall trait mean and β is the regression coefficient for the covariate effect. The expected covariance (Ω) of a quantitative trait between a pair of relatives in a general pedigree can be modeled as

$$\Omega = \sum \Pi_i \sigma_{qi}^2 + 2\Phi\sigma_g^2 + I\sigma_e^2$$

where σ_{qi}^2 is the additive genetic variance due to the ith QTL, Π_i is a matrix whose elements provide the predicted proportion of genes that individuals share IBD at a QTL that is linked to a genetic marker locus, σ_g^2 is the genetic variance due to residual additive genetic factors, Φ is the kinship matrix, σ_e^2 is the variance due to individual-specific environmental effects, and I is an identity matrix. Parameters in this model can be estimated using maximum likelihood and the null hypothesis that the additive genetic variance of the ith QTL is equal to zero can be tested using chi-square statistics derived from likelihood ratio test statistics (Blangero and Almasy, 1997).

Blangero et al. (2000, 2001) provide a detailed review of the assumptions underlying this approach. The primary assumption is that the phenotypes in the pedigrees have a normal multivariate distribution. Allison et al. (1999) studied violations of the normality assumption and found the variance component approach to be robust to some types of nonnormality but not to others. For example, leptokurtosis results in higher than nominal type I error rates (i.e., more false positives than expected for a particular significance level). In general, the new Haseman Elston approach (Elston et al., 2000) is more robust to deviations from normality than the variance component approach (Allison et al., 2000). Feingold (2001) has reviewed the mixed results of attempts to make the variance component approach more robust to deviations from normality.

Comuzzie et al. (1997) applied the variance component approach to identifying QTLs for leptin levels in 10 Mexican-American families ranging in size from 35 to 71 individuals. A single-marker locus was identified with strong evidence for linkage to a QTL that explained about 47% of the variation in serum leptin levels (LOD = 4.95). Thus, the variance component approach was able to detect evidence for a major gene on chromosome 2 that influences human obesity. As a second example, Kissebah et al. (2000) used the variance component linkage approach to carry out a genome scan for QTLs influencing intermediate quantitative risk factor traits for metabolic syndrome. A 10-cM scan was carried out in 2209 individuals from 507 nuclear families. A single marker on chromosome 3q27 was strongly linked to a QTL influencing six different intermediate quantitative traits (LODs = 2.4–3.5). Interestingly, a second marker on chromosome 17p12 was strongly linked to a QTL influencing plasma leptin levels (LOD = 5.0) through an epistatic interaction with the QTL from chromosome 3q27. Here, tests for epistasis were limited to the two chromosomal regions that showed the most evidence for independent main effects and were limited to additive–additive interactions. This study illustrates the usefulness of the variance component approach for testing hypotheses about epistasis. However, this limited set of analyses is likely to overlook additional QTLs whose effects are primarily epistatic.

The advantages of the variance component approach include the ability to handle complex pedigree structures and the ability to test for gene–gene and gene–environment interactions. Mitchell et al. (1997) carried out a simulation study to evaluate the power of the approach to identify epistatic QTLs. Using a linear model, an individual's phenotypic value (y) can be written as $y = \mu + a_1 + a_2 + aa + g + e$, where μ is the overall mean of the trait, a_1 and a_2 are the additive effects of two QTLs, aa is the additive–additive interaction effect, g is the residual polygenic effect, and e is the residual random environmental effect. When applied to simulated data, the variance component approach had poor power to detect epistatic QTLs. It is well known that QTLs that influence quantitative traits primarily through epistatic interactions can be difficult to detect using traditional linkage methods such as variance component analysis (Eaves, 1994; Cheverud and Routman, 1995; Tiwari and Elston, 1998; Cheverud, 2000). Thus, although the variance component approach can directly test for epistasis, the power to detect epistatic QTLs may be very limited.

The SOLAR package (Almasy and Blangero, 1998) provides routines for carrying out variance component linkage analysis in complex pedigrees. Information on obtaining this package is available online at http://www.sfbr.org/sfbr/public/software/solar/index.html.

NONPARAMETRIC METHODS

All of the quantitative trait linkage methods described above use the linear model and thus assume that the quantitative trait is normally distributed. However, in the event that this assumption is violated, there are two alternatives. The first alternative is to apply an appropriate mathematical transformation to restore the distribution to normality. The second alternative is to use a nonparametric approach that is not sensitive to distributional properties. Kruglyak and Lander (1995) provide a nonparametric approach to mapping QTLs that is based on a generalization of the nonparametric Wilcoxin rank-sum statistic. The advantage of this test is that quantitative traits with any distribution can be studied. The disadvantage is that this statistic does not provide a direct estimate of the genetic effect; rather, it simply tests for the presence of a QTL.

The MAPMAKER/SIBS software package provides routines for carrying out nonparametric quantitative trait linkage analysis. Information on obtaining MAPMAKER/SIBS is available online at http://linkage.rockefeller.edu/soft/.

FUTURE DIRECTIONS

Each of the methods presented has reasonable power for identifying genes that have a relatively large independent main effect on interindividual variability in quantitative trait levels. However, identifying major genes that influence quantitative trait levels is only one small part of the genetic architecture of most quantitative traits. None of the methods presented are designed to identify variability or covariability genes. Further, the power of these methods to identify genes whose effects are context dependent is very limited. Why are more powerful methods for detecting interactions not available? The primary reason for this is largely historical in nature. Much of quantitative genetics as we know it today (Falconer and Mackay, 1996; Lynch and Walsh, 1998) can be attributed to the early work of R. A. Fisher (1918) and S. Wright (1932). Fisher's view of genetic architecture was that interindividual variability in quantitative traits was due to many genes, each with independent effects that could be added together. This idea became very popular because it fit nicely with Mendelian genetics and linear models could be used to partition the additive genetic variance. Epistasis or the nonadditive genetic variance was considered a nuisance and a statistical artifact. Wright, on the other hand, emphasized that the relationship between genotype and phenotype is dependent on dynamic interactive networks of genes and environmental factors. This idea hold true today. Gibson (1996) stresses that gene–gene and gene–environment interactions must be ubiquitous given the complexities of intermolecular interactions that are necessary to regulate gene expression and the hierarchical complexity of metabolic networks. Thus, Fisher viewed epistasis as a statistical phenomenon in populations that depends on allele frequencies, while Wright viewed it as a physiological phenomenon resulting from gene effects in individuals that are not dependent on allele frequencies. The debate over whether interindividual variation in quantitative traits is due to the independent additive effects of genes or the interactive effects of gene networks continues today. For the most part, the Fisherian view has been more popular since the theory is relatively easy to understand and apply. However, there is accumulating evidence that gene–gene and gene–environment interactions play an important role in determining interindividual variability in quantitative traits (Schlichting and Pigliucci, 1998; Pigliucci, 2001; Templeton, 2000). As such, it is critical that we develop new methods that are able to identify and characterize QTLs with context-dependent effects.

To address the limitations of the variance component approach for detecting epistasis, Cheverud and Routman (1995) developed a physiological epistasis approach that does not consider allele frequencies. This approach is based on average trait values for each genotype (G) at a locus where the additive (a) and dominance (d) genotypic values are defined as $a = (G_{22} - G_{11})/2$ and $d = G_{12} - (G_{11} + G_{22})/2$. The genotype frequencies do not enter into these equations. As described by Cheverud (2000), the single-locus additive and dominance genotypic values can also be expressed in an unweighted linear regression by $G_{ij} = C + aX_{aij} + dX_{dij}$, where C is the unweighted average of the three genotype values and each X is an indicator variable. For example, a is the coefficient of the genotypic values on the number of 2 alleles such that $X_{a11} = -1$, $X_{a12} = 0$, and $X_{a22} = 1$. The full table of indicator variable values is given by Cheverud (2000). Note that for the single-locus case there are no residuals since there are three genotype values and three parameters in the model. The single-locus model of physiological epistasis is easily extended to two loci. However, with two loci there are nine genotype values and only five parameters so a residual term (e_{ijkl}) is necessary. The four residuals actually specify the additive–additive (aa), additive–dominance (ad), dominance–additive (da), and dominance–dominance (dd) epistatic genotype values. With these four epistatic genotype values, the unweighted linear regression model can be written as $G_{ijkl} = C + a_A X_{aij} + d_A X_{dij} + a_B X_{akj} + d_B X_{dkj} + X_{dij} + a_B X_{akj} + d_B X_{dkj} + aa X_{aij} X_{akl} + ad X_{aij} X_{dkl} + da X_{dij} X_{akl} + dd X_{dij} X_{dkl}$, where C is the unweighted average of all nine genotype values.

As pointed out by Cheverud (2000), these genotypic values measure the effects of genes on phenotypes but they cannot be used to define inheritance because the two alleles that comprise the genotype assort independently during reproduction. Thus, for defining inheritance it is necessary to consider allele frequencies. Cheverud and Routman (1995) illustrate how the genotype values described above contribute to the population-based additive, dominance, and epistatic variance components. The additive genotype values only contribute to the additive variance component. However, the dominance genotype values contribute to both the additive and dominance variance components while the epistatic genotype values contribute to the additive, dominance, and epistatic variance components. This is important because methods such as variance component linkage analysis test for epistasis using only the epistasis variance component and thus do not take into consideration the epistasis effects represented in the additive and dominance variance components (Cheverud and Routman, 1995; Cheverud, 2000). Cheverud (2000) summarized the comparison of statistical tests for epistasis that are based on traditional variance components with those based on genotypic values. For identifying epistatic loci that influence body weight in mice, the physiological epistasis approach was much more powerful than the variance component approach. It should be noted that an important limitation of this approach is that observed values for each of the genotypes must be present. This may present problems for smaller datasets. However, a clear advantage is its simplicity. It is relatively straightforward to implement this method using standard statistical packages such as SAS and S-Plus.

The curse of high dimensionality (Bellman, 1961) is an important limitation of all the methods presented in this chapter. That is, there must be enough data to actually estimate the gene–gene interaction effects. For each additional locus added to a linear model, the number of interaction terms grows exponentially. This can be a problem if the number of loci to be evaluated is relatively large and/or the dataset is small. One approach to dealing with high dimensionality and many interaction terms is to reduce the dimensionality by pooling multilocus genotypes into a smaller number of groups. Nelson et al. (2001) have adopted this approach with the development of the CPM. The CPM simultaneously considers multiple polymorphic loci to identify combinations of genotypes that are most strongly associated with variation in the quantitative trait. In the first step of this procedure, all possible multilocus genotypes are identified and this multilocus genotype space is divided into partitions or groups.

This partitioning serves to collapse the multiple dummy variables needed to encode multiple polymorphisms and their interactions into a single variable with two or more levels. This new independent variable can then be evaluated using linear regression. In the second step, all possible ways of partitioning the multilocus genotypic state space into groups are evaluated, and the partitions that explain the most variation are identified. For two polymorphisms, each with two alleles and three genotypes, there are 255 possible partitions that need to be evaluated.

Application of the CPM to modeling the relationship between 18 diallelic loci from six cardiovascular disease susceptibility genes and interindividual variability in plasma triglycerides identified nonadditive epistatic interactions between multiple loci (Nelson, 2001). Although preliminary, these results are suggestive that the CPM may have reasonable power for identifying nonadditive gene–gene interactions. Nelson et al. (2001) intend this approach to be used for exploratory data analysis and hypothesis generation rather than for formal hypothesis testing. Indeed, Moore et al. (2002a,b) have demonstrated the usefulness of CPM as an exploratory data analysis tool for hypothesis generation. In should be noted that the type I error rate of CPM is currently unknown. However, coupled with permutation testing and cross-validation, it should be possible to use this approach to tests hypotheses about the nonadditivity of gene–gene interaction effects. This approach has not yet been extended to quantitative trait linkage analysis and is limited by its computational intensity. Further, CPM programs have not been made available; however, simpler versions of the method are easily implemented using standard statistical packages such as SAS and S-Plus.

Finally, the search for interacting QTLs in a gene discovery study can be very computationally intensive. For example, the number of orthogonal regression terms needed to describe the interactions among a subset, k, of n biallelic loci is $n!/k!(n-k)!$ (i.e., n choose k) multiplied by 2^k (Wade, 2000). For example, if 10 polymorphisms were measured, 20 parameters would be required to model the main effects (assuming two dummy variables per diallelic locus), 180 parameters to model the two-way interactions, 1920 parameters to model the three-way interactions, 3360 parameters to model the four-way interactions, and so on. Thus, fitting a full model with all interaction terms and then using backward elimination to derive a parsimonious model would be impossible. To address this issue, Carlborg et al. (2000) applied a parallel search strategy called genetic algorithms (Goldberg, 1989) to the identification of combinations of interacting QTLs. In combination with permutation testing (Carlborg and Andersson, 2002), this strategy is a useful alternative to stepwise procedures that first condition on a QTL having a marginal effect. Indeed, genetic algorithms have been very useful in genetic epidemiology for the identification of combinations of polymorphisms associated with complex traits (Congdon et al., 1993; Nakamichi et al., 2001; Moore and Hahn, 2002a,b).

SUMMARY

The quantitative trait linkage analysis methods presented here all have reasonable power to identify major QTLs and can be implemented using freely available computer programs. However, these approaches only address one small part of the genetic architecture of quantitative traits, namely genes that have relatively large independent effect on quantitative trait levels. As our perspective on the genetic architecture of quantitative traits shifts to one where complex gene–gene and gene–environment interactions play a major role, we will need to develop and implement new methods that are capable of identifying such interactions. The design and analysis of any quantitative genetics study should consider that genes might influence trait levels, variability, covariability, and/or dynamics and that each of these effects might be dependent on particular genetic and environmental contexts. Success in identifying QTLs will depend on whether these effects are taken into consideration and will depend on the assumptions of the statistical methods being used.

REFERENCES

Allison DB, Fernandez JR, Heo M, Beasley TM (2000): Testing the robustness of the new Haseman–Elston quantitative-trait loci-mapping procedure. Am J Hum Genet 67:249–252.

Allison DB, Neale MC, Zannolli R, Schork NJ, Amos CI, Blangero J (1999): Testing the robustness of the likelihood-ratio test in a variance-component quantitative-trait loci-mapping procedure. Am J Hum Genet 65:531–544.

Almasy L, Blangero J (1998): Multipoint quantitative-trait linkage analysis in general pedigrees. Am J Hum Genet 62:1198–1211.

Amos CI (1994): Robust variance-components approach for assessing genetic linkage in pedigrees. Am J Hum Genet 54:535–543.

Amos CI, Elston RC, Wilson AF, Bailey-Wilson JE (1989): A more powerful robust sib-pair test of linkage for quantitative traits. Genet Epidemiol 6:435–449.

Amos CI, Zhu DK, Boerwinkle E (1996): Assessing genetic linkage and association with robust components of variance approaches. Ann Hum Genet 60:143–160.

Barnholtz JS, de Andrade M, Page GP, King TM, Peterson LE, Amos CI (1999): Assessing linkage of monoamine oxidase B in a genome-wide scan using a univariate variance components approach. Genet Epidemiol 17(Suppl 1):S49–S54.

Bellman R (1961): Adaptive Control Processes. Princeton, NJ: Princeton University Press.

Berg K (1988): Variability gene effect on cholesterol at the Kidd blood group locus. Clin Genet 33:102–107.

Berg K, Borresen AL, Nance WE (1981): Apparent influence of marker genotypes on variation in serum cholesterol in monozygotic twins. Clin Genet 19:67–70.

Blangero J, Almasy L (1997): Multipoint oligogenic linkage analysis of quantitative traits. Genet Epidemiol 14:959–964.

Blangero J, Williams JT, Almasy L (2000): Quantitative trait locus mapping using human pedigrees. Hum Biol 72:35–62.

Blangero J, Williams JT, Almasy L (2001): Variance component methods for detecting complex trait loci. Adv Genet 42:151–181.

Britton J (1998): Symptoms and objective measures to define the asthma phenotype. Clin Exp Allergy 28(Suppl 1):2–7.

Carlborg O, Andersson L (2002): Use of randomization testing to detect multiple epistatic QTLs. Genet Res 79:175–184.

Carlborg O, Andersson L, Kinghorn B (2000): The use of a genetic algorithm for simultaneous mapping of multiple interacting quantitative trait loci. Genetics 155:2003–2010.

Cheverud JM (2000): Detecting epistasis among quantitative trait loci. In: Wade M, Brodie III B, Wolf J, eds. Epistasis and the Evolutionary Process. New York: Oxford University Press.

Cheverud JM, Routman EJ (1995): Epistasis and its contribution to genetic variance components. Genetics 139:1455–1461.

Comuzzie AG, Hixson JE, Almasy L, Mitchell BD, Mahaney MC, Dyer TD, Stern MP, MacCluer JW, Blangero J (1997): A major quantitative trait locus determining serum leptin levels and fat mass is located on human chromosome 2. Nat Genet 15:273–276.

Congdon CB, Sing CF, Reilly SL (1993): Genetic algorithms for identifying combinations of genes and other risk factors associated with coronary artery disease. In: Proceedings of the Workshop on Artificial Intelligence and the Genome. Chambery.

de Andrade M, Gueguen R, Visvikis S, Sass C, Siest G, Amos CI (2002): Extension of variance components approach to incorporate temporal trends and longitudinal pedigree data analysis. Genet Epidemiol 22:221–232.

Eaves LJ (1994): Effect of genetic architecture on the power of human linkage studies to resolve the contribution of quantitative trait loci. Heredity 72:175–192.

Elston RC, Buxbaum S, Jacobs KB, Olson JM (2000): Haseman and Elston revisited. Genet Epidemiol 19:1–17.

Falconer DS, Mackay TFC (1996): Introduction to Quantitative Genetics. Longman.

Feingold E (2002): Regression-based quantitative trait-locus mapping in the 21st century. Am J Hum Genet 71:217–222.

Fernandez JR, Etzel C, Beasley TM, Shete S, Amos CI, Allison DB (2002): Improving the power of sib pair quantitative trait loci detection by phenotype winsorization. Hum Hered 53:59–67.

Fisher RA (1918): The correlations between relatives on the supposition of Mendelian inheritance. Trans R Soc Edinburgh 52:399–433.

Fulker DW, Cardon LR (1994): A sib-pair approach to interval mapping of quantitative trait loci. Am J Hum Genet 54:1092–1103.

Fulker DW, Cherny SS, Cardon LR (1995): Multipoint interval mapping of quantitative trait loci, using sib pairs. Am J Hum Genet 56:1224–1233.

Gibson G (1996): Epistasis and pleiotropy as natural properties of transcriptional regulation. Theor Popul Biol 49:58–89.

Goldberg DE (1989): Genetic Algorithms in Search, Optimization, and Machine Learning. Reading, MA: Addison-Wesley.

Goldgar DE (1990): Multipoint analysis of human quantitative genetic variation. Am J Hum Genet 47:957–967.

Goldgar DE, Oniki RS (1992): Comparison of a multipoint identity-by-descent method with parametric multipoint linkage analysis for mapping quantitative traits. Am J Hum Genet 50:598–606.

Haseman JK, Elston RC (1972): The investigation of linkage between a quantitative trait and a marker locus. Behav Genet 2:3–19.

Haviland MB, Lussier-Cacan S, Davignon J, Sing CF (1995): Impact of apolipoprotein E genotype variation on means, variances, and correlations of plasma lipid, lipoprotein, and apolipoprotein traits in octogenarians. Am J Med Genet 58:315–331.

Kissebah AH, Sonnenberg GE, Myklebust J, Goldstein M, Broman K, James RG, Marks JA, Krakower GR, Jacob HJ, Weber J, Martin L, Blangero J, Comuzzie AG (2000): Quantitative trait loci on chromosomes 3 and 17 influence phenotypes of the metabolic syndrome. Proc Natl Acad Sci USA 97:14478–14483.

Kruglyak L, Lander ES (1995): Complete multipoint sib-pair analysis of qualitative and quantitative traits. Am J Hum Genet 57:439–454.

Lynch M, Walsh B (1998): Genetics and Analysis of Quantitative Traits. Sinauer.

Mitchell BD, Ghosh S, Schneider JL, Birznieks G, Blangero J (1997): Power of variance component linkage analysis to detect epistasis. Genet Epidemiol 14:1017–1022.

Moore JH (1999): Genetic Analyses of Dynamic Quantitative Traits. Ph.D. Dissertation. University of Michigan.

Moore JH (2001): Improved power of sib-pair linkage analysis using measures of complex trait dynamics. Hum Hered 52:113–115.

Moore JH (2003): Quantitative traits. In: Robinson R, ed., Genetics. New York: Macmillan Science Library.

Moore JH, Hahn LW (2002a): A cellular automata approach to detecting interactions among single-nucleotide polymorphisms in complex multifactorial diseases. Pac Symp Biocomput 2002:53–64.

Moore JH, Hahn LW (2002b): Cellular automata and genetic algorithms for parallel problem solving in human genetics. In: Merelo JJ, Panagiotis A, Beyer H-G, eds. Lecture Notes in Computer Science 2439. Berlin: Springer-Verlag, pp 821–830.

Moore JH, Lamb JM, Brown NJ, Vaughan DE (2002a): A comparison of combinatorial partitioning and linear regression for the detection of epistatic effects of the ACE I/D and PAI-1 4G/5G polymorphisms on plasma PAI-1 levels. Clin Genet 62:74–79.

Moore JH, Smolkin M, Lamb JM, Brown NJ, Vaughan DE (2002b): The relationship between plasma t-PA and PAI-1 levels is dependent on epistatic effects of the ACE I/D and PAI-1 4G/5G polymorphisms. Clin Genet 62:53–59.

Nakagawa K, Ishizaki T (2000): Therapeutic relevance of pharmacogenetic factors in cardiovascular medicine. Pharmacol Ther 86:1–28.

Nakamichi R, Ukai Y, Kishino H (2001): Detection of closely linked multiple quantitative trait loci using a genetic algorithm. Genetics 158:463–475.

Nelson MR, Kardia SL, Ferrell RE, Sing CF (1999): Influence of apolipoprotein E genotype variation on the means, variances, and correlations of plasma lipids and apolipoproteins in children. Ann Hum Genet 63:311–328.

Nelson MR, Kardia SL, Ferrell RE, Sing CF (2001): A combinatorial partitioning method to identify multilocus genotypic partitions that predict quantitative trait variation. Genome Res 11:458–470.

Ott J (1995): Linkage analysis with biological markers. Hum Hered 45:169–174.

Pigliucci M (2001): Phenotypic Plasticity: Beyond Nature and Nurture. Johns Hopkins University Press.

Reilly SL, Ferrell RE, Kottke BA, Kamboh MI, Sing CF (1991): The gender-specific apolipoprotein E genotype influence on the distribution of lipids and apolipoproteins in the population of Rochester, MN. I. Pleiotropic effects on means and variances. Am J Hum Genet 49:1155–1166.

Reilly SL, Ferrell RE, Sing CF (1994): The gender-specific apolipoprotein E genotype influence on the distribution of plasma lipids and apolipoproteins in the population of Rochester, MN. III. Correlations and covariances. Am J Hum Genet 55:1001–1018.

Rice JP (1993): Phenotype definition for genetic studies. Eur Arch Psychiatry Clin Neurosci 243:158–163.

Robertson A (1973): Linkage between marker loci and those affecting a quantitative trait. Behav Genet 3:389–391.

Saccone NL, Kwon JM, Corbett J, Goate A, Rochberg N, Edenberg HJ, Foroud T, Li TK, Begleiter H, Reich T, Rice JP (2000): A genome screen of maximum number of drinks as an alcoholism phenotype. Am J Med Genet 96:632–637.

Schlichting CD, Pigliucci M (1998): Phenotypic Evolution: A Reaction Norm Perspective. Sinauer.

Schork NJ (1993): Extended multipoint identity-by-descent analysis of human quantitative traits: Efficiency, power, and modeling considerations. Am J Hum Genet 53:1306–1319.

Sham PC, Purcell S, Cherny SS, Abecasis GR (2002): Powerful regression-based quantitative-trait linkage analysis of general pedigrees. Am J Hum Genet 71:238–253.

Shete S, Amos CI (2002): Testing for genetic linkage in families by a variance-components approach in the presence of genomic imprinting. Am J Hum Genet 70:751–757.

Shugart YY, Goldgar DE (1997): The performance of MIM in comparison with MAPMAKER/SIBS to detect QTLs. Genet Epidemiol 14:897–902.

Shugart YY, Goldgar DE (1999): Multipoint genomic scanning for quantitative loci: Effects of map density, sibship size and computational approach. Eur J Hum Genet 7:103–109.

Templeton AR (2000): Epistasis and complex traits. In: Wade M, Brodie III B, Wolf J, eds. Epistasis and the Evolutionary Process. New York: Oxford University Press.

Tiwari HK, Elston RC (1998): Restrictions on components of variance for epistatic models. Theor Popul Biol 54:161–174.

Tsuang MT, Faraone SV, Lyons MJ (1993): Identification of the phenotype in psychiatric genetics. Eur Arch Psychiatry Clin Neurosci 243:131–142.

Wade MJ (2000): Epistasis as a genetic constraint within populations and an accelerant of adaptive divergence among them. In: Wade M, Brodie III B, Wolf J, eds. Epistasis and the Evolutionary Process. New York: Oxford University Press.

Weiss KM (1993): Genetic Variation and Human Disease: Principles and Evolutionary Approaches. Cambridge University Press.

Williams JT, Begleiter H, Porjesz B, Edenberg HJ, Foroud T, Reich T, Goate A, Van Eerdewegh P, Almasy L, Blangero J (1999): Joint multipoint linkage analysis of multivariate qualitative and quantitative traits. II. Alcoholism and event-related potentials. Am J Hum Genet 65:1148–1160.

Wright S (1932): The roles of mutation, inbreeding, crossbreeding and selection in evolution. Proc 6th Intl Congr Genet 1:356–366.

Xu X, Weiss S, Xu X, Wei LJ (2000): A unified Haseman–Elston method for testing linkage with quantitative traits. Am J Hum Genet 67:1025–1028.

CHAPTER 10

Advanced Parametric Linkage Analysis

SILKE SCHMIDT

The concept of linkage, introduced in Chapter 1, describes the tendency of two or more loci on a chromosome to cosegregate within families. Linkage analysis examines the joint inheritance of presumed underlying disease (trait) genotypes and genotypes of markers whose position on a chromosome is known. If there is evidence that the disease and the marker do not segregate independently, the unknown disease locus must be within measurable distance of the known marker locus; that is, the two loci are "linked" to each other. The parametric logarithm of the odds (LOD) score is a likelihood-based statistical measure which quantifies the degree of linkage in a pedigree. Note that the disease genotypes can only be inferred from the observed phenotypes of family members. This inference requires the assumption of certain parameters in a genetic model; hence the LOD score method described in this chapter is called "parametric" or "model-based" linkage analysis.

There are four major advantages of model-based linkage analysis over other methods described in this book:

(i) Statistically, it is a more powerful approach than any nonparametric method if the genetic model assumed is approximately correct.

(ii) It utilizes every family member's phenotypic and genotypic information.

(iii) It provides an estimate of the recombination fraction between marker and disease locus.

(iv) It provides a statistical test for linkage and for genetic (locus) heterogeneity.

The genetic model used in parametric linkage analysis includes the following components:

(i) the mode of inheritance (MOI) of the trait (e.g., autosomal dominant, autosomal recessive, X-linked dominant, etc.);

Genetic Analysis of Complex Diseases, Second Edition, Edited by Jonathan L. Haines and Margaret Pericak-Vance

(ii)   the trait and marker allele frequencies;

(iii)   the penetrance values for each possible disease genotype, that is, the probabilities of expressing the disease phenotype given the genotype;

(iv)   the mutation rate of marker loci; and

(v)   sex specificity of recombination fractions.

In practice, model specification typically focuses on components (i)–(iii), while the mutation rate is assumed to be zero and sex-averaged recombination fractions are used. When the genetic model is unknown, as in most genetically complex traits, the LOD score may still be calculated. However, the researcher needs to realize that the LOD score depends not only on the recombination fraction between the two loci of interest but also on all the assumed parameters of the genetic model. Thus, a test of linkage using a parametric approach is a test of all the assumptions of which linkage is only one, and failure to find linkage could be due to a misspecification in any of these parameters; it does not prove lack of linkage. Despite these potential difficulties, LOD score analysis has been successfully applied to several complex traits. It has been used to find single-gene effects in subsets of families, such as in breast cancer (Hall et al., 1990; Easton et al., 1993) and Alzheimer's disease (AD; Pericak-Vance et al., 1991). It has also been applied, in conjunction with nonparametric approaches, to detect more complex linkage signals, such as in type II diabetes (Hanis et al., 1996) and inflammatory bowel disease (Hugot et al., 1996). The purpose of this chapter is to give an overview of the principles of LOD score analysis, to consider the consequences of assuming wrong model parameters, and to provide practical recommendations for LOD score analysis of complex phenotypes, for which it is almost impossible to correctly specify the genetic model.

TWO-POINT ANALYSIS

Chapter 1 explained that the joint inheritance of two loci can be parameterized by the recombination fraction θ, which is a measure of the distance between the two loci expressed as the number of expected crossovers during parental meioses. The parameter θ can take on any value between 0 and $\frac{1}{2}$. The value $\frac{1}{2}$ corresponds to "free recombination," that is, the probability that loci on two completely independent chromosomes recombine in the parental meiosis. The LOD score is simply a ratio of the pedigree likelihood under linkage ($\theta < \frac{1}{2}$) and under free recombination ($\theta = \frac{1}{2}$), where likelihood is defined as the probability of the data as a function of an unknown parameter (here θ). A maximum-likelihood estimator (MLE) of θ is obtained by maximizing the likelihood ratio, that is, by making the actually observed data the most likely to occur. The corresponding maximum $\log_{10}$-transformed likelihood ratio, $Z_{max}(\theta)$, is called the maximum *LOD score* (Barnard, 1949). However, the "odds" in linkage analysis are backward rather than forward odds, and "logarithm of the likelihood ratio" is a less ambiguous description of the LOD score

(Elston, 1997). Formulas for LOD score computation in small pedigrees where recombinant and nonrecombinant meioses can be directly counted were given in Chapter 1 for both phase-known and phase-unknown cases. The overall LOD score for a dataset can be obtained by summing LOD scores (at the same values of θ) from independent pedigrees and is usually reported at $\theta = 0, 0.01, 0.05, 0.1, 0.2, 0.3, 0.4$.

Sequential testing methodology was originally used to derive the critical value of the LOD score test statistic (Morton, 1955). Morton showed that, due to the low prior probability that two loci anywhere on the genome are linked, a LOD score greater than 3.0 is necessary to ensure that the posterior probability of true linkage after declaring the test significant is at least 95%. Today, tests of linkage are generally not carried out in a sequential fashion. However, the rather stringent critical value of 3.0, equivalent to the data being 1000 times more likely under linkage (at the MLE of θ) than under no linkage, has remained the classical threshold for declaring significant linkage. The actual (exact) p-value associated with $Z(\theta) = 3$ was shown to be at most 0.001 (Chotai, 1984). If a large number of informative meioses are observed, likelihood ratio theory can be employed to show that the asymptotic p-value based on a χ^2 [on one degree of freedom (df)] approximation for $2 \times \log_e(10) \times Z(\theta) \approx 4.6 \times Z(\theta)$ is 0.0001 ($X^2 = 4.6 \times 3.0 = 13.8$). Analogous to the critical value for declaring the presence of linkage, the threshold of -2.0 was derived as the critical value for significant evidence against linkage (Morton, 1955). An interval likely to contain the true value of θ for a given dataset is called a support interval in linkage analysis. This interval is conceptually similar to a traditional confidence interval for an unknown parameter but is defined by LOD score units. For example, if the MLE of θ corresponds to a maximum LOD score of $Z_{max}(\theta) = 3.2$, a one-unit support interval would contain all values of θ for which $Z(\theta)$ is at least 2.2 (Ott, 1999, Chapter 4.4).

The classical threshold of 3.0 for establishing significant linkage was derived for single-marker tests. In the context of a genome screen employing many (perhaps 300–500) microsatellite markers, some modifications are necessary; however, there is no general agreement on the most appropriate approach. On the one hand, the problem of multiple testing has to be considered, which would require an increased critical value to control the overall false-positive rate. On the other hand, any correction that ignores the correlations between tests at multiple closely spaced markers (such as the Bonferroni correction) will be overly conservative, thus negatively impacting the power of the analysis. A number of slightly different schemes for declaring "suggestive," "significant," and "highly significant" linkage in the context of a genome screen have been proposed (Thomson, 1994; Lander and Kruglyak, 1995; Haines, 1998). The empirical LOD score thresholds proposed by Haines (1998) are based on a multistaged approach to genome screening: A LOD score of 1 (nominal $p = 0.016$) from a single dataset at the initial screening stage is considered "interesting" enough for inclusion in the follow-up stage; a LOD score of 2 ($p = 0.001$) is considered "very interesting"; and a LOD score of 3 ($p = 0.0001$) would amount to declaring "provisional linkage," awaiting confirmation from an independent dataset in order to be considered highly significant.

More recently, simulation-based methods, which exploit the power and speed of modern computers, have been proposed as an alternative to significance measures based on asymptotic theory. For example, to determine the empirical significance level associated with an observed maximum LOD score in a genome screen, marker genotypes unlinked to any putative disease locus can be simulated for the given pedigree structures and observed disease phenotypes using a program such as SIMULATE (Terwilliger et al., 1993). The same LOD score analysis that was performed on the original dataset is carried out in each replicate. The proportion of replicates with a maximum LOD score equal to or greater than the one observed in the real dataset then approximates the genomewide empirical significance level (*p*-value). It is possible to compute the number of replicates required to obtain a sufficiently accurate estimate of the true (but unknown) significance level for the above situation (Ott, 1999, Chapter 9.7). Depending on the size of the dataset, the number of markers, and the complexity of the analysis, this method of estimating *p*-values may be quite time consuming even with fast computers. As an alternative to assessing individual-marker LOD score significance, approaches for combining multiple correlated *p*-values or test statistics across linked markers to identify a certain region likely to harbor a disease gene have recently been proposed (Zaykin et al., 2002; Hoh and Ott, 2000).

It is interesting to note that even an actually observed LOD score of 3.0 or greater for a single pedigree in a region previously implicated as linked may have to be interpreted with caution. An analysis of a large multigenerational pedigree segregating breast cancer was used to illustrate that it is possible to attain LOD scores of this magnitude on the basis of very few observations when there is a lack, by chance, of observed recombinations with certain markers. In this situation, the LOD score is highly sensitive to further data, and the addition of a single recombinant individual reduced a LOD score of 3.0 at $\theta = 0$ to 1.72 at $\theta = 0.06$ (Skolnick et al., 1984). This emphasizes the importance of follow-up analysis, that is, of saturating a region found to give preliminary evidence for linkage with a higher density of additional markers.

Computationally, the calculation of the LOD score requires sophisticated statistical-genetic software for all but the simplest cases. To specify a genetic model, disease and marker allele frequencies, mode of inheritance, and probabilities of (observed) phenotypes given (unobserved) disease genotypes (the so-called penetrance values) have to be determined. Individuals expressing the disease phenotype due to factors other than the presumed susceptibility gene are called *phenocopies*. The penetrance of phenocopies is sometimes referred to as the *phenocopy rate*, but the same term is also used for the proportion of nongenetic cases among all affected individuals and is therefore best avoided. In some simple cases, disease genotypes can be inferred unambiguously from the observed phenotypes, as illustrated by the following example: Let *D* denote the disease allele and *d* the normal allele. If the disease gene is fully penetrant, phenocopies do not exist, the mode of inheritance is autosomal dominant, and the *D* allele is very rare so that the probability of a *DD* genotype is negligible, an affected person must have genotype *Dd* and an unaffected person must have genotype *dd*. When all family members are genotyped and markers are informative, recombinant gametes can then be directly counted (Chapter 1). With incomplete penetrance and presence of pheno-copies, only genotype probabilities can be computed, and each possible genotype of one pedigree member implies a range of possible genotypes for related members within the constraints of Mendelian inheritance. The general form of the pedigree likelihood function is

$$L(\theta) = \sum_{g} \prod_{i} P(x_i|g_i) \prod_{j} P(g_j) \prod_{k} P(g_k|g_{km}, g_{kf})$$

where g is a vector of genotypes of pedigree members; x_i is the phenotype of the ith individual, with i summing over all pedigree members, j over founders (individuals without parents in the pedigree), and k over nonfounders; and g_{km}, g_{kf} denote geno-types of the kth individual's mother and father, respectively. The algorithms implemented in current software for the efficient computation of this function are described below under Multipoint Analysis.

Example of LOD Score Calculation and Interpretation

We will illustrate LOD score calculation with the example of a linkage study of AD and markers on chromosome 19q. We will use a dataset that includes pedigrees from the original paper reporting linkage of AD to this chromosomal region (Pericak-Vance et al., 1991), which subsequently led to the identification of the apolipoprotein E (APOE) gene as a susceptibility gene for late-onset AD (Corder et al., 1993). Although the mode of inheritance for AD is unknown, we assume an autosomal dominant model allowing for phenocopies and incomplete penetrance in this example. The disease allele frequency is assumed to be $q = 0.001$, which corresponds to a carrier frequency of approximately 0.2% [i.e., $2q(1 - q) + q^2 = (2 \times 0.001 \times 0.999) + 0.001^2 = 0.002$ under Hardy–Weinberg equilibrium]. The population allele frequencies for the marker, D19S246, were calculated from a representative set of unrelated Caucasian individuals (164 chromosomes). The penetrance for gene carriers (f_{DD}, f_{Dd}) is assumed to be 0.80, and the penetrance for noncarriers (phenocopy frequency, f_{dd}) is assumed to be 0.01. With these parameter choices, the proportion of nongenetic cases among all affected individuals is approximately 86% [i.e., $(0.01 \times 0.999^2)/(0.01 \times 0.999^2 + 0.80 \times 0.002) = 0.862$]. For a single-gene disorder, the population prevalence should approximately match the prevalence φ derived from the assumed disease allele frequency and penetrance values [e.g., $\varphi = q^2 \times f_{DD} + 2q(1 - q) \times f_{Dd} + (1 - q)^2 \times f_{dd}$]. However, for a complex and common disease such as AD, with a prevalence of approximately 10% in adults over 75 years of age, it is not realistic to try to achieve this relationship. More appropriate ways to analyze such complex phenotypes with the genetic model-based LOD score approach are discussed later in this chapter.

Table 10.1 presents the two-point linkage analysis for AD and marker D19S246 under the above dominant model. The highest LOD score is 1.57 at $\theta = 0.15$. Using the classical criterion for linkage [$Z_{max}(\theta) \geq 3.0$], one could easily conclude that

TABLE 10.1. Two-Point Linkage Analysis for AD and D19S246

Pedigree	$\theta = 0.00$	$\theta = 0.05$	$\theta = 0.10$	$\theta = 0.15$	$\theta = 0.20$	$\theta = 0.30$	$\theta = 0.40$
401	0.016	0.013	0.010	0.008	0.006	0.002	0.001
701	−0.972	−0.445	−0.242	−0.129	−0.062	−0.004	0.003
736	1.010	0.893	0.773	0.651	0.527	0.282	0.081
757	0.297	0.301	0.286	0.258	0.218	0.123	0.035
763	0.225	0.191	0.158	0.126	0.096	0.045	0.012
794	−0.473	−0.224	−0.092	−0.016	0.026	0.045	0.021
820	0.058	0.108	0.133	0.141	0.135	0.094	0.035
911	0.039	0.030	0.023	0.017	0.012	0.005	0.001
1086	0.163	0.133	0.105	0.081	0.059	0.027	0.007
1207	−0.568	−0.196	−0.063	−0.003	0.023	0.025	0.008
1229	−0.367	−0.217	−0.120	−0.058	−0.022	0.002	0.000
1396	0.230	0.194	0.160	0.128	0.097	0.046	0.012
1399	0.822	0.719	0.611	0.500	0.387	0.179	0.040
1491	0.046	0.056	0.069	0.081	0.086	0.080	0.053
1547	0.081	0.063	0.048	0.035	0.025	0.010	0.002
1592	0.029	0.022	0.017	0.012	0.008	0.003	0.001
1677	−0.325	−0.253	−0.192	−0.141	−0.100	−0.041	−0.010
1682	0.321	0.317	0.297	0.266	0.224	0.127	0.038
1685	0.071	0.066	0.057	0.046	0.034	0.014	0.003
1725	−0.201	−0.168	−0.135	−0.105	−0.078	−0.035	−0.009
1738	−0.297	−0.230	−0.171	−0.122	−0.082	−0.030	−0.006
1743	−0.272	−0.223	−0.180	−0.140	−0.105	−0.048	−0.012
1843	−1.675	−0.960	−0.655	−0.457	−0.315	−0.129	−0.031
1971	0.014	0.011	0.008	0.006	0.004	0.002	0.001
1999	−0.520	−0.351	−0.237	−0.155	−0.098	−0.031	−0.005
2043	0.219	0.190	0.160	0.131	0.103	0.051	0.014
2100	0.693	0.602	0.511	0.420	0.331	0.168	0.046
2120	−0.011	−0.009	−0.007	−0.006	−0.004	−0.002	−0.001
Total	−1.347	0.633	1.335	1.572	1.536	1.011	0.339

there is no significant evidence for linkage. We will use this example to examine more closely the assumptions made in the analysis and their respective impact on the observed result.

EFFECTS OF MISSPECIFIED MODEL PARAMETERS IN LOD SCORE ANALYSIS

The LOD score of 1.57 at $\theta = 0.15$ was obtained using one particular genetic model. For a genetically complex disorder that is likely influenced by multiple susceptibility genes (i.e., is genetically heterogenous) as well as environmental factors and possible gene–gene and gene–environment interaction, this single assumed genetic model is certainly incorrect in at least some aspects. If the assumed genetic model is wrong, both false-positive and false-negative evidence for linkage can be obtained, since the LOD score is a function of both the recombination fraction and the genetic model. There are many factors that contribute to the impact of misspecified genetic parameters on the LOD score, such as the true underlying disease model, the nature of the misspecified parameters, the extent of misspecification, and the pedigree structures. Results from theoretical calculations and simulation studies generally agree that the power to detect linkage is highly sensitive to mode of inheritance (particularly dominant vs. recessive), somewhat sensitive to marker allele frequencies, slightly sensitive to penetrance, and not very sensitive to disease allele frequency. However, the estimation of the recombination fraction may be strongly affected by an error in any genetic parameter (Clerget-Darpoux et al., 1986).

To illustrate the impact of misspecified model parameters on the power to detect linkage or the type I error (false-positive rate), computer simulations using the 28 AD families were performed. Keeping the actual pedigree structure and observed disease phenotypes constant, the program SLINK (Weeks et al., 1990b) was used to simulate disease and marker genotypes in 1000 replicates assuming either complete linkage ($\theta = 0.0$) or no linkage ($\theta = 0.5$) between the disease and marker locus. The disease locus was simulated under the model assumed above for the analysis of the real data: autosomal dominant inheritance with a disease allele frequency $q = 0.001$, penetrances for gene carriers (f_{DD}, f_{Dd}) of 0.80 and for phenocopies (f_{dd}) of 0.01. The marker was simulated using the actually observed allele frequencies of marker D19S246 (12 alleles with frequencies 0.25, 0.02, 0.04, 0.02, 0.03, 0.18, 0.11, 0.06, 0.14, 0.11, 0.02, 0.02). Each replicate was analyzed using the simulated model as well as 11 other models illustrating possible model misspecification: three models with different disease allele frequencies, three models with different penetrance ratios, three models with different modes of inheritance, and two models with different marker allele frequencies. When one parameter was evaluated, the other parameters were generally fixed at their "true" values used in the simulating model; however, when the mode of inheritance was varied, we included an analysis model where the disease allele frequency was adjusted to produce the same disease allele carrier frequency. The effect of misspecified model parameters on the LOD score analysis was measured by the mean maximum LOD score (Z_{max}) and mean estimated MLE (θ) obtained from the 1000 replicates. The results at $\theta = 0.0$ reflect the impact on the power to detect linkage, and those at $\theta = 0.5$ reflect the impact on the type I error.

Impact of Misspecified Disease Allele Frequency

With a true disease allele frequency of 0.001, the data were analyzed using modified frequencies of 0.01, 0.1, and 0.0001, corresponding to disease allele carrier frequencies of 2%, 20%, and 0.02%, respectively (Table 10.2). This represents a wide variation of disease allele frequencies. In the absence of linkage, there is little variation in the mean maximum LOD score (from 0.11 to 0.12). In the presence of complete linkage, the mean Z_{max} decreases slightly when the disease allele frequency is either underestimated or overestimated. In an autosomal dominant model, an increased

TABLE 10.2. Impact of Misspecifying Disease Allele Frequency on LOD Score Analysis

				Generating Model: $q = 0.001, f_{DD} = f_{Dd} = 0.80,$ $f_{dd} = 0.01$				
Analysis Model				$\theta = 0.00$			$\theta = 0.50$	
q	f_{DD}	f_{Dd}	f_{dd}	Mean Z_{max}	Mean MLE (θ)	Proportion with $Z_{max} > 3.0$ (%)	Mean Z_{max}	Mean MLE (θ)
0.001	0.80	0.80	0.01	11.78	0.0	100	0.11	0.40
0.01	0.80	0.80	0.01	11.44	0.0	100	0.12	0.40
0.10	0.80	0.80	0.01	9.95	0.0	99.8	0.11	0.40
0.0001	0.80	0.80	0.01	11.55	0.0	100	0.11	0.40

disease allele frequency can either increase the probability of affected parents being homozygous rather than heterozygous or increase the probability that the disease allele is introduced into the pedigree through married-in individuals rather than a single founder. However, within a reasonable range of plausible values (0.0001–0.10), the impact of misspecifying the disease allele frequency on the LOD score and on the estimated recombination fraction is generally small.

Impact of Misspecified Mode of Inheritance

In general, a misspecified mode of inheritance has a large impact on the LOD score (Clerget-Darpoux et al., 1986). This effect is particularly serious when a dominant disease locus is misspecified as recessive (Table 10.3). In our AD example, when we vary the mode of inheritance, the mean maximum LOD score decreases from 11.78 to 1.64, the power to obtain $Z_{max} \geq 3$ decreases from 100 to 7.5%, and the MLE (θ) increases from 0.0 to 0.10 when the disease allele frequency remains at $q = 0.001$. When the disease allele frequency is adjusted ($q = 0.10$) to obtain approximately the same gene carrier frequency, the decreases in mean Z_{max} and power are less

TABLE 10.3. Impact of Misspecifying Mode of Inheritance on LOD Score Analysis

				Generating Model: $q = 0.001, f_{DD} = f_{Dd} = 0.80, f_{dd} = 0.01$				
Analysis Model				$\theta = 0.00$			$\theta = 0.50$	
q	f_{DD}	f_{Dd}	f_{dd}	Mean Z_{max}	Mean MLE (θ)	Proportion with $Z_{max} > 3.0$ (%)	Mean Z_{max}	Mean MLE (θ)
0.001	0.80	0.80	0.01	11.78	0.0	100	0.11	0.40
0.001	0.80	0.01	0.01	1.64	0.1	7.5	0.11	0.40
0.10	0.80	0.01	0.01	3.94	0.1	75.2	0.11	0.40
0.001	0.80	0.40	0.01	9.75	0.0	99.9	0.12	0.40

pronounced (from 11.78 to 3.94 and from 100 to 75.2%, respectively) but still quite drastic. An intuitive explanation for this observation is that half the time the random transmission of alleles at the disease locus from an unaffected parent to affected offspring is scored as a recombination between the disease and marker locus, because under a dominant model an unaffected parent does not carry the susceptibility allele, but under a recessive model an unaffected parent may carry that allele. In addition, under the recessive model an affected parent will be considered homozygous at the disease locus and is thus uninformative for linkage. The impact of assuming an additive model when the true model is dominant is much less dramatic (Z_{max} decreases from 11.78 to 9.75), demonstrating that the additive model can be viewed as a good "compromise" between dominant and recessive models when the true mode of inheritance is unknown. Table 10.3 shows that in the absence of linkage a misspecified mode of inheritance has little impact on the LOD score.

Impact of Misspecified Disease Penetrances

As long as incomplete penetrance is allowed in the genetic model, misspecifying the exact values for the penetrance has a relatively small impact on the LOD score whether there is linkage or no linkage (Table 10.4). As the ratio of penetrances between disease allele carriers and non–disease allele carriers decreases (from 80 to 40 to 8 in Table 10.4), the mean Z_{max} also decreases (from 11.78 to 9.76 to 4.91, with a power decrease from 100 to 99.9 to 94.5%), since a low ratio decreases the certainty of whether an affected individual is a disease allele carrier or an unaffected individual a noncarrier. In the presence of linkage, most individuals are nonrecombinants, which is why the low ratio results in a decrease in power. When the penetrance of noncarriers is reduced to 0, Z_{max} decreases from 11.78 to 9.64. Thus, failure to allow for the presence of phenocopies is not detrimental as long as there is incomplete penetrance, which permits multiple possible disease genotypes to underlie the observed phenotype. When a Mendelian disorder is analyzed, a nonzero phenocopy rate is sometimes interpreted as a "misdiagnosis parameter," to account for the fact that some family members are diagnosed as affected with less certainty than others and thus may or may not be carriers of the gene that

TABLE 10.4. Impact of Misspecifying Disease Penetrance on LOD Score Analysis

				Generating Model: $q = 0.001, f_{DD} = f_{Dd} = 0.80, f_{dd} = 0.01$				
Analysis Model				$\theta = 0.00$			$\theta = 0.50$	
q	f_{DD}	f_{Dd}	f_{dd}	Mean Z_{max}	Mean MLE (θ)	Proportion with $Z_{max} > 3.0$ (%)	Mean Z_{max}	Mean MLE (θ)
0.001	0.80	0.80	0.01	11.78	0.0	100	0.11	0.40
0.001	0.80	0.80	0.10	4.91	0.0	94.5	0.10	0.40
0.001	0.80	0.80	0.00	9.64	0.1	99.8	0.12	0.40
0.001	0.40	0.40	0.01	9.76	0.0	99.9	0.12	0.40

causes the familial aggregation. In general, in the presence of model uncertainty, choosing a lower penetrance ratio by specifying both a nonzero penetrance of phenocopies and incomplete penetrance for susceptible genotypes (a "weaker model") is sometimes recommended to make LOD score analysis less sensitive to misspecified model parameters (Risch and Giuffra, 1992).

Impact of Misspecified Marker Allele Frequency

In our example dataset, misspecified marker allele frequencies do not have a large impact on the mean Z_{max} for the overall dataset (Table 10.5). This can be explained partly by the relatively small degree of the potential misspecification; in the worst case, the most common allele of a marker with a total of 12 alleles has a frequency of 0.23 and is misspecified as 0.02 (and vice versa). Unlike for allelic association (Chapter 12), the presence of linkage is compatible with the cosegregation of different alleles of a marker in different pedigrees. Therefore, changing the frequency of a particular allele may reduce the LOD score for some families but may increase the LOD score for others. When these effects balance each other, the LOD score for the entire dataset will hardly change at all. However, there are situations when marker allele frequencies can have a substantial impact on the LOD score obtained from a single pedigree (Ott, 1992). In general, a higher proportion of founders without genotype data leads to a greater impact of marker allele frequencies on LOD score analysis, since marker frequencies are used to compute probabilities of marker genotypes for untyped individuals. For example, if two affected cousins share a marker allele in common and not all their parents and grandparents are genotyped, the evidence for linkage increases with the rarity of the allele. If the allele is quite common, there is an increased probability that parents are homozygous for that allele, and thus uninformative for linkage, or that married-in individuals, rather than the same ancestor, may have transmitted the allele, thus diminishing the evidence for linkage. The greatest impact is observed when the marker has only two alleles with very different

TABLE 10.5. Impact of Misspecifying Marker Allele Frequencies on LOD Score Analysis

	Generating Model: $q = 0.001, f_{DD} = f_{Dd} = 0.80, f_{dd} = 0.01$, Marker Allele Frequencies 0.25, 0.02, 0.04, 0.02, 0.03, 0.18, 0.11, 0.06, 0.14, 0.11, 0.02, 0.02				
	$\theta = 0.00$			$\theta = 0.50$	
Analysis Model	Mean Z_{max}	Mean MLE (θ)	Proportion with $Z_{max} > 3.0$ (%)	Mean Z_{max}	Mean MLE (θ)
Same as simulated	11.78	0.0	100	0.11	0.40
Equally frequent alleles	12.09	0.0	100	0.16	0.40
Common/rare alleles reversed	11.84	0.0	100	0.23	0.40

frequencies. In the presence of linkage between disease and marker, misspecification of the rare allele as common would reduce linkage evidence, that is, produce false-negative evidence against linkage. In the absence of linkage, misspecification of the common allele as rare would lead to false-positive evidence for linkage. This probably explains why, in the example dataset (Table 10.5), the mean Z_{max} score under the null hypothesis of no linkage increased from 0.11 (marker alleles as simulated) to 0.16 (equally frequent marker alleles) to 0.23 (most common alleles misspecified as most rare and vice versa).

CONTROL OF SCORING ERRORS

Apart from misspecified model parameters, there are three types of scoring errors that can affect the value of the LOD score: phenotype errors, genotype errors, and misspecified relationships within a pedigree. Phenotype errors occur when an affected individual is misclassified as unaffected or vice versa. Not much can be done from an analysis perspective to minimize the probability of phenotype errors, although their effect on the LOD score can be large. For example, when full penetrance is assumed and an unaffected individual mistakenly classified as affected shows an obligate recombination between a marker and a putative disease locus, the LOD score for a single pedigree might drop from, for instance, 1.5 to $-\infty$ at $\theta = 0$. Therefore, an accurate clinical diagnosis is of utmost importance for any linkage analysis. The program VARYPHEN allows the user to carry out a sensitivity analysis with respect to disease phenotypes. By setting the disease phenotype of each family member in turn to "unknown" and repeating the linkage analysis, a table of resulting changes in LOD scores and estimated θ values is obtained. It is then easy to see which individuals produce the largest changes and are thus the most crucial in terms of diagnostic accuracy (Vieland et al., 1992). Genotype errors, just like missing data, are unavoidable in any real dataset since a machine's or a technician's interpretation of experimental results is an imperfect way of measuring the true marker genotype. For most laboratories, an empirical value of about 1% genotype error is probably a realistic estimate. The introduction of quality control procedures in the genotyping laboratory (Rimmler et al., 1999) is extremely useful for keeping the actual error rate as low as possible. Just like phenotype errors, genotype errors may have an appreciable impact on the LOD score by introducing false recombinants, thus decreasing the evidence for linkage. Several methods can be used to identify genotypes likely to be erroneous after they have been submitted for analysis. The first and most straightforward check is the identification of Mendelian inconsistencies (e.g., the occurrence of an allele in the offspring that is not found in the parents, the occurrence of more than four alleles in a sibship). Software such as PEDCHECK (O'Connell and Weeks, 1998) can be used to identify these inconsistencies, followed by rereading in the laboratory. Another, less commonly employed possibility is to estimate identity-by-state sharing among affected relative pairs using the affected pedigree member (APM) method (see Chapter 11) and to reread any genotypes that display much less sharing than expected

(Weeks et al., 2000). For densely spaced marker genotypes obtained during the fine-mapping stage of a project, rather than the genome screen stage, some additional error-checking approaches can be applied. The program SIMWALK2 (Sobel and Lange, 1996) is often used to estimate the haplotype configurations of multiple linked markers in a pedigree. As an extension of this haplotyping capability, a new "mistyping" option has recently been added to the package and can be used to detect excess recombination in small genomic regions, which typically indicates likely genotype errors. The program SIBMED (Douglas et al., 2000) is particularly suitable for error detection in affected sibpair data without parental genotypes, where haplotyping is more difficult to carry out. It compares the multipoint probability of the sibpair data with the marker in question included versus excluded and pinpoints marker–sibpair combinations with probability ratios that exceed a simulation-based marker-specific threshold.

Seemingly erroneous genotypes in a pedigree may also be due to misspecified relationships between pairs of individuals. A loss of power to detect linkage may be caused by the inclusion of pairs who are less closely related than assumed. For example, two putative full siblings may be unrelated in the case of unknown adoption or sample switches or they may be half siblings in the case of false paternity. False-positive evidence for linkage may be created by sample duplication or incorrect assignment of monozygotic (MZ) twins as full siblings. These types of problems may not be detectable with Mendelian inconsistency checks, particularly if a sample includes only sibpair data without parental genotypes. Once a large number of markers (≥ 50) have been genotyped, software such as RELPAIR (Epstein et al., 2000; Boehnke and Cox, 1997), RELATIVE (Goring and Ott, 1997), or PREST (McPeek and Sun, 2000) can be used to compute the multipoint likelihood of the marker data conditional on certain possible relationships (e.g., MZ twins, full siblings, half siblings, unrelated pair for putative full sibpairs). By comparing the probability of the marker data under the specified versus a different assumed genetic relationship between various relative pairs, the most likely relationship of the pair can be inferred. RELPAIR is able to consider all possible pairs of individuals in the sample, not just those within families, and even allows for the presence of genotyping error. PREST can analyze inbred as well as outbred pedigrees.

GENETIC HETEROGENEITY

Every LOD score analysis should include an examination of the scores for each pedigree. In our AD example, LOD scores ranged from -1.67 to 1.01 (at $\theta = 0.0$), raising the possibility that only some of the pedigrees may be linked to the chromosome 19q13 locus. A complex disorder like AD is very likely to be genetically heterogeneous. In general, genetic locus heterogeneity is said to exist when two or more genes act independently to cause the same clinical disease phenotype or to increase disease susceptibility (see Chapter 1). A study's power to detect linkage can be greatly improved either by increasing sample homogeneity to start with (e.g., by examining only a clinical subtype of the disorder or only patients with a particularly early age of onset) or by teasing out more homogeneous subgroups that may exhibit a stronger linkage signal which is obscured by the heterogeneity in the overall dataset. This type of heterogeneity is likely a major reason for the many failures to replicate linkage findings in complex disease. However, there are many ways in which heterogeneity can be allowed for in the statistical analysis of pedigree data, some of which are described below.

The null hypothesis of homogeneity can be formally tested using two different approaches: the M-test (Morton, 1956; Ott, 1999) or its extension known as the β-test (Risch, 1988) and the admixture test (Smith, 1963). When families can be pre-assigned to several different groups based on certain disease characteristics, such as age at onset, severity of disease, clinical features, or transmission pattern, the M-test can be used to test for different recombination fractions in these k different groups of families, where $\theta_1 \neq \theta_2 \neq \cdots \theta_k$. The first hypothesis ($H_1$) of linkage and homogeneity specifies $\theta_1 = \theta_2 = \cdots \theta_k < 0.5$. Under the second hypothesis of heterogeneity (H_2), $\theta_1 \neq \theta_2 \neq \cdots \theta_k$, the recombination fractions are potentially different in the different groups of families and are estimated separately. Note that H_1 is obtained from H_2 by restricting $k - 1$ parameters ($\theta_2, \ldots, \theta_k$) to be the same as θ_1. To test H_1 against H_2, one computes

$$X^2 = 2 \times \ln(10)\left(\sum_{i=1}^{k} Z_i(\hat{\theta}_i) - Z(\hat{\theta}) \right)$$

where $Z_i(\hat{\theta}_i)$ is the maximum LOD score in each of the family groups and $Z(\hat{\theta})$ is the maximum LOD score over all the pedigrees. Asymptotically, under the assumption of homogeneity (H_1), this statistic follows a χ^2-distribution with $k - 1$ df.

The M-test requires the estimation of a separate θ_i parameter in each of the k groups of families. Alternatively, a Bayesian approach can be taken, in which the θ-values are assumed to follow a β-distribution with two parameters, which are estimated from the posterior distribution of θ given the observed data. This likelihood-based test is known as the β-test and has been shown to be more powerful than the M-test in many situations (Risch, 1988).

The admixture test (Smith, 1963) is used when families cannot be preassigned to different groups based on clinical criteria. In the most commonly employed version of this test, two types of families are assumed, with α denoting the proportion of type I families and $1 - \alpha$ the proportion of type II families. The recombination fraction in type I families is equal to θ_1 and that in type II families is $\theta_2 = 0.5$. The null hypothesis of no linkage and homogeneity is H_0: $\alpha = 0$, $\theta_1 = \theta_2 = 0.5$. The hypothesis of linkage and heterogeneity is H_2: $\alpha < 1$, $\theta_1 < 0.5$, and the hypothesis of linkage and homogeneity is obtained from H_2 by a single restriction, H_1: $\alpha = 1$, $\theta_1 < 0.5$. Under H_2, the likelihood is maximized over α and θ_1. Under H_1, the restricted hypothesis, θ_r is obtained assuming homogeneity ($\alpha = 1$). The test of the hypothesis of homogeneity (H_1 vs. H_2) is carried out by calculating $X^2 = 2$ [ln $L(\alpha, \theta_1) - \ln L(\alpha = 1, \theta_r)$], which asymptotically has a χ^2-distribution with 1 df with probability $\frac{1}{2}$ and is equal to zero with probability $\frac{1}{2}$. Thus, the p-value from a two-sided test has to be halved. Both the M-test and the admixture test can be carried out with the HOMOG program package (Ott, 1986). A computer program to carry out the β-test is available from Dr. Neil Risch. In practice, the admixture test is used more frequently than the M-test and β-test.

A problem with the admixture test concerns significance when the total LOD score (under homogeneity) is less than 3.0, yet a test of homogeneity yields a p-value below the conventional level of 0.05 or 0.01. Declaring the heterogeneity significant would imply the presence of linkage in a subset of the families, even though the usual criterion for significant linkage ($Z \geq 3$) is not met. The more conservative school of thought argues that an overall $Z \geq 3$ ($p \leq 0.0001$) is necessary to reject the null hypothesis of no linkage and that the additional hypothesis of heterogeneity should only be tested if that hypothesis is rejected. A less stringent approach tests linkage and homogeneity simultaneously. In this situation, an overall log-likelihood ratio difference of 3.0 or more between homogeneity with no linkage and heterogeneity with linkage is considered sufficient to declare significant heterogeneity and linkage in a subset of the families. For heterogeneity LOD (HLOD) score, the numerator is maximized over both α and θ and the denominator is the likelihood under H_0: $\alpha = 0$, $\theta = 0.5$. An asymptotic null distribution of the test statistic $X^2 = 4.6 \times \text{HLOD}$, which tests for linkage while allowing for heterogeneity, has been derived (Faraway, 1993). For our sample AD data, the admixture test does not indicate the presence of locus heterogeneity. The HLOD is the same as the previously obtained $Z_{\text{max}} = 1.57$ at $\theta = 0.15$ ($\alpha = 1.0$).

The power of these statistical tests of homogeneity is influenced by a number of factors, including the distance between the linked marker and the disease locus and the amount of linkage information provided by the individual families. A larger recombination fraction between marker and disease gene reduces the power to detect heterogeneity. Larger, more informative families will yield more power than a series of smaller families in discriminating linked and unlinked groups. The main advantage of the admixture test is that it does not require any prior knowledge about possible sources of heterogeneity. If there is a valid prior reason to divide the dataset into more homogeneous subgroups, however, the admixture test is less powerful than the M-test.

Various extensions of the simple admixture model described above have also been implemented in the HOMOG package. For example, HOMOG2 assumes that a disease is caused by either of two loci which are within a measurable distance of each other on the same chromosome and thus are linked to the same marker. Two recombination fractions are then estimated, together with the proportion of families linked to the first locus α, assuming a proportion $1 - \alpha$ to be linked to the second locus. HOMOG3R can be used for a model assuming linkage to one disease locus in some families, linkage to another locus, unlinked to the first (e.g., on a different chromosome), in other families, and a third subset of families unlinked to either of the two disease loci. It has been shown (Schork et al., 1993) that this approach is a good approximation for two-locus linkage analysis, where simultaneous gene mapping of two disease loci is attempted. HOMOGM is the extension of this type of model to an arbitrary number m of disease loci.

In summary, heterogeneity can be a major confounding factor in the gene-mapping process. Once heterogeneity has been confirmed, one must ask what criteria should be used to classify a family as "linked" for molecular follow-up purposes. The parameter estimates from the HOMOG admixture test can be used to calculate the estimated conditional (given marker observations) probability that a family is of the linked type. Even though this is just a statistical estimate derived from a particular set of data, such a probability may be useful to focus laboratory time and effort on the families most likely to harbor a disease gene.

MULTIPOINT ANALYSIS

While data from a genome screen should always initially be analyzed by computing two-point LOD scores, this approach can only extract a limited amount of information from the data. Limitations are imposed both by missing genotypes due to individuals unavailable for blood sampling or unwilling to participate in the study and by individuals uninformative for linkage, that is, those who are homozygous at the marker locus and thus transmit the same allele to their offspring regardless of their underlying genotype at the disease locus. The probability of heterozygous individuals is given by a marker's heterozygosity value, defined as

$$H = 1 - \sum_i p_i^2$$

where p_i is the population frequency of the ith allele, for which estimates are used in practice. Individuals uninformative at one marker locus may contribute linkage information, and missing genotypes may sometimes be inferred if genotypes at flanking markers are taken into account in the analysis, which is the purpose of multipoint analysis.

Multipoint LOD score analysis is an extension of two-point analysis in which linkage of a disease locus is tested not to just a single marker but to an entire map of markers. The map-specific multipoint LOD score using the $\log_{10}$ transformation is most commonly used and is defined as

$$Z(x, \phi_0) = \log_{10}\left(\frac{L(x, \phi_0)}{L(x = \infty, \phi_0)}\right)$$

where $L(x, \phi_0)$ is the likelihood, under an assumed genetic model ϕ_0, that a disease locus is located at a distance x on a fixed map consisting of several markers and $L(x = \infty, \phi_0)$ is the likelihood that the disease locus is not on the map (corresponding to $\theta = 0.5$ in two-point analysis, i.e., no linkage). Usually $Z(x)$ is evaluated at several positions on the map from one end to the other. There are two major advantages to multipoint LOD score analysis. First, as mentioned above, it provides an opportunity to recover linkage information at an originally uninformative locus via haplotype inference. Thus, the linkage results are less sensitive to uninformative or missing genotypes at any single marker. In essence, multipoint analysis can extract more of the total inheritance information from a pedigree. In one simulation study, when a rare dominant disease gene was simulated in the middle of an 18-cM map using a pedigree with two affected fourth cousins, a LOD score of 2.2 (91% of its theoretical maximum) was obtained when 20 markers were analyzed simultaneously. In contrast, the highest two-point LOD score was only 0.83 (34% of its theoretical maximum), and even simultaneous six-marker analysis yielded, at most, a LOD score of 1.74 (72% of the theoretical maximum) (Kruglyak et al., 1996).

Second, the multipoint LOD score approach can be very useful to pinpoint a disease gene location in the fine mapping of a Mendelian disorder. Any true recombinant at some location x will contribute a strongly negative LOD score and will decrease the possibility that this position contains a candidate locus. This significantly narrows the region in which the disease locus can exist, thus helping to better define the minimum candidate region. For a single pedigree, the information conveyed by a multipoint LOD score curve mirrors the cosegregation of multi-marker haplotypes with the disease phenotypes of pedigree members. Therefore, the process of haplotyping, that is, the ordering of genotypes by parental origin (see Chapter 1), can be considered a natural "by-product" of multipoint LOD score analysis. In a given pedigree with evidence for linkage, haplotyping can reveal the critical recombinants, that is, those affected individuals that share only part of the disease-associated haplotype due to a recombination that shortens the region of interest.

Another use of multipoint LOD score analysis for Mendelian disorders is exclusion mapping, where "negative" information is used to exclude regions of the genome from possibly harboring a disease gene. The classical criterion is to exclude all regions with $Z(x) < -2.0$, but it is important to be aware of the effect of heterogeneity and misspecified model parameters on multipoint exclusion mapping. As in two-point analysis, multipoint heterogeneity LOD scores can be computed with the admixture test (e.g., using HOMOG). However, the interpretation of these tests in the multipoint situation is less straightforward because the asymptotic properties of multipoint LOD scores, which are generally multimodal, are not well characterized and the χ^2 approximations are thus questionable. Rather than using p-values, it may be preferable to simply report the observed likelihood ratio between the hypotheses of no linkage and homogeneity versus linkage and heterogeneity. As in two-point analysis, heterogeneity has a major impact on the power to correctly detect linkage using multipoint analysis. With two-point LOD scores, heterogeneity, if not taken into account in the analysis, or other types of model error leading to false apparent recombinations can be partially absorbed by an inflated estimate of the recombination fraction (Risch and Giuffra, 1992). However, in multipoint analysis, the flanking markers prevent the recombination fraction from floating very far, which makes multipoint LOD scores somewhat less robust with respect to genotype and model errors than two-point LOD scores. A typical multipoint LOD score curve in the presence of heterogeneity is shown in Figure 10.1. When heterogeneity is ignored, the region holding the gene of interest is excluded (all families). When heterogeneity is taken into account (linked families), the region is clearly identified.

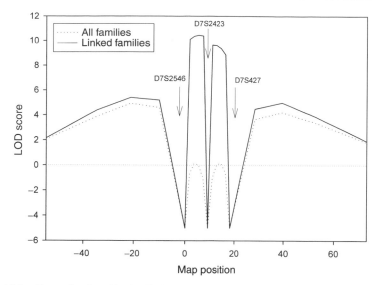

Figure 10.1 Example of multipoint linkage analysis in the presence of genetic heterogeneity for limb–girdle muscular dystrophy. When all families are analyzed, no evidence for linkage to this region is obtained. Examination of linked families after formal heterogeneity testing reveals strong evidence for linkage. (Adapted from Speer et al., 1999.)

Figure 10.1 also illustrates the typical shape of multipoint LOD score curves, which peak in regions between two markers and drop to $-\infty$ at the position of the markers themselves. This is due to the presence of obligate recombinants, which make it impossible for the disease locus to be located exactly at the position of the marker (corresponding to $\theta = 0$ in two-point analysis).

Some researchers believe that multipoint LOD score curves are not as helpful for narrowing the disease gene region for complex disorders as they are for Mendelian phenotypes. The curves often peak throughout extended regions of the genome, and exclusion mapping, using the classical criterion of $Z_{max} < -2.0$, may also be less meaningful. Therefore, it has been suggested to postpone parametric multipoint analysis of complex disorders altogether until a region with preliminary evidence for linkage has been found, that is, to apply the method for follow-up analysis but avoid its use with genome screen data (Ott, 1999). However, some recent work has indicated that the sensitivity of parametric multipoint analysis may have been overestimated and that multipoint LOD scores combined with the admixture test (multipoint HLODs) can be a powerful tool even for linkage analysis of complex traits (see Practical Approaches below).

While multipoint LOD scores are potentially more informative than two-point LOD scores, they will not be of higher absolute value at any given marker locus if the marker itself is already fully informative for linkage. Note that performing a multipoint analysis requires an additional input parameter, a map of ordered markers, and estimated intermarker distances, which means that the linkage test is sensitive to misspecified maps as well as the usual genetic model parameters.

Recent efforts have greatly improved the resolution of these genetic maps, which are based on observed recombination events between polymorphic markers and minimize the required number of recombinations separating the markers to determine the most likely (parsimonious) marker order. A dataset of 146 nuclear Icelandic families containing 1257 meioses was used to develop maps of over 5000 polymorphic microsatellite markers, taking into account sequence information from the physical maps that have been developed as part of the Human Genome Project (Kong et al., 2002). Previously, the Marshfield maps, based on eight three-generation Centre d'Etude du Polymorphisme Humain (CEPH) families containing 188 meioses, had been considered the most comprehensive mapping resource (Marshfield Medical Research Foundation, http://research.marshfieldclinic.org/genetics/).

In multipoint linkage analysis, the computational complexity of pedigree likelihoods of the form (10.1) is greatly magnified, since the joint inheritance of many markers has to be traced through a pedigree to evaluate the LOD score at each point along the marker map, and the number of possible genotypes for untyped pedigree members increases exponentially. Two different algorithms were developed to recursively compute the likelihood terms: the Elston–Stewart "peeling" algorithm (Elston and Stewart, 1971) was implemented in the first generally available software for pedigree likelihood computation (LIPED; Ott, 1974) and continues to be the basis of the current packages FASTLINK (Cottingham et al., 1993), a faster version of the original LINKAGE package (Lathrop et al., 1984), and VITESSE (O'Connell and Weeks, 1995; O'Connell, 2001). The Elston–Stewart algorithm scales linearly in the number of pedigree members and exponentially in the number of markers and is thus typically used for large pedigrees and a moderate number of markers (3–6, depending on the number of untyped pedigree members). VITESSE is the program of choice for simple pedigrees without loops that descend from one founder pair, while FASTLINK is able to handle inbred pedigrees and those with multiple founder pairs.

The Lander–Green algorithm (Lander and Green, 1987) has been implemented in the packages GENEHUNTER (Kruglyak et al., 1996; for GENEHUNTER-PLUS, see Chapter 11), ALLEGRO (Gudbjartsson et al., 2000), and MERLIN (Abecasis et al., 2002), of which MERLIN is superior to both of the other packages in terms of computational speed. The algorithm scales exponentially in the number of pedigree members and linearly in the number of markers and is thus ideally suited for analyzing all the markers for an entire chromosome on small- to moderate-sized pedigrees (in GENEHUNTER, $2n - f \leq 21$, where n is the number of nonfounders and f the number of founders; ALLEGRO and MERLIN are able to handle somewhat larger pedigrees). These programs also implement various types of nonparametric linkage tests (see Chapter 11) and provide other useful statistics of interest, such as the information content of a pedigree and the overall dataset along the marker map, which highlights regions where additional marker genotyping would be most useful.

Another class of computational algorithms is the Markov chain Monte Carlo (MCMC) type, which is a stochastic rather than an exact algorithm. SIMWALK2 (Sobel and Lange, 1996) can be used to compute a random-walk-based approximation of the exact parametric LOD score computed by FASTLINK and VITESSE and is able to analyze simultaneously a large number of markers on pedigrees that are too large for the Lander–Green algorithm. Some packages have begun to implement so-called hybrid algorithms, where pedigrees amenable to exact analysis with the Lander–Green algorithm are processed accordingly, while those too large for exact LOD score computation are analyzed with a SIMWALK2-type MCMC algorithm (SAGE, MERLIN). An excellent resource for downloading most of these computer programs is a website maintained by the Rockefeller University lab, http://linkage.rockefeller.edu, which provides links to all of the program-specific URLs.

PRACTICAL APPROACHES FOR MODEL-BASED LINKAGE ANALYSIS OF COMPLEX TRAITS

For complex traits, susceptibility genes are presumably more common and of more modest effect than Mendelian disease genes, and they are likely to act in concert with multiple other genes and/or environmental factors. Thus, any assumed single-gene mechanism (genetic model) for parametric LOD score analysis is almost certainly incorrect. It has been shown that analysis under a single wrong model [i.e., calculation of "wrod" (wrong LOD) scores] does not inflate the type I error rate of the LOD score method (Williamson and Amos, 1990). However, analyzing the data under multiple models and estimating unknown parameters by maximizing the LOD score over model parameters does lead to grossly inflated significance levels (Weeks et al., 1990a). The question then remains as to how one should choose the model parameters if a parametric LOD score analysis is to be applied to complex trait data.

Traditionally, genetic model parameters have been estimated via segregation analysis. This is a statistically sophisticated technique for estimating the most likely mode of inheritance for a trait from pedigree data that have been collected in a systematic fashion, ideally in such a way that a correction for ascertainment bias can be applied (Khoury et al., 1993). A properly performed segregation analysis of a large dataset can be quite useful for the linkage analyst since it provides maximum-likelihood estimates of the MOI, disease allele frequency and penetrance for susceptible and nonsusceptible genotypes. Due to the improbability of single-gene mechanisms, this type of analysis is only rarely attempted for complex traits, although it has been carried out successfully, for example, to locate highly penetrant genes for early-onset breast cancer (Claus et al., 1991; Easton et al., 1993). Here, the segregation analysis also included the estimation of an age-dependent penetrance function for the gene of interest. Since the ratio of genetic and nongenetic cases is often variable across different age groups, with younger cases typically more likely to be genetic than older cases, the concept of liability classes can be used to flexibly model age-at-onset distributions in parametric linkage analysis, as illustrated by the breast cancer example (Easton et al., 1993; Terwilliger and Ott, 1994). For Mendelian diseases, phenocopies are typically assumed to be absent and age-dependent penetrance is often modeled by assuming a normal age-of-onset distribution with mean and variance estimated from the pedigree data at hand (see Ott, 1999, Chapter 7.5, for more detail).

A much simpler alternative to segregation analysis is to estimate model parameters by making use of epidemiological information (Terwilliger and Ott, 1994). If estimates of population prevalence and the proportion of nongenetic cases among all affected individuals in the population are available, the penetrance for a single-locus model can be determined for an assumed disease allele frequency, because the following relationship between prevalence φ, penetrance of susceptible genotypes f, penetrance of phenocopies f_p, and disease allele frequency q should hold for a dominant disease:

$$\varphi = [q^2 + 2q(1 - q)]f + (1 - q)^2 f_p$$

(an analogous expression holds for recessive traits). While it may seem appealing to use such existing information about disease epidemiology, some guesswork clearly remains and, again, any single-gene mechanism is unlikely to accurately capture the true complex trait etiology. Some other practical applications of parametric LOD score analysis for complex disorders are briefly described below.

Affecteds-Only Analysis

The idea of the affecteds-only ("low-penetrance") analysis is to discard the phenotypic information from unaffected individuals since their phenotype may be due to incomplete penetrance of a single gene, a missing copy of a second susceptibility gene, absence of an environmental trigger, or similar complicating factors. However, one would nevertheless like to incorporate their genotypic information. A cumbersome way of setting up an affecteds-only analysis would be to recode all unaffected people as "unknown." Fortunately, the same effect can be achieved much more efficiently by simply changing the genetic model parameters. Having phenotype "unknown" essentially means that all three possible disease genotypes (DD, Dd, dd) are equally likely to produce the observed phenotype. The implementation of liability classes in LINKAGE and any other programs is such that, for an affected person, the specified penetrance value is interpreted as the probability of being affected given the individual's genotype. For an unaffected person, 1 minus the penetrance value is used for this probability. Thus, if phenocopies are assumed to be absent ($f_p = 0$), it is sufficient to assign very low penetrance values f (e.g., 0.0001) to affected individuals so that, for unaffected individuals, $1 - f_p = 1 - 0 \approx 1 - 0.0001 = 1 - f$, making all three disease genotypes essentially equally likely. Dominant and recessive MOIs only differ by whether the heterozygous disease genotype has penetrance 0 (recessive) or 0.0001 (dominant). The disease allele is commonly assumed to have a frequency of, say, 0.001 for a "generic" dominant and 0.20 for a recessive model. If allowance for phenocopies is desired, the penetrance for nonsusceptible genotypes could be adjusted, relative to the value of 0.0001 for susceptible genotypes, to maintain a penetrance ratio compatible with epidemiological data (Terwilliger and Ott, 1994). Typically, however, phenocopies are assumed to be absent in this type of analysis. Note that the occurrence of phenocopies is essentially a form of heterogeneity (Durner et al., 1996). For our AD data example, an affecteds-only analysis produces $Z_{max} = 3.83$ at $\theta = 0.10$ under a dominant model and $Z_{max} = 1.93$ at $\theta = 0.05$ under a recessive model.

Maximized Maximum LOD Score

Some authors suggest computing a maximized maximum LOD score (MMLS, Greenberg et al., 1998; also called "mod score," Clerget-Darpoux et al., 1986; Hodge and Elston, 1994). They also assume the absence of phenocopies, but rather than using very low penetrance values to discard the phenotypic information provided by unaffected individuals, they recommend the computation of LOD scores under two "simple," single-locus models, one dominant and one recessive, each with 50% penetrance f and with disease allele frequency $q = 1 - \sqrt{(1 - \varphi/f)}$ for the dominant and $q = \sqrt{\varphi/f}$ for the recessive model, where φ denotes the population prevalence (Durner et al., 1999). Then, the maximum of the two LOD scores is corrected for maximization over two models by subtracting 0.3 (Hodge et al., 1997). The MMLS thus corrected was shown to be approximately χ^2-distributed (with 1 df) under the null hypothesis of no linkage, which allows the computation of p-values. It was shown that the MMLS has good power to detect linkage of a complex trait for a variety of simulated pedigree structures and genetic models and can, in fact, be more powerful than a test based on identical-by-descent (IBD) sharing in affected sibpairs or on the GENEHUNTER nonparametric NPL score for affected relative pairs (see Chapter 11; Durner et al., 1999; Abreu et al., 1999). It is informative to report the actual maximum LOD scores under both dominant and recessive models rather than only the MMLS after the correction for maximization over the two models has been applied. Of course, just like the LOD scores from the simple genetic models defined above, affecteds-only LOD scores can also be computed under both a dominant and recessive model. The correction factor is likely to be very similar to the one suggested for the simple models, although the corresponding formal investigations have not been carried out. For our AD data example, a simple dominant model with 50% penetrance and $\varphi = 0.10$ (hence, $q = 0.106$) produces $Z_{max} = 2.30$ at $\theta = 0.05$, and the corresponding recessive model (with $q = 0.447$) produces $Z_{max} = 1.18$ at $\theta = 0.05$.

Heterogeneity LOD

Regardless of whether LOD scores were computed initially based on simple genetic models or an affecteds-only approach, there is a fairly large body of evidence supporting the subsequent calculation of HLOD scores to detect linkage of complex traits, even when the assumed heterogeneity model is incorrect. Most of the relevant studies have been briefly summarized and reviewed (Hodge et al., 2002). In essence, it was shown that the two-point HLOD is a robust and powerful tool to detect linkage even when the assumptions of the HLOD analysis are violated and the obtained estimate of the actual proportion of linked families (α) is strongly biased. In addition, one does not pay much of a price in type I error by using HLODs (Abreu et al., 2002). The statistical properties of the multipoint HLOD have not yet been investigated in as much detail, but there is some indication that results may be similar to those reported for two-point LOD scores (Greenberg and Abreu, 2001). This would imply that earlier conclusions (Risch and Giuffra, 1992) about the greatly increased sensitivity of multipoint LOD scores to misspecified model parameters may have been too pessimistic. Based on a recent study (Greenberg and Abreu, 2001), it appears that a stronger correction than subtracting 0.3 from the two-point MMLS (Hodge et al., 1997) may be needed to appropriately correct for LOD score maximization over two genetic models (dominant and recessive) in multipoint analysis, but more work is needed to arrive at definitive recommendations. For the AD data, the two-point HLODs are 3.82 ($\alpha = 1.0$) and 1.98 ($\alpha = 0.85$) with affecteds-only LOD scores from the dominant and recessive model and 2.30 ($\alpha = 1.0$) and 1.24 ($\alpha = 0.75$) with the LOD scores from the simple models, respectively.

MFLINK

Some researchers have proposed to maximize, not the LOD score, but, independently, the likelihood of the data under linkage and nonlinkage (numerator and denominator of the likelihood ratio) over a limited set of transmission models constrained to produce the correct population prevalence, including dominant and recessive models (Curtis and Sham, 1995). The likelihood under linkage is also maximized over admixture, that is, the proportion of families linked. The authors showed that this test statistic is well behaved under the null hypothesis, that it has good power to detect linkage when it is present, and that its multipoint version is quite robust to errors in the assumed models. The method has been implemented in the software MFLINK.

In practice, the affecteds-only MMLS and HLOD approaches are probably used most frequently for two-point model-based linkage analysis of complex traits, although the MFLINK method certainly has some appeal from a statistical perspective. Multipoint model-based analysis of complex disorders at the initial genome screen stage appears to be applied somewhat less frequently than the corresponding nonparametric approaches popularized by GENEHUNTER. For follow-up analysis of regions with positive linkage signals at the genome screen stage, a multipoint HLOD analysis, assuming both a dominant and recessive model and using either simple genetic models that take into account the population prevalence or the affecteds-only approach, can be recommended. However, possible statistical problems associated with the unknown distribution of multipoint LOD scores should be kept in mind. It is important to evaluate multipoint LOD scores at locations outside the fixed map, which is equivalent to allowing for overestimation of the recombination fraction in the two-point analysis. Multipoint analysis is then essentially the same as testing for linkage to a (potentially more informative) haplotype of markers. A large collaborative study of prostate cancer provides a good example of a parametric multipoint follow-up analysis based on models derived from segregation analysis as well as epidemiological information and implemented through the concept of liability classes (Xu, 2000).

In summary, two-point analysis under two single-locus models (dominant and recessive) can be a powerful way to analyze complex phenotypes. Multipoint parametric LOD scores should be interpreted with more caution and should probably never be the only multipoint analysis method applied to complex disorders. However, recent work indicates that the multipoint HLOD can have more power than the NPL score under a variety of generating models (Greenberg and Abreu, 2001) and may be more robust to misspecified model parameters than previously estimated. Extensions of the parametric LOD score method for jointly analyzing two disease loci have also been proposed and implemented (TLINKAGE; Schork et al., 1993). However, these types of additional complexities are probably better addressed with nonparametric methods based on allele-sharing statistics (Chapter 11).

SUMMARY

Parametric LOD score analysis has been the workhorse of disease gene mapping for many years. It is still the most powerful method of linkage analysis if it is possible to specify the genetic model parameters with reasonable accuracy. For two-point analysis, incorrect specification, particularly of the mode of inheritance, can lead to a dramatic loss of power, as illustrated with simulation studies of AD pedigrees. This has led to the recommendation of computing LOD scores under both a dominant and recessive model. If the maximum over the two models is used as the summary measure of linkage, an appropriate correction for this additional maximization has been derived. Multipoint analysis is potentially able to extract a much greater amount of linkage information from a set of pedigrees but may be more vulnerable than two-point analysis in terms of misspecified model parameters. In practice, model-based multipoint analysis should be complemented by appropriate nonparametric approaches. Despite the difficulty of specifying an approximately correct genetic model, numerous investigations have shown that parametric LOD scores remain a valid and useful tool for modern-day linkage analysis of complex traits.

REFERENCES

Abecasis GR, Cherny SS, Cookson WO, Cardon LR (2002): Merlin—Rapid analysis of dense genetic maps using sparse gene flow trees. Nat Genet 30:97–101.

Abreu PC, Greenberg DA, Hodge SE (1999): Direct power comparisons between simple LOD scores and NPL scores for linkage analysis in complex diseases. Am J Hum Genet 65:847–857.

Abreu PC, Hodge SE, Greenberg DA (2002): Quantification of type I error probabilities for heterogeneity LOD scores. Genet Epidemiol 22:156–169.

Barnard GA (1949): Statistical inference. J R Stat Soc B B11:115–139.

Boehnke M, Cox NJ (1997): Accurate inference of relationships in sib-pair linkage studies. Am J Hum Genet 61:423–429.

Chotai J (1984): On the LOD score method in linkage analysis. Ann Hum Genet 48 (Pt 4):359–378.

Claus EB, Risch N, Thompson WD (1991): Genetic analysis of breast cancer in the cancer and steroid hormone study. Am J Hum Genet 48:232–242.

Clerget-Darpoux F, Bonaiti-Pellie C, Hochez J (1986): Effects of misspecifying genetic parameters in LOD score analysis. Biometrics 42:393–399.

Corder EH, Saunders AM, Strittmatter WJ, Schmechel DE, Gaskell PC, Small GW, Roses AD, Haines JL, Pericak-Vance MA (1993): Gene dose of apolipoprotein E type 4 allele and the risk of Alzheimer's disease in late onset families. Science 261:921–923.

Cottingham RW, Idury RM, Schaffer AA (1993): Faster sequential genetic linkage computations. Am J Hum Genet 53:252–263.

Curtis D, Sham PC (1995): Model-free linkage analysis using likelihoods. Am J Hum Genet 57:703–716.

Douglas JA, Boehnke M, Lange K (2000): A multipoint method for detecting genotyping errors and mutations in sibling-pair linkage data. Am J Hum Genet 66:1287–1297.

Durner M, Greenberg DA, Hodge SE (1996): Phenocopies versus genetic heterogeneity: Can we use phenocopy frequencies in linkage analysis to compensate for heterogeneity? Hum Hered 46:265–273.

Durner M, Vieland VJ, Greenberg DA (1999): Further evidence for the increased power of LOD scores compared with nonparametric methods. Am J Hum Genet 64:281–289.

Easton DF, Bishop DT, Ford D, Crockford GP, Breast Cancer Linkage Consortium (1993): Genetic linkage analysis in familial breast and ovarian cancer: Results from 214 families. Am J Hum Genet 52:678–701.

Elston RC (1997): Algorithms and inferences: The challenge of multifactorial disease. Am J Hum Genet 60:255–262.

Elston RC, Stewart J (1971): A general model for the genetic analysis of pedigree data. Hum Hered 21:523–542.

Epstein MP, Duren WL, Boehnke M (2000): Improved inference of relationship for pairs of individuals. Am J Hum Genet 67:1219–1231.

Faraway JJ (1993): Distribution of the admixture test for the detection of linkage under heterogeneity. Genet Epidemiol 10:75–83.

Goring HH, Ott J (1997): Relationship estimation in affected sib pair analysis of late-onset diseases. Eur J Hum Genet 5:69–77.

Greenberg DA, Abreu PC (2001): Determining trait locus position from multipoint analysis: Accuracy and power of three different statistics. Genet Epidemiol 21:299–314.

Gudbjartsson DF, Jonasson K, Frigge ML, Kong A (2000): Allegro, a new computer program for multipoint linkage analysis. Nat Genet 25:12–13.

Haines JL (1998): Genomic screening. In: Haines JL, Pericak-Vance MA, eds. Approaches to Gene Mapping in Complex Human Diseases. New York: Wiley-Liss.

Haines JL, Terwedow HA, Burgess K, Pericak-Vance MA, Rimmler JB, Martin ER, Oksenberg JR, Lincoln R, Zhang DY, Banatao DR, Gatto N, Goodkin DE, Hauser SL

(1998): Linkage of the MHC to familial multiple sclerosis suggests genetic heterogeneity. The Multiple Sclerosis Genetics Group. Hum Mol Genet 7:1229–1234.

Hall JM, Lee MK, Newman B, Morrow JE, Anderson LA, Huey B, King MC (1990): Linkage of early-onset familial breast cancer to chromosome 17q21. Science 250:1684–1689.

Hanis CL, Boerwinkle E, Chakraborty R, Ellsworth DL, Concannon P, Stirling B, Morrison VA, et al (1996): A genome-wide search for human non-insulin-dependent (type 2) diabetes genes reveals a major susceptibility locus on chromosome 2. Nat Genet 13:161–166.

Hodge SE, Abreu PC, Greenberg DA (1997): Magnitude of type I error when single-locus linkage analysis is maximized over models: A simulation study. Am J Hum Genet 60:217–227.

Hodge SE, Vieland VJ, Greenberg DA (2002): HLODs remain powerful tools for detection of linkage in the presence of genetic heterogeneity. Am J Hum Genet 70:556–559.

Hoh J, Ott J (2000): Scan statistics to scan markers for susceptibility genes. Proc Natl Acad Sci USA 97:9615–9617.

Hugot JP, Laurent-Puig P, Gower-Rrousseau C, Olson JM, Lee JC, Beaugerie L, Naom I, Dupas J-L, Van Gossum A, Groupe d'Etude Therapeutique des Affections Inflammatoires Digestives, Orholm M, Bonaiti-Pellie C, Weissenback J, Mathew CJ, Lennard-Jones JE, Cortot A, Colombel JF, Thomas G (1996): Mapping of a susceptibility locus for Crohn's disease on chromosome 16. Nature 379:821–823.

Khoury MJ, Beaty TH, Cohen BH (1993): Fundamentals of Genetic Epidemiology. New York: Oxford University Press.

Kong A, Gudbjartsson DF, Sainz J, Jonsdottir GM, Gudjonsson SA, Richardsson B, Sigurdardottir S, Barnard J, Hallbeck B, Masson G, Shlien A, Palsson ST, Frigge ML, Thorgeirsson TE, Gulcher JR, Stefansson K (2002): A high-resolution recombination map of the human genome. Nat Genet 31:241–247.

Kruglyak L, Daly MJ, Reeve-Daly MP, Lander ES (1996): Parametric and nonparametric linkage analysis: A unified multipoint approach. Am J Hum Genet 58:1347–1363.

Lander E, Kruglyak L (1995): Genetic dissection of complex traits: Guidelines for interpreting and reporting linkage results. Nat Genet 11:241–247.

Lander ES, Green P (1987): Construction of multilocus genetic linkage maps in humans. Proc Natl Acad Sci USA 84:2363–2367.

Lathrop GM, Lalouel JM, Julier C, Ott J (1984): Strategies for multilocus linkage analysis in humans. Proc Natl Acad Sci USA 81:3443–3446.

McPeek MS, Sun L (2000): Statistical tests for detection of misspecified relationships by use of genome-screen data. Am J Hum Genet 66:1076–1094.

Morton NE (1955): Sequential tests for the detection of linkage. Am J Hum Genet 7:277–318.

Morton NE (1956): The detection and estimation of linkage between the genes for elliptocytosis and the Rh blood type. Am J Hum Genet 8:80–96.

O'Connell JR (2001): Rapid multipoint linkage analysis via inheritance vectors in the Elston–Stewart algorithm. Hum Hered 51:226–240.

O'Connell JR, Weeks DE (1995): The VITESSE algorithm for rapid exact multilocus linkage analysis via genotype set-recoding and fuzzy inheritance. Nat Genet 11:402–408.

O'Connell JR, Weeks DE (1998): PedCheck: A program for identification of genotype incompatibilities in linkage analysis. Am J Hum Genet 63:259–266.

Ott J (1974): Estimation of the recombination fraction in human pedigrees: Efficient computation of the likelihood for human linkage studies. Am J Hum Genet 26:588–597.

Ott J (1986): The number of families required to detect or exclude linkage heterogeneity. Am J Hum Genet 39:159–165.

Ott J (1992): Strategies for characterizing highly polymorphic markers in human gene mapping. Am J Hum Genet 51:283–290.

Ott J (1999): Analysis of Human Genetic Linkage, 3rd ed. Baltimore, MD: Johns Hopkins University Press.

Pericak-Vance MA, Bebout JL, Gaskell PC, Yamaoka LH, Hung WY, Alberts MJ, Walker AP, Bartlett RJ, Haynes CS, Welsh KA, Earl NL, Heyman A, Clark CM, Roses AD (1991): Linkage studies in familial Alzheimer's disease: Evidence for chromosome 19 linkage. Am J Hum Genet 48:1034–1050.

Rimmler JB, Haynes CS, McDowell JG, Stajich JE, Adams CS, Slotterbeck BD, Rogala AR, West SG, Gilbert JR, Hauser ER, Vance JM, Pericak-Vance MA (1999): DataTracker: Comprehensive software for data quality control protocols in complex disease studies. Am J Hum Genet 65(Suppl):A442.

Risch N (1988): A new statistical test for linkage heterogeneity. Am J Hum Genet 42:353–364.

Risch N, Giuffra L (1992): Model misspecification and multipoint linkage analysis. Hum Hered 42:77–92.

Schork NJ, Boehnke M, Terwilliger JD, Ott J (1993): Two trait locus linkage analysis: A powerful strategy for mapping complex genetic traits. Am J Hum Genet 53:1127–1136.

Skolnick MH, Thompson EA, Bishop DT, Cannon LA (1984): Possible linkage of a breast cancer-susceptibility locus to the ABO locus: Sensitivity of LOD scores to a single new recombinant observation. Genet Epidemiol 1:363–373.

Smith CAB (1963): Testing for heterogeneity of recombination fractions values in human genetics. Ann Hum Genet 27:175–182.

Sobel E, Lange K (1996): Descent graphs in pedigree analysis: Applications to haplotyping, location scores and marker-sharing statistics. Am J Hum Genet 58:1323–1337.

Terwilliger, JD, Ott, J (1994): Handbook of Human Genetic Linkage. Baltimore, MD: Johns Hopkins University Press.

Terwilliger JD, Speer MC, Ott J (1993): A chromosome based method for rapid computer simulation in human genetic linkage analysis. Genet Epidemiol 10:217–224.

Thomson G (1994): Identifying complex disease genes: Progress and paradigms. Nat Genet 8:108–110.

Vieland VJ, Hodge SE, Greenberg DA (1992): Adequacy of single-locus approximations for linkage analyses of oligogenic traits. Genet Epidemiol 9:45–59.

Weeks DE, Conley YP, Mah TS, Paul TO, Morse L, Ngo-Chang J, Dailey JP, Ferrell RE, Gorin MB (2000): A full genome scan for age-related maculopathy. Hum Mol Genet 9:1329–1349.

Weeks DE, Lehner T, Squires-Wheeler E, Kaufmann C, Ott J (1990a): Measuring the inflation of the LOD score due to its maximization over model parameter values in human linkage analysis. Genet Epidemiol 7:237–243.

Weeks DE, Ott J, Lathrop GM (1990b): SLINK: A general simulation program for linkage analysis. Am J Hum Genet 47:A204.

Williamson JA, Amos CI (1990): On the asymptotic behavior of the estimate of the recombination fraction under the null hypothesis of no linkage when the model is misspecified. Genet Epidemiol 7:309–318.

Xu J (2000): Combined analysis of hereditary prostate cancer linkage to 1q24-25: Results from 772 hereditary prostate cancer families from the International Consortium for Prostate Cancer Genetics. Am J Hum Genet 66:945–957.

Zaykin DV, Westfall PH, Young SS, KarnoubPL, Wagner MJ, Ehm MG (2002): Missing p, haplotypes, and multiple r: Testing association of statistically inferred haplotypes with discrete and continuous traits in samples of unrelated individuals. Hum Hered 53:79–91.

CHAPTER 11

Nonparametric Linkage Analysis

ELIZABETH R. HAUSER, JONATHAN HAINES, and DAVID E. GOLDGAR

INTRODUCTION

Traditional logarithm of the odds (LOD) score analysis has been highly successful in the mapping of single-gene disorders. However, when the underlying genetic model cannot be specified with any confidence, as is the case for more genetically complex traits, traditional, parametric LOD score analysis loses much of its power, and the results can be misleading (Chapters 1 and 10). Even when a single gene is known to be acting, if there are many ungenotyped individuals in a pedigree, the accurate specification of disease and marker allele frequencies may become critical; even small discrepancies could produce misleading results. In most diseases of late onset, such as Alzheimer's disease, cardiovascular disease, or osteoarthritis, DNA is available only for individuals in the most recent one or two generations and most affected individuals are connected through many ungenotyped individuals. In these cases it is impossible to unambiguously trace the inheritance of particular alleles from the common ancestors. Since a large number of genotypes, for both marker and disease genes, need to be inferred, every specification (however inaccurate) in the genetic model carries a greater weight in the LOD score calculation.

The primary solution to the problem just stated is to utilize procedures that rely less completely on the genetic model specification. Such procedures, less powerful than parametric methods if the model is specified accurately, offer a robust approach when model parameters are less certain. The term *nonparametric* is used generally to describe methods that weaken one or more assumptions of the fully specified genetic model that includes parameters for the disease gene allele frequencies and disease gene penetrance functions. The parameter most likely to be weakened is the penetrance function, either by considering only affected individuals or by reparametrizing the genetic model to include a general indicator of genetic effect without completely specifying the genetic model. The basic premise underlying many of the

Genetic Analysis of Complex Diseases, Second Edition, Edited by Jonathan L. Haines and Margaret Pericak-Vance

methods described in this chapter is that if two related individuals are phenotypically similar (e.g., if both have the same disease, such as hypertension, or both have similar trait values, such as blood pressure), then a genetic marker located nearby to a gene contributing to that trait ought to be similar. The degree of similarity will depend on a number of factors. The two most significant factors are the overall contribution of the particular locus to the trait being studied and the genetic distance between the unknown (or unmeasured) gene influencing the trait and the genetic marker being tested. Thus while we will use the term "nonparametric" to describe these methods, they are not nonparametric in a strict statistical sense; parameters describing the genetic effect, genetic distance between the trait and the marker, and marker allele frequencies are often included.

We will begin by giving some historical background, discuss the fundamental concepts of identity by state, identity by descent, and quantification of genetic effect, and then examine several nonparametric linkage analysis methods for both qualitative and quantitative traits. The goals of this chapter are not to provide an exhaustive description of nonparametric methods, for there are many such methods, but rather to introduce the basic concepts and models and to illustrate how these concepts are incorporated in analytic methods for complex human disease.

BACKGROUND AND HISTORICAL FRAMEWORK

The last 15 years have witnessed dramatic growth in the use of nonparametric linkage analysis in human genetic mapping. This is true from the point of view of theoretical developments and more sophisticated statistical methodology as well as with respect to the successful application of these methods to mapping susceptibility genes for complex human traits. When we refer to complex traits, we typically divide them into two major categories: qualitative or dichotomous traits and quantitative or continuous ones. The qualitative traits most commonly studied are, of course, human diseases in which a person is either clinically affected or unaffected with the disease. The genetic complexity may occur for one or more of the following reasons:

1. The disease may be etiologically heterogeneous, with only a subset of families or individuals affected due to genes conferring high risk.
2. The disease may involve many different genetic loci that act together to cause disease.
3. A gene for the disease may predispose only in the presence of a particular environmental exposure.

Individual differences in quantitative traits such as height, blood pressure, or triglyceride levels are typically the result of several genes together with both specific and random environmental components. These confounding factors in both quantitative and qualitative traits require the use of methods in which the precise mechanisms of disease causation are not required to be known; this has led to the recent interest in the methods described in this chapter.

Although the common use of nonparametric linkage methods is relatively recent for reasons described below, it is important to remember that the concepts that underlie this methodology date back to the early part of this century. Indeed, the first nonparametric method in humans was a sibpair linkage approach in humans described in a seminal paper by Penrose published in 1935 in *Annals of Eugenics*, the predecessor of *Annals of Human Genetics* (Penrose, 1935). Following this paper, the major emphasis in human linkage analysis was focused on the development of likelihood-based methods in which the probability of the observed data is written as a function of the recombination fraction between two loci in the context of a fully specified genetic model, that is, dominant or recessive inheritance and disease allele frequencies. A statistical test is derived by comparing the likelihoods under the two hypotheses of linkage versus free recombination (Fisher, 1935; Haldane, 1934; Haldane and Smith, 1947; Morton, 1955). Aside from a second paper on the sibpair approach by Penrose (1953), the next major developments in sibpair analysis took place in the 1970s. The first such work was the seminal paper by Haseman and Elston (1972), which described an elegant regression-based approach to search for loci influencing quantitative traits, followed by the important work of Suarez et al. (1978), who considered the distribution of identity-by-descent status among siblings for a variety of genetic models and applied this method to the elucidation of the role of the major histocompatibility complex (MHC) in type I (juvenile-onset) diabetes using a set of affected sibpairs.

Thus, until perhaps the mid-1980s, the use of sibpair methodology was largely confined to human leukocyte antigen (HLA) and a few other systems (e.g., Rh blood group), with a relatively high degree of informativeness. With the advent of restriction fragment length polymorphisms (RFLPs), then variable number of tandem repeat (VNTR) markers, and in the 1990s, short tandem repeat (STR) markers (microsatellites), the probability that there will be an informative genetic marker (or set of markers) near any disease gene has steadily increased (see Chapter 6). It is now the case that we have almost 100% coverage of the genome at a resolution of 1 cM with genetic markers of high heterozygosity (>0.70) (Gyapay et al., 1994; Dib et al., 1996). Development of these genetic markers along with algorithmic improvements leading to fast computation of pedigree likelihoods allowed for routine genetic analysis of monogenic disorders with clear Mendelian patterns of inheritance. However, attention is now focused on more common disorders with more complex (or unknown) modes of inheritance. In addition, there is renewed interest in mapping quantitative traits, including those that may serve as surrogates for a disease (e.g., cholesterol levels for cardiovascular disease) or quantitative traits that have a strong genetic component and reflect normal human variation (e.g., height, finger ridge count). The application of nonparametric linkage methods has become standard practice in human genetic analysis of diseases of complex inheritance as well as quantitative traits.

IDENTITY BY STATE AND IDENTITY BY DESCENT

All methods developed for the mapping of complex traits in human relative pairs depend in some way on quantification of the degree to which related individuals "share" alleles at the marker locus or loci under investigation. All such methods can be divided into two basic classes: those that depend on marker alleles shared identical by state (IBS) and those that rely on alleles shared identical by descent (IBD). The simpler of the two to define and understand is identity by state. Two alleles are said to be IBS if they are the same variant of some polymorphic system; that is, two alleles are IBS if they cannot be distinguished by means of a particular method of detection. For example, for an RFLP marker (see Chapter 6), if two individuals both exhibit a band on a gel of a given size when digested with a given restriction enzyme and hybridized to a probe, these two alleles are IBS. Similarly, if two individuals both show the same number of repeat units for an STR marker, they are IBS. Any two individuals, whether related or not, can share zero, one, or two alleles IBS; the two alleles possessed by any one individual can also be IBS, in which case the individual is said to be homozygous at this locus.

Identity by descent, on the other hand, depends not only on whether the alleles appear the same on a gel or by another detection method but also on whether these alleles are derived or inherited from a common ancestor. Clearly, alleles that are shared IBD must also be IBS; however, the converse is not true. That is, alleles that are IBS are not necessarily IBD. Any two unrelated individuals who have the same genotype illustrate this; since they are unrelated, by definition they share no alleles IBD. Figure 11.1 shows two hypothetical examples to illustrate the differences between IBS and IBD.

In Figure 11.1a, in the first genotype configuration (i.e., both siblings are type *13*), it is clear from the parents that each affected child inherited the *1* allele from the father and the *3* allele from the mother. Thus both these alleles are shared IBD as well as IBS. Contrast this with the situation in genotype configuration 3, in which

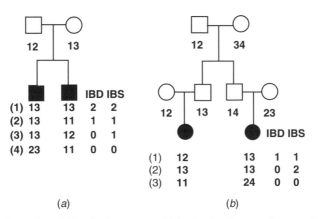

(a)   (b)

Figure 11.1. Examples of identity by state and identity by descent. See text for discussion.

both siblings have a *1* allele and thus share this allele IBS. Inspection of the parental genotypes, however, indicates that in one case this allele was inherited from the father, while for the other sibling it was inherited from the mother. Accordingly, this allele, while shared IBS, is not shared IBD. It is important to point out that in this situation we were only able to distinguish with certainty the origin of each allele because the parents were both genotyped and because the mating type was informative (i.e., the parents were of different heterozygous genotypes). In practice, there will be cases in which one can only estimate the probability that the siblings share a particular number of alleles IBD. In many cases in which the parental genotypes are unknown, these estimates will depend on the frequency of the observed (and unobserved) alleles at the marker locus in question. The more polymorphic the marker, the more precisely we are able to estimate the IBD status.

For example, consider the case of two siblings who are identical homozygotes at a marker locus (i.e., they share two alleles IBS) and the parents are unavailable. If this allele is relatively rare in the population, then it is most likely that both unobserved parents are heterozygous at this locus and thus the two siblings share both alleles IBD. Conversely, if the allele is common, it is likely that at least one parent is homozygous at this locus and thus IBD status is more uncertain. A similar example for cousin pairs is shown in Figure 11.1*b*. In the second configuration, we see that although the two cousins have identical genotypes, they do not share either allele inherited from their common grandparents. Examination of the pedigree structure points out an important principle, namely, that individuals who are related through only one parent (unilineal relationships) can share at most only a single allele IBD. Examples of these are parent–offspring, cousins, half-siblings, and uncle–nephew. Individuals related through both parents (bilineal relationships) can share both alleles IBD. Other than siblings (and monozygotic twins), the most common bilineal relationship is double first cousins, which occur, for example, when two brothers marry two sisters; the offspring of these unions are double first cousins. Other bilineal relationships occur in complex genealogies in which there are regular patterns of intermarriage between related individuals. Table 11.1

TABLE 11.1. Expected Percentage of Affected Pairs Showing 0, 1, or 2 Alleles IBD at a Marker Locus if No Linkage Is Present

Pair Type	Alleles IBD		
	0	1	2
Monozygotic twins	0	0	100
Siblings	25	50	25
Parent–child	0	100	0
Grandparent–grandchild	50	50	0
Half-siblings	50	50	0
Uncle–nephew, etc.	50	50	0
Cousins	75	25	0
Double first cousins	56	38	6

shows the expected proportion of alleles shared IBD for a number of common relationships. These expected proportions are based on the principle of random segregation of gametes during meioses along with the number of meioses separating the two relatives of a pair.

As we saw above, unambiguously determining the IBD status of an affected relative pair is not always possible, even when intervening relatives are genotyped and especially when intervening relatives are not available for genotyping. There are two additional sources of information to help determine relative-pair IBD status at a given marker. The first source of information is from genotypes in additional family members to reduce the number of possible genotypes to consider for untyped or uninformative individuals. As Figure 11.2a illustrates, when the genotypes from the affected sibpair alone are considered, all IBD states are possible due to the large number of parental genotypes that could have resulted in those offspring genotypes. However, when the genotypes from the two unaffected siblings are included, the parental genotypes are clear, $1/2 \times 1/3$, and the affected sibpair share two alleles IBD. The second source of information for IBD status is contained in the IBD states in nearby markers. As Figure 11.2b illustrates, when we consider only marker B, the affected sibs are equally likely to share one or two alleles IBD, since the mother is homozygous. However, when we add genotypes for closely spaced markers A and C, the most likely IBD configuration for marker B is that the affected sibs share one allele IBD. The confidence with which we are able to choose one IBD configuration over another depends on the spacing of the markers. As the distance between the markers increases, the probability of recombination

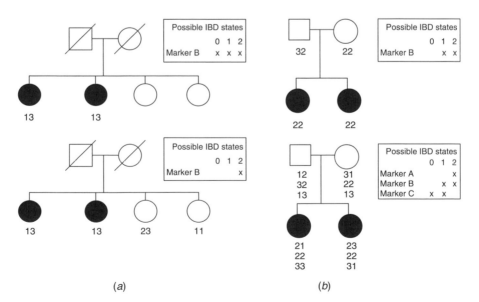

(a)   (b)

Figure 11.2.

increases, decreasing the amount of information that the flanking markers contribute to determination of the IBD state at marker B.

Computer algorithms to allow the incorporation of additional relatives and to perform multipoint analysis to infer IBD states have recently been developed and implemented in several software packages. However, there continues to be a trade-off in the number of individuals in a family that can be analyzed versus the number of markers that can be analyzed. No algorithm can perform exact likelihood calculations on an unlimited number of both individuals and markers. The Lander–Green algorithm (Lander and Green, 1987) and its extensions implemented in GENEHUNTER (Kruglyak et al., 1996; Markianos et al., 2001), Allegro (Gudbjartsson et al., 2000; Jonasson et al., 1995) and MERLIN (Abecasis et al., 2002) can calculate the exact IBD distribution for an arbitrary number of markers but only in moderately sized pedigrees. The computer resources required for analysis with the Lander–Green algorithm increase linearly with the number of markers included in the analysis but exponentially in the number of individuals in the pedigree. This is in contrast to the Elston–Stewart algorithm used for most parametric LOD score analyses, which causes computer time to increase exponentially with the number of markers but linearly with the number of individuals in the pedigree (Elston and Stewart, 1971). The most recently introduced computer program, MERLIN, uses a binary tree algorithm to summarize gene flow through the pedigrees and thereby calculate IBD sharing probabilities for larger pedigrees and denser maps (Abecasis et al., 2002). Nevertheless, there are still pedigrees that may be too large or complicated for calculation of exact likelihoods and approximations will be necessary. In addition, methods to estimate identity by descent have been developed in the context of quantitative trait mapping; however, the IBD estimates obtained from these quantitative trait locus (QTL) packages may be used in any linkage method requiring IBD estimates. Goldgar (1990) first developed a multipoint IBD estimate that averages identity by descent across a chromosomal region. An alternative algorithm implemented in the SOLAR package (Almasy and Blangero, 1998) applies the IBD estimation algorithm of Amos (Amos et al., 1989; Whittemore and Halpern, 1994) at single genetic markers combined with the interval mapping method of Fulker et al. (1995) to estimate the full multipoint likelihood. Because this method does not calculate an exact, fully multipoint likelihood for the pedigree, it is able to handle large, extended pedigrees. In summary, there are a number of ways of estimating IBD proportions for pairs of relatives. Once these estimates are obtained, a variety of statistical analyses may be performed.

MEASURES OF FAMILIALITY

Qualitative Traits

As stated earlier, the ability to detect the effects of a gene influencing a particular trait will depend on the magnitude of the effect of that gene. It may also depend on the overall contribution of genes to the trait.

For dichotomous disease traits, the most natural index of the strength of the genetic/familial component will be some measure of the extent to which relatives of affected individuals also have the disease of interest. The measure most commonly used is called the familial relative risk and is usually denoted by λ_R, where the subscript R denotes the degree of relationship to the index case. Most commonly, we will be concerned with the case of a relative who is a sibling or an offspring of the affected individual, although occasionally we will also be concerned with monozygotic (MZ) twins, half-siblings or other more distantly related relative pairs. For example, if $\lambda_S = 3.0$ for a disease, this indicates that siblings have a threefold increased risk of the same disease compared to the risk to individuals in the general population. Expressed another way, the familial relative risk is the recurrence risk of disease in relatives divided by the population prevalence. For example, a disease in which 6% of siblings of affected individuals are themselves affected and for which the prevalence in the general population is $1/200$, the value of λ_S is $0.06/0.005 = 12.0$. Of course more sophisticated calculations must be performed when the disease is characterized by late and/or variable age at onset, but the concept is essentially the same. Typically, familial risks are estimated using case–control epidemiological studies in which a set of probands with the disease are randomly selected either from a population-based registry or from a consecutive series of patients, whereupon complete and detailed family histories are taken to identify the number and distribution of additional cases of disease in the family members. The same procedure is applied to relatives of age-matched controls, and the rate of disease in each group of relatives is compared, to estimate the familial relative risk, λ.

In considering other relatives, there is a predictable relationship between the λ_R values for different relationships. Following Risch (1990a), for a single locus or a model of several loci acting in an additive manner, we have

$$\lambda_1 - 1 = 2(\lambda_2 - 1) = 3(\lambda_3 - 1) \tag{11.1}$$

where the subscripts refer to unilineal relationships of first (parent–offspring), second (e.g., half-siblings, uncle–nephew), and third degree (cousins). For bilineal relationships such as siblings, the following relationship can be determined between the relative risk to MZ twins, λ_M, and other relationships as follows:

$$\lambda_M = 4\lambda_S - 2\lambda_1 - 1$$

These are general relationships that apply to any genetic model, including both dominant and recessive. However, if the trait does not have a large recessive component, then approximately $\lambda_S = \lambda_1$ and $\lambda_M - 1 = 2(\lambda_S - 1)$, or equivalently $\lambda_M = 2\lambda_S - 1$.

For models in which two or more loci are involved in disease susceptibility, we can distinguish two basic mechanisms of how the loci interact. The first comprises additive models in which the penetrances (i.e., allele-specific risks of disease) associated with the multilocus genotypes across different loci can be modeled as sums of factors for each genotype at each locus. It is important to note that this is different from the locus-specific penetrances themselves being additive; rather the pattern of the multilocus penetrances can be expressed by adding across the genotypes at each locus. For relatively rare genes, this model approximates the situation of genetic heterogeneity in which each locus is sufficient to cause disease on its own. The relationship between the overall familial risk λ_R and that attributable to each locus individually is given by (for two loci)

$$\lambda_R - 1 = \lambda_{R1} - 1 + \lambda_{R2} - 1$$

The second model, the multiplicative or epistatic model, occurs when the multilocus penetrances (or perhaps, more exactly, the ratio of the high- and low-risk penetrances) can act multiplicatively across loci. In this model, individuals generally require the higher risk alleles at both loci to be affected. For a multiplicative model with N loci, for any degree of relationship R, the overall familial risk, λ_R, is equal to the product of the individual locus λ_R values, that is,

$$\lambda_R = \lambda_{R1}\lambda_{R2}\cdots\lambda_{RN}$$

The details of the formulation of these multilocus models are given in papers by Hodge (1981) and Risch (1990a). The important principle is that these different models produce different expectations in observed familial risks as a function of degree of relationship. For the additive model, Eq. 11.1 (for the single-locus case) also may be applied to an arbitrary number of additively acting loci. However, under a multiplicative model, there is a steeper decrease in λ_R values as a function of degree of relationship; for example, the difference in relative risk of disease in cousins compared to parent–offspring is larger for multiplicative models than for additive multilocus models. For a model of N loci acting multiplicatively, each with a common λ_{11} value, the predicted overall λ_1 is λ_{11}^N and the overall $\lambda_2 = (\frac{1}{2})^N(\lambda_{11} + 1)^N$. For example, consider a disease with an overall λ_1 of 8.0 caused by three loci, which contribute equally to the familial risk. Under an additive model, the relationship between λ_2 and λ_1 does not depend on the number of loci involved and can be directly determined from Eq. 11.1 as follows:

$$8 - 1 = 2(\lambda_2 - 1)$$
$$7 = 2\lambda_2 - 2$$
$$\lambda_2 = 4.5$$

Under a multiplicative model, however, it is necessary to first calculate the λ_1 value for each locus; since the three loci are of equal effect, this is simply $8^{1/3} = 2.0$. The expected value for second-degree relationships for each locus is calculated from Eq. 11.1 as 1.5, and the overall λ_2 under the multiplicative model can now be calculated as $(1.5)^3 = 3.375$.

   If estimates of recurrence rates or λ_R values for a number of degrees of relationships are available, these data can be used to get some idea of the number of loci involved and whether it is more likely that these loci are acting in an additive or multiplicative fashion. For example, Risch (1990a) used published data on recurrence risks for relatives of probands diagnosed with schizophrenia to compare a number of models of gene action. Because these studies did not find an increased recurrence risk in siblings compared to offspring, it could be assumed that $\lambda_S = \lambda_1$ and therefore all models could be expressed in terms of a single quantity. All models were constrained to fit the overall value of the observed offspring relative risk of $\lambda_O = 10.0$. These data are reproduced here in Table 11.2. If one assumes that a single locus is responsible for the observed offspring relative risk of 10.0, the predicted risk of MZ twins is 19.0:

$$\lambda_M - 1 = 2(\lambda_1 - 1) = 2(10.0 - 1)$$
$$\lambda_M = 19$$

far less than the observed MZ concordance of 52.1. In addition, we see that the observed values for both second- and third-degree relationships are substantially lower than that predicted under the single-locus model. Thus it is unlikely that susceptibility to schizophrenia is governed by a single locus. Rather, based on these results, the observed familial risks would be consistent with a model of several loci acting epistatically in a multiplicative manner to confer susceptibility to schizophrenia. From Table 11.2, we see that the best fit for the observed data is a multiplicative model consisting of a single major locus with $\lambda_1 = 3$ amid a polygenic background of a large number of loci each with small effect (model IV) or two major loci each with $\lambda_1 = 2$ with a similar polygenic background (model VI).

TABLE 11.2. Multilocus Multiplicative Models for Schizophrenia

Risk Ratio[a]	Observed	Model Prediction[b]						
		I	II	III	IV	V	VI	VII
λ_O	10.0	10.0	10.0	10.0	10.0	10.0	10.0	10.0
λ_S	8.6							
λ_M	52.1	19.0	100.0	75.0	55.6	43.8	56.3	42.2
λ_D	14.2							
λ_H	3.5							
λ_N	3.1							
λ_G	3.3							
λ_2	3.2	5.50	3.16	3.35	3.65	3.95	3.56	3.77
λ_C	1.8	3.25	1.78	1.87	2.03	2.20	1.96	2.07

[a]Subscripts: S, sibling; O, offspring; D, DZ twins; H, half-siblings; N, niece/nephew; G, grandchild; 2, pooled value for all second-degree relationships; C, first cousins.
[b]Definitions of models: I, one locus $\sim\lambda_{10} = 10.0$; II, infinite loci, each with small effects; III, $\sim\lambda_{10} = 2.0$, infinite other loci; IV, $\sim\lambda_{10} = 3.0$, infinite other loci; V, $\sim\lambda_{10} = 4.0$, infinite other loci; VI, $\sim\lambda_{20} = 2.0$, infinite other loci; VII, $\sim\lambda_{10} = \sim\lambda_{20} = \sim\lambda_{30} = 2.0$, infinite other loci.
Source: Risch (1990a). Courtesy of American Journal of Human Genetics.

These results are of practical importance, inasmuch as the ability to map loci for a disease will depend on the magnitude of the effect associated with the locus that contributes most to the overall familial association and on the true way in which the multiple loci interact in causing disease. These concepts are addressed formally in the sections that follow.

Measuring Genetic Effects in Quantitative Traits

For quantitative traits, we no longer characterize an individual as affected or not affected, but by the specific value of some quantitative measurement made on that individual. The values of this measurement will naturally (we hope) vary from individual to individual. Our goal in analyzing such traits is to "explain" this variation in terms of specific components, both genetic and environmental. Before discussing the role of specific genes or specific environmental effects, it is useful to get an overall idea of how much of the observed variation in the trait is due to genetic factors, how much is due to nongenetic familial factors that are shared among family members (common environment), and how much is due to random environmental effects, which are unique to each individual. To do this, we look at the degree to which different categories of related and unrelated individuals have similar values for the quantitative trait of interest. If there is significant evidence of a genetic component, we can begin the process of localizing specific genes that may be responsible for the genetic variation. The methods available for both these tasks are described in the next several sections.

The strength of the genetic component is traditionally measured by a quantity called heritability, denoted h^2. Heritability is defined as the proportion of the total variation in the trait that is due to genetic factors (Chapter 3). Usually, we distinguish between two different types of genetic variation, additive effects and dominance effects. The difference in these two types of variation can be illustrated by considering a locus with two alleles A and B which is associated with a particular quantitative trait. For this locus, individuals will be one of three genotypes: AA, AB, or BB. If the mean of the quantitative trait for the AA group is, for example, 80, and that for the BB is 120, the difference between the two homozygotes is responsible for the additive variance because it shows the independent effects of each allele. The other source of variation in the trait due to this locus is reflected by the mean quantitative trait value associated with the AB heterozygote. If the alleles are acting in an additive manner, then we would expect the mean of the heterozygote individuals to be midway between the two homozygotes, in this case a value of 100.

Any difference between the expected value of 100 and the observed value of the mean for the heterozygotes indicates that one of the two alleles is exerting a stronger influence than the other in the heterozygous state (i.e., is *dominant*). This represents additional variance due to the locus, the dominance variance. In addition to this difference in genotype-specific means, the magnitude of these variance components is influenced by the allele frequencies (hence the genotype frequencies) at the locus of interest. These concepts can be extended to any number of loci. One general case that is often considered is that of an infinite number of loci, each with small effect, which form the basis of the polygenic inheritance model of quantitative traits.

Just as in the case with multilocus disease models, the number of loci will not, in general, be known. However, the magnitude of these effects as a proportion of the total phenotypic variance can be estimated from similarities among relatives, analogous to recurrence risks in discrete traits. For our purpose here the important consideration is that two siblings contribute to half the additive variance and one-fourth of the dominance variance, while parents and offspring contribute to only the additive component of variance. Only bilineal relatives who can potentially share two alleles IBD will include dominance variation; in fact, the proportion of this shared variance is equal to their probability of sharing two alleles IBD.

Heritability (in the broad sense) is defined as $h^2 = (V_A + V_D)/V_T$, where V_T is the total phenotypic variance of the trait. For quantitative traits, the heritability can be estimated from the observed correlations between relatives. For nuclear family data, these consist of the intraclass correlation between siblings ρ_{SS} (estimated from the between- and within-sibship mean squares from analysis of variance) and the interclass correlation (ordinary Pearson product–moment correlation) between parents and offspring, ρ_{PO}. Thus V_A/V_T is estimated as $2\rho_{PO}$, while V_D/V_T is estimated as $4(\rho_{SS} - \rho_{PO})$.

Table 11.3 shows the sibling–sibling and parent–offspring correlations and corresponding estimates of the proportion of total variance due to additive and dominance genetic factors for systolic and diastolic blood pressure.

The estimate of V_D assumes that the increased similarity of siblings compared to parents and children is due to dominance variation; but, depending on the phenotype, it could be the result of higher shared environment among siblings than parent–offspring pairs. Without detailed studies of other relatives, particularly half-siblings, or adoption studies, it is impossible to distinguish true genetic dominance from a common environment. In general, studies to estimate variance components are most easily performed on a random sample of families. One example of such a study is the San Antonio Family Heart Study (Mitchell et al., 1996), which collected a random sample of extended Mexican-American families to examine genetic and environmental determinants of cardiovascular risk factors. This study has resulted in several publications reporting the heritability and linkage results for a variety of quantitative traits [e.g., lipid traits, low- and high-density lipoprotein (LDL, HDL)] (Rainwater et al., 1999; Almasy et al., 1999).

TABLE 11.3. Familial Correlations in Blood Pressure

Trait	σ_{PO}	σ_{SS}	V_A/V_T	V_D/V_T	h^2 (Broad Sense)
Blood pressure					
Systolic	0.237	0.333	0.47	0.38	0.85
Diastolic	0.183	0.265	0.37	0.33	0.70

Source: Cavalli-Sforza and Bodmer (1971), based on work of Miall and Oldham (1963).

Summary of Basic Concepts

We have discussed three basic concepts essential to understanding methods for nonparametric linkage analysis:

1. phenotypic similarity in pairs of relatives, whether the similarity is for a qualitative trait such as disease status or for a quantitative traits, implies genotypic similarity at the disease locus and, by extension, similarity at nearby marker loci;
2. the measurement of genotypic similarity using estimates of identity by descent at one or more linked markers for pairs of relatives; and
3. the summarizing of the effect of a single locus as well as a set of loci as expressed by measures of familiality, either as the recurrence risk to relatives (λ) or the heritability (h^2) of a quantitative trait.

These concepts provide the basis for each of the methods we will discuss below.

METHODS FOR NONPARAMETRIC LINKAGE ANALYSIS

Methods for nonparametric linkage analysis can be distinguished based on four key features: (1) which members of the family are included in the analysis; (2) how phenotypic similarity and estimates of trait locus genetic effect are incorporated; (3) how genotypic similarity at the marker locus is measured and estimated (i.e., by identity by state or identity by descent); and (4) whether the analysis includes IBD estimates at a single marker or at multiple markers. The greatest variability in study design occurs in deciding which family members are included in the analysis and how many markers can be analyzed, as evidenced by the differences in algorithms to estimate identity by descent. We start by examining a few methods for the analysis of qualitative traits and then methods developed for the analysis of qualitative traits. In each section we start by discussing the most basic constellation of relatives, sibling pairs, and then move to methods that handle increasingly more distantly related individuals. In general the implementation of several different methods employing different constellations of family members can be quite useful in the analysis of complex traits, since different methods and different statistics highlight different features of the data (Anderson et al., 1999; Schmidt et al., 2001; Sengul et al., 2001).

Tests for Linkage Using Affected Sibling Pairs (ASPs)

Test Based on Identity by State. The first test proposed for using IBS relationships in sibpairs is that by Lange (1986) in which the observed distribution of pairs sharing zero, one, or two alleles IBS is compared to that expected under the hypothesis of free recombination between disease loci and the marker locus. The test for linkage therefore is a simple chi-square goodness-of-fit test comparing the observed IBS distribution to the expected. The expectations for IBS sharing under the null

TABLE 11.4. Analysis of IBS Sharing Probabilities for Two
Siblings at a Marker with n Equally Frequent Alleles

	Probability of Siblings Sharing Alleles IBS		
Number of Alleles	0	1	2
2	0.03	0.38	0.59
3	0.06	0.48	0.46
4	0.08	0.52	0.40
5	0.10	0.53	0.37
6	0.12	0.53	0.35
8	0.15	0.53	0.32
10	0.16	0.53	0.30
20	0.20	0.52	0.28
Infinite	0.25	0.50	0.25

hypothesis are, of course, a function of the allele frequencies at the marker locus that take into account the distributions of mating types compatible with the observed pair genotypes. Table 11.4 shows the expected distribution of sharing zero, one, and two alleles IBS as a function of the informativeness of the marker (indexed as the number of equally frequent alleles).

Because alleles can be IBS without being IBD, there are more pairs than expected sharing one and two alleles than the $\frac{1}{2}$ and $\frac{1}{4}$ expected for IBD sharing; however, as the number of marker alleles increases, it can be seen that the IBS distribution (at least in nuclear families) approaches the $\frac{1}{4}, \frac{1}{2}, \frac{1}{4}$ expected for IBD relationships. Intuitively, as discussed earlier, this is because for highly polymorphic markers, the probability that one or both parents are homozygous or both are identical heterozygotes (which makes IBD status ambiguous) becomes increasingly small. The power of these methods for siblings and other relative pairs is considered in a paper by Bishop and Williamson (1990), who examined the effect on power of the strength of the genetic component, marker informativeness, and type of relative pair. Although we have denoted these methods "identity by state," it is important to recognize that here one is simply using IBS as a surrogate for the unobservable IBD status. Many of the more sophisticated IBD methods presented in the next section are, in effect, estimating the unknown IBD distribution and thus are similar to the approach presented in this section.

Tests Based on Identity by Descent in ASPs

Simple Tests. In the simplest case, where the numbers of pairs sharing zero, one, or two alleles IBD can be unambiguously determined (i.e., parents are typed and informative), several tests have been developed to test for linkage between a marker locus and a disease. The first such proposed test was a chi-square goodness-of-fit test comparing the observed proportion of sibpairs sharing zero, one, or two alleles IBD with the expected proportions of $\frac{1}{4}, \frac{1}{2}$, and $\frac{1}{4}$ under the hypothesis of no linkage between the marker and disease locus. A related test proposed by Suarez et al. (1978) is to compare only the proportion of sibpairs who share exactly two alleles IBD with the expected proportion of $\frac{1}{4}$. Only differences in which the observed proportion of shared alleles exceeds the expected are considered significant; thus a one-sided test is used. The third test, called the means test, compares the average IBD sharing in the sample of pairs with the expected value of $0.5 = (0 \times \frac{1}{4} + 1 \times \frac{1}{2} + 2 \times \frac{1}{4})/2$ chromosomes. Each of these tests can be written in terms of the observed numbers of pairs with each IBD sharing status:

0	1	2	Total
n_0	n_1	N_2	N

Statistics

1. *Goodness of fit:*

$$\chi^2 = \frac{4(n_0 - N)/4)^2 + 2(n_1 - N/2)^2 + 4(n_2 - N/4)^2}{N}$$

Reject H_0 if χ^2 is greater than $\chi^2_{1-\alpha}$ with two degrees of freedom for size α-test.

2. *Proportion:*

$$T_1 = 4\left(\frac{n_2}{N} - \frac{1}{4}\right)\left(\frac{N}{3}\right)^{1/2} \qquad \text{(one-sided } t\text{-test with } N - 1 \text{ degrees of freedom)}$$

3. *Mean.*

$$T_2 = (2n_2 + n_1 - N)\left(\frac{2}{N}\right)^{1/2} \qquad \text{(one-sided } t\text{-test with } N - 1 \text{ degrees of freedom)}$$

As an example of the application of these three methods, we use data taken from Cox and Spielman (1989) for HLA sharing in sibpairs affected with type I (insulin-dependent) diabetes mellitus (IDDM):

	Alleles or Haplotypes IBD		
	0	1	2
Total pairs: 137			
Observed	10	46	81
Expected	34	69	34

The results of these analyses are shown in Table 11.5.

TABLE 11.5. Results of Simple Sibpair Tests on IDDM Data

Test	Statistic[a]	Degrees of Freedom
Goodness of fit	$\chi^2 = 88.4$	2
Proportions	$T_1 = 9.22$	138
Means	$T_2 = 8.58$	136

[a]T_1 is the proportions test statistic; T_2 is the means test statistic.

Given the overwhelming evidence for the involvement of HLA in IDDM as shown by the data above, all three of the tests described give a highly significant p-value ($p < 10^{-15}$), but in other situations these tests may provide different conclusions regarding the significance of observed IBD data. The choice of the optimal test from the set presented depends on the true underlying genetic model, which is usually unknown. However, studies have shown that the means test performs best under a fairly wide range of genetic models (Blackwelder and Elston, 1985).

Tests Applicable When IBD Status Cannot Be Determined. The tests described above are useful when the IBD status of the entire dataset is known with certainty or when there is a sufficient subset of the entire dataset to permit the subset, where IBD status is unknown, to simply be discarded. In many situations, however, marker genotype data are unavailable on one or both parents of the affected sibpair. This is frequently the case, for example, for disorders with late age of onset, such as Alzheimer's disease or prostate cancer. In these situations, however, information on additional family members often provides some information about the sharing status of the siblings. For this reason, methods that can use all the available data to make inferences about the IBD status of the affected pairs are desirable. The first such method was proposed in a seminal series of papers by Risch (1990b,c) in which the author considered the basic models for general sibpair analysis and the power of these methods. The method, denoted MLS for maximum LOD score method, compares the probability of the observed genotypic data in the affected sibling (or other relative) pairs as a function of the proportion sharing zero, one, or two alleles IBD (Z_0, Z_1, Z_2) with the corresponding probability of the observed marker data under the null hypothesis of no linkage ($Z_0 = 0.25$, $Z_1 = 0.5$, $Z_2 = 0.25$). Numerical search methods are used to find the values of Z_0, Z_1, Z_2 which make the numerator of this ratio (hence the ratio itself, since the denominator is constant) the largest. Put another way, we are finding the values of Z_0, Z_1, Z_2 that are most consistent with the marker genotypes of the affected pairs. This method is designed to use information from affected siblings or other relatives in determining the optimal IBD proportions. To be analogous to parametric analysis (see Chapters 1 and 10), the $\log_{10}$ of the likelihood ratio is taken to produce a LOD score. For example, for the HLA data above, the optimal values of the sharing probabilities (Z_0, Z_1, Z_2) are easily shown to simply be the observed proportions $\frac{10}{137} = 0.073$, $\frac{46}{137} = 0.336$, and $\frac{81}{137} = 0.591$, respectively, since we have complete information in this sample regarding IBD status. This produces a LOD score of

$$\text{LOD} = \log_{10}\left[\frac{0.073^{10}0.336^{46}0.591^{81}}{0.25^{10}0.50^{46}0.25^{81}}\right] = 16.98$$

equivalent to odds of $10^{17} : 1$ in favor of involvement of HLA in this disease.

However, not all values of the Z_i that may be considered in the numerical search are compatible with genetic models. For example, the case of none of the pairs sharing any alleles ($Z_0 = 1$) is not compatible with any Mendelian model of inheritance (in large samples) Holmans (1993) showed that the $\{Z_i\}$ for possible genetic models must satisfy $2Z_0 \le Z_1 \le \frac{1}{2}$, which he called the "genetically possible triangle." Note that this implies that $Z_0 \le 0.25$ and that $Z_2 \ge 0.25$. In this same paper, Holmans showed that restricting the estimation of the $\{Z_i\}$ to satisfy this criterion resulted in a more powerful test of linkage than the test based on unrestricted maximization.

The maximum-likelihood estimates of Z_0, Z_1, Z_2 can be used to make inferences about the contribution of this locus to the overall increased familial risk, as the expected sharing at a marker locus linked to a disease locus can be written in terms of the λ_S value for the disease locus and the distance between this locus and the marker locus (Suarez et al., 1978; Hodge, 1981; Risch, 1990a). These equations can then be solved to estimate the contribution of this locus to the overall familial risk. When there is no recombination between the marker locus and the disease locus—that is, when the marker locus is a candidate locus for the disease—the following equations relate the sharing probabilities and the familial relative risks attributable to a given locus:

$$Z_0 = \frac{1}{4\lambda_S} \qquad \lambda_S = \frac{1}{4Z_0}$$

$$Z_1 = \frac{\lambda_O}{2\lambda_S} \qquad \lambda_O = \frac{Z_1}{2Z_0}$$

$$Z_2 = \frac{\lambda_M}{4\lambda_S} \qquad \lambda_M = \frac{Z_2}{Z_0}$$

When the hypothesized disease gene is located some distance from the marker locus, more complex relationships (see, e.g., Risch, 1990b) hold, although they can be somewhat simplified by assuming that the recombination fraction between the disease locus and marker locus, θ, is not too large (e.g., $\theta < 0.10$) and that the offspring and sibling risks due to the locus are the same (i.e., $\lambda_O = \lambda_S$). If these assumptions can be justified, then the following holds:

$$Z_0 = \frac{1}{4} - \frac{2(\omega - 1)(\lambda_S - 1)}{4\lambda_S}$$

$$Z_1 = \frac{1}{2}$$

$$Z_2 = \frac{1}{4} + \frac{(2\omega - 1)(\lambda_S - 1)}{4\lambda_S}$$

where $\omega = \theta^2 + (1 - \theta)^2$.

In our HLA example, assuming zero recombination between HLA and IDDM susceptibility, we would estimate the sibling risk attributable to this locus to be $\lambda_{S,HLA} = 1/(4 \times 0.076) = 3.3$. If, however, the disease gene is not HLA but is located a distance of $\theta = 0.05$ from the HLA complex, the estimate of $\lambda_{S,HLA}$ increases to 7.1 using the estimate of Z_0. This example illustrates one of the problems of using single-point analysis; the strength of the effect of a given locus and the distance of this locus from the tested marker are confounded. That is, in a genomic search with anonymous STR markers, it will be difficult to distinguish between a relatively weak disease locus located very close to a marker and a locus of larger effect located some distance away. There is no unique estimate of the disease gene location when using one marker at a time.

If we assume that HLA is, in fact, the predisposing locus (or located very close to it), we can use the estimate of 3.3 obtained above to get some idea of how much of the overall familial risk is likely to be accounted for by the HLA system. Since observed values from a variety of studies have estimated the recurrence risk to siblings for IDDM to be 0.06 and the disease has a population prevalence of about 0.004 (Spielman et al., 1980), this yields an overall λ_S value of $0.06/0.004 = 15$. Thus, based on the data, it is clear that other susceptibility loci must be involved in the determination of IDDM. The estimated contribution of the HLA locus is $\log(3.3)/\log(15) = 44\%$ for a multiplicative model and $(3.3 - 1)/(15 - 1) = 16\%$ for an additive multilocus model. The results of a complete genomic search for IDDM susceptibility loci are discussed later in this chapter.

Multipoint Affected Sibpair Methods. As we have seen, the methods discussed in this chapter depend on our ability to distinguish identity by descent from identity by state in affected sibpairs. In addition, it is easy to see that the ability to detect linkage to a disease locus will depend on the distance between the marker locus at which IBD sharing is assessed and a disease locus influencing the disease under study. The use of multiple markers in a defined genetic region addresses both these issues by maximizing the probability that complete or partial IBD information is available in the region spanned by the markers. The first person to examine IBD sharing in nuclear families as a function of marker genotypings at multiply linked loci was Goldgar (1990), who examined the estimation of IBD sharing in a region spanned by genetic markers to map quantitative trait loci. For qualitative traits such as disease phenotypes, the theoretical advantages of using multiple markers (albeit perfectly informative ones) was first discussed by Risch (1990b). More recently, Hauser and Boehnke (1998) have extended the general MLS method for ASPs to multipoint mapping. Risch's method is implemented in the computer program ASPEX and Hauser's method in the computer program SIBLINK.

Handling Sibships with More Than Two Affected Siblings. Although the methods described earlier are oriented toward analyzing a set of affected sibpairs, often other affected siblings are available for study. For methods designed to analyze pairs only, this requires the formation of all possible sibpairs from the sibship. If the sibship contains n affecteds, there are $n(n - 1)/2$ distinct pairs that can be formed for the analysis. The difficulty is that these pairs are not independent, and therefore significance levels based on such data will be inaccurate, resulting in increased type I error rates. For example, using the fact that there are four possible distinct parental alleles in a sibship, if siblings A and B share no alleles IBD, and siblings B and C also share no alleles IBD, we know that siblings A and C must share two alleles IBD. One solution proposed by Hodge (1984) to get around this problem is to weight the contribution of a sibship by $n - 1$ (the number of independent pairs in a sibship of size n). For certain traits, consideration of discordant pairs (one affected, one unaffected) also may provide information for linkage (Olson, 1995a). In this case, one would expect there to be more pairs who share no alleles IBD at the expense of pairs sharing two alleles IBD. However, the power of the discordant pairs under almost all circumstances will be small compared to that for affected pairs. For diseases with variable (or late) age of onset and genes of relatively small effect, discordant pairs will be almost totally uninformative. However, if they are collected and genotyped to contribute additional information to the overall estimation of IBD status in the family, including them in the analysis may be worthwhile (Hauser et al., 1996).

Methods Incorporating Affected Relative Pairs

While studies of affected sibling pairs have some practical advantages, since the methods are conceptually simple and in many cases affected siblings are more abundant than relative pairs, it is often the case that a large variety of affected relative pairs is available for study. Families collected for study often contain both ASPs as well as affected avuncular and cousin pairs and even larger families will have many affected individuals extending across several generations and more distant relationships. The major disadvantage of sibpair analysis is its inability to use additional information that can be extracted from these other affected relatives. Methods that use information from such affected relatives have been developed to take greater advantage of these extended families. These affected relative pair (ARP) methods have proven very useful for linkage studies (St George-Hyslop et al., 1990; Pericak-Vance et al., 1991, 2001; Haines et al., 1996). We discuss four common ARP methods, as well as an approach using the traditional parametric LOD score method that reduces its dependence on the genetic model specification. Some of these methods examine the data by means of comparisons of each possible pair of affected individuals, while others examine all affected individuals simultaneously. We will begin by examining single-point methods that look for linkage between affected status and a single marker and then discuss multipoint methods.

Affected Pedigree Member Analysis. In 1988 Weeks and Lange described the affected pedigree member (APM) method, an extension of earlier work in this field on affected sibling pairs (Weeks and Lange, 1992). The APM analysis requires no knowledge of the underlying genetic model, which can then be arbitrarily complex. Unlike most sibpair methods, the basis of the APM tests for deviation from the expected distribution of IBS, not IBD, relationships. For alleles to be IBS they must simply match in state, most often meaning that each allele is the same size or has the same label (e.g., 117 bp or allele *G*). However, it is not known whether the

allele has been inherited from the same common ancestor (i.e., identity by descent). As discussed earlier in this chapter, to accurately estimate identity by descent, genotypes on all the individuals connecting the affected relative pairs must be available (or completely inferable). The use of identity by state instead of identity by descent eliminated the need for such genotypes for the relatives connecting the affected pair.

Unlike some of the ASP methods that estimated genetic effects as recurrence risk ratios, no assumptions about the underlying genetic etiology of the trait are made. However, the method does depend on three factors: the genotypes of the affected individuals, the marker allele frequencies, and the pedigree relationships of these individuals. The APM statistic for any pair of relatives is:

$$Z_{ij} = \tfrac{1}{4}\delta(G_{ix}, G_{jx})f(P_{Gix}) + \tfrac{1}{4}\delta(G_{ix}, G_{jy})f(P_{Gix}) \tag{11.3}$$
$$+ \tfrac{1}{4}\delta(G_{iy}, G_{jx})f(P_{Giy}) + \tfrac{1}{4}\delta(G_{iy}, G_{jy})f(P_{Giy})$$

where i and j represent the two affected individuals being compared; G_{ix}, G_{iy}, G_{jx}, and G_{jy} represent the maternal (x) and paternal (y) alleles at the marker locus, respectively, of individuals i and j; $\delta(G_{ix}, G_{jx})$, $\delta(G_{ix}, G_{jy})$, $\delta(G_{iy}, G_{jx})$, and $\delta(G_{iy}, G_{jy})$ are 1 if the two alleles (G) match in state or 0 if they do not match in state; and $f(\rho_{Gix})$ and $f(\rho_{Giy})$ represent an arbitrary weighting function of the marker allele frequencies.

Prior to the application of marker allele frequencies, the four possible allele comparisons between these two individuals are equally weighted, so that the maximal value of Z_{ij} is 1 and the minimal value is 0. For example, assuming four equally frequent alleles, if two individuals both have genotype $1/1$, then $Z_{ij} = 1$; if both have genotype $1/2$, then $Z_{ij} = \tfrac{1}{2}$; if one has genotype $1/2$ and the other has genotype $1/3$, then $Z_{ij} = \tfrac{1}{4}$; and if one has genotype $1/3$ and the other has genotype $2/4$, then $Z_{ij} = 0$. A weighting function based on the marker allele frequencies may be introduced to allow for the greater likelihood that rare alleles shared by a pair of relatives indicate linkage than the sharing of a common allele. In this case each of the four possible allele comparisons contributes $\tfrac{1}{4} f(p)$. Three weighting functions are commonly used: $f(p) = 1, f(p) = 1/\sqrt{p}, f(p) = 1/p$, where p is the frequency of an arbitrary allele.

The overall APM statistic for a given family is simply the sum over all pairs:

$$Z = \sum_{i,j} Z_{ij} \tag{11.4}$$

The mean and variance of this statistic are more complicated to calculate, and the reader is referred to Weeks and Lange (1988) for more detail. It should be noted that Eq. 11.4 uses only marker genotypes and marker allele frequencies. The pedigree relationship between the two individuals is used only in the calculation of the mean and variance.

While the APM statistic has been replaced by the generally more powerful SimIBD method (below), its role in the identification of one of the first genes for a complex disease must be acknowledged. Hall et al. (1990) used APM as one of its methods in identification of genetic linkage to early onset breast cancer and subsequent identification of the *BRCA1* gene.

SimIBD Analysis. In 1996 Davis et al. described several new ARP statistics: SimAPM, SimKIN, SimIBD, and SimISO. The SimAPM statistic, an improved version of the APM statistic, still relies on IBS relationships. The latter three attempt to estimate IBD relationships. In the comparison of these three statistics, SimIBD and SimISO were shown to provide the more accurate estimates of IBD sharing, with SimIBD examining sharing between affected relatives, and SimISO examining, in addition, sharing among unaffected relatives. While SimISO may have some applications, the greatest power to detect linkage in genetically complex traits will arise from examining affected individuals, and so the SimIBD measure is generally preferred.

The general approach in SimIBD is to calculate an observed statistic given all the marker and pedigree information and then compare it to a simulated distribution. The simulated distribution is generated by performing many replicates in which the only data generated randomly (within the confines of Mendelian inheritance) are the marker genotypes for the affected individuals conditional on the marker genotypes of the unaffected individuals. The assigned p-value is simply the empirical p-value obtained by comparing the observed statistic to the distribution generated from the replicates of the simulated data. SimIBD depends on three factors: the genotypes of the affected and unaffected individuals, the marker allele frequencies, and the pedigree relationships of these individuals.

The SimIBD statistic for any pair of relatives (parent–child pairs excluded) is

$$Z_{ij} = \sum_{a=1}^{2} \sum_{b=1}^{2} \left(\frac{1}{\sqrt{p}} \right) \alpha_{ia,\,jb} \tag{11.5}$$

where $1/\sqrt{p}$ is the weighting function based on the marker allele frequency p and $\alpha_{ia,jb}$ the probability that the two alleles being compared are IBD.

In Eq. 11.5, $\alpha_{ia,jb}$ equals 1 if the two alleles can be absolutely determined to be IBD and 0 if the two alleles can be absolutely determined not to be IBD. If IBD status cannot be absolutely determined, $\alpha_{ia,jb}$ is estimated given the marker genotypes in the pedigree and will fall between 0 and 1. The estimation procedure uses a recursive algorithm and the reader is referred to Davis et al. (1996) for more detail.

The SimIBD statistic for any pedigree is

$$Z_p = \frac{1}{(r-1)^{1/2}} \left[\sum_{i=1}^{r} \sum_{j=1}^{r} Z_{ij} \right]$$

where r is the number of affected individuals in the pedigree.

The overall statistic across pedigrees is simply the sum of each pedigree statistic:

$$Z = \sum_{p=1}^{m} Z_p$$

The primary advantage of the SimIBD statistic is that it uses IBD information when available and estimates it otherwise. This convention minimizes the amount of potentially misleading information that would make its way into a purely IBS approach, such as the APM method. As with APM analysis, SimIBD does not require any specification of the trait genetic model, making it a desirable method for genetically complex traits. In most situations, it outperforms the APM method, having a lower false-positive rate and higher power. Therefore, SimIBD is recommended over APM analysis.

There are three disadvantages to the SimIBD method. The first is that arbitrarily complex pedigrees cannot be analyzed. The current version of this program can handle simple pedigrees as well as those with one or two marriage loops or consanguineous loops. While this is not a major limitation for most studies, it may severely restrict application of SimIBD in investigations of very rare recessive conditions or inbred populations. The second limitation is that currently no multipoint extension is available. Thus the added information gained by examining nearby markers cannot be utilized. Finally, the program can take substantial amounts of computer time to run. This is due primarily to the simulation aspect of the statistic. Although the observed statistic is calculated only once, the simulated distribution requires hundreds of replicates to achieve reasonable accuracy. We recommend that at least 1000 replicates be performed.

The SimIBD method can be used in several situations:

1. When an extended pedigree structure precludes standard parametric LOD score calculations in single-gene disorders. However, the reduction in computer time may not be substantial.
2. When many affected relatives other than siblings are available in genetically complex traits. Any single pedigree may have only one or a few affected relative pairs, but in aggregate there are many such pairs.
3. When genotypes for unaffected individuals who connect affected individuals are available.

See Table 11.6 for a comparison of the properties, advantages, and disadvantages of SimIBD and other ARP methods.

NPL Analysis. In 1996 Kruglyak et al. introduced an ARP approach they termed NPL (nonparametric linkage) as part of the GENEHUNTER computer program. Like the other ARP methods, the NPL statistic measures allele sharing among affected individuals within a pedigree. The NPL method can analyze the data using a pairwise approach (using the NPL_{pairs} statistic), but it has been extended to provide a simultaneous comparison of alleles in all affected individuals in a pedigree (the NPL_{all} statistic). The NPL approach is inherently multipoint, since it calculates the IBD probability for any given point along the chromosome (the "inheritance distribution") using all available marker data on that chromosome.

TABLE 11.6. Comparison of ARP Methods

Factor	Affected-Only LOD Score	Affected Pedigree Member	SimIBD	Nonparametric Linkage Analysis	Weighted Rank Pairwise Correlation
Parameters trait genetic model	Yes (no penetrance)	No	No	No	No
Recombination fraction	Yes	Marker map only	No	Marker map only	No
Marker allele frequencies	Needed	Needed	Needed	Needed	No
Arbitrary pedigrees	Yes	Yes	Yes (limited size)	Yes (limited size)	Yes
Multipoint analysis	Yes	Yes	No	Yes	Yes
Uses phenotypes from unaffected individuals	No	No	No	No	Yes
Can consider all affected individuals separately	Yes	No	No	Yes	No

The calculation of the NPL statistic can be divided into two parts. The first part, the calculation of the inheritance distribution, is used to estimate the allele sharing among a set of affected individuals, be they just pairs or complete family groupings. This is done as outlined in Lander and Green (1987) and Kruglyak et al. (1996), and the reader is referred to these sources for a detailed description. The second part of the calculation is the individual evaluation of the scoring function that determines whether the inheritance information is indicative of linkage. As mentioned above, two scoring functions are described, NPL_{pairs} and NPL_{all}.

If the inheritance pattern can be determined unambiguously in a pedigree, the resulting NPL_{pairs} statistic is

$$S_{pairs} = \sum S_{ij}$$

where i and j are the two individuals being compared and $S_{ij} = 0, 1, 2$, depending on how many alleles are shared IBD.

In most genetically complex traits, however, the inheritance pattern cannot be determined unambiguously. In this case, the S_{pairs} statistic is an average taken across all possibilities.

To compare S_{pairs} to a statistical distribution, it is normalized

$$Z_{\text{pairs}} = \frac{S_{\text{pairs}} - E(S_{\text{pairs}})}{[\text{var}(S_{\text{pairs}})]^{1/2}}$$

where $E(S_{\text{pairs}})$ and $\text{var}(S_{\text{pairs}})$ are the expectation of the mean and variance of S_{pairs} under the null hypothesis.

The overall statistic across the pedigree is

$$Z = \sum \left(\frac{1}{\sqrt{m}}\right)(Z_{\text{pairs}})$$

where $(Z_{\text{pairs}})_i$ = normalized score for pedigree i

$\qquad m$ = number of pedigrees

The NPL$_{\text{all}}$ statistic is defined as

$$S_{\text{all}} = 2^{-a} \sum_h \left[\prod_{j=1}^{2f} b_j(h)!\right] \tag{11.6}$$

where a = number of affected individuals in pedigree

$\qquad h$ = collection of alleles generated by taking one allele from each affected individual (there are 2^a possible collections)

$\quad 2f$ = total number of founder alleles in pedigree (i.e., total number of different alleles of distinct origin)

$\quad b_j(h)$ = total number of specific founder allele (j) in collection (h)

This statistic is averaged over all feasible inheritance patterns, normalized, and weighted across pedigrees in the same manner as the NPL$_{\text{pairs}}$ statistic.

Statistical significance is determined by comparing the Z score to the standard normal distribution. The use of the standard normal distribution is an approximation, usually a conservative one. In other words, the true p-value is often smaller than the p-value obtained by using a standard normal table. To fix the conservative bias inherent in the NPL statistics based on the standard normal distribution, Kong and Cox (1997) adapted the GENEHUNTER algorithm to provide a likelihood-based statistic presented as a LOD score. The LOD score is less conservative and more powerful.

There are several advantages to the NPL approach. First, it has a friendly and easy-to-use interface. Second, the NPL$_{\text{all}}$ statistic is the only ARP method that can consider all affected relatives simultaneously, rather than as a combination of all possible comparisons of pairs. This implements the intuitively attractive idea that if, for example, five affected individuals in a pedigree all share the same allele IBD, this information should carry more weight than if each of the 15 possible pairs in the same pedigree shared some allele IBD, not necessarily the same allele.

This increased weighting is seen in Eq. 11.6 by including the factorial of $b_j(h)$, rather than $b_j(h)$ itself. However, the NPL$_{all}$ statistic does not perform as well as the NPL$_{pairs}$ statistic in some situations, particularly those involving very common, recessive-acting alleles.

Perhaps the greatest advantage of the NPL statistic is that the data for all markers on a chromosome can be evaluated simultaneously using a multipoint approach. Because GENEHUNTER uses the Lander–Green algorithm for calculating the IBD distribution, the amount of computer time for an analysis increases only linearly with the number of markers included in the analysis. This is in contrast to the Elston–Stewart algorithm used for most parametric LOD score analyses, which causes computer time to increase exponentially with the number of markers.

There is one potential disadvantage to the NPL statistic, related to its inherently multipoint approach. As discussed, the Lander–Green algorithm can incorporate many markers because computer time increases only linearly with the addition of new markers. However, computer time increases exponentially with the number of individuals in the pedigree. Thus NPL is limited to smaller pedigrees. This is dictated by the rule that $2n - f \leq 21$, where n is the number of nonfounders and f is the number of founders. More complicated or extended pedigrees cannot be handled with any confidence, even on very fast workstations. Since the initial release of the GENEHUNTER software, there has been steady improvement in the speed of the algorithm and complexity of the pedigrees that can be analyzed (Gudbjartsson et al., 2000; Markianos et al., 2001). The most recent contender is a program called MERLIN (Multipoint Engine for Rapid Likelihood Inference) that can analyze pedigrees with $2n - f \leq 32$ by using sparse binary trees and by taking advantage of symmetries in pedigree data for untyped individuals (Abecasis et al., 2002).

In general, NPL analysis can be used in several situations:

1. When pedigrees of moderate size are available. Even pedigrees with only a single affected sibpair can be useful, since a large number of markers can be examined simultaneously. However, even with the algorithmic improvements, large pedigrees cannot be examined.
2. When many relatives other than siblings are available for a genetically complex trait. Each pedigree needs only a single affected relative pair.
3. When a large number of linked markers are being (or could be) examined. This is particularly useful if pedigrees have somewhat "sparse" genotyping (i.e., many individuals providing the biological links between affected individuals are not available for genotyping). This situation generally leads to difficulty in generating the IBD status for a single marker, and the problem may be reduced substantially if many linked markers are genotyped.

Likelihood (Parametric LOD Score) Analysis. It may seem contradictory to suggest that parametric LOD score calculations could be used as a "nonparametric" method. However, by appropriately modifying the penetrance parameters, it is possible to approximate a nonparametric method by minimizing the effect of some parameters. This was the first and still perhaps the most widely practiced ARP method. The implementation of the method generally takes two forms. The first amounts to an analysis of only the affected individuals in the pedigree. The second attempts to find a generic genetic model or set of robust models that fit the population prevalence and recurrence risk ratios.

Affecteds-Only Analysis. For genes that moderately increase risk, the penetrance (i.e., the probability of expressing the trait given the gene) is low. If the penetrance parameter is set to zero or to a very small value (e.g., 0.001), the phenotypic information from unaffected individuals is not used, and only phenotype information from affected individuals is used for the trait. Driving this approach is the underlying assumption that the affected phenotype is much more certainly associated with the presence of the trait allele, while the unaffected phenotype is far less certainly associated with the absence of the trait allele.

As an example of how to set up an affecteds-only analysis, we will use the Alzheimer example (Table 11.7). Using the LINKAGE program format, the penetrance matrix on the left is a standard age-at-onset-curve matrix with variable penetrance defined by an age-at-onset distribution (Pericak-Vance et al., 1991). The penetrance matrix on the right mimics this matrix but provides for no penetrance for those individuals who are at risk for Alzheimer's disease (e.g., those who had been assigned a penetrance other than 0 or 1).

Fitting Population Parameters. To implement a genetic model that fits the prevalence and recurrence risk ratios requires knowledge of the λ_R value and an idea of the population prevalence. The following approximate equations can be used to

TABLE 11.7. Penetrance Values for Age-at-Onset Curve Versus Affecteds-Only Analysis

Liability Class[a]	Age-at-Onset Curve			Affecteds Only		
	NN	NA	AA	NN	NA	M
N,A	0.00	1.00	1.00	0.00	1.00	1.00
N40	0.00	0.01	0.01	0.00	0.00	0.00
N50	0.00	0.05	0.05	0.00	0.00	0.00
N60	0.00	0.10	0.10	0.00	0.00	0.00
N70	0.00	0.30	0.30	0.00	0.00	0.00
N80	0.00	0.75	0.75	0.00	0.00	0.00
N90	0.00	0.95	0.95	0.00	0.00	0.00

[a]"Liability class" refers to the phenotype classification of each individual: N, normal; A, affected; NXX, normal with an age at exam in the XX decade.

construct a set of "reasonable" models that fit the epidemiological data:

$$\lambda_S D^2 = qP^2 \qquad \text{for dominantly acting loci}$$

$$4\lambda_S D^2 = q^2 P^2 \qquad \text{for recessively acting loci}$$

where λ_S is the sibling familial risk, D is the disease prevalence, q is the frequency of the allele associated with disease susceptibility, and P is the penetrance (probability of disease) associated with the risk allele.

The risk of disease in noncarriers, t, is given by

$$t = \begin{cases} \dfrac{D - 2qP}{1 - 2q} & \text{for dominant loci} \\ \dfrac{D - q^2 P}{1 - q^2} & \text{for recessive loci} \end{cases}$$

For any known (or assumed) values of λ_S and D, values of q, P, and t can be determined from the equations above. Then traditional LOD score analysis (either two-point or multipoint) can be performed under this limited set of models, with appropriate correction of the significance threshold for the number of models tested (Hodge and Elston, 1994; Abreu et al., 2002).

For example, consider a disease in which the sibling risk due to a hypothesized locus is 4.0 and the disease population prevalence is 1%. This set or parameters would be compatible with the following genetic models:

1. a rare autosomal dominant locus with an allele frequency of 0.0005 for the high-risk allele and penetrances of 89% in individuals who carry at least one copy of the high-risk allele and 0.9% in noncarriers,
2. a common autosomal dominant locus with an allele frequency of 0.01 and a penetrance of 20% in carriers and 0.6% in noncarriers, and
3. an autosomal recessive locus with a disease allele frequency of 0.05.

Individuals who are homozygous for this allele have an 80% chance of disease, compared with 0.8% for heterozygous and normal homozygous individuals.

While these models all produce a familial relative risk of 4 and a disease prevalence of 1%, they differ in several respects. For example, the proportion of all cases due to the proposed risk allele will vary from 39% for model 2 to 20% for model 3 to 9% for model 1. These models will also differ with respect to the parent–offspring risk and in the proportion of families with many cases of disease. Thus, if one also has an idea of the offspring risk, it may be possible to rule out either a dominant or recessive model, and the frequency of extended families with high incidence of disease may help distinguish a rare high-penetrant allele from a more common but lower penetrant gene. In any case, the accurate knowledge of the basic epidemiological profile of the disease in question can aid in choosing a small number of consistent models to use in parametric LOD score analysis of affected sibpairs or other affected relative sets.

The major advantage of the likelihood-based method is that marker genotype information is used from all individuals, and the maximum-likelihood approach captures all the possible marker information and thus generates the best estimates of IBD status. These estimates are used only in conjunction with the most certain of the trait information (i.e., phenotypes of affected individuals). This method can also be used on large and arbitrarily complex pedigrees, and limited (three- to four-marker) multipoint analysis can be performed. It also considers all affected individuals simultaneously, not just in a pairwise fashion. Finally, standard programs such as LINKAGE, VITESSE, and MENDEL can be used for these calculations.

The major disadvantage of the likelihood-based analysis is that the parameters of a genetic model must still be specified. Even if penetrance is set to zero, the inheritance pattern (e.g., autosomal dominant, autosomal recessive, or X linked) must be specified, and misspecification will still generate misleading results. Furthermore, power to detect a locus can be extremely low if the genetic model is grossly misspecified. Thus the results must be interpreted with caution. In addition, the amount of computer time needed for the calculations can become excessive if the pedigrees are large or have marriage or consanguinity loops or if multipoint analysis is undertaken.

Although widely used, affecteds-only analysis only compensates for poor knowledge of penetrance; it is not a satisfying solution to the problems extant in complex traits. Nevertheless, it can be used in several situations when other methods cannot be implemented:

1. in single-gene disorders when the age at onset or other penetrance function cannot be well specified;
2. when a major gene is suspected but cannot be proven;
3. when large or complex pedigrees are being studied, even if the mode of inheritance is unknown; and
4. when multipoint analysis is being contemplated, since other ARP methods (below) may be limited to two-point analysis.

The LOD score approach has the advantages of producing an estimate of the recombination fraction as well as permitting the estimation of other parameters of the genetic model. If the model is even approximately correct, the use of the LOD score method should provide higher power than nonparametric methods (Goldin and Weeks, 1993). Also, under a wrong model, the type I error rate (i.e., the chance of finding significant evidence of linkage when it is not present) is not inflated (Williamson and Amos, 1990). Given that the LOD score method has been shown to be reasonably robust to deviations from the true model, has considerable flexibility in analyzing data from any pedigree structure with any number of affected and unaffected individuals (i.e., does not need to reduce the data to pairs), and makes complete use of available marker information, the use of this method should not be ruled out, even for studies dealing with sibpair data. For further discussion of these issues, the reader is advised to consult Greenberg et al. (1996).

Power Analysis and Experimental Design Considerations for Qualitative Traits

Factors Influencing Power of Sibpair Methods. The power of the affected sibpair method—that is, the probability that a true disease-predisposing locus will be detected by examining a linked marker locus—was first considered in detail in the papers by Risch (1990b,c) for the MLS identity-by-descent method and by Bishop and Williamson (1990) for methods based on identity by state. As demonstrated by Risch (1990b), the power to detect linkage of a trait locus and a marker locus can be expressed solely in terms of the single parameter of sibling relative risk, λ_S, regardless of the underlying genetic model. The power is calculated as a function of the difference in the expected values of the IBD sharing probabilities (Z_0, Z_1, Z_2) if the disease is linked to the marker locus and the values expected under the null hypothesis of no linkage ($Z_0 = 0.25$, $Z_1 = 0.5$, $Z_2 = 0.25$). As we saw above, the expected distribution of (Z_0, Z_1, Z_2) can be written as simple functions of λ_S and the disease–marker locus recombination fraction θ.

For reference, Figure 11.3 (Risch, 1990b) shows the power to detect linkage between a perfectly informative marker locus and a single disease locus that accounts for all the observed familial risk as a function of sibling relative risk λ_S and the recombination fraction between the trait and marker locus for three sample sizes. The

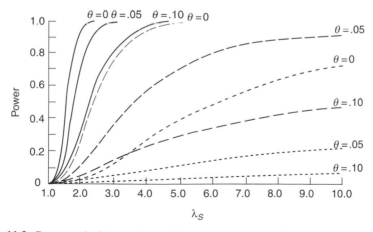

Figure 11.3. Power calculations for MLS sibpair analysis. (From N. Risch, Linkage strategies or genetically complex traits. II. The power of affected relative pairs. American Journal of Human Genetics, 46:229–241 (1990), © 1990. Reproduced with permission of the publisher, The University of Chicago Press.)

dramatic effect of recombination on power is easily seen, particularly for moderate ($N = 100$) and small ($N = 40$) sample sizes, even when λ_S is quite large.

Holmans (1993) showed higher power for the restricted-triangle method described above compared with the unrestricted MLS test for both dominant and recessive models. This gain in power was especially marked for cases of low marker informativeness and/or high recombination fraction between the marker and disease locus ($\theta = 0.1$). Holmans also examined the utility of genotyping parents in sibpair analyses as a function of marker heterozygosity. The results indicated a ratio of required sample size of as much as 1.5 (i.e., not genotyping parents required 50% more pairs) when the marker polymorphism is low. If efficiency is measured by the number of individuals needing to be genotyped, then genotyping parents (which requires twice as much genotyping) could be considered to be inefficient, particularly with markers with heterozygosities of over 0.5. This of course assumes that the desired number of sibpairs needed without genotyping parents can be easily ascertained and that the genotyping error rate is extremely low. Power for the case of interval mapping of a disease locus has been considered by Hauser et al. (1996), who derived conclusions regarding the genotyping of parents and other relatives similar to those of Holmans but with the increased power resulting from a multipoint analysis. It is clear that in designing a study to map a complex disease gene, one should consider the estimated magnitude of the effect of interest (as measured by λ_S), the marker density, the expected genotyping error rate, the availability of parents or other relatives for genotyping, and the relative importance of type I and type II errors. To aid in the design of such studies, the DESPAIR (design of sibpair) program, a part of the statistical analysis for genetic epidemiology (SAGE) package, can be used.

In some situations, however, sibpairs may not be the most efficient sampling strategy for mapping genes involved in complex disease. For example, Goldgar and Easton (1997) compared a variety of nuclear family sampling strategies in an analytical study of a number of two-locus disease models. For the majority of situations examined, affected sibling trios were found to be a more efficient design for mapping disease loci than sibpairs. This was especially true when one of the two loci had been localized/identified and linkage or mutation data were available and incorporated in the analysis. Even accounting for the increased difficulty in ascertaining affected sibling trios, this design (or one more extreme) still proved superior to sibpairs and should be considered for any gene localization effort. Sengul et al. (2001) reviewed and compared a variety of ASP statistics applied to sibship sizes of five affected siblings for a variety of different models. They conclude that the S_{all} statistic performs very well in a variety of situations; however, there can be differences in performance based on the level of informativeness of the marker as well as on the distance between the marker and the disease locus.

Determining the power of a proposed study is clearly important in making decisions about the ability of the study to result in a useful conclusion, and these studies must be done prior to the start of the study. However, it is extraordinarily difficult to perform these power studies. Such studies require knowledge about the overall genetic effect to be observed as well as the role of individual genes, their interactions with each other, and interactions with environmental factors. In addition, particular family structures must be assumed, even before the families are collected; thus the number of ASPs, ARPs, and extended families achieved often is not easy to estimate in advance. The discussion below provides a simple example of power considerations employing nonparametric linkage analysis.

Example of Testicular Cancer. As part of an international effort to map susceptibility loci for testicular cancer, Easton (unpublished) calculated the expected LOD scores for the available sample under a variety of different models for the number and type of loci influencing the trait. All models were constrained to fit the observed sibling risk (λ_S) of 8.0 (e.g., Goldgar et al., 1994) and included one, two, or four loci acting additively or multiplicatively. These models were chosen to span a reasonable range of possibilities, although there are few data that *a priori* favor one model over another. The effects of intermarker distance and heterozygosity were also examined. The data consisted of 77 affected sibpairs, 2 affected half-siblings, 15 affected uncle–nephew pairs, and 6 affected cousin pairs. Table 11.8 shows the expected LOD scores for each situation. While these analyses were done using a parametric analysis that assumed a rare autosomal dominant model with low penetrance, similar conclusions would apply with any of the methods discussed so far in this chapter.

Table 11.8 shows immediately that there is a dramatic decrease in the expected lod score as a function of the number of loci involved. This is to be expected, since

TABLE 11.8. Expected LOD Scores for Mapping Testicular Cancer Susceptibility Loci Using Affected Pairs Under a Variety of Genetic Models and Genome Search Conditions

Marker Density (cM)[a]	Marker Heterozygosity (%)[b]	Single Gene	Heterogeneity[c] Two Genes	Heterogeneity[c] Four Genes	Multiplicative[d] Two Genes	Multiplicative[d] Four Genes
0	100	14.4	3.3	0.8	5.8	1.9
10	90	7.8	2.1	0.4	3.3	1.2
(Multipoint)	80	5.8	1.4	0.3	2.6	0.9
20	90	5.0	1.2	0.3	2.2	0.8
(Multipoint)	80	3.6	0.9	0.2	1.7	0.6
20	90	3.1	0.7	0.2	1.4	0.5
(Two-point)	80	2.1	0.5	0.1	1.0	0.4

[a]Spacing between genetic markers: 0 cM refers to the situation of a disease gene that is a known candidate locus. For the other situations, the disease locus is assumed to lie midway between the two markers. Either a multipoint analysis using the two flanking markers simultaneously or a two-point analysis between the disease locus and each marker is performed.
[b]Informativeness of the marker (i.e., the proportion of individuals heterozygous at a given marker locus).
[c]In the heterogeneity models, most affected pairs are caused by only one of the two loci. The locus specific risks are 4.5 and 2.75 for the two- and four-locus models, respectively.
[d]In the multiplicative models, most pairs are the result of high-risk alleles at two or more loci.

the effect of any single locus is reduced. We note, however, that, in general, if the loci are acting multiplicatively, there is higher power (as measured by the expected LOD score) than if they are acting in an additive fashion, even though in the former case the sibling relative risk due to an individual locus will be reduced (although the relative contribution will be the same). The reason for this was alluded to earlier, when it was observed that in multiplicative models a higher fraction of pairs would be expected to be segregating risk alleles at multiple disease loci, while in additive models each pair may be due to only one of the set of disease susceptibility loci.

It is also apparent that moving from a completely informative marker completely linked to the disease locus ($\theta = 0$) to the more realistic genomic search situation of less than perfect markers with a marker density of 10–20 cM also reduces the expected LOD score considerably. Assuming the usual criterion of a LOD score of 3.0 for detection of linkage (note that an expected LOD score of 3.0 then corresponds to 50% power), one can see that under a realistic genomic search situation, for two or more loci contributing to disease susceptibility, it is unlikely that this sample is of sufficient size to detect linkage to any locus.

Nonparametric Quantitative Trait Linkage Analysis

Early Methods. The first person to address the problem of linkage with a quantitative trait in humans was Penrose (1938), who developed a method of testing by means of the interaction between the marker genotype and the quantitative phenotype in sibpairs from particular parental matings. Penrose's linkage test was designed for ordinal-level phenotypes as opposed to truly continuous ones. Lowry and Schultz (1959) looked at methods for detecting associations between continuous traits and marker loci, while Hill (1975) used a nested analysis-of-variance design (with marker genotype nested within sibship) for detecting and estimating linkage with a quantitative trait. One of the first nonparametric approaches for sibpair mapping of quantitative traits was proposed by Smith (1975).

A method quite different from those discussed above, proposed a generation ago by Haseman and Elston (1972), is still among the most commonly applied methods for examining linkage between a quantitative trait and a marker locus. Haseman and Elston derived the joint IBD distribution for a marker locus and a hypothesized locus determining a quantitative trait in terms of the recombination fraction between the two loci. They used these results to develop a method based on the regression of the squared sibpair difference for the trait on their estimated genetic correlation at the marker locus. Equation 11.7 is the basic relation underlying the Haseman–Elston method. Assuming that the dominance variance is negligible, we have

$$E[Y_j \mid I_{mj}] = \alpha + \pi_{mj}\beta \tag{11.7}$$

where $Y_j = (X_{1j} - X_{2j})^2$ is the squared sibpair difference of the jth sibpair, I_{mj} is the marker information on the jth family, π_{mj} is the estimated proportion of alleles shared IBD by the jth sibpair based on the marker information, and α and β are the coefficients of the linear regression and are equal to

$$\alpha = \sigma_e^2 + 2[\theta^2 + (1 - \theta^2)]\sigma_g^2 \qquad \beta = -2(1 - 2\theta)^2\sigma_g^2 \qquad (11.8)$$

where σ_e^2 and σ_g^2 are the respective environmental and genetic variances of the quantitative trait under study and θ is the recombination fraction between the marker locus and the hypothesized QTL.

A significantly negative estimate of β, the slope of the regression line, indicates linkage between the marker locus and the locus that influences the trait. As before, the basic idea underlying the method is that sibpairs who are IBD at a marker locus will be phenotypically similar for traits influenced by a nearby linked gene. Conversely, sibpairs who do not share genes IBD at the linked marker will tend to be phenotypically different. Originally, the method required IBD status at the marker locus to be determined with certainty, but now the method is generalized to estimate the proportion of alleles shared IBD using maximum-likelihood methods. Amos et al. (1990) have derived computational algorithms for calculating the proportion of alleles at a marker locus shared IBD for an arbitrary relative pair, conditional on the observed marker data in all typed individuals in a pedigree. Amos et al. (1989) also proposed use of a weighted-least-squares (WLS) approach to compensate for the failure of the variance of squared trait values to be constant for each value of IBD sharing when linkage between the trait and marker locus is present, a circumstance that violates an assumption implicit in the use of the regression approach of Haseman and Elston. Amos showed that the use of the WLS approach resulted in higher power for the detection of linkage than the ordinary Haseman–Elston test. This general regression-based approach is implemented in the SIBPAL program of the SAGE package. Recently the Haseman–Elston test has been revised to incorporate additional trait information in the sibpair comparison (Elston et al., 2000).

Although this method was primarily designed for use with quantitative traits, it can also be adapted for use with disease phenotypes by one of two means: transforming disease status to a quantitative susceptibility measure based on sex, age at onset, or other factors (Dawson et al., 1990) or testing for estimated IBD proportion $= 0.5$ in affected sibpairs, but using estimated IBD sharing rather than relying solely on data from pairs with unambiguous IBD status, as done in the simple means test described earlier.

Multipoint Methods. Another method for analyzing quantitative trait loci that uses the concept of IBD is the multipoint IBD method (MIM) proposed by Goldgar (1990). This method differs from the Haseman–Elston procedure in several ways. Most importantly, the method uses sibships, rather than sibpairs, thus avoiding the problems of using multiple, nonindependent sibpairs from the same sibship. Second, as the name implies, the method considers the IBD sharing in a chromosomal region defined by a set of linked genetic markers rather than IBD at a single locus. This method uses a variance-partitioning approach in which the parameter of interest is the proportion of the total genetic variance of the trait that is due to a locus (or loci) in the region spanned by the genetic marker studies. This parameter, P, is estimated by means of a grid search of possible values and is tested by comparing the likelihood of the data under the "best" value of P compared to the likelihood under the null hypothesis $P = 0$. Like all such likelihood ratio tests, the test statistic defined by minus twice the natural logarithm of this likelihood ratio has an approximately chi-squared distribution with one degree of freedom. The method assumes that the dominance variance of the trait V_D is small relative to the additive variance. A more important assumption is that the investigator has some knowledge of the magnitude of V_A, the total additive genetic variance of the quantitative trait under study; thus it is not a "true" nonparametric method. However, as we indicated above, this quantity is easily estimable from the quantitative trait data on parents and offspring via the parent–offspring correlation or, if parental data are unavailable, through the sibling–sibling correlation. Simulation studies (Goldgar, 1990) have shown that this method has higher power than the Haseman–Elston procedure and that the increase in power resulted both from the use of the entire sibship in a statistically meaningful way and from the use of multiply linked markers. An application of this method to a simulated problem involving the elucidation of major genes for a series of related quantitative traits can be found in Lewis and Goldgar (1995).

More recently other investigators have derived interval or multipoint mapping approaches in sibpairs, most of which represent multipoint extensions of the Haseman–Elston method (Fulker and Cardon, 1994; Fulker et al., 1995; Kruglyak and Lander, 1995; Olson, 1995b; Elston et al., 2000). As described in Chapter 9, variance components methods, especially as implemented in the package SOLAR (Almasy and Blangero, 1998), have become very popular tools for modeling genetic effects. As expected, all these efforts demonstrate the higher power for detecting linkage that one obtains from multipoint analysis by making all pairs (or nearly all pairs) informative for determination of IBD status through construction of multi-locus haplotypes. As before, another advantage of the multipoint approach is that it provides some information on the location of the QTL relative to the marker map; this information cannot be estimated using a series of single-marker analyses.

Power and Sampling Considerations for Mapping Quantitative Trait Loci

Just as the power to detect disease loci in affected sibpairs depends on the sibling relative risk λ_S, the power for detecting quantitative trait loci depends on the degree to which the trait is genetically determined (i.e., the trait heritability h^2) and the proportion of this heritability that is due to the specific trait locus to be mapped. Studies have shown that using a random sampling approach and multipoint analysis, these methods can detect, at best, trait loci that account for about 30% of the total phenotypic variance of the trait (Goldgar, 1990; Goldgar and Oniki, 1992; Williams and Blangero, 1999). Given a single trait locus that accounts for a significant fraction of the overall trait variance, it has been shown that the power to detect this locus depends to some extent on the nature of the remaining sources of variation of the trait. For example, Risch and Zhang (1996) showed that the power to detect quantitative trait loci is increased when there is polygenic genetic variation in addition to the major trait locus. That is, it is easier to map a quantitative trait locus that is responsible for 40% of the total phenotypic variance of the trait when most of the remaining 60% arises from other genetic loci rather than when all the remaining variation is due to random environmental effects. Moreover, it has been shown (Goldgar and Oniki, 1992) that any factor that increases the sibling similarity, whether it is additional genetic effects or common environmental effects, results in higher power for QTL detection.

While the quantitative trait methods described thus far, by and large, assume random sampling of sibpairs (or sibships) from the population under study, several investigators have noted that this may not be the most powerful sampling strategy for the identification of loci influencing a quantitative trait. Just as in the case of disease mapping, where the strategy is to sample pairs of affected individuals rather than unaffected, it seems reasonable, in certain cases, to expect to derive advantages from a similar strategy of sampling sibpairs or sibships in which at least one member of the pair has an extreme trait value. The first to address this issue for quantitative traits were Carey and Williamson (1991), who proposed sampling a proband with an extreme value of the trait under study and performing a regression analysis using the other sibling's trait value as the dependent variable and the sibling's IBD sharing status with the proband at the marker locus as the independent variable.

Another approach for increasing the power of these methods is to select sibpairs who are both concordant for extreme values of the trait (e.g., both in the upper or lower 10%) (Cardon and Fulker, 1994) or, as an alternative, to select sibpairs who are "extremely discordant," that is, at opposite ends of the quantitative trait distribution. Risch and Zhang (1995, 1996) showed that among these three choices (single proband with extreme value; extremely concordant pairs, extremely discordant pairs), the extremely discordant sibpair approach (EDSP) was the most powerful and was significantly better than random or single selection for a variety of genetic models. They concluded that mapping of a trait locus that controlled only 10% of the total variation was possible with a reasonable number of extremely discordant sibpairs using the EDSP method.

The major hurdle in this selected sibpair approach is in the difficulty of obtaining such pairs. For example, in their analysis, Risch and Zhang (1996) considered the case of a trait in which 60% of the variance is due to genetic factors composed of 20% due to a dominant major locus with a frequency of the dominant allele of 0.1, with the residual genetic component of 40% due to many genes of smaller effect. Assuming that the major quantitative trait locus is midway between two markers located 20 cM apart (i.e., 10 cM from both markers) and that these markers are informative 80% of the time, it can be shown that a sample size of 212 extremely discordant sibpairs would be required to detect this locus at a significance level of 0.00001 with 80% power. The EDSPs were defined as one member of the pair in the upper 10% and the other in the lower 10% of the overall trait distribution. However, to obtain 212 sibpairs that meet this criterion, an expected total of 12,366 sibpairs would have to be screened for the quantitative trait of interest. If one loosens the criterion to EDSPs in the upper 10% and lower 30%, the sample size required increases to 333, but the number of pairs to be screened is now reduced to 3719. The efficiency of any selection-based approach will depend on the relative costs of sample ascertainment and phenotyping versus the cost of genotyping the selected individuals. If getting the family material and making the phenotypic measurement are difficult and expensive, it might be wasteful to screen such families to obtain the required number of EDSPs when a smaller number of families selected at random or through a single proband would suffice.

EXAMPLES OF APPLICATION OF SIBPAIR METHODS FOR MAPPING COMPLEX TRAITS

So far in this chapter, we have concentrated on presenting an overview of the methods developed to utilize nonparametric linkage methods in mapping genes contributing to complex human qualitative and quantitative traits. However, the reader might well ask whether these methods and sampling schemes, for all their theoretical advantages, have successfully identified genes for any complex human traits through screening of random genomic markers. There has been great progress in several disorders using this approach, primarily in insulin-dependent (juvenile) diabetes, for which in addition to the candidate loci of HLA and insulin, a genomic search for additional IDDM susceptibility locus revealed evidence for at least three other genes (Davies et al., 1994). Of course, many of these linkages have not yet been replicated in independent datasets. All these loci were characterized by relatively small effects (locus-specific λ_S values 1.3–1.5). Given that IDDM is characterized by an overall sibling relative risk of 15, even under a multiplicative model these loci account for only about 60–65% of the overall familial aggregation.

An interesting example of the use of affected sibpair methodology for a disease with a smaller overall familial component but a higher prevalence in the population is provided by the analysis of adult-onset (non-insulin-dependent) diabetes (NIDDM) by Hanis et al. (1996). This study utilized a primary sample of 330 Mexican-American affected sibpairs typed for 490 genetic markers distributed throughout the genome and found evidence (MLS = 3.2) of a susceptibility locus on the terminal portion of the long arm of chromosome 2, designated *NIDDM1*. The maximum effect of this locus (assuming no recombination between the linked marker and the disease) was a locus-specific λ_S value of 1.37. Since the overall sibling relative risk for NIDDM is 2.8, the authors estimate that this locus accounts for 30% of the overall familial risk, assuming a multiplicative model for the interaction between all the loci involved. The original result provided a large region of linkage, not amenable to fine mapping. Using a method that conditions the linkage result at *NIDDM1* on linkage evidence on chromosome 15, the investigators were able to substantially narrow the *NIDDM1* region (Cox et al., 1996) such that association methods (Chapter 12) could be applied. Finally, *CALPAIN 10*, a gene appearing to account for both the linkage and association evidence, was identified as the *NIDDM1* gene (Horikawa et al., 2000).

Another example of a genomewide search for susceptibility loci for a complex disease involves several parallel genomic searches for susceptibility loci for multiple sclerosis (MS), the results of which were published in *Nature Genetics* (Ebers et al., 1996; Haines et al., 1996; Sawcer et al., 1996). Although all three studies found evidence for a susceptibility locus in (or near) the HLA complex (which had been suggested by earlier association studies and biological plausibility), no additional locus was found by more than one of the three studies. Risch (1987) has estimated that less than 30% of the familial aggregation in MS is due to HLA, so it is somewhat surprising that no consistent additional susceptibility loci were identified. This could be because all additional loci are of too small an effect to be consistently detected or, alternatively, because the familial aggregation has nongenetic (e.g., viral) etiology.

For quantitative traits, the success stories for human mapping are somewhat harder to come by; however, the advent of the SOLAR linkage analysis package has allowed for efficient and powerful studies of many quantitative traits, which will certainly lead to identification of genes for complex traits. One of the first reports of linkage to a quantitative trait was reported by Cardon et al. (1994), who reported linkage to chromosome 6 markers for test scores underlying reading ability in a sample of sibships ascertained for a proband with dyslexia. This result was recently replicated in an independent sample by Grigorenko et al. (1997, 2000). SOLAR has been applied to genetic linkage studies of a variety of cardiovascular-related traits, such as the various lipid fractions (Almasy et al., 1999; Rainwater et al., 1999) and obesity (Comuzzie et al., 1997).

ADDITIONAL CONSIDERATIONS IN NONPARAMETRIC LINKAGE ANALYSIS

The application of simple nonparametric linkage methods to complex traits has met with mixed success and thus has increased demand for statistical methods that allow a more comprehensive modeling of the etiology of complex traits. It is clear that etiologic heterogeneity among families affected by a complex disease must be acknowledged to increase power to detect one or a few genes for these diseases. We will discuss two nonparametric linkage analysis methods that begin to address the problem of heterogeneity in this chapter. The weighted pairwise correlation (WPC) allows for heterogeneity by incorporating covariates in the analysis. We also briefly discuss methods for two-locus linkage analysis. The topic of gene–gene interaction and gene–environment interaction is explored in more depth in Chapter 14.

WPC Analysis

In 1994 Commenges described a new nonparametric statistic he termed the WPC statistic based on a score test of homogeneity among strata. In this particular case, the strata are defined by the marker genotypes within each family. If a marker is unlinked to a trait locus, the phenotypes should be homogeneous across these strata. However, if the marker is linked, such homogeneity will no longer exist because certain phenotypes will associate with certain genotypes. By evaluating the IBS status for all pairs of family members, the WPC statistic evaluates the null hypothesis of homogeneity and thus the null hypothesis of no linkage.

It should be noted that unlike the other methods described in this chapter, WPC considers all pairs of individuals, not just affected individuals. In fact, WPC cannot be applied to data on just affected individuals unless they have an associated quantitative trait that varies among affected individuals.

The derivation of the WPC statistic is quite complex, and the reader is referred to Commenges (1994; Commenges and Beurton-Aimar, 1999) for more detail. Unlike the ARP statistics, the WPC method was derived specifically for quantitative traits. Qualitative traits can also be evaluated using a rank or proportional hazard version of the statistic, assuming that some sort of ranking (e.g., by age) can be applied across all individuals. The statistic depends on three factors: the phenotype of each individual, the genotype of each individual, and the pedigree structure (e.g., relationships between individuals).

The WPC statistic for each family is

$$S_L = \sum_{i=1}^{n-1} \sum_{j=i+1}^{n} W_{ij} U_i U_j$$

where W_{ij} is a weight given to the relative pair consisting of individuals i and j. The weight is 0, 1, or 2, depending on the number of alleles shared IBS, and is centered by subtracting the average number of alleles shared by all pairs of the same relationship (e.g., all sibpairs, all cousin pairs, etc.).

The phenotype scores for individuals i and j (U_i, U_j) are the residuals after the expected phenotype has been subtracted under the null hypothesis of no linkage.

When linkage exists, the phenotypes will tend to be similar (e.g., both positive or both negative) if marker sharing exceeds the average ($w_{ij} > 0$). Thus S_L will be positive, and large values of S_L are indicative of linkage. The phenotypes will tend to be dissimilar (e.g., one positive, the other negative) if marker sharing is less than average ($w_{ij} < 0$), making S_L negative. To simplify the score and make it more robust, Commenges replaced U_i and U_j with R_i and R_j, the ranks of the phenotype scores. This resulting weighted rank pairwise correlation (WRPC) statistic is

$$\text{WRPC} = \frac{S_R - E(S_R)}{[\text{var}(S_R)]^{1/2}}$$

where S_R is analogous to S_L except that the ranks of the phenotype scores are used, $E(S_R)$ is the expectation of S_R, and $\text{var}(S_R)$ is the variance of S_R.

The overall WRPC statistic across families is simply the sum across all families (p):

$$\sum_{p=1}^{n} \text{WRPC}_p$$

The WRPC method has several advantages. As with the ARP methods, it does not require any specification of the trait genetic model, making it a desirable method for genetically complex traits. Unlike all ARP methods, it is derived for quantitative traits and has the greatest power when it is used for this purpose. The use of the ranks, however, allows an easy conversion for qualitative traits (assuming the phenotype can be easily ranked), highlighting the flexibility of WPC. In addition, this method inherently allows the incorporation of modifying environmental factors. This is accomplished in forming the phenotype residuals. This approach is quite similar to the method of incorporating covariates in the Haseman–Elston sibpair method (Chapter 9). In fact, if only pairs of siblings are considered, the WPC method simplifies to very nearly the Haseman–Elston sibpair method. The strength of the WPC method is that it uses all relatives, not just those who are affected, thus gaining power by increasing the number of pairs examined. Finally, the calculation of the WRPC statistic is relatively fast.

Some of the strengths of the WPC method are also its weaknesses. For example, the use of both affected and unaffected individuals, while increasing the number of pairs examined, also increases the chance that heterogeneity or misdiagnosis will introduce unwanted variation. In addition, the reliance on ranks requires a ranking scheme. The most commonly used scheme is to rank by age within each trait class (e.g., affected and unaffected). However, this requires both an assumption that age is important and some ranking function (usually a simpler linear function), data that may not be available on many members of the pedigree.

Because WPC uses centered residuals, more than a single pair of relatives is required. With only a single pair, there is no variation, thus no residual, and therefore the statistic is zero. Consequently families with only a single sibpair cannot be used. Additionally the current implementation lumps more distant relationships together into a single category, making the analysis less accurate. As a result, very large pedigrees may produce unexpected and uninterpretable results.

The WPC statistic appears to be anticonservative. That is, when no linkage exists, WPC exceeds the value corresponding to a nominal p-value of 0.05 more often than 5% of the time. Numerous simulation studies (Rogus et al., 1995) indicate that the inflation of this type I error can be as high as 10%.

The WRPC is restricted to examining pairs of individuals at one time. The WRPC method can be used in several situations:

1. When pedigrees of moderate size are available. Pedigrees with only a single sibpair or very large pedigrees cannot be examined with any confidence.

2. When many relatives other than siblings are available in a dataset for a genetically complex trait. Each pedigree must have several relative pairs.

3. When a quantitative trait is being examined. The WPC statistic allows the use of trait information on all individuals without having to arbitrarily define "affected" or "unaffected."

4. When covariates must be included in the analysis. Qualitative covariates can be included by converting them to a linear scale, while a quantitative measure can be used directly.

See Table 11.6 for a comparison of the properties, advantages, and disadvantages of WRPC and other ARP methods.

Two Trait Loci. In this situation, we assume that there are two (or more) loci contributing to the trait being studied, and we want to use genetic marker information to localize these genes. Most often at least one trait locus has been localized, and we want to use this information to help us in the genomic search for one or more additional loci.

The two-locus sibpair method (Dizier and Clerget-Darpoux, 1986) and the MASC (marker association segregation chi-square) method (Clerget-Darpoux et al., 1988; Dizier et al., 1994) were designed to test two candidate loci or explore the interactions between two known loci, rather than to find unidentified genes. For this latter purpose, Knapp et al. (1994) devised a series of two-marker tests and compared the power of these tests based on joint consideration of the IBD distribution at two loci. They assumed that IBD status could be determined with certainty, that there was no recombination between disease and marker, and that the two loci contributed equally to the disease. Under these assumptions, they found that two-locus tests (particularly a test based on the mean two-locus sharing) could provide a substantial increase in power for a variety of models. Cordell et al. (1995) extended the work of Risch and Holmans to define an MLS test for two loci that did not require identity by descent to be known with certainty and incorporated two-locus restrictions on the estimated parameters. They applied this method to sibpairs with IDDM, examining the effect of two loci, while controlling for the effect of HLA at this locus. The results showed that incorporation of the HLA status into the analysis provided increased power for detecting the effect of the additional loci. Recently, in a more general context, Goldgar and Easton (1997), Farrall (1997), and Cox et al. (1999) showed that incorporating linkage and mutation data at known loci provided increased power for a genomic search for other loci. All the aforementioned two-locus methods are typically useful when one or more loci for a disease have already been identified and the aim is to discover additional susceptibility loci. At the present time, they are not generally useful for the simultaneous search for multiple unknown disease loci because of the extreme number of comparisons that have to be made.

SOFTWARE AVAILABLE FOR NONPARAMETRIC LINKAGE ANALYSIS

Many of the methods mentioned above are simple enough to be analyzed using standard statistical packages, or would be if the IBD distribution could be determined with certainty. For others, no user-friendly programs that implement the method are generally available. An excellent reference for linkage analysis software is the list developed by Wentian Li at Rockefeller University (http://linkage.rockefeller.edu/soft/list.html). This website provides direct links or information to obtain most of the statistical analysis software available.

SUMMARY

We have introduced the reader to both the theoretical and practical aspects of the use of sibpairs and more generally relative pairs in the search for genes that either confer increased susceptibility to a qualitative trait (e.g., a disease) or contribute to variation in human quantitative traits. In addition, we have provided a basic understanding of the variety of models of familial aggregation and have shown how these models can be distinguished by means of available epidemiological and statistical data. Other analysis strategies and sampling designs, as alluded to here, may be preferred over relative pair designs and the corresponding nonparametric methods, depending on the characteristics of the phenotype being studied. In particular, methods analogous to those presented here but designed for use in extended pedigrees are presented in Chapter 10. Because, in terms of both theory and implementation, this area is moving forward at a rapid pace, we have concentrated on the underlying principles involved in all these methods rather than specific details of any single approach. Moreover, we have pointed out some of the practical limitations of nonparametric linkage analysis methods and provided the reader wishing to map genes for complex traits some idea of when these methods would be appropriate. These cautions aside, sibpair, relative pair, and related methods are today the most commonly used vehicles for mapping complex human traits. Given the rapid advances in human genome technology and the continuing development of powerful statistical methods, it is uncertain how long this will continue to be the case.

REFERENCES

Abecasis GR, Cherny SS, Cookson WO, Cardon LR (2002): Merlin—Rapid analysis of dense genetic maps using sparse gene flow trees. Nat Genet 30:97–101.

Abreu PC, Hodge SE, Greenberg DA (2002): Quantification of type I error probabilities for heterogeneity LOD scores. Genet Epidemiol 22:156–169.

Almasy L, Blangero J (1998): Multipoint quantitative-trait linkage analysis in general pedigrees. Am J Hum Genet 62:1198–1211.

Almasy L, Hixson JE, Rainwater DL, Cole S, Williams JT, Mahaney MC, VandeBerg JL, Stern MP, MacCluer JW, Blangero J (1999): Human pedigree-based quantitative-trait-locus mapping: Localization of two genes influencing HDL-cholesterol metabolism. Am J Hum Genet 64:1686–1693.

Amos CI, Dawson DV, Elston RC (1990): The probabilistic determination of identity-by-descent sharing for pairs of relatives from pedigrees. Am J Hum Genet 47:842–853.

Amos CI, Elston RC, Wilson AF, Bailey-Wilson JE (1989): A more powerful robust sib-pair test of linkage for quantitative traits. Genet Epidemiol 6:435–449.

Anderson JL, Hauser ER, Martin ER, Scott WK, Ashley-Koch A, Kim KJ, Monks SA, Haynes CS, Speer MC, Pericak-Vance MA (1999): Complete genomic screen for disease susceptibility loci in nuclear families. Genet Epidemiol 17:S467–S472.

Bishop DT, Williamson JA (1990): The power of identity-by-state methods for linkage analysis. Am J Hum Genet 46:254–265.

Blackwelder WC, Elston RC (1985): A comparison of sib-pair linkage tests for disease susceptibility loci. Genet Epidemiol 2:85–98.

Cardon LR, Fulker DW (1994): The power of interval mapping of quantitative trait loci, using selected sibpairs. Am J Hum Genet 55:825–833.

Cardon LR, Smith SD, Fulker DW, Kimberling WJ, Pennington BF, DeFries JC (1994): Quantitative trait locus for reading disability on chromosome 6. Science 266:276–279.

Carey G, Williamson J (1991): Linkage analysis of quantitative traits: Increased power by using selected samples. Am J Hum Genet 49:786–796.

Clerget-Darpoux F, Babron MC, Prum B, Lathrop GM, Deschamps I, Hors J (1988): A new method to test genetic models in HLA associated diseases: The MASC method. Ann Hum Genet 52(Pt 3):247–258.

Commenges D (1994): Robust genetic linkage analysis based on a score test of homogeneity: The weighted pairwise correlation statistic. Genet Epidemiol 11:189–200.

Commenges D, Beurton-Aimar M (1999): Multipoint linkage analysis using the weighted-pairwise correlation statistic. Genet Epidemiol 17(Suppl 1):S515–S519.

Comuzzie AG, Hixson JE, Almasy L, Mitchell BD, Mahaney MC, Dyer TD, Stern MP, MacCluer JW, Blangero J (1997): A major quantitative trait locus determining serum leptin levels and fat mass is located on human chromosome 2. Nat Genet 15:273–276.

Cordell HJ, Todd JA, Bennett ST, Kawaguchi Y, Farrall M (1995): Two-locus maximum lod score analysis of a multifactorial trait: Joint consideration of IDDM2 and IDDM4 with IDDM1 in type I diabetes. Am J Hum Genet 57:920–934.

Cox DW, Spielman RS (1989): The insulin gene and susceptibility to IDDM. Genet Epidemiol 6:65–69.

Cox NJ, Frigge M, Donnelly P, Kong A (in press): A semiparametric model allowing exact calculations of likelihoods implemented in GENEHUNTER. Genet Epidemiol.

Cox NJ, Frigge M, Nicolae DL, Concannon P, Hanis CL, Bell GI, Kong A (1999): Loci on chromosomes 2 (NIDDM1) and 15 interact to increase susceptibility to diabetes in Mexican Americans. Nat Genet 21:213–215.

Davies JL, Kawaguchi Y, Bennett ST, Copeman JB, Cordell HJ, Pritchard LE, Reed PW, Gough SCL, Jenkins SC, Palmer SM, Balfour KM, Rowe BR, Farral M, Barnett AH, Bain SC, Todd JA (1994): A genome-wide search for human type 1 diabetes susceptibility genes. Nature 371:130–136.

Davis S, Schroeder M, Goldin LR, Weeks DE (1996): Nonparametric simulation-based statistics for detecting linkage in general pedigrees. Am J Hum Genet 58:867–880.

Dawson DV, Kaplan NL, Elston RC (1990): Extensions to sib-pair linkage tests applicable to disorders characterized by delayed onset. Genet Epidemiol 7:453–466.

Dib C, Faure S, Fizames C, Samson D, Drouot N, Vignal A, Millasseau P, Marc S, Hazan J, Seboun E, Lathrop M, Gyapay G, Morissette J, Weissenbach J (1996): A comprehensive genetic map of the human genome based on 5,264 microsatellites. Nature 380:152–154.

Dizier MH, Babron MC, Clerget-Darpoux F (1994): Interactive effect of two candidate genes in a disease: Extension of the marker-association-segregation χ^2 method. Am J Hum Genet 55:1042–1049.

Dizier MH, Clerget-Darpoux F (1986): Two-disease locus model: Sibpair method using information on both HLA and Gm. Genet Epidemiol 3:343–356.

Ebers GC, Kukay K, Bulman DE, Sadovnick AD, Rice G, Anderson C, Armstrong H, Cousin K, Bell RB, Hader W, Paty DW, Hashimoto S, Oger J, Duquette P, Warren S, Gray T, O'Connor P, Nath A, Auty A, Metz L, Francis G, Paulseth JE, Murray TJ, Pryse-Phillips W, Risch N (1996): A full genome search in multiple sclerosis. Nat Genet 13:472–476.

Elston RC, Buxbaum S, Jacobs KB, Olson JM (2000): Haseman and Elston revisited. Genet Epidemiol 19:1–17.

Elston RC, Stewart J (1971): A general model for the genetic analysis of pedigree data. Hum Hered 21:523–542.

Farrall M (1997): Affected sibpair linkage tests for multiple linked susceptibility genes. Genet Epidemiol 14:103–115.

Fisher R (1935): The detection of linkage with recessive abnormalities. Ann Eugen 6:339–351.

Fulker DW, Cardon LR (1994): A sib-pair approach to interval mapping of quantitative trait loci. Am J Hum Genet 54:1092–1103.

Fulker DW, Cherny SS, Cardon LR (1995): Multipoint interval mapping of quantitative trait loci. Am J Hum Genet 56:1224–1233.

Goldgar DE (1990): Multipoint analysis of human quantitative genetic variation. Am J Hum Genet 47:957–967.

Goldgar DE, Easton DF (1997): Optimal strategies for mapping complex diseases in the presence of multiple loci. Am J Hum Genet 60:1222–1232.

Goldgar DE, Easton DF, Cannon-Albright LA, Skolnick MH (1994): Systematic population-based assessment of cancer risk in first-degree relatives of cancer probands. J Natl Cancer Inst 86:1600–1608.

Goldgar DE, Oniki RS (1992): Comparison of multipoint identity-by-descent method with parametric multipoint linkage analysis for mapping quantitative traits. Am J Hum Genet 50:598–606.

Goldin LR, Weeks DE (1993): Two-locus models of disease: Comparison of likelihood and nonparametric linkage methods. Am J Hum Genet 53:908–915.

Greenberg DA, Hodge SE, Vieland VJ, Spence MA (1996): Affecteds-only linkage methods are not a panacea. Am J Hum Genet 58:892–895.

Grigorenko EL, Wood FB, Meyer MS, Hart LA, Speed WC, Shuster A, Pauls DL (1997): Susceptibility loci for distinct components of developmental dyslexia on chromosomes 6 and 15. Am J Hum Genet 60:27–39.

Grigorenko EL, Wood FB, Meyer MS, Pauls DL (2000): Chromosome 6p influences on different dyslexia-related cognitive processes: Further confirmation. Am J Hum Genet 66:715–723.

Gudbjartsson DF, Jonasson K, Frigge ML, Kong A (2000): Allegro, a new computer program for multipoint linkage analysis. Nat Genet 25:12–13.

Gyapay G, Morissette J, Vignal A, Dib C, Fizames C, Millasseau P, Marc S, Bernardi G, Lathrop M, Weissenbach J (1994): The 1993–94 Genethon human genetic linkage map. Nat Genet 7:246–339.

Haines JL, Ter-Minassian M, Bazyk A, Gusella JF, Kim DJ, Terwedow H, Pericak-Vance MA, et al (1996): A complete genomic screen for multiple sclerosis underscores a role for the major histocompatability complex. Nat Genet 13: 469–471.

Haldane JBS (1934): Methods for the detection of autosomal linkage in man. Ann Eugen 6:26–65.

Haldane JBS, Smith CAB (1947): A new estimate of the linkage between the genes for color blindness and hemophilia in man. Ann Eugen 14:10–31.

Hall JM, Lee MK, Newman B, Morrow JE, Anderson LA, Huey B, King MC (1990): Linkage of early-onset familial breast cancer to chromosome 17q21. Science 250:1684–1689.

Hanis CL, Boerwinkle E, Chakraborty R, Ellsworth DL, Concannon P, Stirling B, Morrison VA, et al (1996): A genome-wide search for human non-insulin-dependent (type 2) diabetes genes reveals a major susceptibility locus on chromosome 2. Nat Genet 13:161–166.

Haseman JK, Elston RC (1972): The investigation of linkage between a quantitative trait and a marker locus. Behav Genet 2:3–19.

Hauser ER, Boehnke M (1998): Genetic linkage analysis of complex genetic traits by using affected sibling pairs. Biometrics 54:1238–1246.

Hauser ER, Boehnke M, Guo SW, Risch N (1996): Affected-sib-pair interval mapping and exclusion for complex genetic traits: Sampling considerations. Genet Epidemiol 13:117–137.

Hill AP (1975): Quantitative linkage: A statistical procedure for its detection and estimation. Ann Hum Genet 38:439–449.

Hodge SE (1981): Some epistatic two-locus models of disease. I. Relative risks and identity-by-descent distributions in affected sib pairs. Am J Hum Genet 33:381–395.

Hodge SE (1984): The information contained in multiple sibling pairs. Genet Epidemiol 1:109–122.

Hodge SE, Elston RC (1994): Lods, wrods, and mods: The interpretation of lod scores calculated under different models. Genet Epidemiol 11:329–342.

Holmans P (1993): Asymptotic properties of affected-sib-pair linkage analysis. Am J Hum Genet 52:362–374.

Horikawa Y, Oda N, Cox NJ, Li X, Orho-Melander M, Hara M, Hinokio Y, et al (2000): Genetic variation in the gene encoding calpain-10 is associated with type 2 diabetes mellitus. Nat Genet 26:163–175.

Jonasson F, Oshima E, Klintworth GK, Thonar EJM, Smith CF, Johannasson JH (1995): Macular corneal dystrophy in Iceland: A clinical, genealogical and immunohistochemical study of twenty-eight cases. Ophthalmology, under review.

Knapp M, Seuchter SA, Baur MP (1994): Two-locus disease models with two marker loci: The power of affected-sib-pair tests. Am J Hum Genet 55:1030–1041.

Kong A, Cox NJ (1997): Allele-sharing models: LOD scores and accurate linkage tests. Am J Hum Genet 61:1179–1188.

Kruglyak L, Daly MJ, Reeve-Daly MP, Lander ES (1996): Parametric and nonparametric linkage analysis: A unified multipoint approach. Am J Hum Genet 58:1347–1363.

Kruglyak L, Lander ES (1995): A nonparametric approach for mapping quantitative trait loci. Genetics 139:1421–1428.

Lander ES, Green P (1987): Construction of multilocus genetic linkage maps in humans. Proc Natl Acad Sci USA 84:2363–2367.

Lange K (1986): The affected sib-pair method using identity by state relations. Am J Hum Genet 39:148–150.

Lewis CM, Goldgar DE (1995): Screening for linkage using a multipoint identity-by-descent method. Genet Epidemiol 12:777–782.

Lowry DC, Schultz F (1959): Testing association of metric traits and marker genes. Ann Hum Genet 23:83–90.

Markianos K, Daly MJ, Kruglyak L (2001): Efficient multipoint linkage analysis through reduction of inheritance space. Am J Hum Genet 68:963–977.

Mitchell BD, Kammerer CM, Blangero J, Mahaney MC, Rainwater DL, Dyke B, Hixson JE, Henkel RD, Sharp RM, Comuzzle AG, VandeBerg JL, Stern MP, MacCluer JW (1996): Genetic and environmental contributions to cardiovascular risk factors in Mexican Americans. The San Antonio Family Heart Study. Circulation 94:2159–2170.

Morton NE (1955): Sequential tests for the detection of linkage. Am J Hum Genet 7:277–318.

Olson JM (1995a): Multipoint linkage analysis using sibpairs: An interval mapping approach for dichotomous outcomes. Am J Hum Genet 56:788–798.

Olson JM (1995b): Robust multipoint linkage analysis: An extension of the Haseman–Elston method. Genet Epidemiol 12:177–193.

Penrose L (1938): Genetic linkage in graded human characters. Ann Eugen 9:133–138.

Penrose LS (1935): The detection of autosomal linkage in data which consists of pairs of brothers and sisters of unspecified parentage. Ann Eugen 6:133–138.

Penrose LS (1953): The general purpose sib-pair linkage test. Ann Eugen 18:120–124.

Pericak-Vance MA, Bebout JL, Gaskell PC, Yamaoka LH, Hung WY, Alberts MJ, Walker AP, Bartlett RJ, Haynes CS, Welsh KA, Earl NL, Heyman A, Clark CM, Roses AD (1991): Linkage studies in familial Alzheimer's disease: Evidence for chromosome 19 linkage. Am J Hum Genet 48:1034–1050.

Pericak-Vance MA, Rimmler JB, Martin ER, Haines JL, Garcia ME, Oksenberg JR, Barcellos LF, Lincoln R, Goodkin DE, Hauser SL (2001): Linkage and association analysis of chromosome 19q13 in multiple sclerosis. Neurogenetics 3:195–201.

Rainwater DL, Almasy L, Blangero J, Cole SA, VanderBerg JL, MacCluer JW, Hixson JE (1999): A genome search identifes major quantitative trait loci on human chromosomes 3 and 4 that influence cholesterol concentrations in small LDL particles. Arteriosclerosis, Thrombosis & Vascular Biology 19:777–783.

Risch N (1987): Assessing the role of HLA-linked and unlinked determinants of disease. Am J Hum Genet 40:1–14.

Risch N (1990a): Linkage strategies for genetically complex traits. I. Multilocus models. Am J Hum Genet 46:222–228.

Risch N (1990b): Linkage strategies for genetically complex traits. II. The power of affected relative pairs. Am J Hum Genet 46:229–241.

Risch N (1990c): Linkage strategies for genetically complex traits. III. The effect of marker polymorphism on analysis of affected relative pairs. Am J Hum Genet 46:242–253.

Risch N, Zhang H (1995): Extreme discordant sib pairs for mapping quantitative trait loci in humans. Science 268:1584–1589.

Risch N, Zhang H (1996): Mapping quantitative trait loci with extreme discordant sib pairs: Sample size considerations. Am J Hum Genet 58:836–843.

Rogus JJ, Haines JL, Pericak-Vance MA, Harrington DP (1995): An extension of the weighted pairwise correlation statistic (WPC) to test for linkage in the presence of gene-environment interaction. Am J Hum Genet 57:A171.

Sawcer S, Jones HB, Feakes R, Gray J, Smaldon N, Chataway J, Robertson N, Clayton D, Goodfellow PN, Compston A (1996): A genome screen in multiple sclerosis reveals susceptibility loci on chromosome 6p21 and 17q22. Nat Genet 13:464–476.

Schmidt S, Shao Y, Hauser ER, Slifer SH, Martin ER, Scott WK, Speer MC, Pericak-Vance MA (2001): Life after the screen: Making sense of many p-values. Genet Epidemiol 21:S546–S551.

Sengul H, Weeks DE, Feingold E (2001): A survey of affected-sibship statistics for nonparametric linkage analysis. Am J Hum Genet 69:179–190.

Smith CAB (1975): A non-parametric test for linkage with a quantitative character. Ann Hum Genet 38:451–460.

Sobel E, Sengul H, Weeks DE (2001): Multipoint estimation of identity-by-descent probabilities at arbitrary positions among marker loci on general pedigrees. Hum Hered 52:121–131.

Spielman RS, Baker L, Zmijewski CM (1980): Gene dosage and suceptibility to insulin-dependent diabetes. Ann Hum Genet 44:135–150.

St George-Hyslop PH, Haines JL, Farrer LA, Polinsky RJ, Van Broeckhoven C, Goate AM, McLachlan DR, Orr H, Bruni AC, Sorbi S (1990): Genetic linkage studies suggest that Alzheimer's disease is not a single homogenous disorder. Nature 347:194–197.

Suarez BK, Rice JP, Reich T (1978): The generalized sib pair IBD distribution: Its use in the detection of linkage. Ann Hum Genet 42:87–94.

Weeks DE, Lange K (1988): The affected-pedigree member method of linkage analysis. Am J Hum Genet 42:315–326.

Weeks DE, Lange K (1992): A multilocus extension of the affected-pedigree-member method of linkage analysis. Am J Hum Genet 50:859–868.

Whittemore AS, Halpern J (1994): A class of tests for linkage using affected pedigree members. Biometrics 50:118–127.

Williams JT, Blangero J (1999): Power of variance component linkage analysis to detect quantitative trait loci. Ann Hum Genet 63:545–563.

Williamson JA, Amos CI (1990): On the asymptotic behavior of the estimate of the recombination fraction under the null hypothesis of no linkage when the model is misspecified. Genet Epidemiol 7:309–318.

CHAPTER 12

Linkage Disequilibrium and Association Analysis

EDEN R. MARTIN

INTRODUCTION

There are two primary approaches for mapping genes that either cause or increase susceptibility to human disease. The first is the linkage analysis approach, either a parametric logarithm of the odds (LOD) score analysis when the genetic model is known or model-independent affected relative pair analysis when the genetic model is unknown. The linkage approach has been discussed in detail elsewhere in this book. The second approach is to use association analysis. Association studies look for a significantly increased or decreased frequency of a marker allele, genotype, or haplotype with a disease trait than would be expected by chance if there were no association between marker(s) and phenotype. Association analysis is a useful and often necessary tool in identifying disease gene loci, particularly in genetically complex diseases.

Association between genotype and phenotype can be explained either by direct biological action of the polymorphism [e.g., the *APOE-4* allele in Alzheimer's disease (AD)] or by allelic association between the marker and a susceptibility gene. In this chapter we consider *allelic association* to be association occurring between alleles at different loci. Allelic association can occur between linked or unlinked loci. The term *linkage disequilibrium* often is used to refer to allelic associations between linked loci. Associations between alleles at unlinked loci decay rapidly, so allelic associations are typically only found between closely linked loci. For this reason, association studies can play a critical role in the analysis of genetically complex traits, in the evaluation of candidate gene loci, as well as in the fine mapping of a region once linkage studies have indicated a region of interest for follow-up analysis. Certainly, as the Human Genome Project and other initiatives identify and characterize each gene in the human genome as well as millions of

Genetic Analysis of Complex Diseases, Second Edition, Edited by Jonathan L. Haines and Margaret Pericak-Vance
Copyright © 2006 John Wiley & Sons, Inc.

single-nucleotide polymorphisms (SNPs), this approach will be widely used for the identification of complex disease genes.

LINKAGE DISEQUILIBRIUM

Classically, linkage disequilibrium refers to allelic associations that occur between alleles at different loci, whether or not they are linked (e.g., Lewontin, 1988; Weir, 1996). This is sometimes referred to as gametic disequilibrium, since typically linkage disequilibrium is used for associations between alleles from the same gamete. In the human genetics literature, linkage disequilibrium has been used to mean specifically allelic associations that occur between linked loci (Spielman and Ewens, 1996; Crow and Kimura, 1970), which is the definition that we use here.

Measures of Allelic Association

There are many different measures of linkage disequilibrium. All measure deviation from independent assortment of alleles at different loci. Table 12.1 gives the formulas for several common measures. Each depends on allele and haplotype frequencies (i.e., the haplotype frequency is the frequency with which two alleles occur on the same gamete) but in different ways. The disequilibrium coefficient directly measures the deviation of the haplotype frequency from the product of the allele frequencies. The disequilibrium coefficient is zero if there is no allelic association and positive or negative depending on whether the alleles tend to occur together more or less frequently than expected if they were independent. The disequilibrium coefficient can be difficult to interpret because its maximum value depends on allele frequencies. Thus, alternative measures that are standardized to lie between -1 and 1 are often used.

Two standardized measures that are commonly used are the correlation coefficient and Lewontin's D'. In a random-mating population, the count of the number of times that a particular allele at a locus is observed in a sample of unrelated individuals is distributed as a binomial random variable. Therefore, the disequilibrium coefficient D (Table 12.1) is proportional to the covariance of the allele counts at two

---

TABLE 12.1. Measures of Allelic Association for Alleles A and B at Different Loci

Disequilibrium coefficient	$D = \Pr(AB) - \Pr(A)\Pr(B)$
Correlation coefficient	$r^2 = D/\{\Pr(A)[1 - \Pr(A)]\Pr(B)[1 - \Pr(B)]\}$
Lewontin's D'	$D' = \begin{cases} D/\min\{\Pr(A)[1 - \Pr(B)], \Pr(B)[1 - \Pr(A)]\} & \text{if } D > 0 \\ D/\min\{-\Pr(A)\Pr(B), -[1 - \Pr(A)][1 - \Pr(B)]\} & \text{if } D < 0 \end{cases}$

Note: $\Pr(AB)$ = frequency of AB haplotype; $\Pr(A)$ = frequency of A allele; $\Pr(B)$ = frequency of B allele.

loci and the formula for the correlation coefficient r^2 is given in Table 12.1. An advantage of this measure is that it is a standard statistical measure with well-known properties. Lewontin's D' (Table 12.1) is simply the disequilibrium coefficient standardized by its maximum value. The measure is easily interpretable as the proportion of the maximum possible level of association between two loci, given the allele frequencies. As with the disequilibrium coefficient, r^2 and D' are zero if there is no allelic association. The correlation coefficient and Lewontin's D' fall between -1 and 1, and thus the maximum values are not dependent on allele frequencies; however, it is important to recognize that this does not mean that the measures themselves are independent of allele frequencies. Though these measures are often referred to as measures of linkage disequilibrium, they do not explicitly measure linkage. They can be used to measure allelic associations between loci, linked or unlinked. Several other measures of allelic association have been proposed, but these are not covered here.

When haplotype frequencies can be estimated, either through direct observation or through additional family data, we can use estimates of these measures of association to conduct statistical tests for significant allelic associations between different loci. Test statistics can be constructed to test the null hypothesis of no allelic association (e.g., see Weir, 1996). In human genetic studies, it is typically the genotypes at the different loci that are measured in the laboratory, and therefore haplotypes are not directly observed. If one can assume that Hardy–Weinberg equilibrium (HWE) exists, then the expectation–maximization (EM) algorithm (Dempster et al., 1977) can be used to estimate haplotype frequencies (e.g., see Weir, 1996). Alternatively, tests can be based on a composite measure of linkage disequilibrium, which does not require the assumption of HWE (Weir and Cockerham, 1989). To avoid making large-sample approximations, exact tests can be used to test for significant disequilibria between alleles at two loci (Zaykin et al., 1995).

Causes of Allelic Association

There are several forces that can generate allelic associations between loci, including (1) mutation, (2) migration, (3) selection, and (4) genetic drift. After an initial event generates an allelic association, the association begins to decay. The rate of decay is related to the rate of recombination. In general, the more tightly linked two loci are, the slower the decay of association. To illustrate this, consider this hypothetical example in which association is generated by mutation. A new mutation occurs in a gene that results in a disease-causing phenotype. This allele occurs with a particular ancestral haplotype. At the time of the initial mutation, alleles at every marker in the genome are associated with the mutant allele. The chromosome with the mutation is then transmitted to the offspring of the original individual in whom the mutation occurred. Transmission over several generations gives the opportunity for recombination to occur and thus for the rearrangement of the alleles at the marker loci, which breaks up allelic associations. Alleles at marker loci that are farther away from the disease mutation will exchange faster than markers that are closer to the disease mutation. In general, the closer the marker is to the disease gene, the longer the marker allele–disease association will persist.

Consider the disequilibrium coefficient D. For a large, random-mating population, in the absence of forces such as mutation, migration, and selection, the rate of decay of allelic association is dependent on the distance between loci and the number of generations since the allelic association arose. Specifically,

$$D_t = D_0(1 - u)^t$$

where t is the current-generation number, D_t is the current amount of association, D_0 is the association at generation 0, and u is the recombination fraction between loci. Allelic association decays quite rapidly for unlinked loci ($u = 0.5$), but there is little decay if u is close to zero (Fig. 12.1). The slowness of allelic association decay for tightly linked loci makes this a useful mapping tool. However, for tightly linked loci, forces such as recurrent mutation may play a more important role in the decay of allelic association than recombination.

The general rule of thumb is that the stronger the allelic association, the closer the marker is to the disease locus. This is not always the case, however, for several reasons. Mutation rates at the marker locus affect the level of association by increasing the chance that the associated marker allele will change and so seem to represent a different chromosome. New mutations in the disease gene or selection in favor of a particular phenotype can either increase or decrease the level of allelic association.

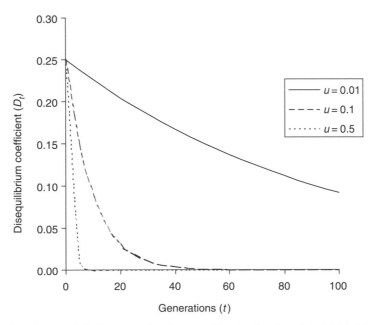

Figure 12.1. Decay of allelic association for recombination fractions (u) of 0.01, 0.1, and 0.5; $D_0 = 0.25$.

Additionally, the frequencies of the marker alleles have an impact on the amount of allelic association that can exist. For example, if the disease susceptibility allele is positively associated with a marker allele whose general population allele frequency is 0.50, then the greatest level of association can be seen when the allele has frequency 1 in the affecteds, which is a 100% increase in the allele frequency. If, however, the positively associated marker allele has frequency 0.8 in the population, then again the greatest level of association is observed when the frequency in affecteds is 1, but this is only a 25% increase. Generally, the potential level of association is greatest when the susceptibility allele is positively associated with a low-frequency marker allele and the chance for substantial association decreases as the positively associated marker allele increases in frequency. It is also important to keep in mind that allelic association is population specific. The level of allelic association between alleles at two loci can differ between populations. It depends on population-specific allele frequencies and critically on population dynamics.

Several studies have examined the relationship between linkage disequilibrium and physical distance. The identification of an increasing number of SNPs and other genetic markers in recent years has enabled detailed studies of linkage disequilibrium in many genomic regions and in different populations. The general result that linkage disequilibrium decays in relation to distance is supported by empirical studies, but the relationship is often not simple for very small distances (see e.g., Terwilliger and Weiss, 1998; Daly et al., 2001; Reich et al., 2001). Markers very close to one another may show very different levels of linkage disequilibrium with other markers in the region; for example, an allele at one marker may be strongly associated with an allele at another locus, while alleles at an immediately adjacent marker may show no association with alleles at the locus (e.g., Martin et al., 2000a; Reich et al., 2001). Not only does linkage disequilibrium vary between markers within a genomic region, but also it is highly variable between regions of the genome and between populations. Some theoretical and empirical studies have suggested that significant linkage disequilibrium will only be found over distances of only a few kilobases (Kruglyak, 1999; Dunning et al., 2000), while others have found linkage disequilibrium spanning distances of 100 kb or larger (Abecasis et al., 2001; Collins et al., 1998). The variability observed highlights the important point that linkage disequilibrium depends on more than physical distance. There is great variability depending on population history, regional mutation rate, and recombination.

Recent studies examining association between SNPs have demonstrated that linkage disequilibrium may be organized in distinct regions of strong linkage disequilibrium, with little association occurring between SNPs in different regions (Daly et al., 2001; Jeffreys et al., 2001). These results suggest that the genome consists of conserved "haplotype blocks" in which little recombination has occurred over time separated by hot spots of recombination. Within these haplotype blocks there is low haplotype diversity, with only a few distinct haplotypes being found. An important implication is that only a few markers may be needed to capture much of the variation within a haplotype block; thus if these haplotype blocks can be defined across the genome, this may improve the efficiency of gene-mapping studies discussed below.

MAPPING GENES USING LINKAGE DISEQUILIBRIUM

Presently, the primary use of association analysis is for fine mapping in candidate regions and testing of candidate genes for biological relevance. Some candidate gene studies attempt to identify direct biological involvement by testing specific functional polymorphisms or mutations. Other studies try to exploit linkage disequilibrium that may exist between marker polymorphisms and disease susceptibility alleles in a gene or small region. Because linkage disequilibrium rarely extends more than 1 cM in general populations and is often found only over much smaller distances, association between the disease and marker alleles signals a significant decrease in the minimum-candidate region. However, this great strength is also its great drawback. Because the effect is so localized, it will be very hard to find against the background of the entire human genome without screening a very dense array of markers. In genetically complex diseases, the further complications of genetic heterogeneity and/or gene–gene interaction may make detection of association even more difficult.

Many early statistical methods of mapping genes using linkage disequilibrium attempted to invoke a simple relationship between linkage disequilibrium and the rate of recombination to estimate the location of a disease locus relative to a genetic marker (Hastbacka et al., 1992; Terwilliger, 1995; Hill and Weir, 1994; Kaplan et al., 1995). These methods achieved some notable successes in mapping genes for Mendelian disorders such as cystic fibrosis (Kerem et al., 1989) and diastrophic dysplasia (Hastbacka et al., 1992, 1994) and have been applied to help localize genes for several other diseases, including Friedreich ataxia (Fujita et al., 1990), myotonic dystrophy (Harley et al., 1991) and Huntington disease (Huntington's Disease Collaborative Research Group, 1993). Methods for linkage disequilibrium mapping can be useful for estimating the location of disease genes when there is a single disease mutation with a strong effect leading to the majority of disease cases in the population. These methods focus on the situation in which chromosomes carrying the disease mutation can be identified, as would be the case for a rare, fully penetrant, recessive disease. Furthermore, estimates of distance depend on accurate knowledge of the history of the disease gene in the population. For these reasons, many early successes with linkage disequilibrium mapping were achieved in isolated populations (e.g., Finnish populations). Isolated populations arising from a small founder group are more homogeneous in disease origin than general populations, since there are likely to be only a few founding individuals who carried the disease mutation. However, even in isolated populations, these methods are restricted to simple Mendelian disorders and have limited utility for disease localization for complex disease genes. In the following sections we discuss several methods for testing for association that are used to identify complex disease genes in general populations.

TESTS FOR ASSOCIATION

For complex diseases, one typically does not know which chromosomes carry the susceptibility allele. Therefore, we cannot directly assess allelic association between marker and disease alleles. Instead, we look for associations between marker alleles and disease phenotype. The phenotype might be a simple dichotomous trait such as affected/unaffected with disease or it might be a quantitative trait such as blood pressure. Initially, we present tests for dichotomous traits.

Tests for association fall into two broad categories, depending on what type of sample is used. Case–control tests use unrelated individuals who are affected (cases) and unaffected (controls). Case–control tests compare allele or genotype frequencies in the cases to the frequency in a set of matched controls. Family-based tests of association use affected individuals and their relatives. Allele frequencies in affected individuals are compared to family-based controls, typically parental controls or unaffected siblings.

Case–Control Tests

In its simplest form, the case–control test is a standard test for the differences between allele or genotype frequencies in the two samples. For balanced data, in which the same number of cases and controls are genotyped, an appropriate test for allele frequency differences is given below:

$$T_{cc} = \frac{\left[\sum_{i=1}^{m} (a_i - u_i) \right]^2}{\sum_{i=1}^{m} (a_i - u_i)^2}$$

where m is the number of marker alleles and a_i and u_i are the number of times the ith allele is found in the cases and controls, respectively. The statistic can be compared to a chi-square statistic with $m - 1$ degrees of freedom (df) for large samples.

It is often difficult to obtain balanced case–control data in any real dataset and there is no need to restrict a study to balanced numbers. A similar case–control statistic can be constructed for unbalanced data, but it takes a slightly more complicated form (see, e.g., Schlesselman, 1982). Essentially, the test compares the observed allele frequencies in the cases and control to those expected under the null hypothesis that the frequencies are equal in cases and controls.

An example of the successful use of a case–control study is the identification of the *APOE-4* allele as the susceptibility gene in late-onset familial and sporadic AD. Table 12.2 presents data on a large study of over 500 AD patients and age and ethnically matched controls. Using standard chi-square analysis, a significant association was found ($p < 0.001$). As indicated in Table 12.2, there appears to be an increase in the *APOE-4* allele and a concomitant decrease in the *APOE-3* allele in AD patients.

Measures of Disease Association and Impact. Case–control studies can also provide estimates of important measures of association and impact. Tests can be based on these measures through either standard hypothesis testing or the

TABLE 12.2. Case–Control Association Studies: *APOE-4* Allele and AD

Allele	Cases	Controls	Total
a. Observed Counts			
APOE-4	240	60	300
Not *APOE-4*	360	340	700
Total	600	400	1000
b. Expected Counts			
APOE-4	180	120	300
Not *APOE-4*	420	280	700
Total	600	400	1000

Note: $\chi^2 = $ (observed $-$ expected)^2/expected $= [(240 - 180)^2/180] + [(60 - 120)^2/120] + [(360 - 420)^2/420] + [(340 - 280)^2/280] = 71.5; p < 0.0001$.

construction of confidence intervals. Several measures are often applied in epidemiological research to quantify the magnitude of the association between the risk factor and the disease (see Table 12.3). The two most commonly used measures of association are the relative risk and the odds ratio. Measures of impact quantify the importance or impact of risk factors on the occurrence of the disease relative to other risk factors for that disease. These measures include the attributable risk and population attributable risk.

The relative risk (RR) is the ratio of the incidence of disease among the "exposed" group, which can be defined as individuals with the susceptibility genotype or with a positive family history, compared to the incidence of disease in the "nonexposed" group, or individuals without the susceptibility alleles. Table 12.4 is a 2×2 table representing the joint distribution of the presence or absence of the genotype and disease in the study population. If the disease represents the occurrence of "newly" diagnosed cases of the disease over a defined period of time, the incidence of disease D among the exposed (i.e., those with a mutation in a susceptibility gene or family history of disease D) is I_e and the incidence of disease D in the unexposed (i.e., absence of a genetic mutation or a family history of disease D) is I_0.

Incidence rates and RRs are derived from cohort study data (prospective studies), where the exposure is defined "before" disease occurs. Study subjects who are disease free at baseline are followed over time and observed for subsequent disease onset. However, as is often the case in genetic linkage studies, cohort data are not available or are impractical to obtain, especially for late-onset and rare disease. Case–control data (retrospective studies), where disease status is established first and exposure subsequently determined, are often utilized to estimate the RR between a study factor and disease. If the data in Table 12.4 were obtained from a case–control study, the marginal totals for the exposed ($a + b$) and nonexposed ($c + d$) groups of the population would not be known, and the incidence rate

TABLE 12.3. Summary of Epidemiological Measures

Measure	Definition
Relative risk (RR)	Measure of association between genetic exposure and disease, the ratio of the incidence of the disease in the "exposed" and "unexposed" groups using prospective/cohort data: $$RR = \frac{I_e}{I_0} = \frac{a(a=b)}{c(c+d)} \quad \text{(see Fig. 12.2)}$$
Odds ratio (OR)	Measure of association between genetic exposure and disease, the ratio of the odds of exposure (genetic susceptibility) among affected subjects or cases to the odds or exposure (genetic susceptibility) among unaffected individuals or controls calculated using data from case–control studies: $$OR = \frac{ad}{bc} \quad \text{(see Fig. 12.2)}$$
Attributable risk (AR)	Measure of impact or the excess risk of disease among genetically susceptible individuals compared to those who are not. The genetic AR is dependent on the proportion of disease due to the susceptibility gene and the magnitude of the relative risk among gene carriers and noncarriers: $$AR = \frac{P_{Aa}[1 - 2q(1 - P_{aa})]}{P_{Aa}(1 - 2q)} \quad \text{(Claus et al., 1996)}$$

could not be calculated directly. In this scenario, an odds ratio (OR) may be calculated to estimate the RR. The OR is defined as the ratio of the odds of exposure among the cases, the subjects diagnosed with disease D, to that among the noncases or controls. Based on the 2×2 table in Table 12.4, the OR is calculated by applying the following formula:

$$OR = \frac{a/c}{b/d} = \frac{a*d}{b*c}$$

TABLE 12.4. 2×2 Contingency Table for Case–Control Analysis

	Disease		
Genetic test	Yes	No	Total
Positive	a	b	$a + b$
Negative	c	d	$c + d$
Total	$a + c$	$b + d$	$a + b + c + d$

Confidence intervals (CIs), which specify the range of the true values for the OR, can be derived by estimating the variance and are calculated from the following equation:

$$CI = (OR) \exp z [\pm (var(\ln OR)]^{1/2}$$

where, for a 95% confidence limit or a significance level of 0.05, $z = 1.96$. The CI for the OR can be approximated using a Taylor series expansion as follows (Hennekens and Buring, 1987):

$$CI = \frac{ad}{bc} \exp\left[\pm z \left(\frac{1}{a} + \frac{1}{b} + \frac{1}{c} + \frac{1}{d} \right)^{1/2} \right]$$

If the CI does not include 1.0 (the null value for a measure of association), the association is statistically significant at the specified level.

In most case–control studies, in the absence of bias, the OR provides a valid estimate of the RR (Hennekens and Buring, 1987). The validity of the RR estimate and the inferences based on this estimate depend on the absence of measurement error, the representativeness of the study population (absence of selection bias), and the absence of an imbalance of other factors related to the disease and the exposure in the exposed and nonexposed groups (confounding bias).

The attributable risk (AR) represents the excess risk of disease in those exposed compared with those unexposed to the factor of interest. The AR depends not only on the magnitude of the relative risk but also on the prevalence of the exposure (the proportion of individuals in a population that are exposed at a specific instant in time, such as the proportion carrying a susceptibility allele) in the study population. The AR reflects the "impact" of the exposure on the occurrence of the disease or how much disease would be "prevented" if that factor were somehow eliminated. The AR can be expressed as a percentage and is calculated (Rothman, 1986) by incorporating the RR (or OR) using the following formula:

$$AR\% = \frac{RR - 1}{RR} \times 100$$

The following equation, for the population attributable risk percent (PAR%), is used to estimate the excess rate of disease in the total study population of exposed and nonexposed individuals that is attributable to the exposure:

$$PAR\% = \left[\frac{(P_e)(RR - 1)}{(P_e)(RR - 1)} + 1 \right] \times 100$$

where P_e is the prevalence of exposure in the population, which can either be estimated in the control group or obtained from another source (Hennekens and Buring, 1987). The AR may also be expressed as the "risk difference," or $I_e - I_0$.

Assessing Confounding Bias. Confounding occurs when an observed association is due to the mixing of effects between the exposure (e.g., family history), the disease, and a third factor (or confounder) that is associated with the exposure and independently affects the risk of developing the disease (Hennekens and Buring, 1987).

Confounding bias in epidemiological studies can be addressed through various statistical analytic strategies to control for such bias. For case–control studies, stratification and logistic regression are the most common methods for assessing an association between the disease and a risk factor while controlling for confounding factors. Survival analyses are often used to analyze data in cohort studies.

Stratified analyses can be useful to control for confounding of either genes or environmental factors. Using this approach, the genetic susceptibility–disease association can be estimated within various levels of the confounding variable. Of particular concern in genetic epidemiology studies is potential confounding by ethnic background of the study participants; such confounding forms the motivation for family-based tests of association.

For example, in a study of Native Americans and type 2 diabetes mellitus, a strong negative association between the Gm haplotype Gm3;5,13,14 from the Gm system of human immunoglobin was found to be confounded by the effect of Caucasian admixture (Knowler et al., 1988). The crude OR for the association between Gm3;5,13,14 and diabetes in a sample of 4920 Native Americans was 0.21 (95% CI = 0.14–0.32) and was statistically significant, suggesting that the absence of this haplotype (or absence of a closely linked gene) is a causal factor in this disease. However, Gm3;5,13,14 is also a marker for Caucasian admixture, and therefore the degree of Native American heritage is a potential confounder for this association. Stratum-specific ORs, according to the degree of Native American heritage, either none, half, or full, revealed that there was no significant relationship between Gm3;5,13,14 and diabetes mellitus. The stratum-specific ORs and 95% CI for full, half, and no Native American heritage were 0.63 (95% CI = 0.34–1.18), 0.69 (95% CI = 0.24–1.96), and 0.75 (95% CI = 0.11–5.32), respectively. Further adjustments for confounding by age resulted in an even weaker association between the marker and disease within these strata. Therefore, the relationship between Gm3;5,13,14 and diabetes mellitus was explained by the inverse relationship between the genetic marker and Caucasian admixture (Knowler et al., 1988).

In designing a case–control study, it is essential to ensure that case and control populations are matched with respect to ethnicity as well as other factors, such as age, which may be associated with affection status or marker–allele frequency. Spurious associations can result because of population stratification (i.e., the existence of multiple population subtypes in what is assumed to be a relatively homogeneous population). The existence of population stratification can lead to significant associations even in unlinked loci or unassociated loci within strata (Table 12.5).

TABLE 12.5. Example of Population Stratification[a]

Population A			Population B			Population C (Mixed)		
	A (0.80)	a (0.20)		A (0.20)	a (0.80)		A (0.50)	a (0.50)
B (0.80)	0.64	0.16	B (0.20)	0.04	0.16	B (0.50)	0.25	0.25
b (0.20)	0.16	0.04	b (0.80)	0.16	0.64	b (0.50)	0.25	0.25

[a]In this example populations A and B have very different allele frequencies for the disease gene A and the unlinked marker B. If the populations mix evenly, then the overall allele frequencies are as seen in population C. If comparisons of the haplotype frequencies are made from population C to either population A or B, the results will be significant even if no linkage exists. For example, if the affected individuals are drawn from population C, they will have an A/B haplotype frequency of 0.25, assuming no association with the disease. If the controls are drawn from population B, the A/B frequency will be 0.04. The erroneous conclusion is that there is an association of the disease with the B allele (0.25 vs. 0.04), since they occur together more often than in population B. In most datasets, it is not possible to determine whether the sample was drawn from one, two, or more populations.

Family-Based Tests of Association

Family-based studies ensure that cases and controls are appropriately matched by using family-based controls. The case–control tests described above are not appropriate in family data since cases and family-based controls are not necessarily independent. Thus, several family-based association tests have been developed.

Early examples of family-based tests for association used family triads (i.e., an affected individual and both parents). In a family triad, there is a transmitted pair of alleles and a nontransmitted pair of alleles. The case genotype is made up of the transmitted pair of alleles and the nontransmitted pair of alleles makes up the family-based control genotype (Fig. 12.2). These tests include the haplotype relative risk (HRR) test (Falk and Rubinstein, 1987), the haplotype-based haplotype relative risk (HHRR) test (Terwilliger and Ott, 1992), the AFBAC method (Thomson, 1995), and the transmission/disequilibrium test (TDT) (Spielman et al., 1993). The HRR, HHRR, and AFBAC approaches were developed to detect association

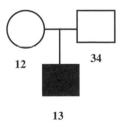

Transmitted pair of alleles   13

Nontransmitted pair of alleles   24

Figure 12.2. Transmitted and nontransmitted alleles in a family triad.

while controlling for possible bias due to population stratification. The TDT approach has the advantage of being valid as a test of linkage as well as association and thus assures that associations detected occur between linked loci. All tests have little power unless linkage and association coexist. The statistical differences between these methods are subtle and are not described here. Reviews of these methods can be found elsewhere (Spielman and Ewens, 1996).

Transmission/Disequilibrium Test. The TDT is one of the most widely used of the family-based association tests. The TDT was originally proposed as a test of linkage in the presence of association (Spielman et al., 1993), meaning that it was intended to verify that significant associations between marker alleles and disease were due to linkage and not some other force such as population admixture or stratification. The TDT can be a more powerful test to detect linkage than the affected sibpair linkage tests, especially when the genetic effect is small (Spielman et al., 1993; Risch and Merikangas, 1996), as is often the case with genetically complex traits. The disadvantage of this approach is that the TDT has little power to detect linkage if allelic association is not present. The TDT examines the number of transmissions of allele *1* or allele *2* from a heterozygous parent to an affected offspring. An example of the TDT is given in Figure 12.3. As a test of linkage the counts needed in Figure 12.3 can come from simplex, multiplex, or even multigenerational family data. The statistical significance can be tested using standard chi-square distribution with 1 df for large samples. The data used in constructing the counts comes only from heterozygous parents. The TDT statistic is as follows:

$$\text{TDT} = \frac{(b - c)^2}{b + c}$$

where *b* is the number of times a parent with genotype *12* transmits allele *1* to an affected offspring and *c* is the number of times a parent with the *12* genotype transmits allele *2* to an affected offspring.

The test statistic then tests for deviations from the expected equal transmission rate into the two categories from the heterozygous parents. Homozygous parents do not need to be scored. This is different from the HRR, HHRR, and AFBAC methods, which use both heterozygous and homozygous parents. A significant result indicates that the marker is linked to the disease locus. However, the test has little power to detect linkage unless there is also allelic association. In fact, if the dataset is composed of only independent family triads, then the TDT can find linkage only in the presence of

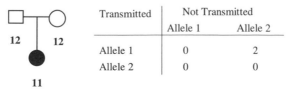

	Transmitted	Not Transmitted	
		Allele 1	Allele 2
Allele 1		0	2
Allele 2		0	0

Figure 12.3. Example of scoring a TDT family.

association. If there were only linkage and no allelic association, then across families there would be no difference between b and c, since the allele coupling with the disease gene in each family is random, preventing the detection of linkage. For larger families, the TDT has some power to detect linkage even without allelic association, though this power is expected to be minimal without significant association.

The TDT can serve as a test of allelic association in addition to linkage (i.e., linkage disequilibrium) if only a single family triad from any family is used in the test. This use of the TDT to test for linkage disequilibrium is critical for narrowing a broad region of interest identified by linkage analysis. Analysis of linkage disequilibrium could potentially identify the markers that are closest to the actual disease susceptibility locus. The TDT is not valid as a test for allelic association if more than a single triad within a family is used since it treats triads independently, which may not be the case if there is linkage. An extension of the TDT, the Tsp test, allows for multiple affected siblings while maintaining validity as a test of allelic association (and linkage). The test properly adjusts for correlations that may occur between transmissions to affected siblings.

The TDT was originally developed to look at biallelic marker systems or situations of alleles that could be readily collapsed because of prior information regarding a known association. With the wide use of multiallelic markers in genetic studies it is necessary to be able to assess association with multiple alleles. There are two approaches that are commonly used for markers with more than two alleles. One is to test for association at each allele versus the others. This provides a test for each marker allele separately; thus multiple testing should be considered when assessing significance of any of these tests. A Bonferroni correction for multiple tests could be applied; the appropriate adjusted significance level is $\alpha/(m-1)$ (Schaid and Rowland, 1998). Alternatively, simulations can be used to obtain appropriate critical values for the statistics (Spielman and Ewens, 1996). The other approach that is used with multiallelic markers is a global statistic that provides an overall assessment of significance. Several global statistics have been proposed. These include the symmetry statistic, the marginal statistic of Bickeboller and Clerget-Darpoux (Bickeboller and Clerget-Darpoux, 1995), the likelihood ratio statistic (extended TDT) of Sham and Curtis (1995), and the marginal statistic with only heterozygous parents (T_{mhet}) of Spielman and Ewens (1996). For a comparison of these tests see Kaplan et al. (1997). In general, T_{mhet} was found to perform well. Table 12.6 gives the definition of T_{mhet}.

The TDT approach was a useful approach in identifying the relationship of the insulin gene in insulin-dependent diabetes mellitus (IDDM) (Spielman et al., 1993) (Table 12.7). The TDT method does not require researchers to go to the expense and effort of recruiting families with multiple affected individuals; indeed, the clinical status of the parents does not have to be known. In some cases, however, these advantages may be outweighed by practical problems. In the diabetes situation, both parents of the affected family members were available for study, making this approach ideally suited to IDDM. However, in the AD/ *APOE-4* example, since AD is a late-onset disease, parental DNA is almost always unavailable, and thus the traditional TDT approach was impossible.

TABLE 12.6. Multiallelic TDT: T_{mhet}^a

	Transmitted Allele			
Nontransmitted Allele	1	2	3	Total
1	n_{11}	n_{12}	n_{13}	$n_{1.}$
2	n_{21}	n_{22}	n_{23}	$n_{2.}$
3	n_{31}	n_{32}	n_{33}	$n_{3.}$
Total	$n_{.1}$	$n_{.2}$	$n_{.3}$	$n_{..}$

$^a T_{\text{mhet}} = \frac{(m-1)}{m} \sum_{i=1}^m \frac{(n_{i.}-n_{.i})2}{(n_{i.}+n_{.i}-2n_{ii})}$, where m is the number of alleles.

Tests Using Unaffected Sibling Controls. An approach that circumvents this difficulty is to use unaffected siblings as controls rather than relying on parental controls. Several test statistics have been proposed (Curtis, 1997; Boehnke and Langefeld, 1998; Spielman and Ewens, 1998). These sibling-based tests compare marker allele frequencies in affected and unaffected siblings. The tests require only a simple affected/unaffected sibpair, (discordant sibpair, DSP), although power may be increased if additional siblings are available. These sibling-based association tests retain the advantages of the TDT in that they provide tests for linkage in the presence of association and are immune to the effects of sampling bias. If only a single DSP from each family is used, then the tests provide tests for association in addition to linkage. For biallelic markers, if there is a single DSP in each family, the tests of Curtis, the DAT, and the Sib-TDT are equivalent. The tests differ in their treatment of multiple alleles and larger sibships. Monks et al. (1998) provide a comparison of these methods.

When larger sibships are available, the sibship disequilibrium test (SDT) (Horvath and Laird, 1998) can be used. The SDT is a sign test based on the sign of the difference in the proportion of times the allele occurs in affected siblings and the proportion of times the allele occurs in unaffected siblings within a sibship. This difference is equally likely to be positive or negative if there is either no allelic

TABLE 12.7. Transmission Disequilibrium Test and Diabetes^a

	Nontransmitted	
Transmitted	*A1*	*A2*
A1	NI	78 (62)
A2	46 (62)	NI

^{a}In this example, the number of *A1* alleles transmitted from a heterozygous parent to the affected (IDDM) child is 78, while the number of *A2* alleles transmitted is 46. If no association exists, the expectation is that each allele would be transmitted 62 times. This result is highly significant. NI, not informative; expected number in parentheses.
Source: Spielman et al. (1993).

association or no linkage between the marker and disease locus, regardless of the size of the sibship. The SDT provides a valid test for linkage and association in sibships of arbitrary size, but it does not take full advantage of the information since it only considers the sign of the difference. An alternative is to use the weighted SDT (WSDT) (Martin et al., 1999), which is based on the magnitude as well as the sign of the difference.

A concern with using unaffected siblings is that one can never really be sure that they will remain unaffected and since they are siblings of affected individuals they are at increased risk. This will bias the test toward the null hypothesis, decreasing the chance of detecting an association. One strategy to help decrease the loss of power is to select unaffected siblings who are older than the age of disease onset in their affected siblings. Alternatively, methods that make inferences about missing parental genotypes where possible without using the phenotype of the unaffected siblings can be used. These methods include the reconstruction-combined TDT (RC-TDT) (Knapp, 1999), which only compares affected siblings to unaffected siblings when parents' genotypes cannot be reconstructed from the sibship, and likelihood-based methods such as those discussed below.

Tests Using Extended Pedigrees. In some disease studies, extended pedigrees containing multiple nuclear families or sibships may be sampled. The tests discussed above are valid as tests of linkage but not valid tests of association when extended pedigree data are used. The difficulty is that contributions of multiple nuclear families or sibships to the statistic are not independent when there is linkage, even if the null hypothesis of no allelic association is true. Thus tests such as the TDT and Sib-TDT that treat nuclear families or sibships, respectively, as independent are not valid; that is, they will not have the correct significance level.

For fine-mapping and candidate gene studies, it is desirable to establish linkage disequilibrium; therefore tests have been developed to test for both linkage and association in extended pedigrees. Several tests for linkage disequilibrium that are appropriate for pedigrees of general structure have been proposed (Martin et al., 2000b, 2001; Abecasis et al., 2000; Rabinowitz and Laird, 2000). The tests properly account for correlations within pedigrees that arise from linkage between disease and marker loci. The test statistics all take a similar form:

$$T = \frac{\left(\sum_{i-1}^{N} X_i\right)^2}{\sum_{i=1}^{N} X_i^2}$$

where X_i is a measure of association for the ith pedigree, $i = 1, \ldots, N$, with mean zero under the null hypothesis. The tests differ in the way that association is measured with pedigrees.

Regression and Likelihood-Based Methods. An appealing property of the TDT and related methods is their simplicity. Alternative methods based on statistical modeling can be employed to expand the uses of the TDT. In family triads or discordant sibpairs, logistic regression can be used to test for linkage disequilibrium (Schaid, 1996, 1999b; Waldman et al., 1999). Regression methods offer several advantages over traditional TDT-like methods. Estimates of the regression parameters provide estimates for allelic or genotypic relative risks. Other genetic or environmental factors can be included in the model to test for gene–gene or gene–environment interactions (Schaid, 1999a). In the regression framework it is straightforward to change the model based on what is known about the mode of inheritance of the disease (Schaid, 1996) to improve statistical power. Similarly, one can incorporate covariates into the model that may be important to take into account when looking for disease genes. A practical difficulty with regression methods is that they are limited to simple pedigree structure (family triads or DSPs) if one is testing for linkage disequilibrium.

Loglinear models have also been used to construct association tests in family triads (Weinberg et al., 1998; Weinberg, 1999). Standard TDT methods treat parental transmissions independently and do not distinguish between maternal and paternal transmissions. The loglinear approach models the probabilities of each possible family triad, with respect to genotype, thus allowing one to test for parent-specific effects. In addition to testing for dependence of affection status on an individual's genotype, loglinear models can be used to test for dependence of affection status on maternal (or paternal) genotype. This allows one to test for important prenatal or other maternal effects. Similarly, loglinear models allow one to test for parent-of-origin effects, which may result from genetic imprinting and may play a role in many complex diseases. Genetic imprinting can lead to differential transmission patterns for marker alleles at loci in linkage disequilibrium with the susceptibility locus.

The likelihood framework is most useful when there is missing parental data. The likelihood ratio test implemented in the computer program Transmit (Clayton, 1999) uses an iterative procedure to test for linkage disequilibrium using transmission from parents to affected offspring even if there is missing genotype data in the parents. Unaffected siblings are used only to improve inference about missing parental genotypes; thus the test is not as sensitive to misclassification of unaffected siblings as methods that compare affected and unaffected siblings. The drawback with making inference about missing data is that the test assumes that the families come from a single, random-mating population, so the test is sensitive to deviations from Hardy–Weinberg equilibrium in the parental population that may arise from population stratification or other forces.

ANALYSIS OF HAPLOTYPE DATA

Traditionally, association tests have been single-locus tests, testing each marker locus independently. Recently there have been several tests proposed that test for association with marker haplotypes composed of alleles at multiple markers. These tests look for association between disease status and a particular combination(s) of alleles at different loci, that is, marker haplotype(s). Intuitively, one might expect that marker haplotypes may provide more information than the markers individually. We often imagine that association arises due to the presence of a susceptibility allele that occurs initially on one, or few, ancestral haplotypes. As discussed earlier, the amount of association between the susceptibility allele and a particular marker allele or marker haplotype depends strongly on the frequency of the associated allele or haplotype. The strongest associations are found when the positively associated allele or haplotype has low frequency in the population. A haplotype necessarily has lower frequency than any of the frequencies for the individual alleles in the haplotype. Thus, there is a potential for greater association when examining haplotypes.

A difficulty to be overcome in testing for haplotype associations is that haplotypes are not generally observed directly. Instead, we typically have information on an individual's genotypes. In some cases, the haplotypes can be inferred with certainty from the genotype, but when the individual is heterozygous at more than one locus, the haplotype assignment is ambiguous. The EM algorithm (Dempster et al., 1977) is often used to accommodate the phase uncertainty. Methods have been proposed to test for haplotype association, allowing for ambiguous haplotypes, in unrelated cases and controls (Zhao et al., 2000; Fallin and Schork, 1999). Genotype information from family members can often aid in haplotype inference. The likelihood ratio test in the Transmit program (Clayton, 1999) provides a family-based approach to testing for haplotype association. The test has the same properties as the single-locus Transmit tests (discussed above), but in addition to assumptions of random mating, the haplotype test assumes no recombination between the markers in the sample.

Another set of pedigree-based association tests is available in the FBAT-tools package (Horvath et al., 2001; Lange et al., 2004), which contains both the FBAT and PBAT programs. As with the PDT and Transmit, FBAT extends the TDT to multiple siblings and entire pedigrees. FBAT has also been extended to handle both dichotomous and quantitative traits. It can examine multiple loci simultaneously, thus handling haplotypic data as well. FBAT considers extended pedigrees as a series of nuclear families. PBAT conditions on the founder genotypes and thus can examine extended pedigrees without breaking them into subpedigrees. PBAT also provides some powerful tools for obtaining power and sample size estimates (Lange et al., 2002). Both FBAT and PBAT can incorporate covariate data, although the methods for incorporating covariates are different.

An additional difficulty with haplotype association tests is that there can be many haplotypes to consider with multiple markers and/or multiple alleles, and often there is no a priori knowledge of which haplotype might be associated. Haplotypes with low frequency will contribute small numbers of observations to the analysis; thus caution must be used in applying and interpreting the results of haplotype tests. Often haplotypes with small numbers can be pooled to provide valid asymptotic tests, but this pooling could reduce power if negatively and positively associated haplotypes are pooled. Typically global as well as individual haplotype tests are produced. Multiple testing should be considered in interpreting many tests for individual haplotypes.

ASSOCIATION TESTS FOR QUANTITATIVE TRAITS

Most diagnoses of complex diseases rely on a number of different clinical criteria. Often quantitative measurements are an important part of these criteria, and identifying genes influencing these quantitative traits could be important for controlling disease, assessing disease risk, and understanding underlying disease mechanisms. Testing for association between genetic markers and a quantitative trait can be used in fine mapping the quantitative trait loci (QTLs) in the same way that we use association tests to look for genes for dichotomous traits.

If a sample of unrelated individuals is available, randomly selected with respect to the quantitative trait, then it is straightforward to test for association between the trait measurement and marker alleles or genotypes. Methods based on general linear models are commonly employed (Searle, 1971). In disease studies, individuals are selected for a genetic study because they have the disease of interest. Therefore, some trait measurements may be different from random members of the population at large. Though this may influence the statistical power of the test, it does not generally affect validity if data are properly normalized.

Association tests for quantitative traits in unrelated individuals suffer from the same potential biases as case–control studies if there is population stratification. Family-based tests can be used to avoid bias and provide an additional test of linkage. Several tests have been proposed for family triads, where the offspring in the triad may have a random or selected trait value (Allison, 1997; Rabinowitz, 1997). If multiple siblings are available, correlations between trait values that may result because of linkage must be taken into account (Monks and Kaplan, 2000). Association tests for quantitative traits can also be used when there is missing parental data, provided multiple siblings are available (Allison et al., 1999). Recently association tests for quantitative traits have been extended to general pedigree structure (Abecasis et al., 2000; Rabinowitz and Laird, 2000; Lange et al., 2002).

ASSOCIATION AND GENOMIC SCREENING

Traditionally, the most common use of association analysis is to test specific candidate loci or markers in a relatively small region. More recently discussions have centered on the use of the association analysis for entire genome scans (Risch and Merikangas, 1996; Sachidanandam et al., 2001). Since allelic association are generally found over only very small distances in general, outbred populations, a very dense array of markers should be screened to give a chance of finding a marker or markers associated with the disease.

The initial success of the human genome project has been the (nearly) complete characterization of the consensus human sequence. This has greatly increased our ability to identify and describe the genomic structure of genes. Perhaps of even more import for disease gene studies is the rapidly developing hapmap data (international HapMap Project 2003). This vast pool of characterized common differences between people greatly increases the power of any targeted or whole genome scan. Multiple technologies now allow a whole genome association (WGA) design to be implemented by genotyping more than 500,000 SNPs with high fidelity and low cost. In such an approach, a dense map of SNPs is genotyped, and alleles, genotypes, or haplotypes are tested directly for association with disease. Estimates suggest that with 500,000 SNPs, about 70–75% of the common variation in the genome will be captured. Thus, WGA is by far the most detailed and complete method of genome interrogation currently possible. WGA has already been validated with the recent identification of the highly significant effect of the T1277C polymorphism in the CFH gene in macular degeneration. This was found simultaneously through WGA (Klein et al., 2005) and targeted locational candidate approaches (Haines et al., 2005; Edwards et al., 2005). It is important to mention, however, that the small sample sizes used for the WGA study were only sufficient because the effect of the T1277C polymorphism was unexpectedly large.

There is no paradigm for the analysis of WGA data. A few recent publications (Lin et al., 2004; Marchini et al., 2005; Skol et al., 2006) have attempted to address very specific analytical issues. There are several potential problems with the use of association analysis in genomic scanning. First, the problem of multiple comparisons arises in this situation if one is testing many markers individually or testing many haplotypes. That is, when so many statistical tests are performed, false-positive results are likely by chance alone unless the usual significance level of 0.05 or 0.01 is modified. Thus use of a significance level that is smaller than the nominal significance level is warranted at each marker or haplotype. The usual Bonferroni correction approach (simply dividing the desired nominal significance level by the number of tests performed) will be too conservative because it assumes each test to be independent of the others. This will not be the case, since many of the markers will be linked and associated with each other. Unfortunately, it is not clear what the appropriate correction needs to be since it depends on the underlying relationship between the markers, although simulation-based methods to establish global significance across many markers taking into account multiple testing can be used (McIntyre et al., 2000).

The third problem is that the association analysis approach rests completely on the assumption that some level of linkage disequilibrium exists. While this may be true in many cases, susceptibility alleles arising from frequent mutation events, existing as extremely old mutations, or arising in regions with very high recombination rates will have little or any detectable linkage disequilibrium. Additionally, allelic heterogeneity, that is, multiple disease mutations within a gene each influence disease susceptibility, can also make detecting association difficult.

SPECIAL POPULATIONS

Without a very dense map of markers, associations may be difficult to detect in general, outbred populations. One way in which one can improve the ability to use linkage disequilibrium to detect disease genes is by studying populations that may be enriched for linkage disequilibrium. So-called special populations, such as isolated founder populations or admixed or inbred populations, can be particularly useful for gene identification through association analysis. This is possible because the linkage disequilibrium tends to extend over larger areas (possibly several centimorgans) of the chromosome than in general, outbred populations. We have previously discussed the use of isolated populations in linkage disequilibrium mapping. Admixed populations, for which disease prevalence is known to differ between mixing populations, have also been used.

Special populations can be especially useful in mapping complex traits. The basic premise is that genetically isolated populations will have fewer genes contributing toward a disease trait, and therefore the effect of each remaining susceptibility gene will be easier to detect. The value of these populations in genetic mapping studies has long been realized. Homozygosity mapping was described early on in the molecular revolution (Lander and Botstein, 1987). These advances have been expanded to include the use of pooling strategies (Sheffield et al., 1994) and the exploitation of the phenomena of linkage disequilibrium together with the isolated inbred nature of these groups through the approach of "shared segment" mapping of a complex phenotype (Houwen et al., 1994; Durham and Feingold, 1997). Thus the great advantage of the special population is its power to detect linkage. However, it must be pointed out that this power comes at the potential cost of specificity. Only one or a few of the entire suite of susceptibility genes may be found, and the effect of this gene or genes may be limited to the special population being studied.

SUMMARY

Allelic associations can arise due to forces such as mutation, selection, admixture, or genetic drift. The decay of association then depends on the degree of linkage between the loci. Allelic associations in the presence of linkage, that is, linkage disequilibrium, can be exploited to aid in the mapping of genetically complex diseases. Both case–control and family-based methods can be used. Tests for association have substantial promise for detection of susceptibility alleles as more finely spaced markers, such as SNPs, are mapped and available for study. In this chapter we have focused on detecting associations for single loci; however, methods of association can also be used to detect complex interactions between different genes. Methods using association to detect gene–gene interactions are reviewed in Chapter 14, and these methods continue to be refined and advanced.

REFERENCES

Abecasis GR, Cookson WOC, Cardon LR (2000): Pedigree tests of transmission disequilibrium. Eur J Hum Genet 8:545–551.

Abecasis GR, Noguchi E, Heinzmann A, Traherne JA, Bhattacharyya S, Leaves NI, Anderson GG, Zhang Y, Lench NJ, Carey A, Cardon LR, Moffatt MF, Cookson WO (2001): Extent and distribution of linkage disequilibrium in three genomic regions. Am J Hum Genet 68:191–197.

Allison DB (1997): Transmission-disequilibrium tests for quantitative traits. Am J Hum Genet 60:676–690.

Allison DB, Heo M, Kaplan N, Martin ER (1999): Sibling-based tests of linkage and association for quantitative trials. Am J Hum Genet 64:1754–1764.

Bickeboller H, Clerget-Darpoux F (1995): Statistical properties of the allelic and genotypic transmission/disequilibrium test for multiallelic markers. Genet Epidemiol 12:865–870.

Boehnke M, Langefeld CD (1998): Genetic association mapping based on discordant sib pairs: The discordant alleles test (DAT). Am J Hum Genet 62:950–961.

Clayton D (1999): A generalization of the transmission/disequilibrium test for uncertain-haplotype transmission. Am J Hum Genet 65:1170–1177.

Collins FS, Brooks LD, Chakravarti A (1998): A DNA polymorphism discovery resource for research on human genetic variation. Genome Res 8:1229–1231.

Crow JF, Kimura M (1970): An Introduction to Population Genetics Theory. New York: Harper & Row.

Curtis D (1997): Use of siblings as controls in case-control association studies. Ann Hum Genet 61:319–333.

Daly MJ, Rioux JD, Schaffner SF, Hudson TJ, Lander ES (2001): High-resolution haplotype structure in the human genome. Nat Genet 29:229–232.

Dempster AP, Laird NM, Rubin DB (1977): Maximum likelihood from incomplete data via the EM algorithm (with discussion). J R Stat Soc B B39:1–38.

Dunning AM, Durocher F, Healey CS, Teare MD, McBride SE, Carlomagno F, Xu CF, Dawson E, Rhodes S, Ueda S, Lai E, Luben RN, Van Rensburg EJ, Mannermaa A, Kataja V, Rennart G, Dunham I, Purvis I, Easton D, Ponder BA (2000): The extent of linkage disequilibrium in four populations with distinct demographic histories. Am J Hum Genet 67:1544–1554.

Durham LK, Feingold E (1997): Genome scanning for segments shared identical by descent among distant relatives in isolated populations. Am J Hum Genet 61:830–842.

Edwards AO, Ritter R, III, Abel KJ, Manning A, Panhuysen C, Farrer LA (2005): Complement factor H polymorphism and age-related macular degeneration. Science 308(5720):421–424.

Falk CT, Rubinstein P (1987): Haplotype relative risks: An easy reliable way to construct a proper control sample for risk calculations. Ann Hum Genet 51:227–233.

Fallin D, Schork N (1999): Haplotype analyses in case-control populations: Factors which influence accuracy of haplotype estimation. Am J Hum Genet 65(A83):439.

Fujita R, Hanauer A, Sirugo G, Heilig R, Mandel JL (1990): Additional polymorphisms at marker loci D9S5 and D9S15 generate extended haplotypes in linkage disequilibrium with Friedreich ataxia. Proc Natl Acad Sci USA 87:1796–1800.

Haines JL, Hauser MA, Schmidt S, Scott WK, Olson LM, Gallins P, Spencer KL, Kwan SY, Noureddine M, Gilbert JR, Schnetz-Boutaud N, Agarwal A, Postel EA, Pericak-Vance MA (2005): Complement factor H variant increases the risk of age-related macular degeneration. Science 308(5720):419–421.

Harley HG, Brook JD, Floyd J, Rundle SA, Crow S, Walsh KV, Thibault MC, Harper PS, Shaw DJ (1991): Detection of linkage disequilibrium between the myotonic dystrophy locus and a new polymorphic DNA marker. Am J Hum Genet 49:68–75.

Hastbacka J, De la Chapelle A, Kaitila I, Sistonen P, Weaver A, Lander E (1992): Linkage disequilibrium mapping in isolated founder populations: Diastrophic dysplasia in Finland. Nat Genet 2:204–211.

Hastbacka J, De la Chapelle A, Mahtani MM, Clines G, Reeve-Daly MP, Hamilton BA, Kusumi K, Trivedi B, Weaver A (1994): The diastrophic dysplasia gene encodes a novel sulfate transporter: Positional cloning by fine-structure linkage disequilibrium mapping. Cell 78:1073–1087.

Hennekens CH, Buring JE (1987): Measures of disease frequency and association. In: Mayrent SL, ed. Epidemiology in Medicine. pp 54–98.

Hill WG, Weir BS (1994): Maximum-likelihood estimation of gene location by linkage disequilibrium. Am J Hum Genet 54:705–714. Erratum, Am J Hum Genet 55(1):217 (1994).

Horvath SM, Laird NM (1998): A discordant-sibship test for disequilibrium and linkage: No need for parental data. Am J Hum Genet 63:1886–1897.

Houwen RHJ, Baharloo S, Blankenship K, Raeymaekers P, Juyn J, Sandkuijl LA, Freimer NB (1994): Genome screening by searching for shared segments: Mapping a gene for benign recurrent intrahepatic cholestasis. Nat Genet 8:380–386.

Huntington's Disease Collaborative Research Group (1993): A novel gene containing a tri-nucleotide repeat that is expanded and unstable on Huntington's Disease chromosomes. Cell 72:971–983.

Jeffreys AJ, Kauppi L, Neumann R (2001): Intensely punctate meiotic recombination in the class II region of the major histocompatibility complex. Nature 29:217–222.

Kaplan NL, Hill WG, Weir BS (1995): Likelihood methods for locating disease genes in nonequilibrium populations. Am J Hum Genet 56:18–32.

Kerem B, Rommens JM, Buchanan JA, Markiewicz D, Cox TK, Chakravarti A, Buchwald M, Tsui LC (1989): Identification of the cystic fibrosis gene: Genetic analysis. Science 245:1073–1080.

Klein RJ, Zeiss C, Chew EY, Tsai JY, Sackler RS, Haynes C, Henning AK, Sangiovanni JP, Mane SM, Mayne ST, Bracken MB, Ferris FL, Ott J, Barnstable C, Hoh J (2005): Complement factor H polymorphism in age-related macular degeneration. Science 308 (5720):385–389.

Knapp M (1999): The transmission/disequilibrium test and parental-genotype reconstruction: The reconstruction-combined transmission/disequilibrium test. Am J Hum Genet 64: 861–870.

Knowler WC, William RC, Pettitt DJ, Steinberg A (1988): GM3;5,13,14 adn type 2 diabetes mellitus: An association in American Indians by genetic admixture. Am J Hum Gen 43:520–526.

Kruglyak L (1999): Prospects for whole-genome linkage disequilibrium mapping of common disease genes. Nat Genet 22:139–144.

Lander ES, Botstein D (1987): Homozygosity mapping: A way to map human recessive traits with the DNA of inbred children. Science 236:1567–1570.

Lewontin RC (1988): On measures of gametic disequilibrium. Genetics 120:849–852.

Lin S, Chakravarti A, Cutler DJ (2004): Exhaustive allelic transmission disequilibrium tests as a new approach to genome-wide association studies. Nat Genet 36(11):1181–1188.

Marchini J, Donnelly P, Cardon LR (2005): Genome-wide strategies for detecting multiple loci that influence complex diseases. Nat Genet 37(4):413–417.

Martin ER, Bass MP, Kaplan NL (2001): Correcting for a potential bias in the pedigree disequilibrium test. Am J Hum Gen 68:1065–1067.

Martin ER, Lai EH, Gilbert JR, Rogala AR, Afshari AJ, Riley J, Finch KL, Stevens JF, Livak KJ, Slotterbeck BD, Slifer SH, Warren LL, Conneally PM, Schmechel DE, Purvis I, Pericak-Vance MA, Roses AD, Vance JM (2000a): SNPing away at complex disease: Analysis of SNPs around APOE in Alzheimer disease. Am J Hum Genet 67:383–394.

Martin ER, Monks SA, Warren LL, Kaplan NL (1999): A weighted sibship disequilibrium test for linkage and association in discordant sibships. Am J Hum Genet 65:A434.

Martin ER, Monks SA, Warren LL, Kaplan NL (2000b): A test for linkage and association in general pedigrees: The pedigree disequilibrium test. Am J Hum Genet 67:146–154.

McIntyre LM, Martin ER, Simonsen K, Kaplan NL (2000): Circumventing the multiple testing problem: A multilocus transmission/disequilibrium test. Genet Epidemiol 19: 18–29.

Monks SA, Kaplan NL (2000): Removing the sampling restrictions from family-based tests of association for a quantitative trait locus. Am J Hum Genet 66:576–592.

Monks SA, Kaplan NL, Weir BS (1998): A comparative study of sibship tests of linkage and/ or association. Am J Hum Genet 63:1507–1516.

Rabinowitz D (1997): A transmission disequilibrium test for quantitative trait loci. Hum Hered 47:342–350.

Rabinowitz D, Laird N (2000): A unified approach to adjusting association tests for population admixture with arbitrary pedigree structure and arbitrary missing marker information. Hum Hered 50:211–223.

Reich DE, Cargill M, Stacey B, Ireland J, Sabeti PC, Richter DJ, Lavery T, Kouyoumjian R, Farhadian SF, Ward R, Lander ES (2001): Linkage disequilibrium in the human genome. Nature 411:199–204.

Risch N, Merikangas K (1996): The future of genetic studies of complex human disorders. Science 273:1516–1517.

Rothman KJ (1986): Modern Epidemiology. Boston/Toronto: Little, Brown and Company.

Sachidanandam R, Weissman D, Schmidt SC, Kakol JM, Stein LD, Marth G, Sherry S, Mullikin JC, Mortimore BJ, Willey DL, Hunt SE, Cole CG, Coggill PC, Rice CM, Ning Z, Rogers J, Bentley DR, Kwok PY, Mardis ER, Yeh RT, Schultz B, Cook L, Davenport R, Dante M, Fulton L, Hillier L, Waterston RH, McPherson JD, Gilman B, Schaffner S, Van Etten WJ, Reich D, Higgins J, Daly MJ, Blumenstiel B, Baldwin J, Stange-Thomann N, Zody MC, Linton L, Lander ES, Altshuler D (2001): A map of human genome sequence variation containing 1.42 million single nucleotide polymorphisms. Nature 409:928–933.

Schaid DJ (1996): General score tests for associations of genetic markers with disease using cases and their parents. Genet Epidemiol 13:423–449.

Schaid DJ (1999a): Case-parents design for gene-environment interaction. Genet Epidemiol 16:261–273.

Schaid DJ (1999b): Likelihoods and TDT for the case-parents design. Genet Epidemiol 16:250–260.

Schaid DJ, Rowland C (1998): Use of parents, sibs, and unrelated controls for detection of associations between genetic markers and disease. Am J Hum Genet 63: 1492–1506.

Schlesselman JJ (1982): Case-Control Studies. Design, Conduct, Analysis. New York/ Oxford: Oxford University Press.

Searle SR (1971): Linear Models. New York: John Wiley & Sons.

Sham PC, Curtis D (1995): An extended transmission/disequilibrium test (TDT) for multi-allele marker loci. Ann Hum Genet 59:323–336.

Sheffield VC, Carmi R, Kwitek-Black AE, Rokhlina T, Nishimura D, Duyk GM, Elbedour K, Sunden SLF, Stone EM (1994): Identification of a Bardet-Biedl syndrome locus on chromosome 3 and evaluation of an efficient approach to homozygosity mapping. Hum Mol Genet 3:1331–1335.

Skol AD, Scott LJ, Abecasis GR, Boehnke M (2006): Joint analysis is more efficient than replication-based analysis for two-stage genome-wide association studies. Nat Genet [Epub ahead of print].

Spielman RS, Ewens WJ (1996): The TDT and other family-based tests for linkage disequilibrium and association [editorial]. Am J Hum Genet 59:983–989.

Spielman RS, Ewens WJ (1998): A sibship test for linkage in the presence of association: The sib transmission/disequilibrium test. Am J Hum Genet 62:450–458.

Spielman RS, McGinnis RE, Ewens WJ (1993): Transmission test for linkage disequilibrium: The insulin gene region and insulin-dependent diabetes mellitus (IDDM). Am J Hum Genet 52:506–516.

Terwilliger JD (1995): A powerful likelihood method for the analysis of linkage disequilibrium between trait loci and one or more polymorphic marker loci. Am J Hum Genet 56:777–787.

Terwilliger JD, Ott J (1992): A haplotype-based "haplotype relative risk" approach to detecting allelic associations. Hum Hered 42:337–346.

Terwilliger JD, Weiss KM (1998): Linkage disequilibrium mapping of complex disease: Fantasy or reality? Curr Opin Biotechnol 9:578–594.

The International HapMap Project (2003): Nature 426(6968):789–796.

Thomson G (1995): Mapping disease genes: Family-based association studies. Am J Hum Genet 57:487–498.

Waldman ID, Robinson BF, Rowe DC (1999): A logistic regression based extension of the TDT for continuous and categorical traits. Ann Hum Genet 63(Pt 4):329–340.

Weinberg C (1999): Allowing for missing parents in genetic studies of case-parent triads. Am J Hum Genet 64:1186–1193.

Weinberg CR, Wilcox AJ, Lie RT (1998): A log-linear approach to case-parent-triad data: Assessing effects of disease genes that act either directly or through maternal effects and that may be subject to parental imprinting. Am J Hum Genet 62:969–978.

Weir BS (1996): Genetic data analysis II: Methods for discrete population genetic data. Sunderland, MA: Sinaur Associates.

Weir BS, Cockerham CC (1989): Complete characterization of disequilibrium at two loci. In: Feldman ME, eds. Mathematical Evolutionary Theory. Princeton, NJ: Princeton University Press, pp 86–110.

Zaykin D, Zhivotovsky L, Weir BS (1995): Exact tests for association between alleles at arbitrary numbers of loci. Genetica 96:169–178.

Zhao JH, Curtis D, Sham PC (2000): Model-free analysis and permutation tests for allelic associations. Hum Hered 50:133–139.

CHAPTER 13

Sample Size and Power

YI-JU LI, SUSAN SHAO, and MARCY SPEER

INTRODUCTION

Before undertaking linkage analysis, it is critical to know whether the available pedigree information is sufficient to allow detection of the gene(s) underlying the trait of interest. Similarly, for association analysis, one needs to consider whether the sample size, allele frequencies of marker loci, and density of the markers are sufficient to detect the association between a marker and the trait of interest. In certain circumstances, estimates can be obtained prior to the collection of samples to develop sampling strategies that will ensure adequate power to detect genetic effects. Such information can be useful in assessing the costs of a study and in determining optimal study design.

The interpretation of power studies is based on fundamental statistical concepts. For a genetic linkage study, the investigator wants to know whether two loci are linked to one another. In this case, the null hypothesis is that the two loci are not linked and the alternate hypothesis is that the two loci are linked. In genetically complex diseases this scheme may be modified so that statistical testing is applied to search for association as opposed to, or in addition to, linkage. Then the question becomes whether the trait allele and an allele at a marker locus occur together more frequently than expected by chance. Following a statistical test, the null hypothesis is either rejected or not. In the latter case, insufficient data are available to reject the null hypothesis. In statistical testing, the alternative hypothesis is never "accepted"; evidence can only be accumulated in support of the alternative hypothesis.

Four possible outcomes of an experiment exist. If the null hypothesis is rejected, the investigator may be correct or incorrect. Similarly, if the null hypothesis is not rejected, the investigator may be correct or incorrect. These four outcomes are shown in Table 13.1. A type I error, quantified as α, occurs when the null hypothesis is incorrectly rejected. This occurrence is known as a false-positive outcome. A type

Genetic Analysis of Complex Diseases, Second Edition, Edited by Jonathan L. Haines and Margaret Pericak-Vance
Copyright © 2006 John Wiley & Sons, Inc.

TABLE 13.1. Relationship of Type I and Type II Errors in Hypothesis Testing

Decision	True Hypothesis	
	H_0: No Linkage	H_a: Linkage
Accept H_0 (no linkage)	$1 - \alpha$	β
Reject H_0 (linkage)	α	$1 - \beta$

II error, quantified as β, occurs when a false null hypothesis is not rejected; this occurrence is known as a false-negative outcome. The power of the study, or the probability of detecting a correct positive result (e.g., linkage), is quantified by $1 - \beta$ (Casella and Berger, 1990).

Power studies in linkage analysis are frequently divided into two general categories, depending on whether the underlying genetic model is known or unknown. In most diseases with a simple Mendelian inheritance pattern, the components of the genetic model are known with enough confidence to allow straightforward calculations that well predict the sample sizes, type I and type II errors, and power of the study. However, in most genetically complex diseases where the mode of inheritance is uncertain, these straightforward approaches can be misleading and often are not the best choice. Although the "answers" about sample size, error frequencies, and power usually would be very desirable to have, they are frequently based on assuming information about the disease that is not known with certainty.

Throughout this chapter, we will discuss various considerations in determining required sample size to design linkage and association studies for complex diseases. Here, we first review two important concepts involved in linkage and association studies, identical-by-descent (IBD) sharing and relative risk to relatives (λ). Identical-by-descent sharing refers to two alleles that originate from the replication of one single ancestral allele. The expected proportion of alleles shared IBD for any particular type of relative pairs has an effect on both linkage and association studies. Figure 13.1 shows the average proportion of genes shared IBD for different types of relative pairs. A general rule of thumb is that the proportion of genes shared IBD between a particular type of relative pair is equal to $1/2n$, where n is the degree of relationship. For instance, genes between the parent–offspring pair share $1/2$ IBD because of their first-degree relationship.

As defined in Chapter 3, the λ-value is a ratio of two risks that describes the increase in risk to a relative of an affected individual compared to the prevalence of the trait in the general population. Most frequently, the following value is reported:

$$\lambda = \frac{\text{trait risk for siblings of affected individual}}{\text{general population trait prevalence}}$$

Type of relationship	Proportion of genes shared identical-by-descent
Siblings	1/2
Parent–offspring	1/2
Half-siblings	1/4
Avuncular	1/4
First cousins	1/8
Grandparent–grandchild	1/4

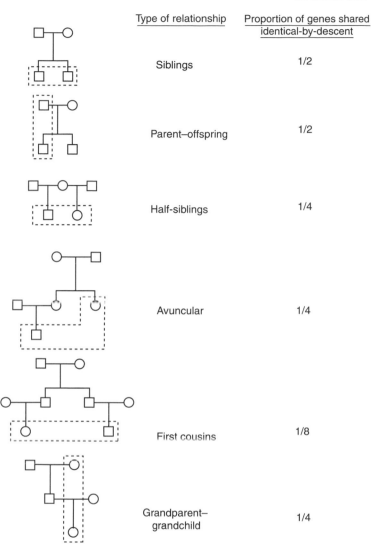

Figure 13.1. Allele sharing between relatives of different types. Note that although sibpairs and parent–child pairs share 50% of their genetic material in common, the type of sharing is different: A parent transmits exactly 50% of his or her genetic material to each offspring, while on average a pair of siblings will share 50% of their genetic material in common— some portion of the 50% from the mother and the remainder from the father.

As λ increases, so does the presumptive genetic contribution to the trait. It should be noted, however, that as the trait incidence in the population increases, overall λ-values will generally decrease.

Several factors are known to affect power to detect linkage or association, regardless of the type of analysis. These include locus heterogeneity, frequency of

TABLE 13.2. Factors Affecting Power to Detect Linkage and/or Association

Factor	Effect
Heterogeneity	When more than one locus leads to the trait phenotype, more families are needed to detect a significant result.
Recombination fraction between trait and marker locus result	The further the marker locus is from the trait locus, the more families are required to detect a significant result.
Marker allele frequency specification	Incorrect specification of marker allele frequencies can lead to an increase in either type I or type II errors.
Phenocopies and diagnostic misclassification	The higher the frequency of phenocopies or misdiagnosis, the more families are required to detect a significant result.
Marker polymorphism	The lower the heterozygosity value of a marker locus, the more families (or genotyped markers) are required to detect a significant result.

phenocopies, marker polymorphisms, recombination between disease and marker locus, and accuracy of the specification of marker allele frequencies. General principles regarding these factors are outlined in Table 13.2.

POWER STUDIES FOR LINKAGE ANALYSIS: MENDELIAN DISEASE

Information Content of Pedigrees

In small pedigrees with all critical individuals available for both phenotyping and genotyping, it is sometimes possible to assess the available information in the pedigree by visual inspection. Guidelines for the contribution of particular offspring to the logarithm of the odds (LOD) score under linkage phase-known conditions are shown in Table 13.3. In an autosomal dominant disease when linkage phase is known and there is no recombination between the disease and marker locus, affected individuals contribute 0.30 to the LOD score. The contribution of unaffected individuals is dependent on the penetrance of the disease allele. As penetrance decreases, so does the contribution of an unaffected individual to the LOD score, because the ability to accurately score these unaffected individuals (who may or may not be gene carriers) as recombinants or nonrecombinants is reduced. Thus, in a disorder that is autosomal dominant and fully penetrant and where linkage phase is known, a minimum of 10 meiotic events (children) are necessary to generate a LOD score of 3.0 when no recombination occurred between the trait and marker locus. Under similar conditions for autosomal recessive disorders, affected individuals contribute 0.60 to the LOD score. In recessive diseases, gametes from both parents contribute linkage information, whereas in dominant diseases only the affected parent contributes linkage information. The contribution of

TABLE 13.3. Approximate Contribution of Affected and Unaffected Offspring to the LOD Score under Dominant and Recessive Inheritance, Assuming Known Linkage Phase, No Recombination between Trait and Marker Locus, and No Phenocopies

	Penetrance		
	1.00	0.90	0.80
Dominant			
Affected	0.30	0.30	0.30
Unaffected	0.30	0.26	0.22
Recessive			
Affected	0.60	0.60	0.60
Unaffected	0.12	0.11	0.10

unaffected individuals is minimal, regardless of penetrance. Thus for a recessive disorder, a minimum of five fully informative nuclear pedigrees with two affected offspring and a marker that demonstrates no recombination with the disease gene are required to obtain a LOD score of 3.0.

In common and genetically complex disorders, the assumption of a single rare disease allele does not hold, making the assumption that the spouse does not carry the disease allele invalid in the autosomal dominant case. Thus, in a complex disease the determination of whether a parent is heterozygous at the trait locus is not straightforward, especially if penetrance of the disease allele is incomplete. Furthermore, given the suspected underlying heterogeneity of most complex diseases (e.g., genetic and/or environmental contributions), estimating the underlying genetic model in advance is virtually impossible.

Computer Simulation Methods

Although small pedigrees are amenable to visual inspection and casual assessment of information content, for most pedigrees obtained in a research setting, such an approach is not feasible. Power assessments thus are frequently done with the assistance of computer programs.

For Mendelian diseases, studies of power are frequently done after an investigator has talked to families willing to participate in the study and has obtained the relevant pedigree information, including pedigree structure, affection status, and which family members are willing to participate in the study. Available computer programs for assessing power in Mendelian disease are based on Monte Carlo methods, which utilize pseudorandom numbers to assign values for analysis to particular situations.

To initiate a simulation process, the investigator must provide to the computer program information about the genetic model of the trait under study, including the degree of dominance for the trait, the penetrance, the frequency of the trait allele, characteristics of the genetic marker (number and frequency of marker alleles and the recombination fraction between the trait and marker loci), the pedigree structure, disease phenotypes for family members, and their availability for the study. The investigator must also provide a value for the extent of heterogeneity. The computer in turn will simulate genetic marker data for the number of replicates specified by the investigator. Each replicate is analyzed and the results are compiled and summarized.

Three computer programs generally available for power studies in Mendelian diseases are SIMLINK (Boehnke, 1986; Ploughman and Boehnke, 1989); SLINK (Ott, 1989; Weeks et al., 1990a) and companion analysis programs MSIM, LSIM, ISIM, and ELODHET; and SIMULATE (Ott and Terwilliger, 1992). All three programs allow simulation of a marker locus unlinked to a disease locus. SLINK and SIMLINK, however, allow the simulation of genetic marker data for a locus linked to a disease gene. These two programs are very similar. One major algorithmic difference between the two programs is in the way they assign genetic marker genotypes given information on the trait phenotypes of members in the pedigree ("conditional" on disease status): SLINK uses a series of risk calculations and SIM-LINK first traverses the entire pedigree, assigning disease genotypes conditional on disease phenotypes. The most critical differences from a practical perspective are that SIMLINK allows a thorough evaluation of the power and type I error rates in one analysis, while multiple analyses must be performed to generate the same information with SLINK. However, SLINK allows the simulated data to be analyzed under a different model than that used to generate the data, which SIMLINK does not. These differences and others are summarized in Table 13.4.

Examples of the process for both conditional and unconditional Monte Carlo simulation are presented in the sections that follow.

Example 1: Unconditional Simulation—Trait and Marker Locus Unlinked. There are numerous uses for simulation of genetic marker data unlinked to a trait marker locus. For instance, investigators may wish to know about the frequency of false-positive results, or they may require assistance in developing genomic screening strategies based on average exclusion from available pedigree material. For this first example, the disease locus was assumed to demonstrate 50% recombination (the "true" recombination fraction) with a four-allele marker with equally frequent alleles; in other words, the disease and marker locus are unlinked to each other. The disease is segregating as an autosomal dominant disorder with complete penetrance of the disease allele, which has frequency of 0.0001 in the general population. Appendix 13.1 shows the process for generating simulated genetic marker genotypes utilizing family 1713 (Fig. 13.2). Note that although the investigator knows the trait phenotypes of members of the pedigree and this information is provided to the computer, it is not used in the calculations since the trait and marker loci are not linked to one another.

Section A in Appendix 13.2 gives the LOD scores associated with the simulated genetic marker genotypes as described in Appendix 13.1, analyzed under the identical genetic model as was input into the simulation (i.e., dominant, fully penetrant disease, disease allele frequency 0.0001, tetra-allelic marker with equally frequent alleles). Note that for this particular replicate the pedigree allows the exclusion of

TABLE 13.4. Practical Differences between Simulation Programs SIMLINK and SLINK

Property	SIMLINK	SLINK
Ability to handle penetrance functions	Often cannot duplicate penetrance function the user would choose to utilize in linkage analysis, especially when using LINKAGE programs. Program simulates and analyzes data under the identical generating model.	Since LINKAGE and its variants are the most frequently utilized linkage analysis packages, penetrance functions can be identical or different between simulation studies and actual data analysis.
Output	Output is thorough and detailed; calculates results assuming both a linked and an unlinked marker, so the investigator can obtain both power and exclusion information from a single run.	Output is less detailed than SIMLINK; additionally, because analysis modules MSIM, LSIM, ISIM, and ELODHET are separate from simulation module SLINK, two separate runs, one for data simulation and then a second for analysis, must be performed to generate results. To obtain both results under the null hypothesis of linkage and nonlinkage, two separate sets of simulations and analyses must be performed. At the end of the analysis runs, a count of the total number of replicates that exceeded specified threshold values is output, providing an intuitive summary of results.
Flexibility	Program does not routinely output replicate pedigree data but can be modified to do so.	Program outputs a very large file with the pedigree replicates in them. These data can be extremely useful if additional studies (e.g., for investigating the robustness of conclusions to model misspecification) need to be performed. Program can simulate genetic marker data given that some individuals have already been genotyped at the marker locus.

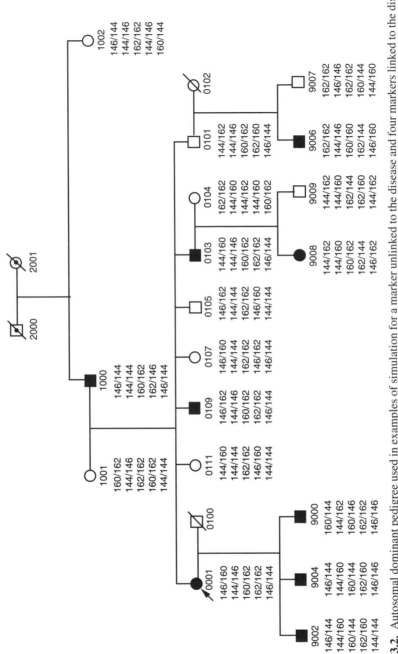

Figure 13.2. Autosomal dominant pedigree used in examples of simulation for a marker unlinked to the disease and four markers linked to the disease locus at 5% recombination. The first set of genotypes under the pedigree individual numbers are those generated from the unconditional simulation in Appendix 13.1. The next four sets of genotype results beneath each individual in the pedigree are the genotypes generated for each of four separate replicates from the simulation that assumes linkage between the disease and marker locus at $\theta = 0.05$. Genotypes are given only for sampled individuals.

approximately 12 cM on either side of the marker for the location of the disease gene using the traditional -2.0 LOD score for exclusion.

Although in this example only one replicate was performed, in practice several hundred—or in many cases up to 1000—replicates usually are performed for such power studies to obtain estimates of the results that are as accurate as possible. By averaging the exclusion region obtained in these series of replicates, the investigator can develop a rational plan for screening the genome using the available pedigree material. For instance, if after performing 1000 replicates of family 1713 as described above, an investigator found that the average exclusion interval for this pedigree was 10 cM, he or she could choose to perform a genomic screen using markers spaced at 20-cM intervals.

Example 2: Conditional Simulation—Trait and Marker Locus Linked. Usually of more interest is the ability to detect true linkage, or the power of the study. In this case, the true recombination fraction between the disease and marker locus would be set at less than 0.50. Simulating genetic marker data conditional on disease status also allows the investigation of the rate of false negatives. Appendix 13.3 shows an example of the process of conditional simulation, first assuming complete penetrance at the disease locus and then allowing for reduced penetrance. Note that there are several algorithms for performing this process; this very simple example simply illustrates the process.

In this example complete penetrance for the trait locus has been assumed. Four separate sets of genotype results, or replicates, were performed assuming a true recombination fraction between the disease and marker locus of 0.05; these are shown on the pedigree in Figure 13.2. The results are summarized in Appendix 13.2, Section B. Note by looking at the results in the pedigree that in replicate 1, by chance, a key individual in the pedigree (individual 1000) is uninformative because he is homozygous at the marker locus. In replicate 2, the pedigree is fully informative with no recombination events between the disease and marker locus, even though the "true" recombination fraction between the disease and marker loci is 5%. In replicate 2, this series of chance events provides the highest attainable LOD score for this particular pedigree. If in the entire series of replicates performed this particular scenario had not occurred, the simulated maximum LOD score would be lower than this one. In replicate 3, there is no recombination between the disease and marker loci, although the pedigree is not fully informative. In this replicate, individuals 0001 and 0103 are homozygous at the marker locus, thereby reducing the attained LOD score even though the highest LOD score still occurs when $u = 0.00$. In replicate 4, the pedigree is fully informative but two recombination events have occurred between the disease and marker locus, so that the best estimate of the recombination fraction in this replicate is 0.15 even though the "true" recombination fraction is 0.05.

Definitions for Power Assessments

The *average LOD score* is calculated as the sum of the LOD scores at a selected value of the recombination fraction divided by the number of replicates. This result is most usefully interpreted in conjunction with its standard deviation, which provides an assessment of the variability of the LOD score. For instance, the average LOD score for family 1713 at $u = 0.05$ is calculated as $(0.66 + 3.86 + 2.47 + 1.30)/4 = 2.07$. This value provides the investigator with an estimate of how much information, on average, should be expected from the available pedigree information at a particular recombination fraction.

The *maximum attainable LOD score* is the highest LOD score obtained at any value of the recombination fraction in the entire set of replicates. In this example, the maximum attainable LOD score is 4.21, which is attained in replicate 2 at $u = 0.0$. This value is very useful in that it allows the investigator to gain insight into the full information content of the available pedigree material, since the true maximum attainable LOD score is the highest potential LOD score when all meioses are fully informative under the assumed model and allele frequencies. Note that the true maximum attainable LOD score may not be attained even in a series of multiplereplicates if, by chance, not all the meioses are informative. In considering the significance of a particular LOD score, it is useful to know whether all the available linkage information has been captured. The observed maximum LOD score of a pedigree may not match the maximum attainable LOD score for a pedigree for several reasons, including lack of informativeness of key individuals and differences in allele frequencies from simulated results and true results, differences between true and simulated penetrances, among others.

The *average maximum LOD score* is the sum of the highest LOD scores obtained in each replicate, regardless of where those scores occur on the LOD score curve, divided by the number of replicates. In this example, the average maximum LOD score is calculated as $(0.84 + 4.21 + 2.71 + 1.59)/4 = 2.34$. Note that the highest LOD score for any replicate need not, and usually will not, occur at the true recombination fraction. There are many caveats associated with the use of this particular measure of information content of a pedigree, the most compelling of which is that it tends to paint too rosy a picture of the true expected outcomes. Note that while the average LOD score is additive across pedigrees, the average maximum LOD score is not, which is an additional disadvantage to using the average maximum LOD score as a measure of pedigree information within a study.

Available pedigree simulation packages will also directly compute the type I and type II error rates and the power of the available material for specific constant values of the LOD score. For instance, the investigator may need to know the probability of obtaining a false positive (a LOD score of 3.0 or greater when the data was simulated assuming free recombination between the disease and marker locus), the probability of obtaining a false negative (a LOD score of -2 or less or some other specified constant value when the disease and marker loci were fixed as linked to one another), and the probability of correctly obtaining a LOD score of 3.0 or more when the disease and marker loci are set as linked to one another. In the example using the data presented in Appendix 13.2, the type I error rate cannot be calculated because only one replicate was performed; the type II error rate is calculated as $\frac{0}{4}$ or 0%, and the power (to obtain a LOD score exceeding 3.0, which represents true linkage) is calculated as $\frac{1}{4} = 25\%$. Here it is important to note that in most cases the investigator will choose to simulate hundreds or even thousands of replicates to obtain estimates of the necessary values (and their standard deviations when appropriate) that are as precise as possible.

Another useful application of computer simulation methods is in determining the empiric p-value associated with a study outcome. By generating genetic marker genotypes under the assumption of free recombination between the disease and marker locus for many replicates of the pedigree structures, the investigator can generate a series of LOD score results that could be obtained from a random genomic screen. By counting all simulated LOD scores exceeding the LOD score from the observed data and dividing by the total number of replicates, the investigator can find the empiric p-value for the observed data (Weeks et al., 1990a). For this application, since the observation of the outcome is so rare, the investigator will typically perform tens of thousands of replicates to obtain a reasonable estimate of the empiric p-value.

Yet another application of computer simulation involves assessing the effect of maximization over models, which may be a more powerful approach than assuming a genetic model when in truth the model is unknown (Hodge et al., 1997). If the true underlying genetic model is unknown, an investigator may choose to analyze several different genetic models (e.g., different degrees of dominance, disease allele penetrance, and frequency) with the same marker data to see which model gives the highest LOD score (Elston, 1989; Greenberg, 1989). In these cases, some correction for multiple comparisons must be performed. This procedure has been illustrated with an example in which chromosome 5 markers were studied in schizophrenia pedigrees (Sherrington et al., 1988). Using three diagnostic schemes that differed in their classification of affected individuals, Sherrington and colleagues found that one of the three models generated a LOD score of 6.5 versus chromosome 5 markers but failed to correct for the fact that three different diagnostic schemes had been used. Weeks et al. (1990b) used computer simulation methods and an approximate reconstruction of the original linkage report to show that this LOD score of 6.5 was inflated by 0.7–1.5 LOD score units.

Hodge et al. (1997) performed a simulation study to assess the magnitude of the increase in type I error when the investigator has maximized both the linkage analysis over penetrance and disease dominance or recessiveness. They developed guidelines for increasing the test criterion under either or both these scenarios. As a general rule of thumb, the critical value of the test statistic should be increased by a LOD value of 0.3 when one of the foregoing parameters is maximized and by 0.6 when both penetrance and dominance/recessiveness are maximized.

POWER STUDIES FOR LINKAGE ANALYSIS: COMPLEX DISEASE

In complex diseases, power studies are not as straightforward as in Mendelian diseases, but they can nonetheless help to guide an investigator in determining whether a particular disorder is amenable to gene mapping. For complex diseases, discussions about power are generally limited to approximate estimates of sample size required to detect linkage making various assumptions about the underlying genetic model. Guidelines for sample size depend on how closely related the sampled individuals (usually affected) are, how tightly linked the disease and marker locus are, how polymorphic the marker system is, and the mode of inheritance of the trait. These issues are frequently interrelated, and reliable estimates for these parameters are often unknown; consequently, developing standard guidelines is difficult. Most linkage analysis methods focus on one single disease locus at a time, so most power studies in complex diseases are based on this premise. However, methods for considering multiple loci at a time have been proposed (Lin, 2000). These issues are discussed briefly. However, it is critically important to recognize that such studies are entirely dependent on the assumptions underlying the genetic model, the validity of which is entirely unknown.

Two of the primary assumptions that must be made about an underlying genetic model involve the frequency of the trait allele and the mode of inheritance of the disease-predisposing locus. The frequency of the trait allele affects power and sample size calculations as well: If the disease allele is rare, then the probability that two affected individuals within a pedigree are IBD at the trait locus is high, whereas if the trait allele is common, the IBD probability decreases. Considered heuristically, the contribution of a distantly related affected relative pair (ARP) such as an affected cousin pair should be greater than the contribution of an affected sibpair when the number and frequency of contributing loci are small. It is a priori more striking if a pair of affected cousins (who share only one-eighth of their genetic material in common by chance alone) share alleles in common than if a pair of affected siblings (who share half of their genetic material in common by chance alone) share alleles in common.

While the true underlying mode of inheritance is rarely known in a genetically complex disease, the investigator can often develop a "gestalt" for the potential underlying genetic factors based on visual inspection of pedigrees. For instance, if a significant proportion of the pedigrees show transmission from affected parent to affected child, these data suggest that an underlying genetic effect may be dominant. Similarly, if carefully screened pedigrees fail to show ARPs other than siblings, a dominance component is unlikely, unless the investigator is willing to assume a dominant allele with low penetrance. As a general rule of thumb, if the underlying trait is recessive, then affected sibpairs are the preferred sampling unit. If the trait is dominant, more distantly related ARPs usually represent the most efficient approach. However, for a dominant trait, the most powerful type of ARP is dependent on the λ-value and on the frequency of the trait. In general, as the frequency of the trait allele decreases, the choice of ARPs should tend toward more distantly affected relatives. For example, for a rare trait allele, the most powerful ARP sampling unit is grandparent–grandchild. However, for a common trait allele, the most powerful ARP sampling unit is affected siblings. For candidate genes, when the recombination fraction between the trait and candidate gene is 0.0, the power of these different types of ARPs is virtually the same regardless of the underlying mode of inheritance.

The majority of linkage analysis methods focus on a single trait locus. The multi-locus disease trait is, in general, more complicated. However, considerable effort has been devoted to studying multilocus disease traits (Lathrop and Ott, 1990; Knapp et al., 1994; Lin, 2000). The program TMLINK (Lathrop and Ott, 1990) in the

LINKAGE package (Lathrop et al., 1984) was the first program to perform two-locus disease trait analysis, but the analysis is generally very computationally intensive. The Markov chain Monte Carlo (MCMC) approach was suggested to reduce the computational burden (Thompson, 1994). Lin (2000) describes in detail how to tackle the two-locus disease model using the MCMC approach (MCLINK-2LOCUS program) and demonstrated its computational feasibility and greater power than the TWLINK method. Here, for simplicity, we focus the discussion by assuming that the investigator is searching for an individual gene that contributes to the underlying phenotype.

Discrete Traits

Affected Sibpairs. Much of the work in assessing power and sample size in complex disease has focused on the use of affected sibpairs. For example, Risch (1990) showed that the power of a sample consisting of affected sibpairs is dependent on λ_R (the ratio of the recurrence risk to a relative of type R of a proband to the prevalence of the disease in the general population) and the recombination fraction between the trait and marker locus. He showed, for instance, that approximately 60 affected sibpairs would provide 80% power to detect linkage in a disorder in which λ_S (where S indicates that the affected relative pair is siblings) is 5.0 when the recombination fraction between the trait and marker locus is 0.0. As expected, a larger sample size is required to detect linkage as the recombination fraction between the trait and marker locus increases. Some estimates of power, sample size, and effects of l are shown in Table 13.5 (see also Chapter 11).

TABLE 13.5. Guidelines on Sample Size for Affected Relative Pairs

A. Approximate Minimum λ_S for Power 0.80 and 0.90 for Specified Number of Affected Sibpairs When $\theta = 0.0$ for Fully Informative Marker

Number of Affected Sibpairs	Power to Detect Linkage	
	80%	90%
60	5.0	7.0
80	3.5	5.0
100	3.0	4.0
200	2.0	2.3

B. Approximate Power to Detect Linkage for Specified Number of Affected Sibpairs and Values for λ_S and θ for Fully Informative Marker

Number of Affected Sibpairs	$\lambda_S = 2.0$			$\lambda_S = 5.0$		
	$\theta = 0.0$	$\theta = 0.05$	$\theta = 0.1$	$\theta = 0.0$	$\theta = 0.05$	$\theta = 0.1$
40	0.05	0.01	0.01	0.4	0.1	0.05
100	0.05	0.18	0.35	0.28	0.6	0.95
300	0.95	0.9	0.4	>0.95	>0.95	>0.95

Source: Risch (1990).

This work was extended by Hauser et al. (1996) for multipoint analysis. As in multipoint linkage analysis for Mendelian disease, the ability to flank a particular interval with markers increases the information in a dataset and the potential to exclude an area of interest. When searching for a locus that confers a λ_S of 2.0, a 10-cM interval flanked with markers with heterozygosity value of 0.75 and a sample size of 200 affected sibpairs can generate a LOD score of 3.0 with a power of 0.51. As the number of affected sibpairs or the λ_S increases, so does the power. Similarly, ability to exclude an interval is dependent on the available sample size, the magnitude of the genetic effect, and the intermarker genetic distance. Table 13.6 summarizes data about sample size used to detect and exclude linkage in a study designed around a multipoint affected sibpair approach.

Inclusion of Unaffected Siblings. The decision about whether to include unaffected siblings in a study is usually based on availability and cost of recruitment and the costs associated with the additional genotyping of these individuals. The advantages of collecting unaffected siblings are many. For instance, the inclusion of the genotypic marker data on unaffected siblings may increase the ability to determine the IBD status of genetic markers in affected sibpairs, which is especially important in late-onset disorders where parents are unavailable for sampling. Additionally, as pointed out by Elston et al. (1996), performing a linkage study utilizing only affected sibpairs is akin to performing a study without controls. Without the unaffected individuals, it is virtually impossible to determine whether a significant result is due to a true linkage or to some other factor, such as meiotic drive at a marker locus. The collection and availability of unaffected siblings also allows for the eventual assessment of penetrance and expression and for considering genes that may potentially modify the phenotype. Although some available methods allow for the

TABLE 13.6. The Power to Detect or Exclude Linkage for Various Sample Sizes N of Affected Sibling Pairs and Contributions of the Locus to the Genetic Effect (λ_S)[a]

	Number of Affected Sibling Pairs[b]								
	100			200			400		
λ_S	p(LOD) > 3.0	%Excl	%All	p(LOD) > 3.0	%Excl	%All	p(LOD) > 3.0	%Excl	%All
1.4	0.02	1.9	1.5	0.06	13.8	10.7	0.23	49.0	43.9
1.6	0.05	10.9	7.9	0.19	43.9	38.1	0.59	82.1	78.5
2.0	0.15	33.6	27.6	0.51	73.1	68.4	0.94	95.7	94.5
3.0	0.45	69.9	64.0	0.90	94.9	93.1	1.00	99.8	99.7

[a]An intermarker distance of 10 cM is assumed between flanking markers, each of which has four equally frequent alleles (heterozygosity = 0.75).
[b]p(LOD) > 3.0, the probability of obtaining a LOD score >3.0 given that the interval contains a disease locus with λ_S as specified; %Excl, the average percentage of a 10-cM interval excluded when no disease locus is present; %All, percentage of intervals in 1000 replicates excluded in their entirely when no disease locus is present.
Source: Hauser et al. (1996).

inclusion of unaffected relatives, they may make the assumption that unaffected relatives are non–gene carriers. If the underlying disease gene has low penetrance, such an assumption may be unrealistic. Most current applications rely exclusively on allele sharing in affected relatives.

Special Problems with Sibpair Data: Considerations of Independence.

Pedigree material, particularly with respect to sibpairs, is usually summarized by simple counts. Some families, however, will have more than one set of affected siblings. The formula for calculating the number of affected sibpairs is $n = m (m - 1)/2$, where n is the total number of affected sibpairs and m is the number of affected siblings. For example, in Figure 13.3a, the parents have three affected children and the number of affected sibpairs that can be formed is $3 \times \frac{2}{2} = 3$ (pairs = individuals [3,4], [3,5], [4,5]). Similarly, the parents in the pedigree Figure 13.3b have five affected children, so $5 \times \frac{4}{2} = 10$ affected sibpairs (pairs = individuals [3,4], [3,5], [3,6], [3,7], [4,5], [4,6], [4,7], [5,6], [5,7], [6,7]) can be created. When multiple affected sibpairs from the same nuclear family are used, the issue of independence is important. Hodge (1984) showed that the pairs formed from more than two affected siblings within a sibship are not independent, but the consequences of this nonindependence have been the subject of considerable debate in the literature. The pedigree in Figure 13.3c illustrates the nonindependence of the affected sibpairs. For instance, once it has been determined that individuals 1 and 2 share one allele IBD and individuals 2 and 3 share two alleles IBD, then the number of alleles (one) that individuals 1 and 3 share is also known. The best mechanism for dealing with this nonindependence is still a matter of intense debate; however, it seems fair to say that it is less problematic the larger the overall sample size. Therefore, for general screening in complex disease, it is accepted practice to assume that the sibpairs are in fact independent. However, once "interesting" areas of the genome have been identified, it is critically important to assess the significance of any nonindependent data, since ignoring it entirely can lead to falsely

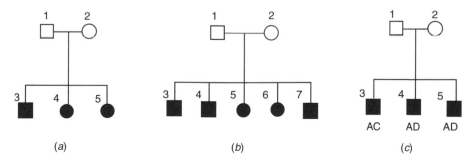

Figure 13.3. (a) Three affected siblings yield 3 nonindependent affected sibpairs. (b) Five affected siblings yield 10 nonindependent affected sibpairs. (c) When the number of alleles shared IBD is known for any two of the three pairs of affected siblings, the identical by descent IBD sharing is known for the third affected sibpair.

high significance levels. Daly and Lander (1996) have suggested a simulation approach to determining the correct empiric *p*-value (see above) of such an "interesting" result for a study in which some families have more than two affected siblings, given information on the sibship sizes and frequencies of the marker alleles.

Affected Relative Pairs of Other Types. In many complex diseases, other types of affected relative pairs will be identified; indeed, trends for different types in a disease may shed some light on the underlying genetic models. For instance, consider a trait in which two different loci are involved. If the disease alleles at the loci are rare and act epistatically, it is unlikely to be observed in affected relative pairs who are more distantly related than siblings. However, if the disease alleles at the loci are relatively common and have low penetrance, the pedigrees may demonstrate affected avuncular or cousin pairs. Because it is unusual for a whole study to be designed around affected relative pairs other than siblings, estimates of sample size for studies specifically geared to other types of ARPs are rare. When other types of ARPs are frequent, the investigator will frequently collect all affected patients and their connecting relatives and utilize an analysis approach other than one dependent on a sibpair design (Chapter 11).

Other Considerations. Some difficulties in the estimates of sample size or power of available pedigree material in the investigation of complex traits warrant consideration. For instance, the presence of phenocopies or genetic heterogeneity will increase the amount of material necessary to correctly identify a linkage (Bishop and Williamson, 1990). Additionally, the lower the heterozygosity of the marker to be tested, the more material will be needed on average to detect a significant effect.

Empiric Examples. Associations for common and genetically complex disorders have been identified in many disorders using the APM (affected pedigree member) method of linkage analysis. Examples include late-onset Alzheimer's disease (Pericak-Vance et al., 1991), which utilized 87 ARPs (including those generated by inference of genotypes in deceased or unavailable individuals utilizing their offspring, both affected and unaffected); early-onset familial breast cancer (Hall et al., 1990), which utilized 69 ARPs; and familial melanoma (Cannon-Albright et al., 1992), which utilized 185 ARPs. In melanoma and breast cancer, these associations were observed in the less frequent Mendelian forms of these common disorders and actually represented true linkages. Since the pedigrees were large enough to permit the identification of the underlying linkage, an empiric assessment of the actual number of ARPs in each of the studies is less meaningful. The linkages were also identified with standard, parametric linkage analysis. The meaning of the significant association results in late-onset Alzheimer's disease were more difficult to unravel. These initial studies were performed using the original version of the APM (Weeks and Lange, 1988), and the resulting effect of the *APOE* gene was found because it was so large. Interestingly, recent extensions to the APM method allow the inference of missing genotypes when possible, and these recent extensions approximate the more powerful IBD methods (Davis et al., 1996). In the late-onset Alzheimer's disease dataset (Pericak-Vance et al., 1991), when ARPs generated by inference of missing genotypes in affecteds are included, the available numbers of ARPs are increased by 29. Clearly, the addition of this much data just by changing the computing algorithm represents a considerable increase in power to detect genetic effects. This scenario—a dataset comprising varying numbers of ARPs of various types—is consistent with many realistic datasets, particularly in late-onset disorders.

Genomic Screening Strategies: One-Stage versus Two-Stage Designs. Interestingly, although much literature is focused on the use of affected sibpairs to study complex disease, it frequently is not the most powerful approach. Elston (1992) and Elston et al. (1996) consider the issue of study design for complex genetic disease in terms of "cost" of the study. Their computing algorithm DES-PAIR is available through the serial analysis of gene expression (SAGE) package. These authors (Elston, 1992; Elston et al., 1996) point out that for a wide range of scenarios in which the genomic screening is done via a two-stage process (sometimes called a grid-tightening approach), the optimal sampling strategy for ARPs is affected grandparent–grandchild pairs because of the limited opportunity for recombination. A two-stage genomic screen involves genotyping a certain number of genetic markers and then following up regions of interest defined by reaching or exceeding a certain predefined significance level. This is done by genotyping markers on either side of the "interesting" markers. Grandparent–grandchild pairs are difficult to find, particularly if the disorder has onset in adulthood. After grandparent–grandchild pairs, the optimal sampling strategies in terms of study cost usually follow the sequence of half-siblings followed by avuncular pairs followed by sibpairs. Affected cousin pairs have varying degrees of utility depending on several considerations. In general, as λ increases, so does the power of affected cousin pairs. Ease of sampling is almost the reverse, however, as sibpairs are generally the easiest to collect.

DESPAIR is valuable for designing the optimal two-stage genomic screening strategy and also for estimating the number of pairs of affected relatives required to attain a certain power. Elston et al. (1996) show that for a disease gene with a λ_S of 2.0 and a cost ratio R of 100 (where cost is defined as the ratio of ascertainment costs for a person to the cost of performing one genotype), the least expensive two-stage genomic screening strategy is to collect 350 affected grandparent–grandchild pairs and genotype 85 markers spaced at 38.8 cM throughout the genome; markers generating a significance level of 0.045 should then be "followed up" by testing two additional markers on either side of the "interesting" marker. Table 13.7 shows additional examples of optimal two-stage designs using different types of ARPs.

Holmans and Craddock (1997) recently reported the results of simulation studies for genomic screening strategies in a complex disease for which nuclear families with one affected sibpair were available. The researchers compared two strategies for maximizing the power of the study and the efficiency of the screen, measured in total number of genotypes performed. The strategies were grid tightening and sample splitting (which involves genotyping only a portion of the total sample in the initial screening). Decisions regarding sample-splitting strategies involved whether to genotype the parents or just the affected sibpairs and whether to genotype the entire sample or half the sample. Overall the investigators concluded that a

TABLE 13.7. Optimal Two-Stage Genomic Screening Strategies for Independent Pairs of Affected Relatives of Various Types, Utilizing a Fully Informative Marker and a Value of 100 for the Ratio of the Cost of Ascertaining a Patient to the Cost of Performing a Genetic Marker Genotype

Affected Relative Pair Type:	$\lambda_i = 2.0$					$\lambda_i = 10.0$				
	Full Sib	Half-sib	Avunc	Cousins	GP-GC	Full Sib	Half-sib	Avunc	Cousins	GP-GC
Number of stage 1 markers	168	147	175	166	85	317	224	279	208	131
Marker spacing, cM	19.6	22.4	18.9	19.9	38.8	10.4	14.7	11.8	15.9	25.2
Number of pairs of ARPs	320	359	362	477	350	26	25	25	32	24
Number of additional makers[b]	2	2	2	2	2	2	2	2	2	2
α^{*}[c]	0.060	0.056	0.059	0.055	0.045	0.096	0.083	0.078	0.07	0.066
Estimated cost[d], $	199,979	203,826	232,093	292,216	143,035	28,231	20,096	23,509	23,695	12,933

[a]Analysis assumes that a single locus accounts for λ_i (a locus-specific λ), where i refers to the type of affected relative pair, although the disease may have an overall λ larger than the locus-specific λ (and this will always be the case for a complex disease in which two or more loci act to define phenotype). Analysis was performed using DESPAIR.

[b]For stage 2 of the genomic screen, this number represents the number of additional markers to genotype on each side of a marker where $p \leq \alpha^{*}$. Note that the total number of markers to genotype for any marker with $p \leq \alpha^{*}$ will be twice the number in this column.

[c]Markers with p-values $\leq \alpha^{*}$ are followed up in stage 2 of genomic screen by genotyping additional markers.

[d]Assuming that each marker genotype costs $1.00.

Source: Elston et al. (1996).

strategy that uses both a grid-tightening and a sample-splitting approach is typically most efficient. From a practical viewpoint, decisions regarding whether to genotype parents will depend on the relative costs of genotyping in the first stage or second stage and on the investigator's knowledge about underlying sample homogeneity (and how important the availability of complete IBD information will be relative to the required genotyping—in other words, whether allele frequencies will make a big difference in the outcome). As always, decisions about strategies will depend on the interplay between correctly detecting a linkage and following up false positives.

Several genomic screens for complex disease have been published. The designs generally consisted of affected sibpairs, but in some cases unaffected siblings and other affected relative types were included. A summary of sampling strategies in selected genomic screens is shown in Table 13.8.

Quantitative Traits

Methods of mapping human quantitative trait loci (QTLs) can be applied to pheno types ranging from height and weight to obesity, blood pressure, "intelligence," and any other trait that is measurable on a continuous scale. The tools for the linkage analysis to detect QTLs have been improved dramatically in the past few years, which allows us to look for loci predisposing to human traits of a quantitative nature. The Haseman–Elston (1972) method utilized a regression model for a trait measured on each sibpair. This approach has been extended to pedigree relationships other than siblings (Amos et al., 1989; Olson and Wijsman, 1993; Elston, 2000). The Haseman–Elston method was implemented in the SIBPAL program of the SAGE package. For discrete traits using ARPs to maximize the chance of detecting linkage, the sampling scheme for Haseman–Elston-based methods is to sample pairs of relatives that are either very different from (extremely discordant) or very similar to (extremely concordant) one another. This approach ensures, to the maximum extent possible, that the family is "segregating" the trait and genes of interest. In other words, if one samples individuals regardless of their trait status (e.g., high value or low value), the investigator may include numerous pedigrees for which the predisposing trait allele is not present in either of the parents.

The other popular linkage analysis method for a quantitative trait is the variance component method, in which the trait value of each individual is modeled by a linear mixed model. The representative program for the variance component method is SOLAR (Sequential Oligogenic Linkage Analysis Routines) (Almasy and Blangero, 1998), which has been widely applied to the real data for the linkage analysis of a quantitative trait (Duggirala et al., 1996; Li et al., 2002). The variance component approach has a very general property, that is, it can accommodate pedigrees of arbitrary size and easily incorporate additional genetic, environmental effects, and interactions. The consideration of sampling schemes in this approach is somewhat different from the Haseman–Elston method. More details are presented below.

TABLE 13.8. Characteristics of Study Design for Selected Published Genomic Screens

Disease	Design	Sample Characteristics	Ref.
Type 1 (insulin-dependent) diabetes mellitus	Affected sibling pairs (ASPs)	Genomic screening in 96 British ASPs. Follow-up of interesting regions in a second set of 102 British ASPs and 84 American ASPs	Davies et al. (1994)
Type 1 (insulin-dependent) diabetes mellitus	Affected sibling pairs	Genomic screening in 61 French and North American Caucasian nonindependent ASPs. Follow-up of interesting regions in additional 253 ASPs of French, North American, and North African Caucasian descent *Note*: Data stratified based on previously identified HLA-DR3 and DR4 associations	Hashimoto et al. (1994)
Multiple sclerosis	Affected sibling pairs	Genomic screening in 100 nonindependent ASPs of Canadian descent. Follow-up of interesting regions in additional 122 ASPs. Unaffected siblings sampled and genotyped to allow inference of missing parental genotypes	Ebers et al. (1996)
Multiple sclerosis	Affected sibling pairs and extended pedigrees with affected relatives of various other types	Genomic screening in 52 families including 81 nonindependent ASPs and 58 pairs of nonsibling affected relatives. Follow-up of interesting regions in additional affected relatives. Follow-up of 23 families, including 45 ASPs and 30 pairs of nonsibling affected relatives	The Multiple Sclerosis Genetics Group (1996)

Extreme Discordant Pairs. The sampling of extremely discordant relative pairs is a potentially powerful approach to identifying genes that predispose to human quantitative traits. Specifically, such relative pairs (usually siblings) should share very few genes IBD at the trait locus (Risch and Zhang, 1995, 1996). Risch and Zhang (1996) have shown that the smallest genetic effect that can be identified using this approach with a reasonable sample size (i.e., <1500) is one that accounts for as little as 10% of the variance. Suppose, for instance, a researcher wants to detect a gene that accounts for 10% of the underlying phenotypic variance, has a trait allele frequency of 0.30, and has no residual correlation by employing a sample scheme that includes pairs of relatives in the top and bottom 10% of the overall distribution. Approximately 1482 sibpairs would be required to detect linkage allowing for a type I error rate of 0.0001 and power of 0.80. Table 13.9 summarizes additional data on required sample size to detect QTLs. In general, the presence of residual correlation between siblings, in which siblings appear to be more phenotypically similar than expected by the presence of a major genetic locus, will tend to decrease the necessary sample size, especially as the heritability decreases. As explained by Risch and Zhang (1996), this reduction in estimated sample size is because phenotypically discordant sibpairs will be more likely to be genetically discordant at the locus of interest. This is important because there is likely to be residual sibling correlation due to other genetic effects for loci with low heritability. With high heritability, most sibling correlation is probably due to that locus, and hence there is unlikely to be a large residual correlation.

TABLE 13.9. Estimated Number of Sibling Pairs Required to Detect a QTL under Various Genetic Models

Allele Frequency	Heritability = 0.10[b]				Heritability = 0.30[b]			
	$p = 0.0$	N	$p = 0.4$	N	$p = 0.0$	N	$p = 0.4$	N
				Additive Model				
0.10	1647	19,120	342	21,441	155	2356	52	2833
0.30	1482	17,357	346	22,336	120	1958	42	2829
				Dominant Model				
0.10	1567	18,153	359	22,320	143	2171	59	3119
0.30	1454	16,777	384	23,957	127	1920	65	3915
				Recessive Model				
0.10	19,984	205,218	849	41,016	18,996	195,204	753	35,514
0.30	2049	22,615	398	21,777	276	3531	185	2943

[a]Assuming a power of 0.80 and a type I error rate of 0.0001, allowing for screening from the top 10% of the distribution and the bottom 10% of the distribution.
[b]N, estimated number of siblings to be screened to obtain the necessary number of sibling pairs fitting the ascertainment criterion; r, the residual correlation.
Source: Risch and Zhang (1996).

Extreme Concordant Pairs. Sampling extremely concordant relative pairs is usually a less effective, though often more practical, ascertainment scheme than sampling extremely discordant pairs. The specific decisions about sampling are more complicated and require the researcher to have at hand information about trait allele frequency, parental mating types, and degree of dominance of the trait (Zhang and Risch, 1996). As a general rule of thumb, however, extremely concordant pairs are more powerful than extremely discordant pairs only when the underlying gene is a rare recessive.

Although in retrospect the collection of extremely discordant or extremely concordant relative pairs rather than pairs with less extreme phenotypes seems intuitive, this idea has served to revolutionize the consideration of human QTLs. There are some potential drawbacks with either approach. To ascertain the extremes of the distribution, hundreds or thousands of families may need to be screened. If the screening tests are expensive, time consuming, or logistically difficult, such an approach may not be feasible. However, if such screening can be undertaken with relative ease, a much smaller dataset can be subjected to the rigors of a genomic screen and ultimate fine mapping of susceptibility genes.

Sampling Consideration for the Variance Component Method. The power of the variance component approach in general varies according to several factors, such as genetic and environmental effects and the resemblance between pedigree members. However, the heritability of the QTL is a key component of power calculation. The greater the QTL effect is, the greater the statistical power is. A QTL with small effect may require considerably larger sample sizes (either larger family size or larger family numbers) to be detectable with comparable power (Williams and Blangero, 1999). The details of sample size and power for the variance component method were discussed in Williams et al. (1999). The residual additive heritability (a^2, the proportion of the total genetic variance that is not explained by the QTL) and the QTL heritability (q^2) are the main parameters for the sample size and power estimations. An example of sample size prediction for 80% power at a LOD score of 3.0 is listed in Table 13.10. For instance, a QTL with 40% heritability under 0.3 residual additive heritability requires 713 sibpairs, or 559 nuclear families of family size 4 (parents and two siblings), to reach 80% power to detect linkage. Overall, additional QTL heritability can reduce the sample size more rapidly than additional residual additive genetic heritability. Collecting parental phenotypes concurrently with sibling data is more efficient than ascertaining additional sibships.

The software program SOLAR (Almasy and Blangero, 1998) has a simulation subroutine to compute the expected LOD score given QTL-specific heritability and pedigree data. Investigators can obtain an estimate of how much information, on average, should be expected from the available pedigree information to assist their study design.

POWER STUDIES FOR ASSOCIATION ANALYSIS

Various statistical methods for testing association between a marker and a trait of interest have been described in detail in Chapter 12. Many papers have been

TABLE 13.10. Number of Sibships and Nuclear Families in Variance Component Linkage Analysis

n_{sibs}	$a^2 = 0.0$		$a^2 = 0.3$	
	$q^2 = 0.1$	$q^2 = 0.4$	$q^2 = 0.1$	$q^2 = 0.1$
Sibships				
2	16,500 (33000)	921 (1842)	14,732 (29463)	713 (1426)
3	5476 (16429)	294 (881)	4697 (14090)	217 (652)
4	2728 (10910)	142 (567)	2270 (9079)	102 (408)
Nuclear families				
2	16,358 (65431)	836 (3342)	13,369 (53474)	559 (2235)
3	5433 (27166)	272 (1361)	4354 (21771)	182 (908)
4	2708 (16246)	134 (801)	2136 (12818)	89 (533)

Note: Number of sampling units and number of individuals (in parentheses) required for 80% power at a LOD score of 3.0/ Entried have been rounded to the nearest integer.

Source: Williams and Blangero (1999).

published addressing the power issues of various association methods (Knapp, 1999b; Kaplan et al., 1997; Teng and Risch, 1999; Schork, 2002). Overall, although case–control design is a cost-effective approach, it suffers the confounding of population substructure ("cryptic stratification") leading to spurious associations. The family-based design becomes a method of choice, because it uses unaffected relatives as controls to circumvent the uncertainty of population stratification. However, the trade-off is that the family-based design costs more than the case–control design, because it requires the collection of DNA samples from family members, which may be difficult to obtain, particularly for late-onset diseases.

Recently, two alternative population-based approaches have been suggested. One is the "genomic controls (GC)" approach (Devlin and Roeder, 1999), which accounts the nonindependence in the case–control samples caused by population stratification. A power study showed that the family-based design using the transmission/disequilibrium test (TDT) method is more powerful when population stratification is substantial, and GC has more power otherwise. Overall, GC is at least comparable to and less expensive than family-based methods (Bacanu et al., 2000). The other method for case–control design is the structured-association method (STRAT) proposed by Pritchard and colleagues (2000), which uses a set of unlinked markers to learn about admixture in the population and applies this information to the association test. Since STRAT requires additional genotyping for the unlinked markers, it is not a less expensive approach than the family-based design unless many candidate genes are studied at the same time. Power studies demonstrated that STRAT has comparable power to the TDT in many settings and may outperform it if conflicting association exists in different subpopulations (Pritchard et al., 2000). Clearly, both the GC and STRAT methods and their extensions provide good alternative association methods to the case–control design. However, unlike family-based designs, few real datasets have used the GC and STRAT methods. Further practice and evaluation of these two approaches are needed.

Below, we discuss sample size and power issues in detail for a few commonly used association methods. A newly developed simulation program for linkage and association analysis is also described.

Transmission/Disequilibrium Test for Discrete Traits

Power studies for the TDT and its various extensions have been performed by either simulation or analytical approaches (Kaplan et al., 1997; Knapp, 1999a; Kaplan and Martin, 2001). Monte Carlo simulation is in general quite accurate, because power is assessed through many simulated pedigrees under different genetic models and recombination rates between the marker and disease locus. However, it is generally time consuming. Various analytical approaches for computing the statistical power of the TDT have been developed (Risch and Merikangas, 1996; Whittaker and Lewis, 1998; Knapp, 1999b; Chen and Deng, 2001). Knapp (1999b) presented an analytical method to estimate the approximate power of the TDT method for samples consisting of nuclear families with either a single affected child or two affected children. The power of the TDT depends on mode of inheritance (e.g., multiplicative, dominant, additive, and recessive), putative disease allele frequency (p), and genotypic relative risk (r). Table 13.11 shows the sample size needed to reach 80% power in TDT with data from singleton families (a single affected child) or affected sibpairs (ASPs), in which the type I error was controlled at 10^{-7} [assuming a TDT genome scan with 500,000 single-nucleotide polymorphisms (SNPs)]. Clearly, the ASP design requires fewer samples than the singleton design. The sample size is also greatly reduced when the genotypic relative risk is high. However, these estimates assume that only one locus contributes to the phenotype so that the attributable risk for that locus is 1.0. In the presence of allelic heterogeneity, the required samples sizes will increase. Also, when the marker locus is not a disease susceptibility locus, the power of the TDT is greatly reduced. Chen and Deng (2001) extended Knapp's (1999b) approach to handle various situations such as mixed family structures, different degrees of linkage disequilibrium between the marker and the disease susceptibility locus, and the existence of allelic heterogeneity. Their approach was implemented in a user-friendly computer program *TDT Power Calculator* (Chen and Deng, 2001).

Kaplan and Martin (2001) formulated an analytical power calculation for a general class of linkage and association tests using nuclear families with multiple affected and unaffected siblings. As we discussed in Chapter 12, various extensions of the TDT (Whittaker and Lewis, 1998; Martin et al., 1997; Abecasis et al., 2000; Lunetta et al., 2000) were proposed after Spielman et al. (1993) introduced the TDT. All of these tests are of the same general form and only differ in how one weights the marker transmission information for affected and unaffected siblings. Similar to other methods, power estimation depends on the genetic model, disease prevalence, and sampling strategy. In summary, Kaplan and Martin found that including unaffected siblings improves the power of the test if the disease prevalence is high. When parental data are available, they recommended using the method from Whittaker and Lewis (1999) and Lunetta (2000).

TABLE 13.11. Sample Size Necessary to Gain 80% Power in TDT with Singletons ($\alpha = 10^{-7}$)

rANDp	Multiplicative		Additive		Recessive		Dominant	
	Singleton	ASP	Singleton	ASP	Singleton	ASP	Singleton	ASP
4.0								
0.01	1056	230	1095	251	4,344,070	724,763	1115	258
0.10	146	48	194	76	5631	1121	231	95
0.5	101	61	218	132	207	94	696	492
0.8	216	158	553	359	259	175	9384	7193
2.0								
0.01	5755	1954	5755	1954	38,654,522	12,404,460	5947	2049
0.10	689	263	689	263	45,071	14,940	949	399
0.5	338	179	338	179	957	424	1839	1108
0.80	634	392	634	392	851	489	21,998	14,328
1.5								
0.01	19,233	7752	18,733	7449	154,174,890	60,955,123	19,755	8012
0.10	2210	939	1755	684	174,694	69,839	2897	1292
0.50	947	484	464	229	3099	1422	4568	2546
0.80	1658	939	698	419	2356	1268	50,826	30,064

Note: r is the genotypic relative risk; p is the frequency of the disease susceptibility allele.
Source: Knapp (1999).

Transmission/Disequilibrium Test for Quantitative Traits

A series of methods have been proposed in the past few years to test association between markers and quantitative traits (see Chapter 12) (Allison, 1997; Fulker et al., 1999; Abecasis et al., 2000). The QTDT program developed by Abecasis and colleagues (http://www.sph.umich.edu/csg/abecasis/QTDT/) supports five association methods from Abecasis et al. (2000), Fulker et al. (1999), Allison (1997), Monks et al. (1998), and Rabinowitz (1997). It has become a useful tool for the allelic association analysis for the quantitative trait.

Sampling strategy generally has a great impact on the statistical test in terms of power. Abecasis et al. (2001) used simulations to evaluate several sampling strategies and the power of the TDT-type association analysis (Monks and Kaplan, 2000) and another approach based on the variance component framework (Abecasis et al., 2000). They demonstrated that proband selection from either end of the trait distribution is the optimal sampling strategy. For single selection (only one affected from each family), the power of extreme proband (EP) selection (i.e., at least one offspring phenotype is either below the bottom or above the top threshold) is consistently greater than that of affected proband (AP) selection (i.e., at least one offspring phenotype exceeds the top-tail threshold). Extreme proband selection requires screening half as many families as AP selection. Thus, for families with one offspring, selection of probands from either end of the distribution appears most practical and advantageous for statistical power.

For sibpair selection, concordant sibpair (CDS), discordant sibpair (DSP), and extreme discordant sibpair (EDAC; the combination of CDS and DSP) methods have been evaluated. Overall, DSPs provide the most dramatic gains in association power. However, ascertainment of DSPs requires phenotype screening of many more families than do other strategies. Thus, selection of sibships on the basis of one EP appears to be a practical, efficient, and powerful sampling design for family-based association studies.

Case–Control Study Design

Case–control study design, a common study design in epidemiology, is increasingly used to test for genetic association. It can precisely locate mutant genes if genetic influence is well established and the study is designed to preclude possible confounding factors and biases.

Consider a biallelic trait-influencing locus with alleles + and −. A test of association for a discrete trait under the case–control design can be described through the odds ratio (OR), $OR = (a \times d)/(b \times c)$, derived from a 2×2 table (Table 13.12). Schlesselman (1982) and others have derived an analytical power function given a proposed sample size. The parameters that affect the power of a case–control design are (1) trait allele frequencies, (2) the OR, (3) marker allele frequency, (4) the strength of LD between the disease locus allele and the marker, and (5) the type I error rate. Given these parameters, one can compute the power of a case–control study under different sample sizes. Schork (2002) extended this calculation to a general format which considers marker density. The study found that the use

TABLE 13.12. Contingency Table for Assessing
Association between Biallelic Locus and Disease

Allelic Status	Cases	Controls	Total Samples
+	A	C	n_+
−	B	D	n_-
Total	n_d	n_n	N

of common marker alleles will reduce power for detecting rare-disease alleles. Matching the allele frequencies between maker and disease locus gains the most power. Sample size will need to increase if a sparse map is used. Similarly, one must use a reasonably dense map (e.g., interlocus map distance $<40\,\text{kb}$) if a moderate sample size is used. Furthermore, there is no guarantee that a marker, no mater how close to the trait locus, will have alleles in LD with a disease allele.

DNA Pooling

As we described at the beginning of this section, family-based designs for TDT-type tests have been the favorite for association analysis. Morton and Collins (1998) demonstrated that the efficiency of a TDT design is at best two-thirds of that of a case–control design. DNA pooling may help increase efficiency. DNA pooling is a technique in which multiple individual DNA samples are pooled together before genotyping. The allele frequencies are then compared between the groups of interest, such as case and control groups. Risch and Teng (1998) compared the power of case–control design with other association designs for DNA pooling. They found that using unrelated controls is more powerful than using related controls (see Table 13.13). For example, affected sibpairs with two unrelated controls are typically 3–3.5 times as efficient as sibpairs with two unaffected siblings.

Genomic Screening Strategies for Association Studies

According to Risch and Merikangas (1996), "the future of the genetics of complex diseases is likely to require large scale testing by association analysis" via the TDT. The demonstrated power of association analyses that exploit linkage disequilibrium effects and the availability of highly dense marker or SNP maps make it tempting to design a genomewide association scan. However, some simulation studies indicate that an extremely large number of markers, perhaps on the magnitude of 500,000 SNPs, would be needed to carry out a genomewide association scan (Kruglyak, 1999). Most recently, Collins et al. (1999) studied several published datasets and discovered that LD can extend as far as several hundred kilobases and argued that a total number of 30,000 SNPs might be enough for an economical genomewide association scan. It is still not clear how successful such efforts can be at the present time for mapping complex diseases.

On the other hand, a "genewide" scanning procedure, that is, scanning within a candidate region, could be more promising for mapping complex disease genes

TABLE 13.13. Number of Families Required to Detect Linkage Disequilibrium for Sibships with $r = 1, 2, 3$ Affected and 2 Unaffected Siblings/Unrelated Controls Using Pooling

	Number of Families Required to Detect LD for Sibships with r Affected and 2 Unaffected Siblings			Number of Families Required to Detect LD for Sibships with r Affected Siblings Compared with 2 Unrelated Controls		
p-Value	$r = 1$	$r = 2$	$r = 3$	$r = 1$	$r = 2$	$r = 3$
			Dominant			
0.05	534	227	147	207	66	36
0.20	357	247	248	158	73	51
0.70	4317	4357	5638	2204	1158	819
			Recessive			
0.05	59,234	14,555	4810	28,820	5015	1325
0.20	1498	494	237	712	154	55
0.70	271	196	206	160	72	49
			Multiplicative			
0.05	2032	992	605	872	265	121
0.20	655	361	258	300	102	52
0.70	642	478	471	352	152	93
			Additive			
0.05	1213	569	351	502	154	74
0.20	526	318	258	238	90	52
0.70	990	765	760	530	231	141

for a disease locus that has been established through other means. A hybrid, two-stage design that employs a linkage scan in the first stage followed by an association scan in the implicated regions using high-density SNPs thus becomes both attractive and feasible. In the first stage, power is maximized to detect as many true positive signals as possible, with the trade-off of some false positives. In the second stage, highly dense SNP markers within candidate regions are used to increase precision. Also in the second stage, use of a lower significance threshold to minimize false positives is recommended. This kind of two-stage design is more promising than a single-scan approach. With new technology development and the establishment of public SNP databases, as more SNPs become available and genotyping becomes less expensive, single genomewide association scans will become more feasible. Until then, a combined approach may be more practical.

Simulation of Linkage and Association Program

Simulation of Linkage and Association (SIMLA; Bass et al., 2002) is a program used to generate pedigree data. What distinguishes SIMLA from other simulation programs is that it is possible to specify varying levels of both linkage and LD among markers and between markers and disease loci, so one can simultaneously study linkage and association in extended pedigrees. Among the user-specified conditions are the location and prevalence of up to 10 disease loci, a map of one or more makers, the size of a dataset, and the number of replicates. The output files can be used as input into various genetic analysis packages. It is a useful tool to allow the researcher a greater amount of flexibility in specifying test conditions and further assessing the power and sample size for the study. The program is available at http://wwwchg.mc.duke.edu under the "software" link.

SUMMARY

Power studies in genetic linkage analysis are invaluable in helping to design studies to identify disease genes. In Mendelian diseases, the study strategy often involves confirming that the material collected is sufficient to detect evidence for linkage with reasonable power. In complex genetic diseases, however, the strategy is more difficult to define and depends critically on factors inherent in the underlying genetic model. Since the underlying genetic model for the disease is virtually unknown in almost all cases, traditional applications of power studies in Mendelian disease are not directly applicable to complex disease.

For Mendelian diseases, computer simulation programs are available to determine the power of an available sample to detect linkage, the type I error rate, and the type II error rate. The Monte Carlo methods employed in these programs can be used for a wide breadth of applications, including assessing potential increases in error rates associated with maximization of LOD scores over disease model parameters and estimating empiric p-values.

These approaches can also be utilized to generate some broad generalizations about power and sample size in complex diseases, as long as investigators understand that all results are dependent on the assumptions they are willing to make about how the disease works. Programs such as DESPAIR are valuable in providing estimates of necessary sample size under an assumed genetic model and also for evaluating various genomic screening strategies.

Optimal study design plays a critical role for successful mapping of complex disease. As part of a study design, investigators should decide on sampling schemes, marker selection, and analysis methods according to the particular situation at hand.

One of the most valuable outcomes anticipated from the plethora of genomic screens currently under development for various complex diseases will be the ability to compare different genomic screening strategies, sample designs, and type I and type II error rates. The availability of LOD scores, p-values, or other statistics from these genomic screens will allow important cross-study comparisons for a variety of diseases whose underlying models are different. The compilation of these empiric experiences will provide important insights into design and analysis of genetic studies in complex human disease.

APPENDIX 13.1: EXAMPLE OF MONTE CARLO SIMULATION ASSUMING THAT TRAIT AND MARKER LOCI ARE UNLINKED TO EACH OTHER

Step I: Assign Genotypes to Founders

Decision Rules

If $0.00 \leq$ random number < 0.25, assign allele *144*.
If $0.25 \leq$ random number < 0.50, assign allele *146*.
If $0.50 \leq$ random number < 0.75, assign allele *160*.
If $0.75 \leq$ random number < 1.00, assign allele *162*.

Process. A random number is selected to generate each of the two alleles of the founder The paternally derived allele is arbitrarily assigned to be the "left" allele and the maternally derived allele is therefore the "right" allele. For instance, for individual 2000 a random number of 0.24 is selected (Table 13.14). Since this value is less than 0.25, assign the left allele as *144*. Next, a random number of 0.34 is selected. Since this value is equal to or more than 0.25 and less than 0.50, assign the right allele as *146*. The process continues for each of the founders.

Note that genotypes are simulated regardless of whether the individual is available for study. In this case, individuals 2000, 2001, 0100, and 0102 are deceased and unavailable for genotyping. Although their genotypes are simulated for use by the computer program, the simulated genotypes are not printed into the output file. These genotypes are listed as the first set of marker genotypes in Figure 13.2.

Step 2: Transmit Alleles Throughout Family According to Mendel's First Law of Segregation

Decision Rules

If $0 \leq$ random number < 0.50, then transmit paternally derived (left) allele.
If $0.50 \leq$ random number < 1.0, then transmit maternally derived (right) allele.

TABLE 13.14. Example of Genotype Assignment for Founders

Individual	Random Number for Allele *1*	Random Number for Allele *2*	Assigned Genotype
2000	0.24	0.34	*144/146*
2001	0.33	0.04	*146/144*
1001	0.55	0.78	*160/162*
0100	0.03	0.19	*144/144*
0104	0.82	0.99	*162/162*
0102	0.87	0.47	*162/146*

TABLE 13.15. Example of Genotype Assignment for Nonfounders

| Individual | Random Number to Assign Allele Transmitted From | | Genotype |
	Father	Mother	
1000	0.53	0.87	*146/144*
1002	0.57	0.59	*146/144*
0001	0.01	0.19	*146/160*
0111	0.73	0.18	*144/160*
0109	0.11	0.83	*146/162*
0107	0.12	0.33	*146/160*
0105	0.27	0.90	*146/162*
0103	0.67	0.44	*144/160*
0101	0.93	0.99	*144/162*
9002	0.12	0.37	*146/144*
9004	0.32	0.91	*146/144*
9000	0.48	0.42	*160/144*
9008	0.16	0.01	*144/162*
9009	0.40	0.83	*144/162*
9006	0.71	0.18	*162/162*
9007	0.64	0.29	*162/162*

Because the simulation here is performed without regard to whether the individual is affected with the disease, the selection of the marker allele that is transmitted is independent of the disease status of the parents.

Process. For example, in individual 1000 the first random number selected is 0.53, which is utilized to select which of the two alleles was transmitted from individual 2000, his father. Since 0.53 exceeds 0.50, the right allele (*146*) is selected. The second random number selected is 0.87, which is utilized to select which of the two alleles was transmitted from his mother, individual 2001. Since 0.87 exceeds 0.50, again the right allele (*144*) is transmitted from the mother. The assignment of genotypes for the remainder of the nonfounders in the pedigree is demonstrated in Table 13.15.

Note that the nonrandom transmission of alleles (meiotic drive) can be simulated here if the decision rules are changed. For instance, if any random number ≤ 0.70 causes the program to select the paternally derived allele, then there is a bias in transmission in favor of the paternally derived allele.

APPENDIX 13.2: EXAMPLE LOD SCORE RESULTS FOR PEDIGREE IN FIGURE 13.2

The simulation model is described in the text. Note that a LOD score of -99.99 at $\theta = 0.0$ represents an impossible likelihood (the occurrence of a recombination event when the recombination fraction between the disease and marker locus is 0.0) and is often represented in text as $-\infty$.

Replicate	$\theta = 0.00$	$\theta = 0.05$	$\theta = 0.10$	$\theta = 0.15$
A. True recombination fraction between the disease and marker locus is 0.50:				
1	-99.99	-4.30	-2.59	-1.66
B. True recombination fraction between the disease and marker locus is 0.05; italicized LOD scores are the highest LOD score for the calculated θ-values:				
1	-99.99	0.66	0.81	0.84
2	*14.21*	3.86	3.49	3.10
3	*2.71*	2.47	2.21	1.95
4	-99.99	1.30	1.58	*1.59*

APPENDIX 13.3: EXAMPLE OF SIMULATION OF GENETIC MARKER GENOTYPES CONDITIONAL ON TRAIT PHENOTYPES ALLOWING FOR COMPLETE AND REDUCED PENETRANCE

A. Complete Penetrance

For this example, a small nuclear pedigree is utilized to illustrate the concepts (Fig. 13.4); this simple example underscores the complexity of these conditional simulation problems. For this example, the trait locus is transmitted as an autosomal dominant with complete penetrance. The frequency of the trait allele d is 0.001 and the frequency of the normal allele D is 0.999. In this model, there are no phenocopies. Note that to simplify the example, the consideration of unknown phenotypes is not discussed. The marker locus has three alleles with frequency 0.40, 0.30, and 0 30, and it demonstrates 5% recombination with the trait locus.

Step 1: Calculate genotypic probabilities for founders, and calculate the probabilities of genotypes given known phenotype of a founder. The frequencies of the four possible genotypes at the trait locus are calculated assuming Hardy–Weinberg equilibrium as follows:

$$p(DD) = 0.999 \times 0.999 = 0.998001$$
$$p(Dd) = 0.999 \times 0.001 = 0.000999$$
$$p(dD) = 0.001 \times 0.999 = 0.000999$$
$$p(dd) = 0.001 \times 0.001 = 0.000001$$

Because the penetrance of the trait allele has been specified as 1.00, the following conditional probabilities for the phenotype given the genotype are known:

$$p(\text{affected}|DD) = 0.00 \quad p(\text{unaffected}|DD) = 1.00$$
$$p(\text{affected}|Dd) = 1.00 \quad p(\text{unaffected}|Dd) = 0.00$$

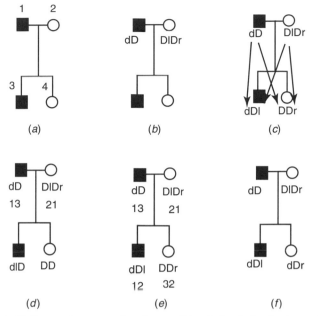

Figure 13.4. Pedigrees demonstrating the conditional simulation in Appendix 13.3.

$$p(\text{affected}|dD) = 1.00 \qquad p(\text{unaffected}|dD) = 0.00$$
$$p(\text{affected}|dd) = 1.00 \qquad p(\text{unaffected}|dd) = 0.00$$

The conditional probabilities for the genotype given the phenotype must be calculated, and these calculations can be done utilizing the Bayes rule. So, for example,

$$\frac{p(DD|\text{unaffected}) = (0.99801) \times (1.0)}{(0.998001 \times 1.0) + (0.000999 \times 0.0) + (0.000999 \times 0.0) + (0.000001 \times 0.0)} = 1.0$$

In summary (see also Fig. 13.5):

	Conditional Probability	Cumulative Probability
$p(DD|\text{unaffected})$	1 000000	1 000000
$p(Dd|\text{unaffected})$	0 000000	1 000000
$p(dD|\text{unaffected})$	0.000000	1.000000
$p(dd|\text{unaffected})$	0.000000	1.000000
$p(DD|\text{affected})$	0.000000	0.000000
$p(Dd|\text{affected})$	0.499750	0.499750
$p(dD|\text{affected})$	0.499750	0.999500
$p(dd|\text{affected})$	0.0005 00	1.000000

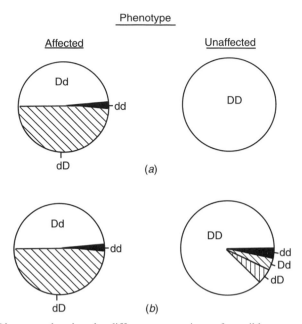

Figure 13.5. Diagrams showing the different proportions of possible genotypes underlying the disease phenotype. (*a*) Complete penetrance. The most likely disease genotypes for an affected individual are heterozygous (*dD, Dd*). However, an affected individual can be homozygous at the disease locus (*dd*). The probability for a founding affected individual to be of genotype *dd* is calculated from disease allele frequencies provided by the investigator. Since penetrance is complete, an unaffected individual can only be of disease genotype *DD*. (*b*) Reduced penetrance. Note that the probabilities for the underlying genotypes at the disease locus remain unchanged for an affected individual regardless of the penetrance. However, the genotypes for unaffected individuals at the disease locus when reduced penetrance is allowed are different from those when penetrance is complete. Specifically now an unaffected individual has a defined probability of being a disease gene carrier in the heterozygous state (*dD, Dd*) or in the homozygous state (*dd*).

And the decision rules for assigning genotypes given phenotypes are as follows. Given that the founder's phenotype is affected:

> If $0.0 \leq$ random number < 0.000000, then genotype $=>DD$.
> If $0.000000 \leq$ random number < 0.499750, then genotype $=>Dd$.
> If $0.499750 \leq$ random number < 0.999500, then genotype $=>dD$.
> If $0.999500 \leq$ random number < 1.000000, then genotype $=>dd$.

Note here that if an individual's phenotype is affected, then the genotype *DD* is impossible. Given that the founder's phenotype is unaffected:

> If $0.0 \leq$ random number ≤ 1.00, then genotype $=>DD$.

Note that because penetrance is complete, all phenotypically asymptomatic founders must not carry the disease allele and are therefore always of genotype DD.

Next, select random numbers for use in assigning trait genotypes to founders. For individual 1, random number 0.8731 is selected so the trait genotype is dD; for individual 2, random number 0.3215 is selected so the trait genotype is DD (Fig. 13.4b).

Step 2: Assign trait locus genotypes to offspring given their phenotypes and the parental trait locus genotypes. Since the parental mating type is dD by DD, there are two possible genotypes for offspring which occur with equal frequency: dD and DD. This probability is designated as P(genotype of offspring | parents). For the sake of simplicity, we will assume that the unaffected mother always transmits her left allele (designated as D_1) to offspring. Since she is homozygous for the normal allele, which allele at the trait locus the mother transmits to her children is irrelevant of their trait phenotype and trait genotype. In practice, this process is also accomplished via selection of random numbers.

Decision rules for assignment of genotypes are dependent upon these Mendelian transmission rules and on the correspondence between the phenotype and genotype (i.e., the penetrance). If the offspring phenotype is affected, then:

Possible Offspring Genotypes (g)	$P(g$|parents)	Conditional Probability	Cumulative Distribution
DD_1	½	$P(DD$|affected$) = 0.0$	0.0
dD_1	$1-2$	$P(Dd$|affected$) = 0.0$	1.0

If the offspring phenotype is unaffected, then:

Possible Offspring Genotypes (g)	$P(g$|parents)	Conditional Probability	Cumulative Distribution
DD_1	½	$P(DD$|unaffected$) = 1.00$	1.0
dD_1	$1-2$	$P(Dd$|unaffected$) = 0.10$	1.0

So the decision rules are established as follows. Given that the phenotype of the offspring is affected:

If $0.0 \leq$ random number < 0.00, then genotype $=>DD_1$.
If $0.00 \leq$ random number < 1.00, then genotype $=>dD_1$.

(Note that if the offspring is affected, the probability of genotype DD is zero.) Given that the phenotype of the offspring is unaffected:

If $0.0 \leq$ random number < 1.00, then genotype $=>DD_1$.

(Note that an unaffected individual is much more likely to have genotype DD than Dd.) In both cases, the unaffected, noncarrier mother could have transmitted either the left or right copy of her D allele. To simplify the example, we will presume that we have established—also by selection of random numbers—that the mother has transmitted her left allele, which is designated as D_1.

To assign genotypes for individuals 3 and 4, select random numbers. For individual 3, random number 0.7604 is selected so the assigned genotype is dD_1. For individual 4, random number 0.9773 is selected so the assigned genotype is DD_1 (Fig. 13.4c).

Step 3: Assign marker genotypes (and therefore linkage phase) to founders. For founders, genotypes are assigned according to Hardy–Weinberg expectations based on allele frequencies provided by the user. The linkage phase, however, is set by determining which of the alleles is associated with the disease allele, which is again performed by selecting random numbers.

Decision Rules

If $0 \leq$ random number < 0.40, then allele $=>1$.

If $0.40 \leq$ random number < 0.70, then allele $=>2$.

If $0.70 \leq$ random number ≤ 1.00, then allele $=>3$.

For individual 1, for the left allele, the random number is 0.33; for the right allele, the random number is 0.87. So, the genotype for this individual is *13*. Since his trait genotype is dD, the phase for the trait and marker loci is established as d_1/D_3, so the disease allele is established to be "segregating" with the *1* allele.

For individual 2, for the left allele, the random number is 0.62; for the right allele, the random number is 0.19. So, the genotype for this individual is *21*. Since her trait genotype is DD, the phase for the trait and marker loci is established as D_2/D_1 where the *2* allele is segregating with the left copy of D (Fig. 13.4d). Note, however, that this individual does not carry the disease allele. Since she is homozygous for the normal allele at the trait locus, she is uninformative for linkage analysis.

Step 4: Assign marker genotypes to offspring conditional on trait phenotypes and established frequency of recombination between the trait and marker locus. Because the model has been established such that the trait and marker loci are linked and our simulation has determined the disease locus genotype for each individual, we know which marker allele the affected father has transmitted to each of his offspring *unless* there has been a recombination event between the trait and marker locus. (We further earlier assigned the transmission of the D_1 disease–marker allele combination to both children from the mother.) The Monte Carlo simulation process will allow us to mimic the effects of recombination by establishing decision rules for transmission of marker allele from affected parent to child who inherited the d allele from him. We recall that the user has specified that the true recombination fraction between the trait and marker loci is 0.05.

If $0 \leq$ random number < 0.95, then transmit d_1 (a nonrecombinant gamete).

If $0.95 \leq$ random number ≤ 1.0, then transmit d_3 (a recombinant gamete).

Similarly, if the affected father transmitted the D allele to an offspring, the decision rules would be:

If $0.0 \leq$ random number < 0.95, then transmit d_3 (a nonrecombinant gamete).

If $0.95 \leq$ random number ≤ 1.00, then transmit d_1 (a recombinant gamete).

For individual 3, random number 0.7 is selected, so he receives allele *1* from his affected father; for individual 4, random number 0.98 is selected, so she receives allele *3* from her affected father (Fig. 13.4*e*).

The identical process is utilized to determine which of the maternal alleles is transmitted. For individual 3, the selected random number is 0.93 and for individual 4, the selected random number is 0.09, so each has inherited a nonrecombinant gamete from the mother; in other words, each is assigned the *2* allele at the marker locus.

B. Reduced Penetrance

Step 1: Calculate genotypic probabilities for founders, and calculate the probabilities of genotypes given known phenotype of a founder. The first step proceeds identically in assigning probabilities for trait genotypes for founders; however, since the penetrance of the trait allele is 0.90, the conditional probabilities for the genotype given the known phenotype are

$$p(\text{affected}|DD) = 0.00 \qquad p(\text{affected}|dD) = 1.00$$
$$p(\text{unaffected}|DD) = 1.00 \qquad p(\text{unaffected}|dD) = 0.00$$
$$p(\text{affected}|Dd) = 1.00 \qquad p(\text{affected}|dd) = 1.00$$
$$p(\text{unaffected}|Dd) = 0.00 \qquad p(\text{unaffected}|dd) = 000$$

Step 2: Simulate trait genotypes of founders. Again, the conditional probabilities for the genotype given the phenotype must be calculated.

In summary (see also Fig. 13.5*b*):

	Conditional Probability	Cumulative Distribution
$p(DD|\text{unaffected})$	1.000000	1.000000
$p(Dd|\text{unaffected})$	0.000000	1.000000
$p(dD|\text{unaffected})$	0.000000	1.000000
$p(dd|\text{unaffected})$	0.000000	1.000000
$p(DD|\text{affected})$	0.000000	0.000000
$p(Dd|\text{affected})$	0.499750	0.499750
$p(dD|\text{affected}\}$	0.499750	0.999500
$p(dd|\text{affected})$	0.000500	1.000000

The two sets of decision rules for assigning genotypes to given phenotypes are as follows. Given that the founder's phenotype is affected:

If $0.0 \leq$ random number < 0.000000, then genotype $=> DD$.

If $0.000000 \leq$ random number < 0.499750, then genotype $=> Dd$.

If $0.499750 \leq$ random number < 0.999500, then genotype $=>dD$.
If $0.999500 \leq$ random number ≤ 1.000000, then genotype $=>dd$.

Note here that if an individual's phenotype is affected, then the genotype DD is impossible. Given that the founder's phenotype is unaffected:

If $0.0 \leq$ random number ≤ 0.999798, then genotype $=>DD$.
If $0.999799 \leq$ random number < 0.999899, then genotype $=>Dd$.
If $0.999899 \leq$ random number < 0.999999, then genotype $=>dD$.
If $0.999999 \leq$ random number ≤ 1.000000, then genotype $=>dd$.

Next, select random numbers for use in assigning trait genotypes to founders. For individual 1, the random number 0.9007 is selected so the trait genotype is $dD;$ for individual 2, the random number 0.6581 is selected so the trait genotype is D_lD_r.

Step 3: Assign trait locus genotypes to offspring given their phenotypes and the parental trait locus genotypes. Since the established parental mating type is unchanged from the preceding example, the two possible offspring genotypes remain the same. We will also make the same simplifying assumption that the mother transmits the left allele.

If the offspring phenotype is affected:

Possible Offspring Genotypes (g)	$P(g|parents)$	Conditional Probability	Cumulative Distribution
DD_1	½	$p(DD|affected) = 0.0$	0.0
dD_1	½	$p(Dd|affected) = 1.00$	1.0

If the offspring phenotype is unaffected, then:

Possible Offspring Genotypes (g)	$P(g|parents)$	Conditional Probability	Cumulative Distribution
DD_1	½	$p(DD|affected) = 1.00$	0.91
dD_1	½	$p(Dd|affected) = 0.10$	1.00

So the decision rules are established as follows. Given that the phenotype of the offspring is affected:

If $0.0 \leq$ random number < 0.00, then genotype $=>DD$.
If $0.00 \leq$ random number ≤ 1.00, then genotype $=>Dd$.

Note that if the offspring is affected, the probability of genotype DD is zero. Given that the phenotype of the offspring is unaffected:

If $0.0 \leq$ random number < 0.91, then genotype $=>DD$.
If $0.91 \leq$ random number ≤ 1.00, then genotype $> Dd$.

Note that an unaffected individual is much more likely to have genotype DD than Dd.

Again, for simplicity, we assume that the mother has transmitted her left allele, which is designated as D_1.

To assign genotypes for individuals 3 and 4, select random numbers. For individual 3, random number 0.3371 is selected, so the assigned genotype is dD_1. For individual 4, random number 0.9802 is selected, so the assigned genotype is dD_1 (Fig. 13.4f).

From here, the assignment of marker alleles to founders and transmission to offspring proceeds as described in the full penetrance example.

REFERENCES

Abecasis GR, Cookson WOC, Cardon LR (2000): Pedigree tests of transmission disequilibrium. Eur J Hum Genet 8:545–551.

Abecasis GR, Noguchi E, Heinzmann A, Traherne JA, Bhattacharyya S, Leaves NI, Anderson GG, Zhang Y, Lench NJ, Carey A, Cardon LR, Moffatt MF, Cookson WO (2001): Extent and distribution of linkage disequilibrium in three genomic regions. Am J Hum Genet 68:191–197.

Allison DB (1997): Transmission-disequilibrium tests for quantitative traits. Am J Hum Genet 60:676–690.

Almasy L, Blangero J (1998): Multipoint quantitative-trait linkage analysis in general pedigrees. Am J Hum Genet 62:1198–1211.

Amos CI, Elston RC, Wilson AF, Bailey-Wilson JE (1989): A more powerful robust sib-pair test of linkage for quantitative traits. Genet Epidemiol 6:435–449.

Bacanu SA, Devlin B, Roeder K (2000): The power of genomic control. Am J Hum Genet 66:1933–1944.

Bass M, Martin E, Hauser E (2002): Software for simulation studies of complex traits: SIMLA. Am J Hum Genet 71:569.

Bishop DT, Williamson JA (1990): The power of identity-by-state methods for linkage analysis. Am J Hum Genet 46:254–265.

Boehnke M (1986): Estimating the power of a proposed linkage study: A practical computer simulation approach. Am J Hum Genet 39:513–527.

Cannon-Albright LA, Goldgar DE, Meyer LJ, Lewis CM, Anderson DE, Fountain JW, Hegi ME, Wiseman RW, Petty EM, Bale AE, Olopade OI, Diaz MO, Kwiatkowski DJ, Piepkorn MW, Zone JJ, Skolnick MH (1992): Assignment of a locus for familial melanoma, MLM, to chromosome 9p13–p22. Science 258:1148–1152.

Chen WM, Deng HW (2001): A general and accurate approach for computing the statistical power of the transmission disequilibrium test for complex disease genes. Genet Epidemiol 21:53–67.

Collins A, Lonjou C, Morton NE (1999): Genetic epidemiology of single-nucleotide polymorphisms. Proc Natl Acad Sci USA 96:15173–15177.

Daly MJ, Lander ES (1996): The importance of being independent: Sib pair analysis in diabetes. Nat Genet 14:131–132.

Davis S, Schroeder M, Goldin LR, Weeks DE (1996): Nonparametric simulation-based statistics for detecting linkage in general pedigrees. Am J Hum Genet 58:867–880.

Devlin B, Roeder K (1999): Genomic control for association studies. Biometrics 55: 997–1004.

Duggirala R, Stern MP, Mitchell BD, Reinhart LJ, Shipman PA, Uresandi OC, Chung WK, Leibel RL, Hales CN, O'Connell P, Blangero J (1996): Quantitative variation in obesity-related traits and insulin precursors linked to the OB gene region on human chromosome 7. Am J Hum Genet 59:694–703.

Elston RC (1989): Man bites dog? The validity of maximizing LOD scores to determine mode of inheritance. Am J Med Genet 34:487–488.

Elston RC (1992): Designs for the global search of the human genome by linkage analysis. In: Proceedings of the XVIth International Biometric Conference, Hamilton, New Zealand, pp. 39–51.

Elston RC (2000): Introduction and overview. Statistical methods in genetic epidemiology. Stat Methods Med Res 9:527–541.

Elston RC, Guo X, Williams LV (1996): Two-stage global search designs for linkage analysis using pairs of affected relatives. Genet Epidemiol 13:535–558.

Fulker DW, Cherny SS, Sham PC, Hewitt JK (1999): Combined linkage and association sibpair analysis for quantitative traits. Am J Hum Genet 64:259–267.

Greenberg DA (1989): Inferring mode of inheritance by comparison of LOD scores. Am J Med Genet 34:480–486.

Hall JM, Lee MK, Newman B, Morrow JE, Anderson LA, Huey B, King MC (1990): Linkage of early-onset familial breast cancer to chromosome 17q21. Science 250: 1684–1689.

Haseman JK, Elston RC (1972): The investigation of linkage between a quantitative trait and a marker locus. Behav Genet 2:3–19.

Hauser ER, Boehnke M, Guo SW, Risch N (1996): Affected-sib-pair interval mapping and exclusion for complex genetic traits: Sampling considerations. Genet Epidemiol 13:117–137.

Hodge SE (1984): The information contained in multiple sibpairs. Genet Epidemiol 1: 109–122.

Hodge SE, Abreu PC, Greenberg DA (1997): Magnitude of type I error when single-locus linkage analysis is maximized over models: A simulation study. Am J Hum Genet 60:217–227.

Holmans P, Craddock N (1997): Efficient strategies for genome scanning using maximum-likelihood affected-sib-pair analysis. Am J Hum Genet 60:657–666.

Kaplan NL, Martin ER (2001): Power calculations for a general class of tests of linkage and association that use nuclear families with affected and unaffected sibs. Theor Popul Biol 60:193–201.

Kaplan NL, Martin ER, Weir BS (1997): Power studies for the transmission/disequilibrium tests with multiple alleles. Am J Hum Genet 60:691–702.

Knapp M (1999a): A note on power approximations for the transmission/disequilibrium test. Am J Hum Genet 64:1177–1185.

Knapp M (1999b): The transmission/disequilibrium test and parental-genotype reconstruction: The reconstruction-combined transmission/disequilibrium test. Am J Hum Genet 64:861–870.

Knapp M, Seuchter SA, Baur MP (1994): Two-locus disease models with two marker loci: The power of affected-sib-pair tests. Am J Hum Genet 55:1030–1041.

Kruglyak L (1999): Prospects for whole-genome linkage disequilibrium mapping of common disease genes. Nat Genet 22:139–144.

Lathrop GM, Lalouel JM, Julier C, Ott J (1984): Strategies for multilocus linkage analysis in humans. Proc Natl Acad Sci USA 81:3443–3446.

Lathrop GM, Ott J (1990): Analysis of complex diseases under oligogenic models and intrafamilial heterogeneity by the LINKAGE program. Am J Hum Genet 47:A188.

Li YJ, Scott WK, Hedges DJ, Zhang F, Gaskell PC, Nance MA, Watts RL, et al. (2002): Age at onset in two common neurodegenerative diseases is genetically controlled. Am J Hum Genet 70:985–993.

Lin S (2000): Monte Carlo methods for linkage analysis of two-locus disease models. Ann Hum Genet 64:519–532.

Lunetta KL, Faraone SV, Biederman J, Laird NM (2000): Family-based tests of association and linkage that use unaffected sibs, covariates, and interactions. Am J Hum Genet 66:605–614.

Martin ER, Kaplan NL, Weir BS (1997): Tests for linkage and association in nuclear families. Am J Hum Genet 61:439–448.

Monks SA, Kaplan NL (2000): Removing the sampling restrictions from family-based tests of association for a quantitative-trait locus. Am J Hum Genet 66:576–592.

Monks SA, Kaplan NL, Weir BS (1998): A comparative study of sibship tests of linkage and/or association. Am J Hum Genet 63:1507–1516.

Morton NE, Collins A (1998): Tests and estimates of allelic association in complex inheritance. Proc Natl Acad Sci USA 95:11389–11393.

Olson JM, Wijsman EM (1993): Linkage between quantitative trait and marker loci: Methods using all relative pairs. Genet Epidemiol 10:87–102.

Ott J (1989): Statistical properties of the haplotype relative risk. Genet Epidemiol 6:127–130.

Ott J, Terwilliger JD (1992): Assessing the evidence for linkage in psychiatric genetics. In: Mendlewicz J, Hippius H, eds. Genetic Research in Psychiatry. Berlin: Springer-Verlag.

Pericak-Vance MA, Bebout JL, Gaskell PC, Yamaoka LH, Hung WY, Alberts MJ, Walker AP, Bartlett RJ, Haynes CS, Welsh KA, Earl NL, Heyman A, Clark CM, Roses AD (1991): Linkage studies in familial Alzheimer's disease: Evidence for chromosome 19 linkage. Am J Hum Genet 48:1034–1050.

Ploughman LM, Boehnke M (1989): Estimating the power of a proposed linkage study for a complex genetic trait. Am J Hum Genet 44:543–551.

Pritchard JK, Stephens M, Rosenberg NA, Donnelly P (2000): Association mapping in structured populations. Am J Hum Genet 67:170–181.

Rabinowitz D (1997): A transmission disequilibrium test for quantitative trait loci. Hum Hered 47:342–350.

Risch N (1990): Linkage strategies for genetically complex traits. II. The power of affected relative pairs. Am J Hum Genet 46:229–241.

Risch N, Merikangas K (1996): The future of genetic studies of complex human disorders. Science 273:1516–1517.

Risch N, Teng J (1998): The relative power of family-based and case–control designs for linkage disequilibrium studies of complex human diseases: DNA pooling. Genome Res 8:1273–1288.

Risch N, Zhang H (1995): Extreme discordant sib pairs for mapping quantitative trait loci in humans. Science 268:1584–1589.

Risch N, Zhang H (1996): Mapping quantitative trait loci with extreme discordant sib pairs: Sample size considerations. Am J Hum Genet 58:836–843.

Schlesselman JJ (1982): Case-Control Studies. Design, Conduct, Analysis. New York/ Oxford: Oxford University Press.

Schork NJ (2002): Power calculations for genetic association studies using estimated probability distributions. Am J Hum Genet 70:1480–1489.

Sherrington R, Brynjolfsson J, Petursson H, Potter M, Dudleston K, Barraclough B, Wasmuth J, Dobbs M, Gurling H (1988): Localization of a susceptibility locus for schizophrenia on chromosome 5. Nature 336:164–167.

Spielman RS, McGinnis RE, Ewens WJ (1993): Transmission test for linkage disequilibrium: The insulin gene region and insulin-dependent diabetes mellitus (IDDM). Am J Hum Genet 52:506–516.

Teng J, Risch N (1999): The relative power of family-based and case–ontrol designs for linkage disequilibrium studies of complex human diseases. II. Individual genotyping. Genome Res 9:234–241.

Thompson EA (1994): Monte Carlo likelihood in the genetic mapping of complex traits. Philos Trans R Soc Lond B Biol Sci 344:345–350.

Weeks DE, Lange K (1988): The affected-pedigree member method of linkage analysis. Am J Hum Genet 42:315–326.

Weeks DE, Lehner T, Squires-Wheeler E, Kaufmann C, Ott J (1990a): Measuring the inflation of the LOD score due to its maximization over model parameter values in human linkage analysis. Genet Epidemiol 7:237–243.

Weeks DE, Ott J, Lathrop GM (1990b): SLINK: A general simulation program for linkage analysis. Am J Hum Genet 47:A204.

Whittaker JC, Lewis CM (1998): The effect of family structure on linkage tests using allelic association. Am J Hum Genet 63:889–897.

Whittaker JC, Lewis CM (1999): Power comparisons of the transmission/disequilibrium test and sib-transmission/disequilibrium-test statistics. Am J Hum Genet 65:578–580.

Williams JT, Blangero J (1999): Power of variance component linkage analysis to detect quantitative trait loci. Ann Hum Genet 63:545–563.

Williams JT, Van Eerdewegh P, Almasy L, Blangero J (1999): Joint multipoint linkage analysis of multivariate qualitative and quantitative traits: I. Likelihood formulation and simulation results. Am J Hum Genet 65:1134–1147.

Zhang H, Risch N (1996): Mapping quantitative-trait loci in humans by use of extreme concordant sib pairs: Selected sampling by parental phenotypes. Am J Hum Genet 59:951–957. Erraratum, Am J Hum Genet 60:748–750 (1997).

CHAPTER 14

Complex Genetic Interactions

WILLIAM K. SCOTT and JOELLEN M. SCHILDKRAUT

INTRODUCTION

In the study of the etiology of complex traits, there is a need for analytic strategies that can determine the cause of the discordance between genetic susceptibility and expression of the trait. Genetic models of complex traits need to account for genetic heterogeneity and interactions with other genes and the environment. In previous chapters, analytic strategies have focused on identification of genes that *individually* increase or decrease the occurrence or expression of complex traits. However, such approaches have little power to identify factors that have no individual effect on the development of a trait but influence expression of the trait through interactions with other genetic or environmental factors. Such effects will only be identified by the use of methods that consider the simultaneous effects of multiple genetic and environmental factors. A combination of genetic and epidemiological approaches in the study of complex genetic interactions has the potential for success in investigating this multifactorial nature of many common diseases.

There are several ways in which genes and environment may interact to influence the development of a complex disease. Genetic susceptibility can influence the risk of disease by itself; it may exacerbate the effect of an environmental risk factor; or the risk factor may exacerbate the genetic effect (Ottman, 1990). Therefore, study designs that consider both genetic and environmental factors are necessary to evaluate interactions between genotype or family history and environmental influences, thereby enhancing the ability to uncover genetic influences on disease (Ottman, 1990). Since exposures to some environmental factors that influence genetic risks are modifiable, the discovery of such relationships has important public health implications. This chapter reviews methods applied to investigate complex interactions between susceptibility genes and environmental factors in family- or population-based genetic epidemiology studies.

Genetic Analysis of Complex Diseases, Second Edition, Edited by Jonathan L. Haines and Margaret Pericak-Vance
Copyright © 2006 John Wiley & Sons, Inc.

EVIDENCE FOR COMPLEX GENETIC INTERACTIONS

In general, complex diseases lack one-to-one correspondence of genotype and phenotype. In other words, the genetic basis of complex traits cannot be explained by detecting one or two mutations in a gene, segregating in Mendelian fashion to all affected family members. Complex traits are multifactorial and are influenced by multiple genes and/or environmental factors. Expression of complex phenotypes is therefore influenced by genetic heterogeneity, the effects of modifying genes and gene–gene interaction, and gene–environment interaction. Such complex effects complicate the task of identifying complex disease genes using traditional linkage and association methods discussed in earlier chapters. Therefore, consideration of complex interactions is essential in the design and implementation of successful studies of multifactorial traits.

Genetic Heterogeneity

Genetic heterogeneity exists when several genes are associated with the same disease (Chapter 2). Genetic heterogeneity may also be due to variations in the same genes that are associated with differences in disease risk (allelic heterogeneity). Genetic heterogeneity might affect the ability to identify genetic and non-genetic factors associated with the risk of disease, since etiologically distinct subgroups might not be distinguished by phenotype alone. Ignoring the possibility that different genetic factors could cause the same phenotype might result in weaker estimates of linkage or association between important genetic risk factors and disease (Hodge, 1994; Newman et al., 1995; Slattery et al., 1995).

Alzheimer's disease (AD) is an example of a complex trait related to multiple genes. Several genes influence the risk of developing AD, and there is suggestive evidence that environmental factors play a role as well. Age at onset is one factor that distinguishes subgroups of AD. Genes for rare early-onset familial AD (i.e., onset before the age of 60 years) are found on chromosomes 1, 14, and 21 (St George-Hyslop et al., 1990; Goate et al., 1991; Levy-Lahad et al., 1995; Rogaev et al., 1995; Sherrington et al., 1995). The more common late-onset form of AD (i.e., onset after age 60) has been associated with the apolipoprotein E (APOE) locus on chromosome 19 (Pericak-Vance et al., 1991; Corder et al., 1993; Saunders et al., 1993). Allelic variation in the *APOE* gene is differentially associated with the risk of AD, with increased risk at a decreased age at onset associated with the *APOE*E4* allele and a decreased risk of AD associated with the *APOE*E2* allele (Pericak-Vance and Haines, 1995).

Unlike the three early-onset AD loci, APOE accounts for a significant proportion of the genetic risk of late-onset AD. Therefore, any study searching for additional late-onset genes must control for the influence of APOE on risk of AD. The existence of at least three additional late-onset genes has been suggested by significant evidence for linkage to chromosomes 9 (Pericak-Vance et al., 2000; Farrer et al., 2001), 10 (Kehoe et al., 1999; Myers et al., 2000; Ertekin-Taner et al., 2000), and 12 (Pericak-Vance et al., 1997; Scott et al., 2000; Mayeux et al., 2002). Notably, each of these studies considered the evidence for linkage conditional on the effect of APOE, and results indicate that some effects, such as that on chromosome 12, may be independent of the effect of the *APOE*E4* allele.

Gene–Gene Interaction (Epistasis)

An additional level of complexity for genetic traits is the interaction of multiple genes to determine the phenotype. The degree and type of genetic interaction strongly influences the chance of detecting genes through linkage analysis (Risch, 1990a,b). Modifier genes and polygenic traits are examples of gene–gene interaction. The presence of a modifier gene alters the main effect of a primary (or major) gene for the trait. Because the major gene exerts an influence on the trait regardless of the presence of the modifier, such effects might still be detectable if the effect of the modifier gene is not considered. In contrast, polygenic traits require the simultaneous presence of variations in multiple genes and consideration of the interaction is crucial to detecting the genetic factors involved (Lander and Schork, 1994; Cordell and Todd, 1995). Polygenic inheritance is difficult to demonstrate in humans and complicates genetic mapping because no single locus is solely responsible for producing the trait (Lander and Schork, 1994).

Examples of gene–gene interaction illustrate the need for a combination of analytic approaches to understand the complexity of disease susceptibility. One approach to detecting gene–gene interaction is to select loci with main effects on development of disease and examine them for potential interaction. For example, analysis of affected sibpairs with insulin-dependent diabetes mellitus (IDDM) detected linkage to human leukocyte antigen (HLA) genes (Thomson, 1994; Cordell and Todd, 1995). In addition to HLA, several other loci have been linked or associated with IDDM (Cordell and Todd, 1995). Once identified, these loci have been examined for interaction using two-locus linkage models or case–control association tests. Two-locus linkage models suggest a potential multiplicative interaction between *IDDM1* (HLA) and *IDDM2* (INS, on chromosome 11p) and genetic heterogeneity between *IDDM1* (HLA) and *IDDM4* (on chromosome 11q) (Cordell et al., 1995). Similarly, a case–control association study approach has been used to examine potential gene–gene interaction between *NAT1* and *NAT2* enzyme genetic polymorphisms, important in N- or O-acetylation of xenobiotics, in determining colorectal cancer risk (Bell et al., 1995). Although the *NAT1*10* allele is associated with a 1.9-fold [95% confidence interval (CI) = 1.2–3.2; $p = 0.009$] increased risk of colorectal cancer, the association with this *NAT1* variant is found to be stronger among individuals with *NAT2* alleles associated with rapid acetylation of some drugs and carcinogens [odds ratio (OR) = 2.8; 95% CI = 1.4–5.7; $p = 0.003$]. While subsequent studies failed to find an association with *NAT1*, *NAT2*, or the interaction, these studies may not have had enough statistical power to detect the effect (Brockton et al., 2000). Studies that aim to detect novel complex interactions or replicate previously reported ones must have a much larger sample size than studies designed to detect only main effects (Smith and Day, 1984).

A second approach to accounting for gene–gene interaction is to conduct a search for additional disease loci conditional on the genotype at a known genetic risk factor. This conditional analysis approach was used in a whole-genome screen to identify additional non-insulin-dependent diabetes (NIDDM) loci while conditioning on the known effect of a previously described locus on chromosome 2 (*NIDDM1*). When controlling for the effect of *NIDDM1* and allowing for an epistatic interaction, Cox and colleagues (1999) discovered that evidence for linkage to chromosome 15 in a sample of Mexican-Americans increased from a logarithm of the odds (LOD) score of 1.3–4.0.

A third approach is to consider many genetic factors at once, allowing statistical model-building or data reduction methods to select the most significant factors. Methods for these analyses include classification and regression trees (CARTs) (Wilcox et al., 1999; Province et al., 2000) and multifactor dimensionality reduction (MDR) (Ritchie et al., 2001). These approaches are similar in motivation, in that many exposures are evaluated for association with disease, and the pattern that best predicts affection status is selected as the solution. One drawback to such analyses is that the statistical significance of the solution is difficult to calculate (although empiric *p*-values can be obtained) and many patterns may have similar predictive value; therefore, several complex patterns may perform similarly in predicting disease status.

Gene–Environment Interaction

In addition to genetic heterogeneity and gene–gene interaction, it is important to consider the influence of environmental factors in the study of complex genetic traits. Once a major genetic or environmental effect has been identified, gene–environmental interactions can be explored. The examination of candidate genes in the study of disease–exposure associations, or gene–environment interactions, can reveal effects of an environmental factor on the risk of disease that would be concealed if genetic susceptibility were ignored (Hwang et al., 1994). Similarly, consideration of an established environmental risk factor may facilitate the identification of novel genetic factors. Six plausible models of gene–environment interaction have been described (Yang and Khoury, 1997):

1. Disease risk is increased only in the presence of the susceptibility genotype and the environmental risk factor.
2. Disease risk is increased by the environmental risk factor alone but not by the genotype alone.
3. Disease risk is increased by the genotype in the absence of the environmental risk factor but not by the risk factor alone.
4. Disease risk is increased by both genotype and environmental risk factor; joint effects may be additive, superadditive, multiplicative or supermultiplicative.

5. Disease risk is reduced by the genotype and is not affected by the environmental factor alone, but disease risk is increased in the presence of the genotype and environmental factor.

6. Disease risk is reduced by the genotype, is increased by the environmental factor, and is increased by the presence of both.

These models are admittedly simple, considering the effect of only one gene and one environmental factor; interactions are likely to be much more complex, involving multiple genes, environmental factors, genetic heterogeneity, and heterogeneity of exposure (Yang and Khoury, 1997). However, these models provide a framework for discussing potential study designs and approaches to the analysis of complex interactions.

There are several examples of gene–environment interaction in the literature. Models for carcinogenesis are illustrative of complex traits with multiple risk factors and etiologic pathways, since it is thought that interactions between environmental factors and inherited polymorphisms of genes are important in the multistage process of carcinogenesis. For example, the previously discussed interaction between the *NAT1*10* allele and *NAT2* fast acetylator alleles in increasing the risk of colorectal cancer was not widely replicated by other studies. However, analysis of 212 men with colorectal cancer and 221 controls from the Physicians' Health Study found that although an interaction between *NAT1* and *NAT2* could not be shown to increase colorectal cancer risk, a gene–gene–environmental interaction was observed when red meat consumption was considered (Chen et al., 1998). In this case, men with *NAT1* and *NAT2* rapid acetylator alleles that consumed greater than one serving of red meat per day had 5.8 times the risk of colorectal cancer (95% CI = 1.1–30.6) compared to men with rapid acetylator alleles who consumed at most half of a serving of red meat per day. As well, interactions between environmental chemical exposures and N- and O-acetylation phenotypes encoded by two loci, *NAT1* and *NAT2*, have been associated with an increase in the risk of bladder cancer (Risch et al., 1995; Hein et al., 2000).

ANALYTIC APPROACHES TO DETECTION OF COMPLEX INTERACTIONS

It may be necessary to combine several analytic approaches to explore genetic heterogeneity and interactions between other genes and the environment. The integration of genetic and epidemiological methods in the study of complex genetic interactions has the potential of providing a successful approach for investigating the multifactorial nature of many common diseases (Weeks and Lathrop, 1995).

Modeling effects of several loci and environmental factors simultaneously may reveal important information about biological interactions between disease determinants. Genetic approaches to assessing complex genetic interactions are discussed below. For more complete information on specific genetic analyses, consult other chapters in the text.

Segregation Analysis

Although segregation analysis is limited to the investigation of modes of inheritance, the development of statistical methods using mixed genetic models (containing Mendelian and polygenic parameters) has allowed the estimation of genetic parameters for several complex traits (Weeks and Lathrop, 1995). Mixed models contain transmission probabilities, gene frequencies, and penetrance parameters for Mendelian models and heritability, sample means, and variances for polygenic models. Regressive models, which condition the phenotype of each individual on the phenotype of preceding relatives, have been shown to be useful in the analysis of both qualitative (categorical) and quantitative traits (Bonney et al., 1989; Demenais and Lathrop, 1993). Regressive models incorporate a major gene effect as well as other sources of familial correlation (Bonney, 1984; Demenais et al., 1992). Segregation analysis for determining major gene effects using regressive models (Bonney, 1984) incorporates covariates such as age, sex, and specific genetic polymorphisms to assess the effects of complex genetic interactions. However, like all segregation analyses, this approach is very sensitive to sample size and to ascertainment bias; moreover, it generally requires substantial computer resources for complex models with many factors (see Chapter 3).

Combining segregation and linkage analyses can be a powerful modeling technique (Bonney et al., 1988). Together, these analytic approaches have been used to detect genetic interactions and improve LOD scores when genotype-dependent effects of covariates (such as environmental exposures) are incorporated into the linkage analysis (Bonney et al., 1988; Weeks and Lathrop, 1995; Gauderman and Faucett, 1997). While in theory combined segregation and linkage analysis can be a powerful approach to evaluate complex genetic models, certain drawbacks limit the utility of this approach. These include the need to correct for the ascertainment scheme, which may not be possible when sampling strategies are complex, computational constraints that occur when maximizing the likelihood of the model over many parameters, the difficulty in modeling genetic heterogeneity (thought to be the norm in complex traits), and the nature of the method itself, which is better adapted for the analysis of quantitative traits in simple random samples.

Linkage Analysis

Linkage analysis can be used to evaluate complex interactions in several different ways. If the goal of the analysis is to identify new regions of linkage while controlling for the effects of known genetic or nongenetic factors, stratified or conditional linkage analysis may be used. The simultaneous effect of multiple loci may be evaluated using two-locus modeling. More complicated interactions may be evaluated using more elaborate statistical models, such as those employed in variance component analysis (Chapter 9). The goal of the analysis dictates which of these approaches is potentially most useful.

Stratified and Conditional Linkage Analysis. The use of linkage analysis in the study of complex genetic traits has produced mixed results. Many genomic regions have been linked to complex diseases, but relatively few susceptibility genes have been identified. The progression from linkage to gene identification is complicated by the wide intervals of suggestive linkage normally detected in complex diseases. These wide intervals are often the result of genetic and phenotypic heterogeneity, in which only a fraction of the total sample is linked to the region of interest. By narrowing the definition of the disease or by stratifying the patient population by a particular characteristic, the chances for unraveling the genetic etiology of a complex disease may be improved. Perhaps the simplest method for doing so is to stratify the dataset into more homogeneous subsets for parametric or nonparametric linkage analysis. For example, focusing on early-age-at-onset cases, such as with AD (St George-Hyslop et al., 1990; Schellenberg et al., 1992) and breast cancer (Hall et al., 1990), has proved to be advantageous and has allowed the identification of several causative genes (Goate et al., 1991; Futreal et al., 1994; Miki et al., 1994; Levy-Lahad et al., 1995; Rogaev et al., 1995; Sherrington et al., 1995).

While stratifying the sample a priori into more homogeneous subsets has facilitated the identification of genes in AD and breast cancer, for many diseases, it is unclear what the optimal stratification scheme might be. In these cases, a post-hoc subsetting strategy might be helpful. One such strategy, ordered subset analysis (Ghosh et al., 2000), allows the study of a set of families rank ordered by a continuous characteristic, such as mean age at onset, LOD scores at a particular marker, or other phenotypic characteristics. Linkage analysis is then performed in progressively larger subsets (starting with the top-ranked family and increasing the size of the subset by 1 each time). Permutation testing is then used to determine if any increased evidence for linkage in a subset is statistically significant. A similar approach was used for identification of linkage to chromosome 17 in breast cancer; 23 families were ordered by age at onset and families were added (in order of increasing age at onset) to the total LOD score until the score began to decrease (Hall et al., 1990). The results indicated that the linkage to the region that eventually was shown to contain *BRCA1* was strongest in the 8 families with the earliest mean ages of onset.

One potential drawback to stratified analysis is that some statistical power may be lost due to reducing the sample size being analyzed. Conditional nonparametric linkage analysis approaches, such as that implemented in the GENEHUNTER-PLUS program package (Kong and Cox, 1997), attempt to avoid this issue by weighting each family's contribution to the linkage statistic by the covariate (e.g., linkage results at another marker). When examining two genetic loci, the family-specific weights may be constructed to test for genetic heterogeneity (e.g., results at locus 2 become stronger when giving more weight to families with negative results at locus *1*) or epistasis (results are stronger at locus 2 when giving more weight to families linked to locus *1*). Statistical significance of the results may be assessed using simulation and permutation testing—weights are shuffled among the families many times (e.g., 1000 replicates) and LOD scores calculated in each replicate.

The significance level for the weighted LOD score is then the percentage of the time the permuted LOD score exceeds the weighted LOD score. This approach has been successfully used to identify potential interaction between loci on chromosomes 2 and 15 in NIDDM (Cox et al., 1999) and to detect potential phenotypic heterogeneity in linkage to chromosome 12 in AD (Scott et al., 2000).

A limitation to both stratified and conditional linkage analysis approaches is that the conditioning is based on attributes of the family, rather than an individual; for example, mean age at onset or linkage scores at another locus. Using these methods to incorporate information on known disease genes or measured environmental factors requires that individual information somehow be summarized over the entire family. For example, in the study mentioned above, to adjust for known genetic effect of *APOE*E4* while examining families for linkage to chromosome 12, nonparametric linkage analysis was performed weighting the family-specific scores by the proportion of affected individuals in the family that carried the *APOE*E4* allele (Scott et al., 2000). Moreover, many weighting schemes will reduce the power of the analysis by simply stratifying the sample into a smaller subset possessing the factor of interest (e.g., families with positive LOD scores at a particular locus). For these reasons, approaches that specifically model linkage simultaneously at multiple loci (such as two-trait-locus linkage models) may be more powerful.

Two-Locus Models.   Once two disease loci are linked, the joint effect of the loci may be studied using parametric or nonparametric two-locus models. Parametric two-locus linkage models, such as those implemented in the program TMLINK (Lathrop and Ott, 1990), require that parameter estimates be provided for the two loci and for their proposed interaction. Such estimation is often difficult in complex traits, and few examples of parametric two-locus models exist in the complex disease literature. An early example was the two-locus linkage analysis of the HLA region on chromosome 6 and the myelin basic protein on chromosome 18 in a set of Finnish families with multiple sclerosis (Tienari et al., 1994). In this study, the two-trait-locus/two-marker-locus LOD score was 10.26, substantially higher than the LOD scores obtained when each locus was analyzed separately (3.1 and 6.75). The sum of the two single-locus LOD scores (9.85) was not significantly different from the two-trait-locus/two-marker-locus LOD score, suggesting that the joint effect of these two loci is additive. While a potentially powerful approach, the original parametric two-trait-locus/two-marker-locus algorithm implemented in TMLINK is limited to considering two unlinked marker loci—the calculation of an exact LOD score with more markers would likely render the analysis computationally intractable. To address this limitation, a Markov chain Monte Carlo (MCMC) method has been implemented in a program called MCLINK-2LOCUS to calculate approximate LOD scores for parametric two-trait-locus models using fairly dense marker maps and general pedigrees (Lin, 2000). Initial comparisons between the two approaches appear to show that MCLINK-2LOCUS is faster, requires fewer computer resources, and can analyze more marker data than TMLINK with little loss of precision. However, the approach still requires that a parametric two-locus model be specified, and this may limit the application of either method.

Just as parametric LOD score approaches have been extended to multiple loci, two-locus nonparametric allele-sharing methods [such as affected sibpair and affected relative pair (ARP) analyses] have been developed. Initially, nonparametric approaches, such as the extension of the marker-association-segregation-chi-square (MASC) method (Dizier et al., 1994) and the likelihood ratio test introduced by Knapp et al. (1994), were restricted to the two-trait-locus/two-marker-locus approaches developed for parametric LOD scores. However, Cordell et al. (1995) extended the multipoint maximum LOD score (MLS) affected sibpair method developed by Risch (1990a) to two-trait loci; this approach allowed the consideration of many markers in two unlinked regions. The approach was generalized to include two linked trait loci and the algorithm was implemented in the TWOLOC program package (Farrall, 1997). This approach allows the consideration of epistatic and heterogeneity models for interaction between the two loci. A comparison of the MLS obtained under each model allows the selection of a potential method of interaction between the two loci (Cordell et al., 1995). In the additive model, the locus-specific risk ratios are added together to determine the joint effect. One potential limitation of this approach is the difficulty in assessing the statistical significance of the potential interaction (Farrall, 1997); simulation approaches, such as those described by Cordell et al. (1995) are one potential approach to assessing significance. Another limitation of affected sibpair two-locus models is that they ignore information from other ARPs that might be available in the dataset; when such family structures are available for analysis, the use of recently developed two-locus ARP approaches may be more powerful.

Two of these methods examine allele sharing at multiple unlinked disease loci in small or moderately sized pedigrees. These approaches allow multipoint analysis of dense marker maps using the Lander–Green algorithm. The extension of the MLS approach described above to ARPs allows the calculation of MLS curves across the genome for a series of single- and multilocus models, allowing for analysis conditional on previously identified genes or a "conditional search" of the genome given results at other locations (Cordell et al., 2000). Under a multiplicative model, this approach is limited to the examination of about three loci at one time; under an additive model many more loci could be examined simultaneously. A similar extension of the NPL approach implemented in GENEHUNTER implements scoring functions S_{pairs} and S_{all} that examine allele sharing at two unlinked disease loci; however, this method may only analyze relatively small families with 11–12 meioses (Strauch et al., 2000). A third approach extends the weighted pairwise correlation (WPC) method to ARPs; however, this extension is limited to two unlinked marker loci, computes allele-sharing estimates based on identity by state, and does not allow the use of a multipoint marker map (Zinn-Justin and Abel, 1998). However, it does allow a more general pedigree structure than the other two nonparametric two-locus ARP approaches.

Multivariable Regression Models. Several approaches have employed the use of regression models to assess interactions. For example, approaches to quantitative trait locus mapping such as the Haseman–Elston algorithm and variance component models (see Chapter 9) may be extended to incorporate covariates and interaction terms. For example, the variance component model implemented in the program SOLAR (Almasy and Blangero, 1998) allows the analysis of large, extended families and can incorporate identical-by-descent (IBD) sharing at a marker locus, household effects, measured environmental exposures, additional genes, and interactions into a single statistical model. For common diseases, a threshold model may be applied to allow the analysis of a dichotomous trait; however, the underlying quantitative trait must be normally distributed for the threshold model to be valid. For affected sibpairs, an extension of the mean test of the Haseman–Elston algorithm allows a test of gene–environment interactions in multipoint linkage analysis of siblings concordant and discordant for an environmental exposure (Gauderman and Siegmund, 2001). Two similar approaches to evaluating interactions in ARPs in small families have been implemented in the program LODPAL (Olson, 1999) and in programs written for the STATA package (Holmans, 2002). Both programs use conditional logistic models to examine the joint effect of IBD sharing at a marker locus, covariates, and their interactions.

Association Analysis

The application of epidemiological methods to the study of genetics in complex diseases has been increasing. Population- and family-based association designs can be used to test models for the relationship between genetic susceptibility and environmental risk factors. In particular, the case–control method has emerged as a powerful approach for addressing the role of genetic factors and their interactions with other genes and environmental factors (Khoury and Beaty, 1994). Additionally, family-based case–control studies, case-only studies, and a prospective cohort approach offer certain advantages.

Case–Control Design. Using a case–control approach, subjects are selected according to their disease status. Cases are defined as those affected with the particular disease of interest. The controls, although at risk for developing the disease (i.e., old enough to develop a disease with a characteristically late onset or the appropriate sex for developing a disease such as ovarian cancer), are unaffected with the disease. The case and control subjects are then compared with respect to the proportion having the exposure or characteristic of interest. In genetic epidemiology studies, the cases and controls are classified according to the presence or absence of the genotype (or family history) as well as the presence or absence of an additional risk factor, such as an environmental factor or a second gene.

To achieve valid case–control comparisons, controls should be selected so that they represent the same source population as the cases. Thus the selection of controls should include the same exclusion or restriction criteria applied to the cases. Another important issue in selecting controls is the source of the subjects. The source could be the hospital from which the cases were identified (a hospital-based case–control design), the general population (a population-based case–control design), or friends, neighbors, or relatives of cases. Each set offers advantages and disadvantages that must be considered in the context of the source of the case selection, the nature and goals of the study, and the type of information obtained (Rothman and Greenland, 1998). To avoid introducing secular influences of disease risk that differ in the cases versus controls, it is preferable to choose control subjects during the same time period as the cases.

Association studies that use the case–control method provide a complementary strategy for detection of disease susceptibility genes and can be used for identifying such genes once the region of the chromosome has been determined by linkage (Khoury and Beaty, 1994). In this case, candidate genes are selected based on their function *and* location, rather than simply by their functional relevance. In the candidate gene approach, the frequency of the disease susceptibility gene or genetic marker of interest in a group of affected individuals or cases is compared to the frequency in a group of controls or cases (Khoury and Beaty, 1994). This strategy can be expanded to permit examination of complex genetic interactions by analyzing the joint effects of genes on disease susceptibility. As with all case–control studies, selection of an appropriate control group, such as matching on ethnicity and other factors, is important to avoid spurious associations due to variation in allelic frequencies in different populations. Spurious associations can also result from confounding due to recent admixture and selection or drift between unlinked loci (Lander and Schork, 1994; Weeks and Lathrop, 1995). Methods of detecting and correcting for this bias in case control studies have been recently described (reviewed in Pritchard and Donnelly, 2001). These approaches rely on the genotyping of many polymorphic markers in the cases and controls and determining if subpopulations may be identified by sets of alleles at these marker loci. If such subpopulations are identified, the marker data may then be used to adjust for confounding by genetic background in tests of association at candidate gene loci (Pritchard et al., 2000).

Approaches to analysis of case–control data include methods that can detect and adjust for bias due to confounding. Confounding occurs when an observed association is due to the mixing of effects between the exposure (e.g., family history), the disease, and a third factor (or confounder) that is associated with both the exposure and disease (Rothman and Greenland, 1998). For case–control studies, stratification and Mantel–Haenszel chi-square analyses, as well as logistic regression, are the most common methods for assessing an association between the disease and a risk factor while controlling for confounding factors. Stratified analyses can be useful to control for confounding of either genes or environmental factors. Using this approach, the genetic susceptibility–disease association can be estimated within various levels of the confounding variable such as age or race. Using Mantel–Haenszel estimation, a uniform odds ratio from data stratified according to a potential confounding variable can be calculated (Rothman and Greenland, 1998).

As strata become numerous and in the presence of a need to control for more than a few variables, stratified analyses become cumbersome, sample sizes decrease, and the results are difficult to describe and summarize. In addition, using a stratified analytic approach that involves the categorization of a continuous variable can introduce a degree of arbitrariness and perhaps result in loss of information. A multi-variable approach allows for the simultaneous control of several factors to determine which ones are independent effects in predicting disease.

Multivariable analysis involves construction of a mathematical model to describe the association between a genetic exposure and disease and can incorporate variables, either continuous or categorical, that confound or modify the genetic effect. In many genetic epidemiological studies, particularly case–control studies, where the outcome of interest is a binary variable, such as affected versus unaffected, the most appropriate analytic approach is logistic regression analysis (Rothman and Greenland, 1998). In the logistic model, the risk of developing a disease is expressed as a function of independent predictor variables such as genotype as well as other known risk factors, either genetic or environmental, which could potentially confound or modify the disease–gene association (see Chapter 12).

Stratification and logistic regression can be also used to assess possible interaction. Examining stratum-specific ORs allows for a preliminary investigation of possible interactions between risk factors for a disease. In its simplest form, such an analysis takes the form of a 2×4 table, where each subject is classified as exposed or not exposed on each of two factors (Yang and Khoury, 1997). One indication that two factors interact would be notable differences in the association between the genetic exposure and the risk of the disease in the presence versus the absence of the other factor. Such a situation is depicted in Table 14.1, where exposure to risk factor M modifies the effect of genotype AA in the risk of developing D. In this case, an OR of 15.5 is observed among those with the AA genotype who had been exposed to M while an OR of 1.5 appeared among those with the AA genotype who had not been exposed to M. To assess the interaction statistically, one must decide on a multiplicative or additive model to describe the relationship between risk factors for disease. For either model, an OR of interaction may be calculated from the ORs to detect interaction between the two factors. To assess interaction on a multiplicative scale, $OR_i = [OR_{ge}]/[OR_g \times OR_e]$ (Goldstein and Andrieu, 1999). When $OR_i \neq 1$, the risk factors depart from a multiplicative model and therefore may be said to interact on a multiplicative scale. For the data in Table 14.1, $OR_i = 15.5/1.5 = 10.33$, which is indicative of a multiplicative interaction. To assess interaction on an additive scale, $OR_i = [OR_{ge}]/[OR_g + OR_e - 1]$ (Goldstein and Andrieu, 1999). When $OR_i \neq 1$, the effects of the risk factors depart

TABLE 14.1. An Example of Evaluation of Gene–Environment Interaction Using the 2 × 4 Table

Genotype AA?	Exposure to M?	OR
Yes	Yes	15.5
Yes	No	1.5
No	Yes	1.0
No	No	1.0 (referent)

from an additive model and therefore interact on an additive scale. It has been argued that additivity is a better definition of independent effects (Rothman and Greenland, 1998). For more complicated analyses, such as continuous exposures, or multiple factors, multiplicative and additive models may be fitted using logistic regression, from which ORs of interaction may be calculated (Thompson, 1994).

Examples of gene–environment interaction in case–control studies demonstrate the importance of taking the assessment of interactions into account in the design of a case–control study. The synergistic relationship between the glutathione *s*-transferase θ (*GSTT1*) genotype and cigarette smoking in the etiology of prostate cancer serves as an example of such an interaction. The risk of prostate cancer due to both factors was equivocal, with some studies suggesting links between smoking and *GSTT1* and prostate cancer, and others failing to find such associations. Given the potential role of metabolic enzymes (such as *GSTT1*) in metabolizing xenobiotic compounds into toxic metabolites, one hypothesis for the divergent results for this gene, smoking, and prostate cancer was that the underlying genetic susceptibilities to cigarette smoking differed across studies. Kelada and colleagues (2000) studied whether interactions between *GSTT1* genotype and cigarette smoking increased the risk of prostate cancer. In their study of 276 cases and 499 controls, no main effect of smoking was observed, but an increased risk of prostate cancer was found in individuals with nondeleted *GSTT1* genotypes (OR = 1.61). When considered together, an interaction between smoking and *GSTT1* genotype was detected; individuals with both risk factors were 2.03 times as likely to have prostate cancer as those with neither risk factor nor a deletion in *GSTT1*.

Case-Only Studies. It is possible to study certain aspects of the joint effects of genetic and environmental factors without using control subjects (Khoury and Flanders, 1996; Goldstein and Andrieu, 1999). The case-only study design is an approach that lends itself to the investigation of interactions between genes and the environment as well as gene–gene interactions and does not require a control group. The case-only approach is more appropriate than a case–control design for the investigation of the interaction between an exposure and a rare genetic risk factor on risk of disease. This is because the case–control approach would be very costly considering the addition of genotyping costs for the controls that would yield a low number of genetic risk factor carriers.

The case-only approach may improve power to detect interactions and is valid when the disease is rare and there is a strong a priori reason to believe that the genotype and the exposure are independent of each other in the population (Piegorsch et al., 1994; Yang et al., 1997). Several studies have examined the impact of a violation of this assumption and have found it to produce grossly inflated type I error (Albert et al., 2001). A preliminary assessment of the independence of the two factors can be performed by determining the independence of the susceptibility genotype and the exposure status (or the modifying allele) among a set of unaffected controls (in a case–control study) or a set of unaffected relatives (in a family study) who are genotype or mutation positive but do not have the disease of interest. However, even when a test of independence is first carried out in a sample of controls, the bias may persist (Albert et al., 2001). Thus, the case-only design may be useful for assessing departures from multiplicative interaction when there is strong empirical evidence for the independence between the factors, but estimates and tests should be interpreted cautiously in the absence of such evidence.

When the factors are independent and the disease is rare, the interaction terms can be estimated by calculating an OR for the association between the susceptibility genotype and the modifying factor (Khoury and Flanders, 1996; Goldstein and Andrieu, 1999). This case-only interaction OR is interpreted in a similar fashion to the OR_i on a multiplicative scale calculated from an unmatched case–control study. The case-only interaction OR can be estimated using a standard logistic regression model (Khoury and Flanders, 1996; Goldstein and Andrieu, 1999).

An illustration of a case-only design is a study where the interaction between oral contraceptive use and *BRCA1* and *BRCA2* mutation carriers was examined among women with breast cancer (Ursin et al., 1997). Women included in this study were diagnosed with breast cancer before the age of 40 and were of Ashkenazi Jewish heritage. All subjects genotyped for *BRCA1* and *BRCA2* were included in the study. In this study, long-term oral contraceptive use before a first full-term pregnancy was associated with an elevated risk of being classified as a *BRCA1* or *BRCA2* mutation carrier (OR = 7.8), suggesting that oral contraceptive use increased the risk of breast cancer more among mutation carriers compared to noncarriers (Ursin et al., 1997).

Cohort Studies. The "gold standard" of epidemiological study designs is the longitudinal cohort study. Rather than selecting study subjects on the basis of their disease status and looking backward to a possible cause, study subjects are selected on the basis of their exposure status, are disease free at the start of the follow-up period, and are followed forward in time to onset of disease. In the context of genetic epidemiology, the exposure can be defined as family history of disease or a susceptibility genotype. Some of the advantages of a cohort design are the ability to establish a temporal relationship between exposure and disease, the suitability for the study of rare exposures, and the ability to study multiple effects of a single exposure. Cohort studies are more costly and time consuming than case–control studies and are not efficient for the investigation of relatively rare disease outcomes such as cancer or diseases with late onset, such as AD. The analysis of interactions in prospective cohort studies is performed using methods similar to those described for case–control studies. Unlike case–control studies, prospective data may be used to provide direct estimates of the relative risk (RR) and to calculate measures of impact, such as attributable risk (see Chapter 12 for more details).

Other multivariate analytic methods, including survival analysis techniques, may also be employed to assess familial aggregation. In the study of diseases with variable age of onset, it is also important to account for the age of those at risk for developing the disease. Survival analyses, applied in the analysis of cohort data, account for varying lengths of observation of study subjects. These techniques are particularly appropriate for late-onset disease, where censoring of data (when a family member dies as a result of a competing cause or is lost to follow-up before the disease is diagnosed) is important, since these methods weight the contribution of older family members more than that of younger members.

Age-of-onset distributions of relatives of cases and controls are often examined to help determine recurrence risk patterns and the influence of age and age-specific risk of disease (Schildkraut et al., 1989a,b; Claus et al., 1991; Khoury and Beaty, 1994). The Cox (1972) proportional-hazards model is one survival method that takes into account the unequal lengths of time or variable age at onset in family members at risk for developing the disease of interest. The proportional-hazards model describes the relation of the independent variable (disease status) to the natural logarithm of the incidence rate of disease rather than to the odds of disease (Rothman and Greenland, 1998). The following equation describes the proportional-hazards model:

$$\ln[\text{incidence rate } (t)] = a(t) + b_1 X_1 + \cdots + b_n X_n$$

where $a(t)$ = baseline incidence rate expressed as a function of age (or time)
$X_1, \ldots, X_n$ = independent variables, such as family history, status of probands (affected vs. unaffected), genotype, or other risk factors for disease under study
$b_1, \ldots, b_n$ = respective coefficients for each independent variable

Proportional-hazards analysis allows for simultaneous inclusion of other factors in the model to assess the independent effects, control for confounding, and test for interaction. The coefficients from the proportional-hazards model are estimates of the incidence density ratio and are analogous to the coefficients from the logistic regression model.

One example of the use of survival analyses is in a study investigating the genetic relationship between ovarian, breast, and endometrial cancers (Schildkraut et al., 1989b). The data analysis was based on information about cancer in relatives of cancer cases and controls from large population-based case–control study. By employing the proportional-hazards model, age-adjusted RRs and 95% confidence intervals for breast, ovarian, and endometrial cancers among mothers and sisters of women diagnosed with cancer at these three sites (or cases) were calculated. The reference group for all comparisons consisted of the mothers and sisters of population controls.

Using the proportional-hazards model, the age-specific incidence rates for relatives of cases and controls were compared under the assumption that the hazard ratio for case and control relatives is constant throughout the risk period. In this example, $a(t)$ is the incidence rate expressed as a function of age or age at death (for those without cancer) to age at diagnosis of cancer (for diagnosed with cancer) in mothers and sisters. At each point in time, women who developed cancer of the breast, ovary, or endometrium were compared to a risk set comprising all women who survived at least as long and did not develop cancer. The variable (X_1) included in this model was the case–control status of the proband. The result demonstrated there were significantly elevated age-adjusted RRs for ovarian cancer (RR = 2.8; 95% CI = 1.6–4.9) and breast cancer (RR = 1.6; 95% CI = 1.1–2.1) among relatives of ovarian cancer probands and for breast cancer (RR = 2.1, 95% CI = 1.7–2.5) and ovarian cancer (RR = 1.7; 95% CI = 1.0–2.0) among relatives of breast cancer probands. Relatives of endometrial cancer probands had an elevated RR for endometrial cancer only (RR = 2.7; 95% CI = 1.6–4.8). The RRs that were derived from proportional-hazards models were then used in an analysis in which a multivariate polygenic threshold model was chosen to estimate heritability and to determine genetic relationships by estimating the genetic correlation among traits. The heritability estimates and genetic correlations derived from the multivariate polygenic threshold model were consistent with the results of the proportional-hazards analysis.

Family-Based Case–Control Studies. This design is a special case of a matched case–control study (see Chapter 12). Since cases and controls (usually unaffected siblings or cousins) are well matched for ethnic background, concerns about potential bias of the case-unrelated control design due to population stratification are eliminated. In addition, this design is potentially more efficient for estimating gene–environment interactions, particularly when the genetic factor is rare (Witte et al., 1999). The data may be analyzed with a conditional logistic regression model, incorporating interaction terms to test for interactions. When multiple affected or unaffected siblings from the same family are used, analysis methods for correlated data can be employed to correct for correlations among cases and controls (Siegmund et al., 2000).

While the stratification approach may seem to be an effective way of exploring interactions in case–parent triads, this approach is limited to qualitatively comparing results in triads with exposed probands to triads without exposures. Some authors have suggested that it is possible to test for interaction by comparing allelic transmission rates in exposed and unexposed triads (Schaid, 1999). However, Umbach and Weinberg (2000) have shown that if the population being studied consists of two nonmating subpopulations with different exposure and disease frequencies, transmission rate comparisons are invalid for detecting interactions and may lead to erroneous conclusions. In the presence of gene–environment interaction, the conditional genotype distributions of exposed and unexposed cases given parental genotypes are expected to differ. A test of the null hypothesis of equal distributions can be obtained by fitting a loglinear model to observed transmission counts in a table of all possible parental mating types, stratified by the case's exposure status (Umbach and Weinberg, 2000). The model-based test of interaction remains valid even when a genotype–exposure association is due to population admixture. The test assumes absence of genotypic differences on survival and conditional (on parental genotypes) independence of the genotype and environmental exposure.

Higher Order Interactions and Data Reduction. The identification of interactions in studies of complex diseases is hindered by two primary factors: the lower power of available samples to detect interactions compared to main effects (Smith and Day, 1984) and the limitations of commonly used statistical methods for case–control data for handling higher order interactions (Hosmer and Lemeshow, 2000). As increasing numbers of risk factors (genetic or environmental) are considered, the possible number of exposure-level combinations increases exponentially. The application of the traditional methods described in this chapter, such as logistic regression, becomes problematic in commonly available sample sizes since the high number of exposure combinations quickly results in cells containing no observations. Stepwise logistic regression deals with this combinatorial problem by entering only those factors into the model that have a statistically significant marginal or main effect. Factors with purely interactive effects will be missed. It is apparent that traditional methods such as logistic regression will not have the flexibility or the power to detect high-order interactions (Templeton, 2000). To address this issue, data reduction techniques such classification and regression trees (Province et al., 2000), combinatorial partitioning (Nelson et al., 2001), and multifactor dimensionality reduction (Ritchie et al., 2001) have been applied to genetic data.

The CART methods partition a dataset into more homogeneous subsets with varying risks for a potential outcome (e.g., risk of a complex disease or expression level for a quantitative trait locus). Given a set of covariates, the CART procedure selects the covariate most strongly associated with the outcome and partitions the dataset into two parts (or "daughter nodes"). This procedure is repeated with each daughter node and the remaining variables until no additional splits into significantly more homogeneous subsets may be made. This approach will identify a set of covariates that together define a class of individuals at strong risk of the disease under investigation. This approach has been applied to the study of interactions in case–control samples (Province et al., 2000) and family-based linkage studies (Shannon et al., 2001). CART algorithms for case–control data are available in the S-Plus software package. Software for fitting regression trees to linkage data has also been developed (Shannon et al., 2001).

The combinatorial partitioning method (CPM) (Nelson et al., 2001) was developed to consider all possible genotypic interactions underlying quantitative traits. This method considers multiple loci to determine a combination of genotypes that explains the greatest amount of variation in the trait. In the first step, subsets of measured loci are selected for analysis and the resulting multilocus genotypes are determined. For example, for two loci each with three genotypes there are nine two-locus genotype combinations. In the second step, all possible ways of partitioning the genotypes are evaluated for association with the trait. Sets explaining a significant amount of variation in the trait are retained for validation. In step 3, the prediction error of each model is estimated using 10-fold cross-validation. An example of CPM may be found in a study of polymorphisms in the angiotensin-converting enzyme (*ACE*) and *PAI-1* genes and plasma t-PA and PAI-1 levels (Moore et al., 2002). In this study, no significant main effects of the *ACE* and *PAI-1* polymorphisms were detected in samples of African-American and Caucasian men and women. The application of CART methods to the data would not have detected any interactions, since no main effects existed upon which to partition the sample. However, the application of CPM to the data detected a significant correlation between the *ACE I/D* and *PAI-1* *4G/5G* genotypes and plasma t-PA and PAI-1 levels in Caucasian men and women and African-American women (Moore et al., 2002). The statistical significance of this interaction cannot be evaluated, as CPM is intended to be a hypothesis-generating approach; validation of the results in an independent sample using hypothesis testing is necessary (Nelson et al., 2001; Moore et al., 2002).

The MDR method (Ritchie et al., 2001, 2003) is a data reduction approach, motivated by CPM, that may be used to evaluate interactions in qualitative traits. In the first step, subsets of measured loci are selected for analysis and the resulting multilocus genotypes are determined. In the second step, the case–control ratio is estimated for each multilocus genotype. In the next step, genotypes are classified as either "high-" (if the case–control ratio exceeds a threshold, often set at 1) or "low-"risk sets, reducing the number of dimensions from n to 1. In the final step, the prediction error of each model is estimated using 10-fold cross-validation. Once an optimal model is identified, the null hypothesis that the factors do not predict the disease endpoint can be tested by permuting the data and repeating the classification or prediction to estimate an empirical p-value. As with CPM, MDR has the ability to detect complex interactions in the absence of significant main effects. For example, a study of several candidate genes in a case–control study of sporadic breast cancer failed to identify a single significant main effect using logistic regression, yet detected a four-way interaction among polymorphisms in the *COMT*, *CYP1A1*, and *CYP1B1* genes (Ritchie et al., 2001). Software for using the MDR to evaluate gene–gene and gene–environment interactions in case–control and discordant sibpair designs has recently been developed (Hahn et al., 2003), with extensions underway for more complex sampling designs.

A common problem for all these methods is the difficulty in interpreting the results. Each of these approaches identifies a pattern of exposures that is significantly associated with disease; however, the "best" solution may not be that much more significant from other potential solutions, some of which may be more biologically plausible. Therefore, the results of these data reduction analyses must be examined in more detail to determine if a more parsimonious model exists. One method for examining results is pruning (Ripley, 1996). This method is often applied in CART analyses and involves removing variables one at a time to determine the most parsimonious model that fits nearly as well as the original full model.

Potential Biases

As discussed in more detail in Chapter 12, genetic epidemiology studies are subject to several biases. These include confounding bias, selection bias, and measurement error.

Confounding bias results when the relationship between a risk factor (such as a susceptibility allele at a candidate gene) and an outcome (such as disease status) is altered by an unmeasured third factor (such as ethnic background) that is associated with both the risk factor and outcome. Considering these confounding variables in matched study designs or data analysis is critical in generating unbiased results.

Choosing a study sample that is not representative of the underlying population causes selection bias. To provide valid estimates of risk, a study sample should be a random sample of the population being studied, and cases and controls should both be selected from the same population. When the genetic risk factor is associated with both disease status and the probability of selection for the study sample, biased estimates of the relative risk may result.

Measurement error, such as misclassification of the susceptibility genotype or of an environmental exposure, can occur regardless of the study design. In studies that use family history in place of an unknown genetic factor, it is possible that genetic susceptibility will be misclassified. Such misclassification, if nondifferential with respect to an environmental risk factor, will cause a dilution in the estimate of association between the genetic susceptibility factor and the disease (Ottman, 1990). The attenuation in the magnitude of the association makes it more difficult to detect an association that actually exists. However, nondifferential misclassification does not induce a spurious association between the risk factor and disease. Therefore, if an association is detected in the presence of nondifferential misclassification, the true value for the measure is at least as great as what is observed.

Another important aspect to consider when evaluating any genetic study is the generalizability of the results or how findings in the study population are applicable to other populations. Generalizability depends on how the study population is selected or defined. Replication of the results in other populations is one way to address this question directly.

CONCLUSION

Combining epidemiological and genetic approaches can unravel the complexities of gene–gene and gene–environment interactions. Results of epidemiological studies have helped define syndromes and have aided in the identification of phenotypes associated with specific genetic syndromes.

Genetic epidemiological research illustrates that it is important that research of complex multifactorial traits address the characterization of how genes and the environment interact. Discovery of major genetic effects allows exploration of these avenues of research and can further our understanding of complex genetic traits. Advances in molecular technology promise to provide powerful tools for identifying genetic and environmental factors in the etiology of complex diseases.

REFERENCES

Albert PS, Ratnasinghe D, Tangrea J, Wacholder S (2001): Limitations of the case-only design for identifying gene-environment interactions. Am J Epidemiol 154:687–693.

Almasy L, Blangero J (1998): Multipoint quantitative-trait linkage analysis in general pedigrees. Am J Hum Genet 62:1198–1211.

Bell DA, Stephens EA, Castranio T, Umbach DM, Watson M, Deakin M, Elder J, Hendrickse C, Duncan H, Strange RC (1995): Polyadenylation polymorphism in the acetyltransferase 1 gene (NAT1) increases risk of colorectal cancer. Cancer Res 55: 3537–3542.

Bonney GE (1984): On the statistical determination of major gene mechanisms in continuous human traits: Regressive models. Am J Med Genet 18:731–749.

Bonney GE, Dunston GM, Wilson J (1989): Regressive logistic models for ordered and unordered polychotomous traits: Application to affective disorders. Genet Epidemiol 6: 211–215.

Bonney GE, Lathrop GM, Lalouel JM (1988): Combined linkage and segregation analysis using regressive models. Am J Hum Genet 43:29–37.

Brockton N, Little J, Sharp L, Cotton SC (2000): N-acetyltransferase polymorphisms and colorectal cancer: A HuGE review. Am J Epidemiol 151:846–861.

Chen J, Stampfer MJ, Hough HL, Garcia-Closas M, Willett WC, Hennekens CH, Kelsey KT, Hunter DJ (1998): A prospective study of N-acetyltransferase genotype, red meat intake, and risk of colorectal cancer. Cancer Res 58:3307–3311.

Claus EB, Risch N, Thompson WD (1991): Genetic analysis of breast cancer in the cancer and steroid hormone study. Am J Hum Genet 48:232–242.

Cordell HJ, Todd JA (1995): Multifactorial inheritance in type 1 diabetes. Trends Genet 11:499–504.

Cordell HJ, Todd JA, Bennett ST, Kawaguchi Y, Farrall M (1995): Two-locus maximum LOD score analysis of a multifactorial trait: Joint consideration of IDDM2 and IDDM4 with *IDDM1* in type I diabetes. Am J Hum Genet 57:920–934.

Cordell HJ, Wedig GC, Jacobs KB, Elston RC (2000): Multilocus linkage tests based on affected relative pairs. Am J Hum Genet 66:1273–1286.

Corder EH, Saunders AM, Strittmatter WJ, Schmechel DE, Gaskell PC, Small GW, Roses AD, Haines JL, Pericak-Vance MA (1993): Gene dose of apolipoprotein E type 4 allele and the risk of Alzheimer's disease in late onset families. Science 261:921–923.

Cox DR (1972): Regressin models and life tables (with discussion). J R Stat Soc B 34:187–220.

Cox NJ, Frigge M, Nicolae DL, Concannon P, Hanis CL, Bell GI, Kong A (1999): Loci on chromosomes 2 (*NIDDM1*) and 15 interact to increase susceptibility to diabetes in Mexican Americans. Nat Genet 21:213–215.

Demenais F, Lathrop M (1993): Use of the regressive models in linkage analysis of quantitative traits. Genet Epidemiol 10:587–592.

Demenais FM, Laing AE, Bonney GE (1992): Numerical comparisons of two formulations of the logistic regressive models with the mixed model in segregation analysis of discrete traits. Genet Epidemiol 9:419–435.

Dizier MH, Babron MC, Clerget-Darpoux F (1994): Interactive effect of two candidate genes in a disease: Extension of the marker-association-segregation χ^2 method. Am J Hum Genet 55:1042–1049.

Ertekin-Taner N, Graff-Radford N, Younkin LH, Eckman C, Baker M, Adamson J, Ronald J, Blangero J, Hutton M, Younkin SG (2000): Linkage of plasma Aβ42 to a quantitative locus on chromosome 10 in late-onset Alzheimer's disease pedigrees. Science 290: 2303–2304.

Farrall M (1997): Affected sibpair linkage tests for multiple linked susceptibility genes. Genet Epidemiol 14:103–115.

Farrer LA, Bowirrat A, Friedland RP, Warasaka K, Adams JC, Korczyn A, Baldwin CT (2001): Identification of multiple loci for Alzheimer Disease in an inbred Israeli-Arab community. Am J Hum Genet 69(suppl):200.

Futreal PA, Liu Q, Shattuck-Eidens D, Cochran C, Harshman K, Tavtigian S, Bennett LM, Haugen-Strano A, Swensen J, Miki Y, Eddington K, McClure M, Frye C, Weaver-Feldhaus J, Ding W, Gholami Z, Soderkvist P, Terry L, Jhanwar S, Berchuck A, Iglehart JD, Marks J, Ballinger DG, Barrett JC, Skolnick MH, Kamb A, Wiseman R (1994): BRCA1 mutations in primary breast and ovarian carcinomas. Science 266: 120–126.

Gauderman WJ, Faucett CL (1997): Detection of gene-environment interactions in joint segregation and linkage analysis. Am J Hum Genet 61:1189–1199.

Gauderman WJ, Siegmund KD (2001): Gene-environment interaction and affected sibpair linkage analysis. Hum Hered 52:34–46.

Ghosh S, Watanabe RM, Valle TT, Hauser ER, Magnuson VL, Langefeld CD, Ally DS, Mohlke KL, Silander K, Kohtamaki K, Chines P, Balow JJ, Birznieks G, Chang J, Eldridge W, Erdos MR, Karanjawala ZE, Knapp JI, Kudelko K, Martin C, Morales-Mena A, Musick A, Musick T, Pfahl C, Porter R, Rayman JB (2000): The Finland–United States investigation of non-insulin-dependent diabetes mellitus genetics (FUSION) study. I. An autosomal genome scan for genes that predispose to type 2 diabetes. Am J Hum Genet 67:1174–1185.

Goate A, Chartier-Harlin MC, Mullan M, Brown J, Crawford F, Fidani L, Giuffra L, Haynes A, Irving N, James L, Mant R, Newton P, Rooke K, Roques P, Talbot C, Pericak-Vance MA, Roses A, Williamson R, Rossor M, Owen M, Hardy J (1991): Segregation of a missense mutation in the amyloid precursor protein gene with familial Alzheimer's disease. Nature 349:704–706.

Goldstein AM, Andrieu N (1999): Detection of interaction involving identified genes: Available study designs. J Natl Cancer Inst Monogr:49–54.

Hahn LW, Ritchie MD, Moore JH (2003): Multifactor dimensionality reduction software for detecting gene–gene and gene–environment interactions. BioInformatics, in press.

Hall JM, Lee MK, Newman B, Morrow JE, Anderson LA, Huey B, King MC (1990): Linkage of early-onset familial breast cancer to chromosome 17q21. Science 250: 1684–1689.

Hein DW, Doll MA, Fretland AJ, Leff MA, Webb SJ, Xiao GH, Devanaboyina US, Nangju NA, Feng Y (2000): Molecular genetics and epidemiology of the *NAT1* and *NAT2* acetylation polymorphisms. Cancer Epidemiol Biomarkers Prev 9:29–42.

Hodge SE (1994): What association analysis can and cannot tell us about the genetics of complex disease. Am J Med Genet 54:318–323.

Holmans P (2002): Detecting gene–gene interactions using affected sibpair analysis with covariates. Hum Hered 53:92–102.

Hosmer D, Lemeshow S (2000): Applied Logistic Regression. New York: John Wiley & Sons.

Hwang SJ, Beaty TH, Liang KY, Coresh J, Khoury MJ (1994): Minimum sample size estimation to detect gene-environment interaction in case–control designs. Am J Epidemiol 140:1029–1037.

Kehoe P, Wavrant-De Vrieze F, Crook R, Wu WS, Holmans P, Fenton I, Spurlock G, Norton N, Williams H, Williams N, Lovestone S, Perez-Tur J, Hutton M, Chartier-Harlin MC, Shears S, Roehl K, Booth J, Van Voorst W, Ramic D, Williams J, Goate A, Hardy J, Owen MJ (1999): A full genome scan for late onset Alzheimer's disease. Hum Mol Genet 8:237–245.

Kelada SN, Kardia SL, Walker AH, Wein AJ, Malkowicz SB, Rebbeck TR (2000): The glutathione S-transferase-mu and -theta genotypes in the etiology of prostate cancer: Genotype-environment interactions with smoking. Cancer Epidemiol Biomarkers Prev 9:1329–1334.

Khoury MJ, Beaty TH (1994): Applications of the case–control method in genetic epidemiology. Epidemiol Rev 16:134–150.

Khoury MJ, Flanders WD (1996): Nontraditional epidemiologic approaches in the analysis of gene-environemnt interaction: Case-control studies with no controls. Am J Epidemiol 144:207–213.

Knapp M, Seuchter SA, Baur MP (1994): Two-locus disease models with two marker loci: The power of affected-sib-pair tests. Am J Hum Genet 55:1030–1041.

Kong A, Cox NJ (1997): Allele-sharing models: LOD scores and accurate linkage tests. Am J Hum Genet 61:1179–1188.

Lander ES, Schork NJ (1994): Genetic dissection of complex traits. Science 265:2037–2048.

Lathrop GM, Ott J (1990): Analysis of complex diseases under oligogenic models and intra-familial heterogeneity by the LINKAGE program. Am J Hum Genet 47:A188.

Levy-Lahad E, Wasco W, Poorkaj P, Romano DM, Oshima J, Pettingell WH, Yu CE, Jondro PD, Schmidt SD, Wang K, Crowley AC, Fu YH, Guenette SY, Galas D, Nemens E, Wijsman EM, Bird TD, Schellenberg GD, Tanzi RE (1995): Candidate gene for the chromosome 1 familial Alzheimer's disease locus. Science 269:973–977.

Lin S (2000): Monte Carlo methods for linkage analysis of two-locus disease models. Ann Hum Genet 64:519–532.

Mayeux R, Lee JH, Romas SN, Mayo D, Santana V, Williamson J, Ciappa A, Rondon HZ, Estevez P, Lantigua R, Medrano M, Torres M, Stern Y, Tycko B, Knowles JA (2002): Chromosome-12 mapping of late-onset Alzheimer disease among Caribbean Hispanics. Am J Hum Genet 70:237–243.

Miki Y, Swensen J, Shattuck-Eidens D, Futreal PA, Harshman K, Tavtigian S, Liu Q, Cochran C, Bennett LM, Ding W, Bell R, Rosenthal J, and 33 others (1994): A strong candidate for the breast and ovarian cancer susceptibility gene BRCA1. Science 266:66–71.

Moore JH, Lamb JM, Brown NJ, Vaughan DE (2002): A comparison of combinatorial partitioning and linear regression for the detection of epistatic effects of the ACE I/D and PAI-1 4G/5G polymorphisms on plasma PAI-1 levels. Clin Genet 62:74–79.

Myers A, Holmans P, Marshall H, Kwon J, Meyer D, Ramic D, Shears S, Booth J, DeVrieze FW, Crook R, Hamshere M, Abraham R, Tunstall N, Rice F, Carty S, Lillystone S, Kehoe P, Rudrasingham V, Jones L, Lovestone S, Perez-Tur J, Williams J, Owen MJ, Hardy J, Goate AM (2000): Susceptibility locus for Alzheimer's disease on chromosome 10. Science 290:2304–2305.

Nelson MR, Kardia SLR, Ferrell RE, Sing CF (2001): A combinatorial partitioning method (CPM) to identify multi-locus genotypic partitions that predict quantitative trait variation. Genome Res 11:458–470.

Newman MF, Croughwell ND, Blumenthal JA, Lowry E, White WD, Spillane W, Davis RD, Glower DD, Smith LR, Mahanna EP, Reves JG (1995): Predictors of cognitive decline after cardiac operation. Ann Thorac Surg 59:1326–1330.

Olson JM (1999): A general conditional-logistic model for affected-relative-pair linkage studies. Am J Hum Genet 65:1760–1769.

Ottman R (1990): An epidemiologic approach to gene-environment interaction. Genet Epidemiol 7:177–185.

Pericak-Vance MA, Bass MP, Yamaoka LH, Gaskell PC, Scott WK, Terwedow HA, Menold MM, Conneally PM, Small GW, Vance JM, Saunders AM, Roses AD, Haines JL (1997): Complete genomic screen in late-onset familial Alzheimer disease: Evidence for a new locus on chromosome 12. JAMA 278:1237–1241.

Pericak-Vance MA, Bebout JL, Gaskell PC, Yamaoka LH, Hung WY, Alberts MJ, Walker AP, Bartlett RJ, Haynes CS, Welsh KA, Earl NL, Heyman A, Clark CM, Roses AD (1991): Linkage studies in familial Alzheimer's disease: Evidence for chromosome 19 linkage. Am J Hum Genet 48:1034–1050.

Pericak-Vance MA, Grubber J, Bailey LR, Hedges D, West S, Kemmerer B, Hall JL, Saunders AM, Roses AD, Small GW, Scott WK, Conneally PM, Vance JM, Haines JL (2000): Identification of novel genes in late-onset Alzheimer disease. Exp Gerontol 35:1343–1352.

Pericak-Vance MA, Haines JL (1995): Genetic susceptilbility to Alzheimer disease. Trends Genet 11:504–508.

Piegorsch WW, Weinberg CR, Taylor JA (1994): Non-hierarchical logistic models and case-only designs for assessing susceptibility in population-based case–control studies. Stat Med 13:153–162.

Pritchard JK, Donnelly P (2001): Case-control studies of association in structured or admixed populations. Theor Popul Biol 60:227–237.

Pritchard JK, Stephens M, Rosenberg NA, Donnelly P (2000): Association mapping in structured populations. Am J Hum Genet 67:170–181.

Province MA, Arnett DK, Hunt SC, Leiendecker-Foster C, Eckfeldt JH, Oberman A, Ellison RC, Heiss G, Mockrin SC, Williams RR (2000): Association between the alpha-adducin gene and hypertension in the HyperGEN Study. Am J Hypertens 13: 710–718.

Ripley BD (1996): Pattern Recognition and Neural Networks. Cambridge: Cambridge University Press.

Risch A, Wallace DM, Bathers S, Sim E (1995): Slow N-acetylation genotype is a susceptibility factor in occupational and smoking related bladder cancer. Hum Mol Genet 4:231–236.

Risch N (1990a): Linkage strategies for genetically complex traits I. Multilocus models. Am J Hum Genet 46:222–228.

Risch N (1990b): Linkage strategies for genetically complex traits II. The power of affected relative pairs. Am J Hum Genet 46:229–241.

Ritchie MD, Hahn LW, Moore JH (2003): Power of multifactor dimensionality reduction for detecting gene–gene interactions in the presence of genotyping error, missing data, phenocopy, and genetic heterogeneity. Genet Epidemiol, in press.

Ritchie MD, Hahn LW, Roodi N, Bailey LR, Dupont WD, Parl FF, Moore JH (2001): Multifactor-dimensionality reduction reveals high-order interactions among estrogen-metabolism genes in sporadic breast cancer. Am J Hum Genet 69:138–147.

Rogaev EI, Sherrington R, Rogaeva EA, Levesque G, Ikeda M, Liang Y, Chi H, Lin C, Holman K, Tsuda T (1995): Familial Alzheimer's disease in kindreds with missense mutations in a gene on chromosome 1 related to the Alzheimer's disease type 3 gene. Nature 376:775–778.

Rothman KJ, Greenland S (1998): Modern Epidemiology. Philadelphia: Lippincott-Raven.

Saunders AM, Strittmatter WJ, Schmechel D, George-Hyslop PH, Pericak-Vance MA, Joo SH, Rosi BL, Gusella JF, Crapper-MacLachlan DR, Alberts MJ (1993): Association of apolipoprotein E allele epsilon 4 with late-onset familial and sporadic Alzheimer's disease. Neurology 43:1467–1472.

Schaid DJ (1999): Case-parents design for gene-environment interaction. Genet Epidemiol 16:261–273.

Schellenberg GD, Bird TD, Wijsman EM, Orr HT, Anderson L, Nemens E, White JA, Bonnycastle L, Weber JL, Alonso ME, Potter H, Heston LL, Martin GM (1992): Genetic linkage evidence for a familial Alzheimer's disease locus on chromosome 14. Science 258:668–671.

Schildkraut JM, Myers RH, Cupples LA, Kiely DK, Kannel WB (1989a): Coronary risk associated with age and sex of parental heart disease in the Framingham Study. Am J Cardiol 64:555–559.

Schildkraut JM, Risch N, Thompson WD (1989b): Evaluating genetic association among ovarian, breast, and endometrial cancer: Evidence for a breast/ovarian cancer relationship. Am J Hum Genet 45:521–529.

Scott WK, Grubber JM, Conneally PM, Small GW, Hulette CM, Rosenberg CK, Saunders AM, Roses AD, Haines JL, Pericak-Vance MA (2000): Fine mapping of the chromosome 12 late-onset Alzheimer disease locus: Potential genetic and phenotypic heterogeneity. Am J Hum Genet 66:922–932.

Shannon WD, Province MA, Rao DC (2001): Tree-based recursive partitioning methods for subdividing sibpairs into relatively more homogeneous subgroups. Genet Epidemiol 20:293–306.

Sherrington R, Rogaev EI, Liang Y, Rogaeva EA, Levesque G, Ikeda M, Chi H, Lin C, Li G, Holman K, Tsuda T, Mar L, Foncin JF, Bruni AC, Montesi MP, Sorbi S, Rainero I, Pinessi L, Nee L, Chumakov I, Pollen DA, Brookes A, Sanseau P, Polinsky RJ, Wasco W, Da Silva HAR, Haines JL, Pericak-Vance MA, Tanzi RE, Roses AD, Fraser PE, Rommens JM, St George-Hyslop PH (1995): Cloning of a gene bearing missense mutations in early-onset familial Alzheimer's disease. Nature 375:754–760.

Siegmund KD, Langholz B, Kraft P, Thomas DC (2000): Testing linkage disequilibrium in sibships. Am J Hum Genet 67:244–248.

Slattery ML, O'Brien E, Mori M (1995): Disease heterogeneity: Does it impact our ability to detect dietary associations with breast cancer? Nutr Cancer 24:213–220.

Smith PG, Day NE (1984): The design of case–control studies: The influence of confounding and interaction effects. Int J Epidemiol 13:356–365.

St George-Hyslop PH, Haines JL, Farrer LA, Polinsky RJ, Van Broeckhoven C, Goate AM, McLachlan DR, Orr H, Bruni AC, Sorbi S (1990): Genetic linkage studies suggest that Alzheimer's disease is not a single homogenous disorder. Nature 347:194–197.

Strauch K, Fimmers R, Kurz T, Deichmann KA, Wienker TF, Baur MP (2000): Parametric and nonparametric multipoint linkage analysis with imprinting and two-locus-trait models: Application to mite sensitization. Am J Hum Genet 66:1945–1957.

Templeton AR (2000): Epistasis and complex traits. In: Wolf JB, Brodie N, Wade M, eds. Epistasis and the Evolutionary Process. Oxford University Press, pp 41–57.

Thompson WD (1994): Statistical analysis of case–control studies. Epidemiol Rev 16:33–50.

Thomson G (1994): Identifying complex disease genes: Progress and paradigms. Nat Genet 8:108–110.

Tienari PJ, Terwilliger JD, Ott J, Palo J, Peltonen L (1994): Two-locus linkage analysis in multiple sclerosis (MS). Genomics 19:320–325.

Umbach DM, Weinberg C (2000): The use of case-parent triads to study joint effects of genotype and exposure. Am J Hum Genet 66:251–261.

Ursin G, Henderson BE, Haile RW, Pike MC, Zhou N, Diep A, Bernstein L (1997): Does oral contraceptive use increase the risk of breast cancer in women with BRCA1/BRCA2 mutations more than in other women? Cancer Res 57:3678–3681.

Weeks DE, Lathrop GM (1995): Polygenic disease: Methods for mapping complex diesase traits. Trends Genet 11:513–519.

Wilcox MA, Smoller JW, Lunetta KL, Neuberg D (1999): Using recursive partitioning for exploration and follow-up of linkage and association analyses. Genet Epidemiol 17:S391–S396.

Witte JS, Gauderman WJ, Thomas DC (1999): Asymptotic bias and efficiency in case–control studies of candidate genes and gene-environment interactions: Basic family designs. Am J Epidemiol 149:693–705.

Yang Q, Khoury MJ (1997): Evolving methods in genetic epidemiology. III. Gene-environment interaction in epidemiologic research. Epidemiol Rev 19:33–43.

Yang Q, Khoury MJ, Flanders WD (1997): Sample size requirements in case-only designs to detect gene-environment interaction. Am J Epidemiol 146:713–720.

Zinn-Justin A, Abel L (1998): Two-locus developments of the weighted pairwise correlation method for linkage analysis. Genet Epidemiol 15:491–510.

CHAPTER 15

Genomics and Bioinformatics

JUDITH E. STENGER and SIMON G. GREGORY

INTRODUCTION

The primary goal of this chapter is to give the reader insight into how bioinformatics tools and genomic resources and databases can best be applied to research aimed at identifying genes contributing to susceptibility of complex diseases. The chapter describes a strategy to identify candidate genes and to unravel the mechanism by which an associated gene may contribute to a disease phenotype. The chapter's contents will provide background to the origin of the available genomic data, a description of the various public databases and analysis tools, and how different sources of public and private data can be analyzed synergistically to help advance a research project. Most research questions can be approached from multiple directions using a variety of tools. The authors direct genetics researchers to some of the most reliable and useful bioinformatics resources that are available at the time of this writing.

Era of the Genome

On April 25, 1953, a short letter was published in *Nature* revealing to the world that the secondary structure of our genetic material was comprised of a double helix containing two complementary strands of DNA (Watson and Crick, 1953). The authors contended that the double-stranded helix could both accommodate the building blocks of life, nucleic acids (Pauling and Corey, 1953), and provide a guide to genetic inheritance by serving as a template for replication. This enormous breakthrough provided the groundwork for scientists to decipher the genetic code and uncover the mechanisms of transcription and translation. In April 2003, a mere 50 years after Watson and Crick's discovery, an international consortium of publicly funded scientists accomplished an equally extraordinary achievement, the complete

Genetic Analysis of Complex Diseases, Second Edition, Edited by Jonathan L. Haines and Margaret Pericak-Vance

decoding of the human genome sequence. This accomplishment signified the true beginning of the "genomic era" and the dawn of genomic medicine (Collins et al., 2003).

The speed at which the vast amount of human sequence data was generated can be attributed to the evolution of strategies and techniques developed to map and sequence organisms such as bacteria (Kohara et al., 1987), yeast (Olson et al., 1986), and the nematode worm (Coulson et al., 1986). The availability of such an evolutionary diverse collection of species, with the addition of mouse (Mouse Genome Sequencing Consortium, 2002) and other complex multicellular organisms, has also enabled comparisons to be made at a nucleotide level. The generation and analysis of sequence from these complex organisms could not have been possible without commensurate advances in computational biology and bioinformatics. Because of their relative simplicity, the sequence of nucleic acids and proteins is amenable to computer-aided assembly and analysis. Their nucleotide and amino acid components are represented by letters of 4 and 20 elements, respectively, that can be treated as simple character strings that can be interrogated for matching of known patterns such as genes or matching between species on a comparative level.

The first genomes that were characterized at the nucleotide level were relatively small by current standards, bacteriophage ϕX174, 5 kb (Sanger et al., 1977, 1978) and bacteriophage λ, 48 kb (Sanger et al., 1982), but they provided the underlying techniques and strategies that are being used for the more complex organisms currently being studied. Chain termination sequencing, developed by Sanger et al., is highly sensitive and robust and has been amenable to biochemical optimization, producing long, accurate sequence reads, and also to automation, which was necessary for large-scale application of the technique. However, terminator sequencing is not capable of generating single reads of greater than 200–300 nucleotides, limited in part by the sequence production itself and partly by the ability to separate the sequence by gel electrophoresis at single-base resolution (even today sequencing read lengths approaching 1 kb are rare).

Assembly of larger tracts of DNA therefore required the development of methods to reassemble a consensus sequence from multiple individual reads. Two approaches were adopted for this: first, the construction of physical maps, which incorporated previously available cytogenetic data and genetic and radiation hybrid maps, and, second, the use of the information gained from each individual sequence read to order and orient each segment relative to overlapping neighbors. Both of these approaches required the development of advanced computer programs to make the task possible on all but the smallest scale.

MAPPING THE HUMAN GENOME

The human genome is contained within 22 autosomes (1–22, numbered largely according to size) and two sex chromosomes, X and Y. The initial size estimate of the genome (3,200,000,000 bp, or 3200 Mb) was based largely upon cytometric measurements (Morton, 1991) and has since been revised to 2900 Mb in light of the higher resolution human draft and finished chromosome analyses (International Human Genome Mapping Consortium, 2001; Dunham et al., 1999; Hattori et al., 2000; Deloukas et al., 2001; Heilig et al., 2003; Hillier et al., 2003; Skaletski et al., 2003; Mungall et al., 2003). The construction of a variety of maps of the human genome provided a means by which all the features could be ordered and partitioned, therefore providing an important step toward understanding and characterizing the sequence contained within it.

Cytogenetic Mapping. The treatment of metaphase chromosome spreads with trypsin digestion and Giemsa staining creates differential chromosome banding patterns [by the generation of light (R-bands) and dark (G-bands)], therefore permitting a regional division of a chromosome. The characterization of these chromosomal banding patterns provided the basis for much of the early categorization of chromosome aberrations (duplications, deletions, and translocations) that were associated with clinical phenotypes (Pinkel et al., 1988; Tkachuk et al., 1990; Dauwerse et al., 1990). The maximum genomewide resolution was, however, limited to an 850 genomewide banding pattern (Bickmore et al., 1989). A dramatic improvement upon this resolution was achieved by hybridizing labeled probes containing specific sequences to detect their location on metaphase chromosome by autoradiographic or fluorescent detection techniques [fluorescence in situ hybridization (FISH)] (Pinkel et al., 1986). This technique typically utilizes cloned DNA as the template for the generation of a fluorescently labeled polymerase chain reaction (PCR) probe for hybridization to a metaphase spread of genomic DNA and the signal emitted by the fluorescent nucleotide contained within the probe being detected by using epifluorescence microscopy. Initially, the location of the probe relative to the metaphase banding pattern provided an approximate map position for the sequence represented by the probe. Pairs of markers, labeled with different fluorochromes, could be simultaneously placed relative to the cytogenetic banding and also ordered with respect to each other. The use of pairs of differentially labeled markers in combination with a third reference marker enabled FISH to be applied to chromosomal DNA in a less condensed state (in interphase nuclei). Although no banding pattern can be obtained in interphase DNA, the decondensed state of the chromatin relative to metaphase chromosomes means that increased levels of resolution could be obtained as probes were better separated. An interprobe distance of 1–5 Mb can be resolved using metaphase FISH, 0.1–1.0 Mb by interphase FISH (Wilke et al., 1994), and 5 kb by FISH using mechanical pretreatment to extend DNA into fibers (Heiskanen et al., 1994).

Genetic Mapping

Genetic maps utilize the likelihood of recombination between adjacent markers during meiosis to calculate intermarker genetic distances and from this to infer a physical distance. The closer two landmarks are together on a chromosome, the less likelihood there is of a recombination event occurring between, with the opposite being true for markers that are further apart (see Chapter 1). The calculation of distance, and therefore the metric upon which the genetic map is based, is the length of the chromosomal segment that, on average, undergoes one exchange with a sister chromatid during meiosis, the morgan (M). Therefore, a 1% recombination frequency is equivalent to 1 cM, and, since the human genome covers 3000 cM and contains approximately 3,000 Mb, 1 cM is approximately equivalent to 1 Mb. However, recombination is known to be nonrandom, which can lead to a level of inaccuracy (Dib et al., 1996) in inferring physical distances from measurements of genetic recombination.

The limitation of the primary genetic maps, that is, the lack of availability of polymorphic markers between which genetic distances could be calculated, was ameliorated in part by the use of restriction fragment length polymorphisms (RFLPs) (Kan and Dozy, 1978). However, the first such map (Donis-Keller et al., 1987) was limited in its usefulness due to RFLPs having a maximum heterozygosity of 50% and the low level of resolution of the 403 characterized polymorphic markers (including 393 RFLPs) that covered the genome. The identification of microsatellite markers (containing di-, tri-, or tetranucleotide repeats) greatly facilitated the generation of genetic maps. They were proven to be widely distributed throughout the genome, showed allelic variation (Litt and Luty, 1989; Weber and May, 1989), and were amenable to PCR amplification (Saiki et al., 1988) by sequence-tagged site screening (Olson et al., 1989). In a relatively short period of time a number of genetic maps were published with increasing marker density and resolutions, culminating in the deCODE genetic map that contains 5136 markers genotyped across 1257 meioses (Kong et al., 2002).

Genetic Map Resources. Genetic maps are crucial elements in the development of a linkage study and public databases of genetic markers and maps can aid in selecting genotyping markers for performing a genomic screen. The Cooperative Human Linkage Center (http://gai.nci.nih.gov/CHLC/), now maintained at the National Cancer Institute, integrates several types of genetic maps showing the location of CHLC markers along with information on markers. It includes the ABI Prism medium- and high-density maps and the Genetic Annotation Initiative (GAI) single-nucleotide polymorphism (SNP) marker maps from the Cancer Genome Anatomy Project, a project that characterizes genetic variation in genes important in cancer. The CEPH–Foundation Jean Dausset site at the Centre d'Etude du Polymorphisme Humain (CEPH) (http://www.cephb.fr/) contains the CEPH genotype database, integrated maps containing CEPH-Généthon mapping data, the CEPH Chromosome 21 mapping resource, and the CEPH cDNA resource. The mapping information is based on a yeast artificial chromosome (YAC) contig map of the genome (Chumakov et al., 1995). The CEPH site contains Genexpress data and a map of expressed sequence tags (ESTs). Detailed information on markers is also available. The CEPH genotype database contains RFLP, variable-number tandem repeat (VNTR) data, and microsatellite marker data submitted from laboratories around the world. The Quickmap database and browser are available to the community by anonymous ftp. The Marshfield Center for Medical Genetics site (http://research.marshfieldclinic.org/genetics/) contains accurate, well-maintained, error-checked genetic maps. There is significant data on human DNA poly-morphisms, a comparison of genetic and physical maps, and links for the mapped polymorphisms to GenBank genomic sequences. The Marshfield site also includes a Build Your Own Map feature and a Search for Markers feature. For researchers involved in genomic screening, this site is a valuable marker resource.

The Location Data Base (LDB) site (http://cedar.genetics.soton.ac.uk/public_html/LDBmain.html) provides summary maps that integrate genetic, radi-ation hybrid, cytogenetic, and physical maps for all chromosomes. LDB2000 fea-tures sequence-based integrated maps of the human genome. However, much of the data are somewhat outdated. GENLINK (www.genlink.wustl.edu/) features TelDB—telomeric markers and maps. Detailed meiotic maps are also provided. Resources are provided to integrate genetic and physical maps. There are 2215 refer-ences in the literature citation database and user-specified queries of the database can be performed. This site also provides access to the Généthon database and links to the Genome Data Base (GDB, see below).

Genetic Marker Resources. The GDB (www.gdb.org/) is the most complete resource for genetic markers. The site is a detailed repository of information on gen-etic markers; alternate names; flanking regions; primer sequences; allele sizes; allele frequencies; genetic, cytogenetic, and physical mapping information; and evidence for cytogenetic location. This site is under the management of the Hospital for Sick Children in Toronto and hosts the HUGO chromosome pages. Additionally, the GDB has completely regenerated its comprehensive maps, integrating all available mapping information.

Radiation Hybrid Mapping

The utilization of somatic cell hybrids to maintain human genomic fragments, such as whole chromosomes or chromosomal regions, permits the generation of another form of mapping resource, the radiation hybrid map. The modification of a tech-nique that fragmented human chromosomes by irradiation which were then rescued by fusion to rodent cells (Goss et al., 1975) prompted Cox et al. (1990) to propose that radiation hybrid (RH) mapping could be applied to the construction of long-range maps of mammalian chromosomes.

The premise of the technique is similar to that of the genetic map, that is, the more closely related two markers are related within the genome, the less likelihood there is of a radiation-induced break in between them in a reference panel of cell lines, and hence the less likely is their segregation to different chromosomal locations based on association of the markers to different sets of fragments. As the presence of two markers within a radiation fragment gives no indication to their physical distance, a panel of RHs was required. By estimating the frequency of breakage, and thus the distance between two markers, it is possible to determine their order. The unit of map distance is the centiray and represents 1% probability of breakage between two markers for a given radiation dose. Unlike the level of information garnered from a genetic marker, which may or may not be informative within a varying number of meioses, the RH marker is either positive or negative for a DNA fragment, effectively digitizing PCR results. Any amplifiable single-copy sequence can therefore be placed in a RH map. The RH mapping technique has been used for the construction of high-resolution gene maps using ESTs (Schuler et al., 1996; Deloukas et al., 1998) and has also been used to supplement the construction of chromosome physical maps via sequence-tagged site (STS) localization (Mungall et al., 1996).

Radiation Hybrid Software Resources. A number of RH mapping packages are available to researchers (see the RH mapping information page, http://compgen. rutgers.edu/rhmap/), including servers established at some of the institutions constituting the international RH mapping consortium.

The National Center for Biotechnology Information (NCBI) genemap99 home page (www.ncbi.nlm.nih.gov/genemap99) provides links to the two RH mapping servers which allow the user to map STSs relative to the respective center's RH maps: the Stanford G3 RH server (www-shgc.stanford.edu/RH/index.html) and the Wellcome Trust Sanger Institute's gene map server (www.sanger.ac.uk/ Software/RHserver/RHserver.shtml) for the GB4 RH panel. The Sanger Institute's site allows the user to input the results of a marker typed against the RH screening panel for analysis in relation to the international integrated Gene Map described above. Users can find which markers on the site's RH panels are closest to a STS submitted. In addition to the G3 panel, the Stanford server assays the higher resolution TNG marker set. Both sites make available maps for all chromosomes and provide detailed instructions. Other servers for RH map marker ordering can be found at the Technion in Israel (www.cs.technion.ac.il/Labs/cbl/CGI/rh-wizard.pl) and at the Massachusetts Institute of Technology's Whitehead Institute (http:// www-genome.wi.mit.edu/cgi-bin/contig/rhmapper). The collaborative Radiation Hybrid mapping database (http://corba.ebi.ac.uk/RHdb/) is a database of raw data used in constructing RH maps. This includes marker data, scores, experimental conditions, and extensive cross-references.

Physical Mapping

The generation of a physical map relies upon the construction of an ordered and orientated set of clone-based contigs. Staden coined the term "contig" to refer to a contiguous set of overlapping segments that together represent a consensus region (Staden, 1980). These segments can be sequences or bacterial clones, whose relationship is defined by information in common between the overlap of pairs of segments. Pairwise comparisons of datasets associated with each segment define the overlaps. Similarities that are statistically significant indicate the presence and sometimes the extent of overlap. Cloned contigs are the most convenient route for the sequence generation of larger genomes. They present a means of coordinating physical mapping and, because of the way in which they are constructed, provide an optimal set of clones (the tile path) for sequencing.

YAC Maps. The main benefit of using YACs for the constructing of a physical map is that the insert size (up to 2 Mb) results in coverage of large regions of the genome with relatively few clones. Yeast artificial chromosomes have been utilized to construct a physical map across a candidate gene region (Green and Olson, 1990), in addition to chromosome-specific (Chumakov et al., 1992; Foote, 1992) and genome-wide maps (Chumakov et al., 1995; Hudson et al., 1995). Though STS content mapping is the most frequently used method to generate YAC contigs, techniques such as repeat mediated fingerprinting, either by *Alu*-PCR (Coffey et al., 1992) or by repeat content hybridization (Cohen et al., 1993), have also been used. The advantages of using YACs are, however, offset by the relative difficulty of constructing YAC libraries, analysis of the cloned DNA, in addition to many YAC clones having been found to be chimeric (contain fragments derived from noncontiguous parts of genomic DNA being cloned) (Green et al., 1991; Bates et al., 1992; Slim et al., 1993). Rather than being used as a primary sequence resource, YACs have generally become used to support the construction of detailed landmark maps and to underpin sequence-ready bacterial clone maps (Collins et al., 1995; Bouffard et al., 1997) by STS content mapping. In these cases the YACs have been sequenced directly.

Bacterial Clone Maps. In contrast to YACs, bacterial clone libraries are easier to make, the cloned DNA is more easily manipulated, chimerism is low (Shizuya et al., 1992; Ioannou et al., 1995), and the recombinant DNA is more easily purified. An important factor influencing the construction of a physical map utilizing bacterial clones is the availability of genomic resources. Current bacterial clone contig construction utilizes large-insert P1-derived artificial chromosome (PAC) (Ioannou, 1995) and bacterial artificial chromosome (BAC) (Shizuya et al., 1992) libraries. Each BAC or PAC clone typically contains an insert of 100–300 kb and maps are constructed from a >10-fold genomic clone coverage.

A combination of two strategies was used for the construction of sequence-ready bacterial maps of the human genome, the hierarchical strategy and the whole-genome fingerprinting approach. The hierarchical strategy was based upon the utilization of well-characterized publicly available markers, the majority of which were used to construct genetic and RH maps, at a target density of 15 markers/Mb along the length of a chromosome. The whole-genome fingerprinting approach relied upon the *in silico* assembly of fingerprints generated from the restriction digest of large-insert bacterial clones from total genomic PAC or BAC libraries. The assimilation of both sets of data generated an estimated >99% coverage of the coding (euchromatic) portion of the human genome and acted as the resource for the generation of high-quality finished sequence data. The improved resolution of markers on the physical map enables correct ordering of markers with respect to their locations on the RH map and permits the separation of markers previously binned within the same genetic interval on the genetic map.

The next phase in the evolution of physical map construction was driven by the availability of ordered genomic sequence. Conservation of sequence and long-range order between organisms that are sufficiently closely related means that the genome of one species can act as the template upon which a physical map of another can be built and, in doing so, elucidate the homologous relationship between them (Thomas et al., 2002). The success of the comparative physical mapping approach was demonstrated by construction of a clone map of the mouse genome using the assembled human genome sequence as a template (Gregory et al., 2002). In this study, human genomic sequence was used to align stringently assembled BAC fingerprint contigs by matching mouse BAC end sequences (BESs) to their corresponding locations in the human genome. The availability of BESs from a highly redundant fingerprint assembly of BAC clones greatly simplified the process of contig assembly. This permitted the rapid construction of a physical map covering 99% of the 2500-Mb mouse genome compared to the physical map of the human genome. The same approach could be adopted for any genome where there is sufficient sequence homology to allow alignment of BESs (or equivalent sequence tags) plus sufficient homology between the template genome and the genome under study. The approach both has important applications for genomes where the full genome sequence is anticipated and also (perhaps even more importantly) is a cost-effective way to provide access to regions of a genome for which there are no plans to generate genomic sequence on any scale.

Online Physical Map Resources. The two main types of clone-based maps are derived from BACs and YACs. The Washington University in St. Louis (WUSTL) Genome Sequencing Center provides both a human genome project BAC-based clone map and a map of accession numbers (http://genome.wustl.edu/projects/human/index.php?fpc_get=1). WUSTL synchronizes its layout files with periodic freezes of the Human Genome Project Working Draft Sequence (http://genome.ucsc.edu/) to provide more up-to-date information. The maps are available for download from the web page. In addition, WUSTL has a WebAce site for the physical map of the human genome released with the draft sequence (International Human Genome Mapping Consortium, 2001). The whole-genome BAC map is enriched with regional map data. An overview of this map can be found at http://genome.wustl.edu/cgi-bin/ace/GSCMAPS.cgi.

France's CEPH-Généthon (www.cephb.fr/ceph-genethon-map.html) had constructed the first physical map of the entire human genome using a bank of YAC clones containing seven haploid human genome equivalents (Chumakov et al., 1995). Its maps and navigation tool Quickmap are available for download. Quickmap, which is also usable through CEPH's online server (www.cephb.fr/quickmap.html), enables retrieval of extensive information on 35,000 YACs, STSs, Alu-PCR hybridization, and L1/THE fingerprint and sizing data.

The Whitehead/MIT YAC map (http://www-genome.wi.mit.edu/cgi-bin/contig/phys_map) of the human genome presents an integrated map where genetic and RH maps provided the global framework to which 25,000 STSs were mapped (Hudson et al., 1995). In addition to CEPH-Généthon's and Whitehead/MIT's global mapping efforts, there had been numerous chromosome-specific sites available online. However, many of these are no longer maintained as their data have been incorporated into integrated sites.

The Unified Data Base (http://bioinformatics.weizmann.ac.il/udb/) features integrated maps for all chromosomes. The UDB map incorporates genomic sequencing information from NCBI's clones and contigs. The chromosomal position of a contig is determined based on the markers' UDB positions using e-PCR (ftp:// ncbi.nlm.nih.gov/pub/schuler/e-PCR/). The UDB map currently includes genes, STSs, and EST clusters, with further links to the Genome Database, GeneCards (http://bioinformatics.weizmann.ac.il/cards/), UniGene (www.ncbi.nlm.nih.gov/ UniGene/Hs.Home.html), and NCBI's Human Genome Sequencing.

Physical Mapping and Whole-Genome Shotgun Sequencing. While the construction of a physical map, and therefore a clone-by-clone approach, proved successful for the generation of human sequence, are physical maps required given the contribution of a whole-genome shotgun (WGS) approach to sequencing complex organisms? The main advantages of WGS are that the production of data is very rapid, can be highly automated, avoids cloning biases of BAC systems, and is very cost-effective. The assembly inherent from the sequence alignment also provides important mapping information which is unbiased by additional experimental mapping systems or procedures. While it can be contended that WGS in isolation has disadvantages that prevent completion of either the map or finished sequence of a large genome, the possibility of combining the advantages of both approaches has been explored. A hybrid strategy emerged from the *Drosophila* project and has since been adopted for the mouse genome. Sevenfold WGS coverage was generated from subcloned plasmids of varying sizes which, when assembled with BESs, generated 96% coverage of the euchromatic portion of the mouse genome. A tiling path of BAC clones from the physical map is currently being used for directed finishing of the draft genomic sequence. The physical map helped to assemble the sequence scaffold, while the WGS data increased the rate of clone-based finishing (Mouse Genome Sequencing Consortium, 2002).

Given that WGS sequence data can accurately place BACs via their BESs within the sequence assembly, are de novo bacterial clone maps actually required? While BES localization within a WGS assembly facilitates a more optimal minimum tiling path selection, overlaps within fingerprinting contigs can link sequence assemblies (as reported by the assembly of the mouse WGS sequence; Mouse Genome Sequencing Consortium, 2002). The overlaps generated by fingerprint analysis may also be able to resolve errors in the genomic assembly where, for example, low-copy repeats may have resulted in a compression of the sequence assembly. The proven success of assembling genomewide physical maps, the cost of constructing a >10-fold genomic BAC library, and the ease with which genomewide fingerprint databases can be assembled have led to the construction of several genomic fingerprint databases (Marra et al., 1999; Tao et al., 2000; McPherson et al., 2001; Gregory et al., 2002). While genomewide fingerprint maps will facilitate the large-scale characterization of many varied species, the construction of small region-specific sequence-ready maps will continue to be important for detailed interspecies sequence comparisons (Thomas et al., 2002).

Public Data Repositories and Genome Browsers

The availability of large tracts of human genomic sequence has necessitated the development of databases (genome browsers) that provide a framework upon which the enormous amount of data associated with the human genome can be stored and displayed. The three main databases—the NCBI browser (http://www.ncbi.nlm.nih.gov/mapview/map_search.cgi?taxid = 9606); Ensembl, developed at the Wellcome Trust Sanger Institute and the European Bioinformatics Institute (http://www.ensembl.org/Homo_sapiens/); and the University of California Santa Cruz (UCSC) genome browser (http://genome.cse.ucsc.edu/), developed by Jim Kent and David Haussler—each contain information pertaining to physical maps, chromosome-specific sequence assemblies, aligned mRNAs and ESTs, cross-species homologies, SNPs, and repeat elements. The development of generic genome browsers, such as those hosted by Ensembl, makes possible the rapid identification of homologous sequences between comparative organisms and in doing so assists in identifying conserved features that may be of some functional significance.

Although each of the three sites contains the same human genome assembly (which is in turn based upon curated chromosome-specific tiling path files of bacterial clones), they continue to use their own analysis pipelines for automated annotation so that predictions of gene structure for novel genes may differ depending on the program used. There are also differences between the various annotation features that users can opt to display. There is considerable overlap between the three resources in their ability to access the most commonly desired features; consequently each genome browser is capable of providing a comprehensive overview of a region of interest. However, there are slight differences that exist between the NCBI, UCSC, and Ensembl annotations that, depending on the needs of the researcher, may determine which database is used for primary data or for specific data retrieval.

National Center for Biotechnology Information. The NCBI has been at the forefront of bioinformatics research for over a decade. Its map view of the human genome has an integrated sequence and mapping resource that is comprehensive and easy to use. As part of the NCBI data model, Entrez provides easy access to many other bioinformatics resources (see www.ncbi.nlm.nih.gov/Sitemap/index.html) by integrating multiple sequence databanks with information on loci and information from PubMed.

The map viewer features an options box that allows the user to have some control over the display. The user can choose a number of different maps for comparison. These include the localization of FISH mapped clones, the locations of sequence-ready contigs, the placement of individual GenBank entries, the placement of STSs and SNPs, and annotated genes that have been mapped to the genomic contigs. The NCBI map viewer also includes ideograms of the G-banding pattern for each chromosome, BAC clones mapped to G-bands using FISH, the Mitelman map of cancer-related chromosomal breakpoints (Mitelman et al., 1997), the cytogenetic location of genes, and the morbid map of disease genes in the Online Mendelian Inheritance in Man (OMIM, www.ncbi.nlm.nih.gov/omim/). Genetic linkage maps include the Marshfield map, the Genethon map, physical mapping data from genemap99, and the more recent and accurate DeCode map (Kong et al., 2002).

Annotation Within UCSC and Ensembl. The two other most useful repositories of human and model organism sequence data are the UCSC and Ensembl genome browsers. These sites use a different approach from the NCBI regional display by using an analysis pipeline to annotate the complete genome by combining computational gene prediction tools such as Genscan with publicly submitted experimental evidence including EST, cDNA, and mRNA data. These data are combined into a baseline annotation of the entire human genome and include location information for SNPs, sequence contigs, BAC clones, microsatellite markers, ESTs, predicted and confirmed gene structures, and numerous other features. Each site displays a linear representation of a segment of the genome annotated with a ruler to show the position in base pairs at the top and allows the user to select among a wide variety of data sources displayed as marks on linear tracks that directly correspond to the position above. They both enable the user to choose among features and select the level of detail seen. The sites are flexible in that the researcher can zoom in and out to view the data. The other main advantage to these systems is that additional annotation can be added to the display using the Distributed Annotation System (Dowell et al., 2001), allowing annotation of a local version of the database with project-specific data.

UCSC Genome Browser. The UCSC sequence map, or Golden Path, provides the user with the ability to view previous genome assemblies so that previous sequence assemblies and annotation can be viewed in the correct context, since genomic assemblies can vary significantly from one build to the next. The complete sequence data and tiling path files are available from www.genome.ucsc.edu. A particularly useful tool at the site is the Blast-Like Alignment Tool, or BLAT (Kent et al., 2002). This extremely fast heuristic alignment program easily enables the user to extract a complete genomic sequence with the exons highlighted when a cDNA, mRNA, or protein sequence is used as the query.

Ensembl Project. Ensembl is a joint Sanger Institute and European Bioinformatics Institute project that displays annotated public genomic sequence data in a manner similar to the UCSC browser, is viewable through the Ensembl website at www.ensembl.org, or the Ensembl databases can be downloaded and installed locally for specialized queries. Source code and extensive documentation are also available. Each gene is assigned a unique Ensembl gene identifier, which remains constant through each successive Ensembl gene build. The gene prediction methods and other automated features used in Ensembl provide a simple yet thorough first-pass analysis of the human genome, which can greatly aid researchers searching for genes in a genomic region.

Using Ensembl. Ensembl is by far the most informative of the three sites in terms of flexibility by virtue of the provision of a data extraction tool called EnsMart. EnsMart provides extensive filtering to generate user-defined subsets of data. Users are provided with a considerable list of features allowing them to retrieve data from many different fields. These data can be extracted in html file, text file, or MS Excel file format and downloaded in either compressed or uncompressed form. Ensembl also has several different portals through which data can be queried or extracted. Among a variety of entry points, MarkerView displays comprehensive information associated with, for example, genetic markers including marker source, synonyms, flanking sequence, and genetic, RH, and physical map location; Gene-View contains links to the physical map location, gene description (from SWISS-PROT), transcript description and supporting evidence, ortholog prediction, and SNP information. The SNP data associated with an annotated gene is displayed in; GeneSNPView is a very useful tool for displaying SNPs contained within the region of a gene that may have structural significance to the transcript, that is, within the coding region of the gene, or which is important to transcription, that is, the 5′ and 3′ untranslated regions, exon boundaries, and putative promoter regions.

Other genome databases can be found at the Rosalind Franklin Centre for Genomic Research website (http://www.hgmp.mrc.ac.uk/GenomeWeb/human-gen-db-genome.html). Included among these is the Genome Channel at Oak Ridge National Laboratory (http://compbio.ornl.gov/channel/index.html), which allows for retrieval of genomic data for multiple organisms through Java viewers for both sequence and features. A variety of graphics windows, text windows, and summary reports provide the underlying information and evidence for genes and other features. Annotated features include repetitive DNA, simple repeats, CpG islands, Poly-A sites, BAC ends, STS markers, GenScan and Grail genes and exons, and GenBank gene features. Pop-up menu buttons provide summaries, sequence data, similarity data, and source information.

SINGLE-NUCLEOTIDE POLYMORPHISMS

Most differences between individuals, at the nucleotide level, can be attributed to allelic sequence variation. The characterization of sequence differences and comprehension of how these genomic variations affect the expression and function of genes will be crucial for the study of molecular alterations in human disease. While sequence variation has previously been used for genomewide linkage and disease gene mapping studies (leading to the identification of many disease-causing genes), the association of SNPs with genes, either by mapping or as causal sequence variants, promises to be a valuable method in the future for identifying genes involved in complex diseases.

Approximately 90% of the allelic differences existing within the human genome can be attributed to SNPs, the remainder being insertions or deletions (Collins et al., 1998). A comparison of any two diploid genomes is estimated to identify one SNP per 1.3 kb which has an allele frequency of >1% [International SNP Map Working Group (ISNPMWG), 2001]. The prevalence of SNPs in the genome, their existence as biallelic variants, and their stability through inheritance make them amenable to large-scale high-throughput analyses. Therefore, SNPs will be applied to several research areas, including (1) large-scale genome analysis of linkage disequilibrium and haplotype patterns, (2) genetic analysis of simple and complex disease states, and (3) genetics and diversity of human populations.

SNP Discovery

Initially, de novo candidate SNPs were identified by the alignment of STSs and ESTs to available genomic sequence (Wang et al., 1998; Picoult-Newberg et al., 1999; Irizarry et al., 2000; Deutsch et al., 2001). The clone-based strategy used by the Human Genome Project for the large-scale production of human genomic sequence contributed to a dramatic increase in the SNP numbers by allowing identification of novel SNPs within sequence overlaps between minimum tile path clones (Taillon-Miller et al., 1998; Dawson et al., 2001). A more directed approach to SNP discovery was initiated by sequencing DNA from population-specific individuals (Mullikin et al., 2000; Altshuler et al., 2000). Two to fivefold redundant shotgun sequence coverage was generated from 1.5-kb small-insert library clones and the resultant sequences were aligned to each other in clusters. As the Human Genome Project progressed, these assemblies and additional shotgun sequence data were aligned to available genomic sequence to identify more SNPs. The total number of SNPs identified using the strategies outlined above culminated in the ISNPMWG constructing a SNP map of the human genome which contained 1.42 million candidate SNPs (ISNPMWG, 2001). A proportion of the candidate SNPs were validated experimentally during the project, confirming that >90% were real SNPs. The SNPs identified by The SNP Consortium (TSC) (http://snp.cshl.org, Marshall et al., 1999) and the Human Genome Project had generated a SNP density of 1 SNP per ~1.9 kb of available sequence.

Akin to the rationalization that was required to establish a unique set of ESTs, a database was established to generate a nonredundant collection of candidate SNPs (dbSNP) (Sherry et al., 2001, http://www.ncbi.nlm.nih.gov/SNP/index.html). Currently dbSNP contains 10.4 million human SNP entries which have been condensed into a nonredundant set of 4.8 million SNPs, that have been validated to date (build 125). Localization of these unique SNPs within a recent human sequence assembly (build 34) yields a SNP density of 1 per 1.3 kb.

Many different platforms (see Chapter 6) have been developed for SNP analysis which are based upon four basic allele-specific assays types: (1) hybridization with allele-specific probes, (2) oligonucleotide ligation, (3) single-nucleotide primer extension, and (4) enzymatic cleavage. Many of the techniques have been developed further and automated in commercial systems. The range of formats used include colorimetric microtiter-plate-based assays (Taqman by Applied Biosystems or Invader assay by Third Wave Technologies) or fluorometric methods of detecting SNP alleles that have been separated by gel electrophoresis (Applied Biosystems) and fluorometric assay of targets hybridized to oligonucleotides immobilized in a microarray chip format (Affymetrix) or immobilized via beads on the ends of array light-sensing glass fibers (Illumina).

Utilizing SNPs

Single-nucleotide polymorphisms may be utilized for population genetic studies to identify an association between a SNP allele and a specific phenotype. Ultimately the goal of such a study is to identify the causal variant, the mechanism by which the variant has its functional effect. The functional variant will have maximal predictive value in future individual tests, and the gene involved may encode a target protein or mRNA for possible therapeutic intervention. Functional variants may be assayed for using two approaches. The direct approach requires prior availability of a candidate functional variant (e.g., a SNP which alters the encoded protein sequence in a nonconservative way, thus affecting function). The variant is then tested by genotyping a population of defined phenotype and comparing the frequency of one allele with the frequency in a population of matched controls (a case–control study). In the absence of a candidate functional variant, the indirect approach can be taken, in which available SNPs within specific genes (candidate gene association studies) or throughout the genome (genomewide association studies) can be used to test the same populations.

An aid to the indirect approach is the identification of allele-specific sequence variants and generation of a map of common combinations of specific alleles (or haplotype patterns) that have been largely conserved during the recent population expansions. Among other factors, it is believed that the regions of conserved local haplotype patterns have been maintained by the absence of ancestral recombination within each region. Identification of these conserved segments is facilitated by the availability of SNPs identified by the ISNPMWG. Pairs of alleles can be statistically quantified to determine whether recombination has occurred between them, in which case they are said to be in equilibrium, or if the alleles share evolutionary cosegregation and are therefore in linkage disequilibrium (LD). The generation of an LD map does not require the analysis of related individuals (by comparison to the genetic map), only that they share a common evolutionary history (although inclusion of pedigrees allows direct determination of the phase between SNP alleles and facilitates definition of long-range haplotypes). It is hoped that the generation of a map of common haplotype patterns (HapMap) will facilitate the identification of common diseases by indirect association studies as described above (Couzin et al., 2002; Harris et al., 2002).

The availability of genome sequence with annotated gene structures provides the means to search for candidate functional variants. Build 125 in dbSNP contains >120,000 SNPs that have been localized to exons, untranslated regions, or noncoding regions adjacent to genes (introns or flanking sequence) (Table 15.1).

TABLE 15.1. SNP Totals Contained Within or
Adjacent to Coding Features

SNP Count	Functional Classification
397,161	Gene region
48,277	Synonymous
60,763	Nonsynonymous
619,857	Untranslated region
3,289,086	Intron
950	Splice site

Single-nucleotide polymorphisms localizing within a coding feature (cSNPs) have the greatest potential to affect the structure and function of the gene. The characterization of allelic variants enables conclusions to be drawn as to whether a specific allele may have an effect upon the amino acid sequence. Slightly less than half of the SNPs localizing to coding sequences result in a synonymous change (no change in the amino acid sequence because of codon redundancy) while the remaining SNPs result in a nonsynonymous change. Nonsynonymous changes are further classified as to whether the resultant amino acid has similar biological properties to the "normal" allele, in which case the change is conservative, or if the biological properties of the amino acid are different, and then the change is nonconservative. While the molecular significance cSNPs have upon protein structure and function has previously been reported (Chasman et al., 2001; Sunyaev et al., 2001; Wang and Moult, 2001), the effects that SNPs have in noncoding sequences such as splice junctions (Pan et al., 2002; Khan et al., 2002), folding of mRNAs (Shen et al., 1999), and promoter function (Knight et al., 1999; Hijikata et al., 2000) have also been described.

The identification of SNPs that show an allelic influence on the functioning of proteins, particularly of drug-metabolizing enzymes, promises a bright future for the optimization of clinical therapeutics. Associating inherited variations with pharmacological responsiveness provides a basis for the possible development of personalized medicine which will improve the efficacy of drug treatments and decrease the side effects experienced by the individual (Pfost et al., 2000).

Computational SNP Resources

Single-nucleotide polymorphisms are annotated features integrated with genetic and physical maps into the whole-genome sequence assemblies at UCSC, NCBI, and Ensembl. HGBASE, the Human Genetic Bi-Allelic SEquences database (hgbase. interactiva.de/), records polymorphisms in genetic regions to facilitate SNP-based genotype–phenotype association studies. Repositories of SNP information are available at the SNP database (www.ncbi.nlm.nih.gov/SNP/) and the SNP Consortium database (snp.cshl.org/data/). An extremely high-quality repository of SNP information, although limited to those found in the Asian population, is JSNP (http://snp.ims.u-tokyo.ac.jp/). This site has extensive annotation, including allele frequency data. Two other very useful resources that link to SNP information are NCBI Locus Link (www.ncbi.nlm.nih.gov/LocusLink/), which indexes all the SNPs found within a gene, and the MIT-Whitehead Institute site (www-genome. wi.mit.edu/). Information on allele frequencies may be found at the Marshfield site, the CEPH site, and the Duke Center for Human Genetics site (wwwchg.mc. duke.edu/software/allele.html).

MODEL ORGANISMS

While BLAST alignment of human mRNA, EST, and protein sequences to predicted coding structures provides a primary level of support, predicted features can also be supported by alignments with sequences from other organisms (http://wit.integrated genomics.com/GOLD/) for comparison (comparative sequence analysis) (Birney et al., 2001). The identification of sequences that are conserved between species is important because sequences that contain elements that are potentially functional are more likely to retain their sequence than nonfunctional segments under the constraints of natural selection during evolution. The evolutionary distance between species is an important consideration. Sequence comparisons between closely related species may facilitate the identification of gene structures and regulatory elements, but if the evolutionary distance between the species is relatively small, these sequences may be obscured by nonfunctional sequence conservation. Therefore a variety of species, including more distantly related species, might be required to identify potential functional sequences using the comparative approach.

The identification of conserved sequences by comparative analysis has focused on the identification of noncoding regions (Hardison et al., 1993; Koop et al., 1994; Hardison et al., 1997) and protein coding regions (Makalowski et al., 1996; Ansari-Lari et al., 1998; Jang et al., 1999) between human and mouse genomes. The alignment of sequences from multiple organisms has also been used to identify upstream regions that may affect gene expression (Gottgens et al., 2000). While comparative sequence analysis may not identify all control regions associated with a gene, conserved regions may be identified that would be candidates for further experimental investigation (Pennacchio et al., 2001).

PEDANT (http://pedant.gsf.de) is a web-based tool for comparative, functional, and structural genomics that holds data processed from the analysis of 293 complete and draft genome sequences. PEDANT generates databases by processing these genomes through a comprehensive, automated analysis pipeline and includes comprehensive tools to compare genomes that include genome comparison tables (Frishman et al., 2003)

The finished sequences of the mouse, rat, zebrafish, fugu, mosquito, *C. elegans*, and *C. briggsae* have been released to the public and are available on the UCSC genome browser, Ensembl, and the NCBI map viewers. The Mouse Ensembl genome annotation site has features and navigational tools that are virtually indistinguishable from the human Ensembl site discussed at length above. The other leading resource is the Mouse Genome Informatics site at the Jackson Laboratory (www.informatics.jax.org/). This site offers maps and mapping data, data on genotypes and phenotypes, strains and polymorphisms, information on molecular probes and segments, and comparative maps and homology data. Whole-genome homology maps depict overview of human orthology information plotted against the MGD consensus genetic map for the mouse. The Jax site also features the Oxford Grid query form, which enables homology information to be retrieved in the form of a grid display, thereby providing an overview of homology between the genomes of two species (chosen among the human, mouse, rat, cat, pig, cow, and sheep). The phenotype database enables possible candidate gene selection based on similar phenotypes between human and mouse. Other useful tools for doing comparative and structural genomes include the mouse sequencing project (www.hgsc.bcm.tmc.edu/mouse), the mouse BAC mapping project (sequence.aecom.yu.edu/mouseDB/mousePUB/mouse_welcome_all.hts) that allows you to search for markers and clones, the mouse EST project (genome.wustl.edu/est/mouse_esthmpg.html), and the Whole Mouse Catalog (http://www.MURIDAE.COM/wmc/).

IDENTIFYING CANDIDATE GENES BY GENOMIC CONVERGENCE

"Genomic convergence" is an approach that assembles data gathered from different methods to help identify candidate genes in complex traits. This approach has been used successfully to identify candidate susceptibility genes for Parkinson's disease and Alzheimer's disease (Hauser et al., 2003; Li et al., 2003). Genomic convergence integrates multiple sources of genomic data such as gene expression data and linkage analysis date to help prioritize candidate genes for association analysis.

For example, initial genomic screening usually identifies multiple significant linkage peaks scattered throughout the genome. The first bioinformatics task is to generate a list of genes that can be assigned to the region between the markers flanking the peak logarithm of the odds (LOD). It is advisable to extend this region to include a few genes that lie outside of but near the peak in either direction. It is not uncommon to have several hundred genes under a single linkage peak; therefore it is neither practical nor cost effective to test all genes in a more detailed analysis. A traditional approach for researchers to shorten the list of candidate genes is to assess genes for biological plausibility. This approach is limited by its requirement that such functional data exist and that rather extensive knowledge of the phenotype and biochemical pathways for the disease in question is available. Fortunately, there are several pathway databases, albeit far from complete, to assist researchers with their understanding of biological pathways besides literature searching in PubMed.

Many laboratories are using gene expression analysis to identify genes whose level of expression is significantly altered in diseased tissue as compared to normal tissue. While there are limitations to these methods, genes with perturbed expression are reasonable candidates. However, it is possible that the critical gene may not be overexpressed, and if the researcher is using microarrays, the desired gene may not even be represented on the chip. Another problem with gene expression approaches is that many genes with altered expression are not involved with the disease phenotype, resulting in false positives. The researcher will want to generate a list of the perturbed genes and use data-mining techniques to extract as much necessary and relevant information on these genes as possible. This list will likely still be too long to follow up each gene.

This is where genomic convergence becomes useful. Now armed with two long lists of genes, one generated through linkage analysis and the other generated from gene expression data, that are reasonable for follow-up, the researcher can prioritize candidates for association analysis by taking the intersection of the two sets. If these lists are long enough to make manual inspection impractical or if the laboratory plans on employing this strategy on a routine basis, a short script [in Perl (Stein, 2001) or another language] should be written to quickly find the overlapping genes. With luck the gene associated with the disease is among those listed.

The genomic convergence strategy can include pathway databases (Krishnamurthy et al., 2003) to extend the list of prioritized candidate genes. Clustering methods can be used to bin genes with altered expression levels into related groups. Often the data will direct the researcher to one or more regulatory, signal transduction, or biochemical pathways. Pathway databases can help the researcher identify genes that are associated with a pathway that is not identified through gene expression. The researcher should then determine if any of these genes are on the list generated from linkage analysis or microarray analysis and follow up with association analysis. The problem is that there still is a good chance that the researcher will not find the smoking gun.

DE NOVO ANNOTATION OF GENES

When attempting to identify candidate disease genes, it may be necessary to perform de novo analysis of novel genes in the region of interest. Gene-finding programs used in Ensembl, UCSC, and NCBI have identified thousand of putative genes, a large proportion of which have no known function. Sequence alignment and other analysis tools are, therefore, essential in determining putative gene function. Although many molecular biologists are adept at utilizing online sequence analysis tools such as BLAST and FASTA (Pearson and Lipman, 1988), interpreting the results is often difficult.

The biocomputing initiative at the Pittsburgh Supercomputer Center (PSC) is a leading resource in the training of biologists in the sequencing of macromolecules. In addition to providing excellent on-site workshops, the PSC offers comprehensive online tutorials on the sequencing of nucleic acids and proteins. Practical tutorials are available on retrieving, reformatting, finding homologous genes and proteins, performing multiple-sequence alignments, and using the GCG Wisconsin package (available through Accelrys) of sequence analysis tools.

Without a thorough understanding of the algorithms, scoring parameters, and statistics used by database searching programs, one may make inappropriate choices and interpretations. Fortunately, the PSC website also provides tutorials that deal with these issues and lists a comprehensive bibliography. Scores of other online tutorials in sequence analysis on the web as well as numerous journal articles and books are available on the subject for those who want to understand the underlying algorithms and statistical issues in detail. Any researcher that anticipates using alignment methods extensively or who needs an optimal alignment should read the classic papers describing the use of dynamic programming for global (Needleman and Wunsch, 1970) and local (Smith and Waterman, 1981) alignment as well as papers by Henikoff and Henikoff (1992, 2000) on the BLOSUM Scoring matrices.

Software Suites

The GCG Wisconsin Package has been the gold standard in sequence analysis, with its popular SeqWeb, an online graphical user interface (GUI). In addition to several enhancements, SeqWeb 2.0 includes new programs for searching and finding an optimal alignment among all six reading frames for a protein sequence and for gene finding and pattern recognition.

For those who have limited financial resources there is an open-source alternative to the Wisconsin Package, the European Molecular Biology Open Software Suite (www.uk.embnet.org/Software/EMBOSS) for sequence analysis. The EMBOSS software repertoire is very comparable to the Wisconsin Package. EMBOSS can be downloaded, installed, and run on most UNIX computers. Unlike GCG's SeqWeb, EMBOSS lacks a GUI, but many bioinformaticians find command-line operation preferable to SeqLab and SeqWeb 1.0.

Another popular commercially available bioinformatics software suite is Vector NTI from Informax (www.informaxinc.com/products/vectornti/vector_suite. html). The suite contains programs for data management, mapping and illustration, primer design and analysis, strategic recombinant design, protein and nucleic acid sequence analysis, multiple-sequence alignment, and ContigExpress for sequencing project management and fragment assembly.

Online Sequence Analysis Resources

The BLAST server at NCBI (www.ncbi.nlm.nih.gov/BLAST), the preeminent online tool for searching for homologous sequences in GenBank by performing local alignments, was completely revamped in 2002. In addition to performing nucleotide and protein queries, "translated BLAST" searches (blastx, tblastn, tblastx) are possible as are searches against specialized databases (Human Genome, Microbial Genomes, etc.). Other features include the MegaBLAST service for comparison of large sets of long sequences, search page optimization for nearly exact matches to short query sequences (nucleotide and protein), and taxonomy reports for PSI and PHI BLAST searches.

The Biology Workbench (http://workbench.sdsc.edu/) requires users to register and enter a password to access a point-and-click interface for rapid access of biological databases and analysis tools. Tools are available for protein and nucleic acid sequence analysis, alignment, and structural analysis.

Understanding Molecular Mechanisms of Disease

Once a gene associated with the disease is found, the next step is to identify the molecular mechanism at work. Some basic bioinformatics programs included in the GCG Wisconsin Package (available from Accelrys) or the European Molecular Biology Open Software Suite (www.emboss.org) can assist in this task. Pairwise sequence alignment between the associated gene in the affected individual and the wild type may identify a mutation that will either truncate the protein product or alter its amino acid sequence. If there is no truncation, the function domain that contains the amino acid sequence must be identified by comparing it to the profile databases or using a tool such as Cn3D (www.ncbi.nlm.nih.gov/Structure/ CN3D/cn3d.shtml) to view the three-dimensional structure in the Protein Data Bank (PDB; www.rcsb.org/pdb/). However, nothing obviously disruptive to the structure may be evident even if you have an overt mutation. Due to the degenerative nature of the genetic code, it is also possible to identify a SNP that does not alter the encoding amino acid. It is also possible that the identified SNP is not within an exon. The next step is to see if the SNP alters the splice junction or is in a regulatory region. Two tools to aid this procedure are DNABIND (Mrazek and Kypr, 1992) and PromoterWise, part of the Wise2 package at EBI (www.ebi.ac.uk/Wise2/). These are designed to help identify promoters by allowing for translocation and inversions when comparing sequences.

Assigning Gene Function

Characterization of the functional product of a gene is not achieved directly by the identification of translated amino acid sequences contained within the open reading frame of the coding sequence. Within a genomic context, the final sequence and structure of an mRNA and the encoded protein may be influenced by priming from multiple promoters, splice variation within the coding exons, or the existence of alternative polyadenylation sites. Posttranslational processing can also result in modification of a protein product. While *in vivo* and *in vitro* studies within model organisms by chemical mutagenesis and gene targeting can identify gene function by generating an observable phenotype, it may not always be clear how a disruption of the target gene has given rise to a particular effect within a complex network of gene interactions. Alternatively, protein function may be predicted by in silico structural analysis. The assignment of a new function to a novel protein at a nucleotide sequence level comparison, however, may fail as BLAST analysis within the PDB was shown to only find 10% of the known relationships (Brenner et al., 1998). While iterative PSI-BLAST (Altschul et al., 1997) is more sensitive, relationships are still missed. A convenient place to start is the ProSAL site (xray.bmc.uu.se/sbnet/ prosal.html). This Protein Sequence Analysis Launcher facilitates sending a protein sequence to numerous protein analysis servers which are grouped according to six major categories: local sequence alignment; global sequence alignment; domains, families, motifs, and functions; properties; predictions; and tertiary structure.

An alternative approach is the sequence-to-function method, which uses pairwise sequence or motif alignment to derive significant homologies between proteins and hence suggests similarity of function. While these methods are powerful, they are not ideally suited to identify loss or gain of function during protein evolution and encounter difficulties when assigning function as protein databases become more diverse (Skolnick and Fetrow, 2000). Protein and gene family-specific databases are invaluable tools for predicting the structure and function of proteins translated from genomic or cDNA sequences. Often distantly related sequences prohibit the detection of homology by global sequence alignment. The presence of domains, patterns, or motifs can facilitate familial identification so that searching a protein sequence against any of several family-specific databases can be used to detect function. There are a number of these family-specific databases. These include Prosite (http://www.expasy. ch/prosite), a data library of biologically significant sites and patterns used to reliably identify to which known family of protein a sequence belongs; Blocks (http:// www.blocks.fhcrc.org/), a library of highly conserved regions in groups of proteins represented in the Prosite library; and Prints (http://bioinf.man.ac.uk/fingerPRINTS can/), a data library of conserved motifs used to characterize a protein family. Finger-prints can encode protein folds and functionalities with greater flexibly and more power than can single motifs since motif neighbors can provide the mutual context whereas blocks and prints detect local regions of similarity; ProDom, Pfam, and ProfileScan (http://hits.isb-sib.ch/cgi-bin/PFSCAN) are used to detect global similarity; The ProDom (http://protein.toulouse.inra.fr/prodom/current/html/ home.php) protein domain database uses recursive PSI-BLAST search to automatically compile a database of homologous domains. Pfam (http://www.sanger.ac.uk/ Software/Pfam/) is a data library of multiple-sequence alignments containing hidden Markov model representation of complete domains. Interpro Scan (http:// www.ebi.ac.uk/interpro/), an integrated search in Prosite, Pfam, ProDom, Smart, Prints, and Swiss-Prot + Trembl, is highly recommended.

In many cases, however, even in these well-characterized families, the catalytic component may be recognizable but the specific substrate binding properties may be difficult to determine. Additional protein domains (encoded by separate exons) that are required for function but localize to other regions of the protein are less readily elucidated by homology alone. For this, direct experimental approaches are required to determine the substrate and products in the appropriate biochemical pathway. For example, protein-binding assays using yeast two-hybrid systems can identify interacting binding proteins. Knockouts, or natural mutants, may be investigated to determine the biochemistry of the altered phenotype in some detail. For example, a defective enzyme may result in accumulation of abnormally high levels of substrate, and comparison of normal versus mutant systems will reveal candidates as possible substrates.

Alternatively, the possible function of a protein may be suggested by comparison of three-dimensional structure to proteins of known function. Since the tertiary structure of proteins of common function is likely to be more conserved than their primary structures (amino acid sequences), attempts have been made to classify groups of proteins based on structural and phylogenetic relationships (e.g., SCOP; Murzin et al., 1995). A second approach describes proteins according to their structural characteristics, such as class of architecture and fold type. In practice, both approaches are used, initially grouping proteins according to their sequence homology and then by their structural descriptors (Thornton et al., 1999).

The application of the sequence to structure-to-function approaches aims to determine the structure of a protein and then to identify the functionally important residues. Ab initio folding can be used to predict a native structure based on domains contained within the protein. A process known as threading utilizes a known structure as a template upon which proteins of up to 500 residues can be molded. These three-dimensional structures can then be used to infer function by analysis of internal or external residues, the shape and molecular composition of the protein, or the juxtaposition of individual groups. The prediction of protein folds, their three-dimensional structure and function, is, however, primarily reliant upon experimental evidence, either as a basis for modeling or as support for a prediction. X-ray crystallography and nuclear magnetic resonance spectroscopy are methods by which these proteins structures have been experimentally determined.

Determining the cellular and tissue localization of the protein can gain additional information. The colocalization of proteins in a highly tissue-specific pattern may provide evidence for some level of protein interaction. The fusion of the sequence encoding a novel protein to the sequence of a reporter molecule in a shuttle vector can be used to determine cellular localization if the construct can be introduced into a physiologically relevant cell line. This work may be followed up, for example, by manipulation of the construct and introduction into embryonic stem cells in order to create a transgenic animal model where the gene is under control of the endogenous promoter. This would enable investigation of the expression of the gene presumably in response to physiologically natural intracellular and extracellular signals. The cellular distribution of the signal molecule should, therefore, reflect the distribution of the natural gene product. Data from colocalization experiments may be correlated with protein–protein interaction studies, and possibly analysis (e.g. by mass spectrometry) of the components of copurified complexes, to build a picture of the interactions between specific proteins

LOOKING BEYOND GENOME SEQUENCE

Although scientists have learned a great deal about genomic biochemistry in the last 50 years, it has only been in recent years that instrumentation and technology have developed sufficiently to make serious attempts at studying the human transcriptome (the entirety of all mRNA species of a cell under defined conditions). Since the publication of the first edition of this book (Haines, 1998), there has been an explosion of gene expression data generated by hybridization-based nucleic acid array analysis using commercial and customized spotted microarrays or "gene chips" [(Schena et al., 1995; DeRisi et al., 1997; Wodicka et al., 1997), reviewed in Hacia et al. (1998)], in addition to serial analysis of gene expression (SAGE) technology, which allows for the detection of unknown ESTs (Velculescu et al., 1985; Saha et al., 2002, www.ncbi.nlm.nih.gov/dbEST/). Gene chip microarray technology is a very powerful method for analyzing messenger RNA levels in a particular tissue under a particular circumstance. Laboratory metadata generated by an expression microarray experiment require meticulous and detailed analysis and the variability of signal-to-noise ratio poses a number of statistical challenges; nevertheless consideration of these data alongside linkage analysis [e.g. "genomic convergence" (Hauser et al., 2003)] can aid in the identification of candidate disease susceptibility genes. Cluster analysis of coexpressed messenger RNA, which can provide insight into affected pathways, together with the use of resources such as the *Kyoto Encyclopedia of Genes and Genomes* [KEGG (Goto et al., 1997)], can identify additional candidate genes for further study.

Other Databases

Morbidity. Online Mendelian Inheritance in Man (www3.ncbi.nlm.nih.gov/Omim/) is a database of human genes and genetic disorders that also includes their cytogenetic map locations. It is also now part of the NCBI Entrez integrated data retrieval system.

Another useful data source is the GeneCards encyclopedia (bioinformatics. weizmann.ac.il/cards/) maintained at the Weizmann Institute. GeneCards is an electronic database of biological and medical information that includes human genes, their products, and their involvement in diseases.

The Human Gene Mutation Database (HGMD, archive.uwcm.ac.uk/uwcm/mg/hgmd0.html) at the Institute of Medical Genetics in Cardiff is an up-to-date and comprehensive reference source for the spectrum of inherited human gene lesions and thus only includes nuclear genes (Krawczak, 1997). It also now includes polymorphisms, for which there is a convincing association with a disease phenotype.

MitoMap (www.mitomap.org) is a compendium of polymorphisms and mutations of the human mitochondrial DNA at the University of California, Irvine. It includes the mitochondrial DNA sequence and information on mitochondrial genome structure and function, pathogenic mutations and their clinical characteristics, population-associated variation, and gene–gene interactions.

Proteins. For proteins, the same NCBI Entrez website used for access to GenBank can be used to access the protein sequence database, while the annotated European protein sequence database, SWISS-PROT, can be found at http://www.ebi.ac.uk/swissprot/. One must exercise some caution when working with the protein sequence data since it contains translated DNA sequence data as well as experimentally determined protein sequences.

For help in keeping up with rapidly expanding databases, it is a good idea to register with Swiss Shop (expasy.cbr.nrc.ca/swiss-shop) and the Sequence Alerting System (www.bork.embl-heidelberg.de/Alerting). Swiss Shop uses a user-defined query to scan, on a weekly basis, for newly added proteins related to a field of interest and returns the results via email. The Sequence Alerting System searches several databases on a daily basis for new homologues to a sequence of interest and will also inform you by email if it has detected a new relative.

Biological Systems. To truly understand the complexity of a biological system, research scientists need to take a multidimensional view of the overall picture. To facilitate our ability to synthesize all of this information, visualization tools are needed to represent this information in a two-dimensional view to be able to be displayed on a computer screen. Toward this end bioinformaticians are working on solutions to graphically represent the complexity of biochemical pathways while incorporating data on expression levels and the state of the cell. Cytoscape is one such solution that is available for use and download at this time (http://www.cytoscape.org). Cytoscape is a very elegant software platform developed by a collaborative team of researchers from the Institute of Systems Biology in Seattle, the Whitehead Institute at MIT, and the Memorial Sloan-Kettering Cancer Center in New York. This tool also features the ability to illustrate alternative pathways and expression levels in response to perturbations such as knockouts of a particular enzyme (Ideker et al., 2001).

ViMAc (Luyf et al., 2002) is another flexible tool for the visualization of expression data integrated into a genomewide metabolic map. This presents the data very differently than Cytoscape and is also freely available to the academic community, but through the request of the authors.

SUMMARY

It is anticipated that substantial efforts will result in all of the genes in the human genome being investigated in detail. These studies will result in a fuller understanding of the specificity and range of biochemical structures and functions that are encoded in the human genome sequence. In general, there is likely to remain a distinction between the study of functions encoded at the DNA level, which affect gene expression via transcriptional control, and the study of functions reflected at the protein level following translation, taking into account posttranslational modifications (processes which are largely genetically determined). Without a genic catalogue, functional studies are necessarily limited to the investigation of a specific target—a gene, a protein, or a disease. These approaches are an essential part of fully interpreting the genome as they provide a means by which hypotheses can be experimentally tested and which produce valid and supplementary results. However, the production of a complete gene catalogue (if completion can indeed be measured or achieved) will provide the raw material for modeling whole systems. The extensive use of computational biology to suggest how such complex systems are made up of their interacting components will, in itself, enable predictions to be made of the system model. These predictions can be tested both to determine the validity of the modeled system and to test the success of the methods used to derive the system. Advances in genome informatics, in particular, will continue to empower geneticists, even those with minimal training in bioinformatics and computational biology, with the ability to locate, visualize, query, filter, analyze, mine, and download the complete finished sequence of the human genome, including known polymorphisms and an ever-expanding list of model organisms, symbionts, and pathogens in the public domain.

To fully understand systems biology, we need to find a way to deal with the numerous difficult analytical problems that genetics presents. Continued refinement of visual imaging, simulations, analysis methodologies, and integration of phenotypic, medical history, vital signs and other biometrics, pedigrees, genomic, transcriptomic, proteomic, metabolic and laboratory metadata on experimental conditions and parameters will greatly assist in unraveling the function of the thousands of unknown genes and is essential in identifying the environmental and genetic factors contributing to the complex diseases that plague our society. Computational methods and bioinformatics will become increasingly important in the acquisition, storage, and interpretation of data in the research laboratory as we begin to understand the complexities of the interactions between an organism, its symbionts, its parasites, and its environment in influencing the health of the individual.

A more complete knowledge of biochemical processes will yield a better understanding of complex disease and how it should be treated. At present, our knowledge is primarily based on monogenic diseases. As the problem is reduced to a single gene, hypotheses for function can be tested by biochemical assays, protein structural studies, experimental knockouts, or the study of naturally occurring mutants. The approach to complex disease centers on a similar approach, that is, trying to identify the one or few dominant genetic factors that contribute the most significant effect to the overall phenotype. However, there is a realization that these genetic factors may not fully explain the observed phenotype and that a proportion of the remaining factors may not be identified. In these instances, a comprehensive knowledge of the systems involved will be more informative than the approach of complex disease genetics, in both how the phenotype arises and how it might be possible to intervene more effectively. This is potentially a true long-term value of the genome sequence and its interpretation in a biochemical, biological, and genetic context for the advancement of medicine in the future.

ACKNOWLEDGMENTS

We would like to thank Jason Stajich, who helped in developing the scope of this chapter, as well as Margaret Pericak-Vance and Jonathan Haines for their patience and Frank Zhao for his comments and assistance in the preparation of the manuscript.

REFERENCES

Altschul SF, Gish W, Miller W, Myers EW, Lipman DJ (1990): Basic local alignment search tool. J Mol Biol 215:403–410.

Altschul SF, Madden TL, Schaffer AA, Zhang J, Zhang Z, Miller W, et al (1997): Gapped BLAST and PSI-BLAST: A new generation of protein database search programs. Nucleic Acids Res 25:3389–402.

Altshuler D, Pollara VJ, Cowles CR, Van Etten WJ, Baldwin J, Linton L, et al (2000): An SNP map of the human genome generated by reduced representation shotgun sequencing. Nature 407:513–516.

Ansari-Lari MA, Oeltjen JC, Schwartz S, Zhang Z, Muzny DM, Lu J, et al (1998): Comparative sequence analysis of a gene-rich cluster at human chromosome 12p13 and its syntenic region in mouse chromosome 6. Genome Res 8:29–40.

Bates GP, Valdes J, Hummerich H, Baxendale S, Le Paslier DL, Monaco AP, et al (1992): Characterization of a yeast artificial chromosome contig spanning the Huntington's disease gene candidate region. Nat Genet 1:180–187.

Bickmore WA, Sumner AT (1989): Mammalian chromosome banding—An expression of genome organization. Trends Genet 5:144–148.

Birney E, Bateman A, Clamp ME, Hubbard TJ (2001): Mining the draft human genome. Nature 409:827–828.

Bouffard GG, Idol JR, Braden VV, Iyer LM, Cunningham AF, Weintraub LA, et al (1997): A physical map of human chromosome 7: An integrated YAC contig map with average STS spacing of 79 kb. Genome Res 7:673–692.

Brenner M, Lampel K, Nakatani Y, Mill J, Banner C, Mearow K, et al (1990): Characterization of human cDNA and genomic clones for glial fibrillary acidic protein. Brain Res Mol Brain Res 7:277–286.

Chasman D, Adams RM (2001): Predicting the functional consequences of non-synonymous single nucleotide polymorphisms: Structure-based assessment of amino acid variation. J Mol Biol 307:683–706.

Chumakov I, Rigault P, Le Gall I, Cohen D, et al (1995): A YAC contig map of the human genome. Nat Genome Directory 377(Suppl):174–297.

Chumakov IM, Le Gall I, Billault A, Ougen P, Soularue P, Guillou S, et al (1992): Isolation of chromosome 21—Specific yeast artificial chromosomes from a total human genome library. Nat Genet 1:222–225.

Coffey AJ, Roberts RG, Green ED, Cole CG, Butler R, Anand R, et al (1992): Construction of a 2.6-Mb contig in yeast artificial chromosomes spanning the human dystrophin gene using an STS-based approach. Genomics 12:474–484.

Cohen D, Chumakov I, Weissenbach J (1993): A first-generation physical map of the human genome. Nature 366:698–701.

Collins FS, Brooks, LD, Chakravarti A (1998): A DNA polymorphism discovery resource for research on human genetic variation. Genome Res 8:1229–1231.

Collins FS, Morgan M, Patrinos A (2003): The Human Genome Project: Lessons from large-scale biology. Science 300(5617):286–290.

Collins JE, Cole CG, Smink LJ, Garrett CL, Leversha MA, Soderlund CA, et al (1995): A high-density YAC contig map of human chromosome 22. Nature 377:367–379.

Coulson A, Sulston J, Brenner S, Karn J (1986): Toward a physical map of the genome of the nematode *Caenorhabditis elegans*. Proc Natl Acad Sci USA 83:7821–7825.

Couzin J (2002): Human genome. HapMap launched with pledges of $100 million. Science 298:941–942

Cox DR, Burmeister M, Price ER, Kim S, Myers RM (1990): Radiation hybrid mapping: A somatic cell genetic method for constructing high-resolution maps of mammalian chromosomes. Science 250:245–250.

Dauwerse JG, Kievits T, Beverstock GC, van der Keur D, Smit E, Wessels HW, et al (1990): Rapid detection of chromosome 16 inversion in acute nonlymphocytic leukemia, subtype M4: Regional localization of the breakpoint in 16p. Cytogenet Cell Genet 53:126–128.

Dawson E, Chen Y, Hunt S, Smink LJ, Hunt A, Rice K, et al (2001): A SNP resource for human chromosome 22: Extracting dense clusters of SNPs from the genomic sequence. Genome Res 11:170–178.

Deloukas P, Matthews LH, Ashurst J, Burton J, Gilbert JG, Jones M, et al (2001): The DNA sequence and comparative analysis of human chromosome 20. Nature 414:865–871.

Deloukas P, Schuler GD, Gyapay G, Beasley EM, Soderlund C, Rodriguez-Tome P, Hui L, Matise TC, McKusick KB, Beckmann JS, Bentolila S, Bihoreau M, Birren BB, Browne J, Butler A, Castle AB, Chiannilkulchai N, Clee C, Day PJ, Dehejia A, Dibling T, Drouot N, Duprat S, Fizames C, Bentley DR, et al (1998): A physical map of 30,000 human genes. Science 282:744–746.

DeRisi JL, Iyer VR, Brown PO (1997): Exploring the metabolic and genetic control of gene expression on a genomic scale. Science 278:680–686.

Deutsch S, Iseli C, Bucher P, Antonarakis SE, Scott HS (2001): A cSNP map and database for human chromosome 21. Genome Res 11:300–307.

Dib C, Faure S, Fizames C, Samson D, Drouot N, Vignal A, et al (1996): A comprehensive genetic map of the human genome based on 5,264 microsatellites. Nature 380:152–154.

Donis-Keller H, Green P, Helms C, Cartinhour S, Weiffenbach B, Stephens K, et al (1987): A genetic linkage map of the human genome. Cell 51:319–337.

Dowell RD, Jokerst RM, Day A, Eddy SR, Stein L (2001): The distributed annotation system. BMC Bioinform 2(1):7. Erratum, October 10 (2001).

Dunham I, Shimizu N, Roe BA, Chissoe S, Hunt AR, Collins JE, et al (1999): The DNA sequence of human chromosome 22. Nature 402:489–495.

Foote S, Vollrath D, Hilton A, Page DC (1992): The human Y chromosome: Overlapping DNA clones spanning the euchromatic region. Science 258:60–66.

Frishman D, Mokrejs M, Kosykh D, Kastenmuller G, Kolesov G, Zubrzycki I, Gruber C, Geier B, Kaps A, Albermann K, Volz A, Wagner C, Fellenberg M, Heumann K, Mewes HW (2003): The PEDANT genome database. Nucleic Acids Res 31(1):207–211.

Goss SJ, Harris H (1975): New method for mapping genes in human chromosomes. Nature 255:680–684.

Goto S, Bono H, Ogata H, Fujibuchi W, Nishioka T, Sato K, Kanehisa M (1997): Organizing and computing metabolic pathway data in terms of binary relations. Pac Symp Biocomput 175–186.

Gottgens B, Barton LM, Gilbert JG, Bench AJ, Sanchez MJ, Bahn S, et al (2000): Analysis of vertebrate SCL loci identifies conserved enhancers. Nat Biotechnol 18:181–186.

Green ED, Olson MV (1990): Chromosomal region of the cystic fibrosis gene in yeast artificial chromosomes: A model for human genome mapping. Science 250:94–98.

Green ED, Riethman HC, Dutchik JE, Olson MV (1991): Detection and characterization of chimeric yeast artificial-chromosome clones. Genomics 11:658–669.

Gregory SG, Sekhon M, Schein J, Zhao S, Osoegawa K, Scott CE, et al (2002): A physical map of the mouse genome. Nature 418:743–750.

Hacia JG, Brody LC, Collins FS (1998): Applications of DNA chips for genomic analysis. Mol Psychiatry 3(6):483–492.

Haines JL (1998): In: Haines JL, Pericak-Vance MA, eds. Using Public Database in Approaches to Gene Mapping in Complex Human Diseases. New York: Wiley-Liss.

Hardison R, Slightom JL, Gumucio DL, Goodman M, Stojanovic N, Miller W (1997): Locus control regions of mammalian beta-globin gene clusters: Combining phylogenetic analyses and experimental results to gain functional insights. Gene 205:73–94.

Hardison R, Xu J, Jackson J, Mansberger J, Selifonova O, Grotch B, et al (1993): Comparative analysis of the locus control region of the rabbit beta-like gene cluster: HS3 increases transient expression of an embryonic epsilon-globin gene. Nucleic Acids Res 21:1265–1272.

Harris RF (2002): Hapmap flap. Curr Biol 12:R827.

Hattori M, Fujiyama A, Taylor TD, Watanabe H, Yada T, Park HS, et al (2000): The DNA sequence of human chromosome 21. Nature 405:311–319.

Hauser MA, Li YJ, Takeuchi S, Walters R, Noureddine M, Maready M, Darden T, Hulette C, Martin E, Hauser E, Xu H, Schmechel D, Stenger JE, Dietrich F, Vance J (2003): Genomic convergence: Identifying candidate genes for Parkinson's disease by combining serial analysis of gene expression and genetic linkage. Hum Mol Genet 12(6):671–677.

Heilig R, Eckenberg R, Petit JL, Fonknechten N, Da Silva C, Cattolico L, Levy (2003): The DNA sequence and analysis of human chromosome 14. Nature 421(6923):

Heiskanen M, Karhu R, Hellsten E, Peltonen L, Kallioniemi OP, Palotie A (1994): High resolution mapping using fluorescence in situ hybridization to extended DNA fibers prepared from agarose-embedded cells. Biotechniques 17:928–929, 932–933.

Henikoff S, Henikoff JG (1992): Amino acid substitution matrices from protein blocks. Proc Natl Acad Sci USA 89(22):10915–10919.

Henikoff S, Henikoff JG (2000): Amino acid substitution matrices. Adv Protein Chem 54:73–97.

Hijikata M, Ohta Y, Mishiro S (2000): Identification of a single nucleotide polymorphism in the MxA gene promoter (G/T at nt-88) correlated with the response of hepatitis C patients to interferon. Intervirology 43:124–127.

Hillier LW, Fulton RS, Fulton LA, Graves TA, Pepin KH, Wagner-McPherson C, et al (2003): Nature 424(6945):157–164.

Hudson TJ, Stein LD, Gerety SS, Ma J, Castle AB, Silva J, et al (1995): An STS-based map of the human genome. Science 270:1945–1954.

Ideker T, Thorsson V, Ranish JA, Christmas R, Buhler J, Eng JK, Bumgarner R, Goodlett DR, Aebersold R, Hood L (2001): Integrated genomic and proteomic analyses of a systematically perturbed metabolic network. Science 292(5518):929–934.

International Human Genome Mapping Consortium (2001): A physical map of the human genome. Nature 409:934–941.

International SNP Map Working Group (2001): A map of human genome sequence variation containing 1.42 million single nucleotide polymorphism. Nature 409:928–933.

Ioannou PA, Amemiya CT, Garnes J, Kroisel PM, Shizuya H, Chen C, et al (1994): A new bacteriophage P1-derived vector for the propagation of large human DNA fragments. Nat Genet 6:84–89.

Irizarry K, Kustanovich V, Li C, Brown N, Nelson S, Wong W, et al (2000): Genome-wide analysis of single-nucleotide polymorphisms in human expressed sequences. Nat Genet 26:233–236.

Jang W, Hua A, Spilson SV, Miller W, Roe BA, Meisler MH (1999): Comparative sequence of human and mouse BAC clones from the mnd2 region of chromosome 2p13. Genome Res 9:53–61.

Kan YW, Dozy AM (1978): Polymorphism of DNA sequence adjacent to human beta-globin structural gene: Relationship to sickle mutation. Proc Natl Acad Sci USA 75:5631–5635.

Kent WJ (2002): BLAT—The BLAST-like alignment tool. Genome Res 12(4):656–664.

Khan AS, Wilcox AS, Hopkins JA, Sikela JM (1991): Efficient double stranded sequencing of cDNA clones containing long poly(A) tails using anchored poly(dT) primers. Nucleic Acids Res 19:1715.

Knight JC, Udalova I, Hill AV, Greenwood BM, Peshu N, Marsh K, et al (1999): A polymorphism that affects OCT-1 binding to the TNF promoter region is associated with severe malaria. Nat Genet 22:145–150.

Kogelnik AM, Lott MT, Brown MD, Navathe SB, Wallace DC (1998): MITOMAP: A human mitochondrial genome database—1998 update. Nucleic Acids Res 26(1):112–115.

Kohara Y, Akiyama K, Isono K (1987): The physical map of the whole E. coli chromosome: Application of a new strategy for rapid analysis and sorting of a large genomic library. Cell 50:495–508.

Kong A, Gudbjartsson DF, Sainz J, Jonsdottir GM, Gudjonsson SA, Richardsson B, Sigurdardottir S, Barnard J, Hallbeck B, Masson G, Shlien A, Palsson ST, Frigge ML, Thorgeirsson TE, Gulcher JR, Stefansson K (2002): A high-resolution recombination map of the human genome. Nat Genet 31(3):241–247.

Koop BF, Hood L (1994): Striking sequence similarity over almost 100 kilobases of human and mouse T-cell receptor DNA. Nat Genet 7:48–53.

Krawczak M, Cooper DN (1997): The Human Gene Mutation Database. Trends Genet 13:121–122.

Krishnamurthy L, Nadeau J, Ozsoyoglu G, Ozsoyoglu M, Schaeffer G, Tasan M, Xu W (2003): Pathways database system: An integrated system for biological pathways. Bioinformatics 19(8):930–937.

Li YJ, Oliveira SA, Xu P, Martin ER, Stenger JE, Scherzer CR, Hauser MA, et al (2003): Hum Mol Genet 12(24):3259–3267. Erratum October 21 (2003).

Litt M, Luty JA (1989): A hypervariable microsatellite revealed by in vitro amplification of a dinucleotide repeat within the cardiac muscle actin gene. Am J Hum Genet 44:397–401.

Luyf AC, de Gast J, van Kampen AH (2002): Visualizing metabolic activity on a genome-wide scale. Bioinformatics 18(6):813–818.

Makalowski W, Zhang J, Boguski MS (1996): Comparative analysis of 1196 orthologous mouse and human full-length mRNA and protein sequences. Genome Res 6:846–857.

Mrazek J, Kypr J (1992): DNABIND: An interactive microcomputer program searching for nucleotide sequences that may code for conserved DNA-binding protein motifs. Comput Appl Biosci 8(4):401–404.

Marra M, Kucaba T, Sekhon M, Hillier L, Martienssen R, Chinwalla A, et al (1999): A map for sequence analysis of the *Arabidopsis thaliana* genome. Nat Genet 22:265–270.

Marshall, E (1999): Drug firms to create public database of genetic mutations. Science 284:406–407.

McPherson JD, Marra M, Hillier L, Waterston RH, Chinwalla A, Wallis J, et al (2001): A physical map of the human genome. Nature 409:934–941.

Mitelman F, Mertens F, Johansson B (1997): A breakpoint map of recurrent chromosomal rearrangements in human neoplasia. Nat Genet 15 (Spec No):417–474.

Morton NE (1991): Parameters of the human genome. Proc Natl Acad Sci USA 88: 7474–7476.

Mouse Genome Sequencing Consortium (2002): Initial sequencing and comparative analysis of the mouse genome. Nature 420:520–562.

Mullikin JC, Hunt SE, Cole CG, Mortimore BJ, Rice CM, Burton J, et al (2000): An SNP map of human chromosome 22. Nature 407:516–520.

Mungall AJ, Edwards CA, Ranby SA, Humphray SJ, Heathcott RW, Clee CM, et al (1996): Physical mapping of chromosome 6: A strategy for the rapid generation of sequence-ready contigs. DNA Seq 7:47–49.

Mungall AJ, Palmer SA, Sims SK, Edwards CA, Ashurst JL, Wilming L, Jones MC, et al (2003): Nature 425(6960):805–811.

Murzin AG, Brenner SE, Hubbard T, Chothia C (1995): SCOP: A structural classification of proteins database for the investigation of sequences and structures. J Mol Biol 247:536–540.

Needleman SB, Wunsch CD (1970): A general method applicable to the search for similarities in the amino acid sequences of two proteins. Mol Biol 48:443–453.

Olson M, Hood L, Cantor C, Botstein D (1989): A common language for physical mapping of the human genome. Science 245:1434–1435.

Olson MV, Dutchik JE, Graham MY, Brodeur GM, Helms C, Frank M, et al (1986): Random-clone strategy for genomic restriction mapping in yeast. Proc Natl Acad Sci USA 83:7826–7830.

Pan SS, Han Y, Farabaugh P, Xia H (2002): Implication of alternative splicing for expression of a variant NAD(P)H:quinone oxidoreductase-1 with a single nucleotide polymorphism at 465C > T. Pharmacogenetics 12:479–488.

Pauling L, Corey RB (1953): Nature 171:346.

Pearson WR, Lipman DJ (1988): Improved tools for Biological Sequence Comparison. Proc Nat Acad Sci 85:2444–2448.

Pennacchio LA, Olivier M, Hubacek JA, Cohen JC, Cox DR, Fruchart JC, et al (2001): An apolipoprotein influencing triglycerides in humans and mice revealed by comparative sequencing. Science 294:169–173.

Pfost DR, Boyce-Jacino MT, Grant DM (2000): A SNPshot: Pharmacogenetics and the future of drug therapy. Trends Biotechnol 18:334–338.

Picoult-Newberg L, Ideker TE, Pohl MG, Taylor SL, Donaldson MA, Nickerson DA, et al (1999): Mining SNPs from EST databases. Genome Res 9:167–174.

Pinkel D, Landegent J, Collins C, Fuscoe J, Segraves R, Lucas J, et al (1988): Fluorescence in situ hybridization with human chromosome-specific libraries: Detection of trisomy 21 and translocations of chromosome 4. Proc Natl Acad Sci USA 85:9138–9142.

Pinkel D, Straume T, Gray JW (1986): Cytogenetic analysis using quantitative, high-sensitivity, fluorescence hybridization. Proc Natl Acad Sci USA 83:2934–2938.

Saha S, Sparks AB, Rago C, Akmaev V, Wang CJ, Vogelstein B, Kinzler KW, Velculescu VE (2002): Using the transcriptome to annotate the genome. Nat Biotechnol 20(5): 508–512.

Saiki RK, Gelfand DH, Stoffel S, Scharf SJ, Higuchi R, Horn GT, et al (1988): Primer-directed enzymatic amplification of DNA with a thermostable DNA polymerase. Science 239:487–491.

Sanger F, Air GM, Barrell BG, Brown NL, Coulson AR, Fiddes CA, et al (1977): Nucleotide sequence of bacteriophage phi X174 DNA. Nature 265:687–695.

Sanger F, Coulson AR, Friedmann T, Air GM, Barrell BG, Brown NL, et al (1978): The nucleotide sequence of bacteriophage phiX174. J Mol Biol 125:225–246.

Sanger F, Coulson AR, Hong GF, Hill DF, Petersen GB (1982): Nucleotide sequence of bacteriophage lambda DNA. J Mol Biol 162:729–773.

Schena M, Shalon D, Davis RW, Brown PO (1995): Quantitative monitoring of gene expression patterns with a complementary DNA microarray. Science 270:467–470.

Schuler GD, Boguski MS, Stewart EA, Stein LD, Gyapay G, Rice K, White RE, Rodriguez-Tome P, Aggarwal A, Bajorek E, Bentolila S, Birren BB, Butler A, Castle AB, Chiannilkulchai N, Chu A, Clee C, Cowles S, Day PJ, Dibling T, Drouot N, Dunham I, Duprat S, East C, Hudson TJ, et al (1996): A gene map of the human genome. Science 274(5287):540–546.

Shen LX, Basilion JP, Stanton VP, Jr (1999): Single-nucleotide polymorphisms can cause different structural folds of mRNA. Proc Natl Acad Sci USA 96:7871–7876.

Sherry ST, Ward MH, Kholodov M, Baker J, Phan L, Smigielski EM, et al (2001): dbSNP: The NCBI database of genetic variation. Nucleic Acids Res 29:308–311.

Shizuya H, Birren B, Kim UJ, Mancino V, Slepak T, Tachiiri Y, et al (1992): Cloning and stable maintenance of 300-kilobase-pair fragments of human DNA in *Escherichia coli* using an F-factor-based vector. Proc Natl Acad Sci USA 89:8794–8797.

Skaletsky H, Kuroda-Kawaguchi T, Minx PJ, Cordum HS, Hillier L, Brown LG, Repping S, et al (2003): Nature 423(6942):825–837.

Skolnick J, Fetrow JS (2000): From genes to protein structure and function: Novel applications of computational approaches in the genomic era. Trends Biotechnol 18:34–39.

Slim R, Le Paslier D, Compain S, Levilliers J, Ougen P, Billault A, et al (1993): Construction of a yeast artificial chromosome contig spanning the pseudoautosomal region and isolation of 25 new sequence-tagged sites. Genomics 16:691–697.

Smith TF, Waterman MS (1981): Identification of common molecular subsequences. J Mol Biol 147:195–197.

Staden R (1980): A new computer method for the storage and manipulation of DNA gel reading data. Nucleic Acids Res 8:3673–3694.

Stein LD (2001): Using Perl to facilitate biological analysis. Methods Biochem Anal 43:413–449.

Sunyaev S, Ramensky V, Koch I, Lathe W 3rd, Kondrashov AS, Bork P (2001): Prediction of deleterious human alleles. Hum Mol Genet 10:591–597.

Taillon-Miller P, Gu Z, Li Q, Hillier L, Kwok PY (1998): Overlapping genomic sequences: A treasure trove of single-nucleotide polymorphisms. Genome Res 8:748–754.

Tao Q, Chang Y-L, Wang J, Huaming C, Islam-Faridi MN, Scheuring C, Wang B, Stelly, DM, Zhang H-B (2001): Bacterial artificial chromosome-based physical map of the rice genome constructed by restriction fingerprint analysis. Genetics 158:1711–1724.

Thomas JW, Prasad AB, Summers TJ, Lee-Lin SQ, Maduro VV, Idol JR, et al (2002): Parallel construction of orthologous sequence-ready clone contig maps in multiple species. Genome Res 12:1277–1285.

Thornton JM, Orengo CA, Todd AE, Pearl FM (1999): Protein folds, functions and evolution. J Mol Biol 293:333–342.

Tkachuk DC, Westbrook CA, Andreeff M, Donlon TA, Cleary ML, Suryanarayan K, et al (1990): Detection of bcr-abl fusion in chronic myelogeneous leukemia by in situ hybridization. Science 250:559–562.

Velculescu VE, Zhang L, Vogelstein B, Kinzler KW (1995): Serial analysis of gene expression. Science 270(5235):484–487.

Wang DG, Fan JB, Siao CJ, Berno A, Young P, Sapolsky R, et al (1998): Large-scale identification, mapping, and genotyping of single-nucleotide polymorphisms in the human genome. Science 280:1077–1082.

Wang Z, Moult J (2001): SNPs, protein structure, and disease. Hum Mutat 17:263–270.

Watson JD, Crick FHC (1953): Molecular structure of nucleic acids: A structure for deoxyribose nucleic acid. Nature April 25(4356).

Wilke CM, Guo SW, Hall BK, Boldog F, Gemmill RM, Chandrasekharappa SC, et al (1994): Multicolor FISH mapping of YAC clones in 3p14 and identification of a YAC spanning both FRA3B and the t(3;8) associated with hereditary renal cell carcinoma. Genomics 22:319–326.

Wodicka L, Dong H, Mittmann M, Ho M-H, Lockhart DJ (1997): Genome-wide expression monitoring in *Saccharomyces cerevisiae*. Nat Genet 15:1359–1367.

CHAPTER 16

Designing a Study for Identifying Genes in Complex Traits

JONATHAN L. HAINES and MARGARET A. PERICAK-VANCE

INTRODUCTION

Disease gene discovery in humans has a long history, predating even the identification of DNA as the genetic molecule (Watson and Crick, 1953) and the determination of the number of human chromosomes (Ford and Hamerton, 1956; Tjio and Levan, 1956). In fact, as early as the 1930s some simple statistical methods had been developed (Bernstein, 1931; Fisher, 1935a,b). However, these methods were severely limited in their application. Not only were genetic markers lacking (the ABO blood type was one of the few that had been described), but these methods were restricted to small nuclear pedigrees, perhaps including grandparents. Any calculations had to be done by hand, of course, making analysis very laborious.

There were two hurdles to be overcome before human disease gene discovery could be performed routinely. First, appropriate statistical methods were lacking, as were ways of automating the laborious calculations of the statistics. Second, sufficient genetic markers to cover the entire human genome had to be developed. Morton (1955), building on the work of Haldane and Smith (1947) and Wald (1947), described the use of maximum-likelihood approaches in a sequential test for linkage between two loci. He used the term *LOD score* (for logarithm of the odds of linkage) for his test. This score is the basis of most linkage analyses being performed today, and it represents a milestone in human disease gene discovery. However, the complex calculations had to be done by hand, severely limiting the use of this approach. In 1971, Elston and Stewart (1971) described a general approach for calculating the likelihood of any nonconsanguineous pedigree. This was extended by Lange and Elston (1975) to include pedigrees of arbitrary complexity. Soon thereafter the first general-purpose computer program for linkage in

Genetic Analysis of Complex Diseases, Second Edition, Edited by Jonathan L. Haines and Margaret Pericak-Vance

455

humans, LIPED (Ott, 1974), was described. Thus the first of the two major hurdles had been overcome.

By the mid-1970s there were 40–50 red cell antigen and serum protein polymorphisms available as genetic markers. A few markers could be arranged into initial linkage groups, but these markers covered only approximately 5–15% of the human genome. In addition to this limited coverage, genotyping these polymorphisms was labor intensive, time consuming, and often quite technically demanding. This remaining hurdle was crossed with the description of restriction fragment length polymorphisms (RFLPs) by Botstein et al. (1980). Not only are these markers easier to genotype in a standard manner, but also they occur with great frequency throughout the genome, opening up the remaining 85–95% of the human genome for the first time.

With these tools in place, the field of human disease gene discovery blossomed. The first successful linkage using RFLPs was reported (Gusella et al., 1983), locating the Huntington disease gene to chromosome 4p. This was the beginning of the approach to disease gene identification often termed *positional cloning*. It is noteworthy that these early successes were all diseases inherited in a simple manner: autosomal dominant, autosomal recessive, or X linked (i.e., Mendelian inheritance). Although confounding factors such as genetic heterogeneity, variable penetrance, and phenocopies may exist for Mendelian traits, it is generally possible with a known genetic model in hand to determine the best and most efficient approach to identifying the responsible gene. The success of these tools is apparent since over 1600 Mendelian traits now have at least one identified molecular genetics lesion.

However, the inheritance patterns for traits such as the common form of Alzheimer's disease, multiple sclerosis, insulin-dependent diabetes, and hypertension (with the exception of some rare subtypes) do not fit any simple genetic explanation, making it far more difficult to determine the best approach to identifying the unknown underlying effect. In addition to the confounding factors involved in single-gene disorders such as genetic heterogeneity and phenocopies, gene–gene and gene–environment interactions must be considered when a complex trait is dissected.

With the attention now being paid to genetically complex traits, it is important to change our thinking about what, exactly, we are trying to find. No longer are we simply searching for the one or few rare mutations in a single gene that cause a rare and devastating disease. We are now searching for multiple alterations in one or more genes that alone or in concert (and we do not know which) either increase or decrease the risk of developing a trait. Such alterations may not be rare at all but could be common polymorphisms. One etiological explanation for this is the common disease variant hypothesis. This hypothesis supposes that the risk for common diseases, which span multiple ethnic groups, arises from evolutionarily old variants that have had substantial time to spread throughout the human population. The alternative multiple rare-variant hypothesis supposes that risk arises from a larger number of rare variants in one or more genes, perhaps occurring more recently. Unfortunately testing these two hypotheses requires different analytical and molecular approaches as described below.

COMPONENTS OF A DISEASE GENE DISCOVERY STUDY

Each genetically complex trait has its own peculiarities that require special attention. However, a guiding paradigm can be applied to most situations. Originally the general paradigm that emerged for application to single-gene disorders was called positional cloning. With the sequencing of the human genome, cloning is no longer a necessary step; thus we prefer to describe the process as disease gene discovery. This classical disease gene discovery approach consists of identifying families, collecting blood samples, genotyping markers, and performing analyses for initial localization, refining the genetic localization to define the minimum-candidate region, identifying and testing genes within this region, and ultimately finding the causal mutation. This process is essentially linear, although there are some steps that are recursive in nature, particularly in the genotyping, initial localization, and fine-mapping steps (Fig. 16.1). This linear property has made it easier to apply disease gene discovery as a general approach to many different Mendelian traits.

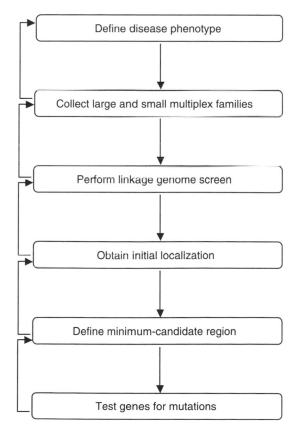

Figure 16.1.

In contrast to the classical approach, the current approaches to finding genes for common and genetically complex traits are not linear, and many steps are works in progress, subject to further defining, refining, or replacement. Figure 16.2 is a diagrammatic interpretation of the steps as we see them today. Each of these steps has its own peculiarities and key factors that must be considered. Particularly for complex traits, the answers and paths that define the best approach will be trait specific. This fact is perhaps underappreciated, and it contrasts strongly with the classical disease gene discovery approach. Thus the amount of confusion surrounding the issue of how to approach complex traits is not entirely surprising.

This section discusses the steps in Figure 16.2. The purpose is not a detailed description of each step but a review of the major points of consideration and a guide to the chapters that cover these points in more detail.

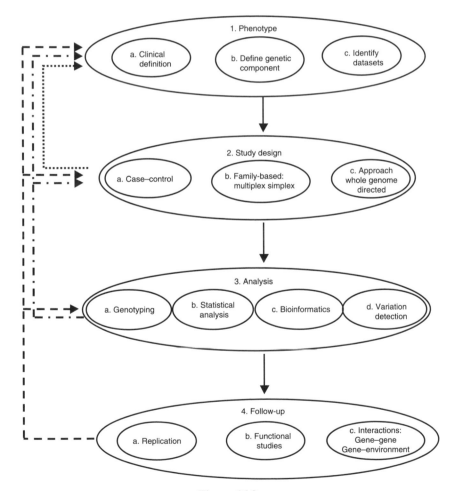

Figure 16.2.

Define Phenotype

The first step in any disease gene discovery process is to know what phenotype is to be studied. This may sound trivial, but it is often overlooked in the rush to move forward. There are three aspects that need to be considered: clinical definition, determining that a trait has a genetic component, and identification of datasets.

Clinical Definition. It is not enough to define a trait as Huntington disease or diabetes. In the former case there can be wide variation in the symptoms, with some only psychological or very mild motor disturbances, and the age at examination plays a critical role. In the latter case there are known subtypes [insulin-dependent (IDDM) vs. non-insulin-dependent (NIDDM)] as well as age effects. In addition, blood glucose levels (a quantitative trait) are strongly associated with the diabetes (a qualitative trait) and could perhaps be used as a surrogate measure. One critical role of the clinician in the study design is to assess the various diagnostic procedures and tools and determine which ones best define a consistent phenotype. In addition, dissecting genetically complex diseases usually requires large datasets to supply enough power to unravel genetic effects. For this reason participant ascertainment often extends across multiple centers. It is critical when this occurs that consensus diagnostic criteria be established a priori and used by all centers. For example, the establishment of a consensus diagnostic scheme played an important role in a successful complex disease linkage study in late-onset familial Alzheimer's disease (AD) (McKhann et al., 1984).

The phenotype assignment must be done in a rigorously consistent fashion. Even a few misassigned phenotypes may have major negative implications in the analysis, possibly leading to both false-positive and false-negative results. Thus which data will be used to assign the trait status must be determined. Must clinical records be obtained and reviewed for consistency on every patient? Is the report of a patient's relative sufficient? Is a note in a medical record sufficient? Or is direct examination for the study required? Are biomarkers (e.g., antibody titers, protein levels) or laboratory tests (e.g., electroencephalogram, electrocardiogram, magnetic resonance imaging) available and known or suggested to correlate with the trait? The goal is to minimize sources of error and uncertainty. This is discussed in more detail in Chapter 2.

Determining That Trait Has a Genetic Component. It is critical that as much as possible be known about the genetic basis of a trait for linkage studies before the expensive and time-consuming process of data collection and genotyping is begun. That a trait "runs in families" is insufficient evidence, since this phenomenon can occur for several reasons, including common environmental exposure and biased ascertainment as well as a true genetic disposition. There are numerous lines of evidence that can be examined, including family studies, segregation analysis, twin studies, adoption studies, heritability studies, and population-based risks to relatives of probands. For most traits being contemplated some such data already exist in the literature. A thorough and considered review of this literature may provide most of the necessary information and point out any missing data. The data may not only indicate the strength of the genetic effect on the trait but also give some indication of the underlying genetic model. For example, there may be obvious evidence of a single "major" gene, such as in Huntington disease, or multiple genes interacting in complex ways, such as in multiple sclerosis (Sadovnick et al., 1996). This is discussed in more detail in Chapter 3.

Identification of Datasets. It is helpful to identify early on what potential datasets exist or can be collected. Do large extended families exist or are most cases apparently sporadic? Are large cohort or case–control studies available? Are there repositories of multiplex families with the necessary clinical data available? Are there existing clinical networks or large specialty clinics available? The answers to these questions may determine what study designs are feasible for the trait under study. These issues are discussed in Chapters 2 and 3.

Develop Study Design

This step and step 1 are not independent. Review of the available data may indicate that a trait as originally defined has little, or even no, evidence of a genetic component. However, there may be strong evidence that a particular subform of the trait is strongly genetic. For example, there had for many years been debate about the role of genetics in AD. Over time it became increasingly clear that a subform of AD with onset before age 65 existed and is, although rare, strongly influenced by the action (within families) of a single gene. By restricting further ascertainment and analysis to families with such early-onset AD, three different single genes have now been identified (Goate et al., 1991; Levy-Lahad et al., 1995; Rogaev et al., 1995; Sherrington et al., 1995).

The exact approach to the disease gene discovery process should be outlined as completely as possible before the project gets under way. With the clinical phenotype in hand, it is possible to determine the best strategy for defining what type of dataset to collect. Ascertainment is perhaps the longest and most labor-intensive step in the entire process. It is imperative that the ascertainment of families proceed with careful consideration of the wishes of the participating families and that their rights of participation/nonparticipation and confidentiality be protected. There are many sources of potential ascertainment, such as support groups, hospital clinics, and private referrals. These issues are discussed in detail in Chapter 4.

Determination of the type(s) of individuals to collect (e.g., using case–control or family-based approaches, see below) is based on the knowledge of phenotype and any evidence for the genetic model. For example, if a single major gene is suspected, large extended families are likely to prove most efficient. For a more complex genetic model, small families, such as sibpairs, may prove most efficient, whereas if a trait phenotype could be defined in numerous ways, a restrictive set of clinical criteria may limit the available family material to a single type.

It is also important to have some sense of the sample size required to identify the genes being sought. For single-gene disorders, formal power studies using standard simulation programs are possible because the underlying genetic model can be assigned with reasonable confidence. For genetically complex traits, however, the underlying genetic model is usually unknown; thus simulation studies can only give general guidelines to the potential detection of genes. Chapter 13 discusses sample size and power analyses.

Case–Control Studies. In many instances the most easily accessible sample is one with single affected cases matched to unaffected controls. This has been a workhorse approach in epidemiological studies for years but has only recently been adopted for disease gene discovery. It has been shown (Lee, 2004) that such studies have more power per genotyped individual than family-based studies, but they are more sensitive to violations of the underlying assumptions. For example, undiscovered sample stratification, lack of Hardy–Weinberg equilibrium, or poor matching between cases and controls can all create misleading results. This is discussed in more detail in Chapters 12 and 14.

Family-Based Studies. Family-based studies include large extended families (most likely resulting from a Mendelian acting single gene), smaller affected sibpair or affected relative pair families (useful for both linkage and association studies), and discordant sibpair studies (useful for association studies). While not as powerful per genotype as case–control studies, they are more robust to violations of underlying sample structure, particularly population stratification. Family-based studies are discussed in more detail in Chapters 9–14.

Approach. There are two general, but not mutually exclusive, ways to approach disease gene discovery. The first is to take a whole-genome screening approach. Genomic screening can be defined as testing the chosen trait for linkage to polymorphic markers spread throughout the genome. A good genomic screen will attempt to cover the entire human genome using markers evenly spaced across a given genetic map. It does not require any knowledge of the function of any genes or of the biology of the trait in question. In fact, practitioners tend to revel in the fact that they (and most everyone else) know very little about the functional relationship between genes and the trait.

Genomic screening was first applied using linkage analysis. While a few early screens were done using RFLPs (Tsui et al., 1985), most linkage screens have been done using microsatellite markers. Most screens have been done with sets of markers at an average spacing of 10 cM ($\sim$350–450 markers) using one of a few mapping sets such as those from Marshfield (http://research.marshfieldclinic.org/genetics/Map_Markers/maps/IndexMapFrames.html), Decode (Kong et al., 2002), and Applied Biosystems. It must be remembered that any claims of a particular spacing are based on an average spacing with a large standard deviation. Linkage screening using high-density panels of single-nucleotide polymorphisms (SNPs) (5000–10,000) have recently been undertaken (John et al., 2004). More detail on genotyping is given in Chapter 6.

Under certain situations the special design of homozygosity mapping can be used. In this approach, single affected individuals can be used for disease gene localization, under the assumption that each person has inherited the same ancestral chromosomal segment surrounding the gene in question. Genomic screening can be performed in essentially the same manner as for other types of families.

It has been proposed (Risch and Merikangas, 1996) that the current linkage genome screening approach using anonymous microsatellite markers will be supplanted by a genome screening approach using numerous SNPs and testing directly for allelic, genotypic, or haplotypic association. There remains debate on whether SNPs should be concentrated within coding sequences or more broadly represent regulatory and evolutionarily conserved sequences (Botstein and Risch, 2003). This approach is now technically feasible but it bears a bit of scrutiny. There remains substantial debate over the number of SNPs necessary to obtain good coverage of the genome. Despite the downward trend on per-genotype costs, performing such a screen is still extremely expensive. In addition, the real bottleneck is not genotype generation but data analysis. Potential genotypic errors must be identified and tested, a more difficult task without family structures. The undefined linkage disequilibrium relationships between SNPs must be taken into account. Finally, the problem of what p-value constitutes a true rare event must be solved.

In contrast to the genomic screening approach, a directed genomic screening approach may be used. By directed genomic screening we simply mean that the polymorphisms chosen for analysis are not spread either randomly or evenly across the genome but instead are concentrated in certain areas based on additional information. The additional information that defines these candidate regions can come from several sources.

Functional candidate regions are defined by the genes that reside within them. If something is known of the biology of the trait, then genes affecting that biology become candidates. For example, multiple sclerosis is an autoimmune disease in which the myelin sheaths around nerves are attacked and often destroyed. This information suggests that certain genes, such as the human leukocyte antigen (HLA) genes, T-cell receptor genes, and the myelin basic protein gene are prime candidates for analysis. The strength and weakness of this approach arises from the confidence placed on the role of these genes. If the evidence is strong that a direct role is played, only one or a few such genes may need to be tested. If the evidence is more circumstantial, then many genes may have an equal chance of being involved, and not much has been gained over a random genomic screen.

At the phenotypic level, what is known of the physiology may suggest a scheme for stratification (subsetting) of the data such that a single gene (or at least a reduced set of genes) may be responsible. For example, for many years neurofibromatosis was considered a single disorder, manifesting both peripheral and auditory nerve tumors. Later it became clear that the peripheral and auditory tumors rarely occurred together and likely represent two different forms. This eased the linkage analysis, with each form being linked to different chromosomes (Barker et al., 1987; Rouleau et al., 1987).

Additional information can come from the rapidly growing field of gene expression profiling (Chapter 7). Complementary DNA microarray and serial analysis of gene expression (SAGE) experiments are but two ways of measuring the expression of genes at the RNA level.

Analysis

Genotyping. In most cases, collecting biological samples is done simultaneously with family ascertainment (Chapter 4). This usually consists of obtaining 10–40 mL of blood, from which DNA is extracted. However, other options, such as initiating cell lines, using buccal washes, finger sticks, or extracting DNA from pathological tissue specimens, may be preferable under some circumstances. These methods are discussed in Chapter 5.

The specific approaches toward genotype generation are discussed in Chapter 6. It is important to note that genotype generation should not move forward independently of analysis, since many laboratory errors, such as sample mix-ups or contamination, may be detected only in the analytical process. Data generation also takes very different tacks depending on the goal. For an initial genomic screen, high-throughput, highly automated genotyping using a standard set of markers is desirable. However, for follow-up or saturation genotyping, a less automated approach may prove more efficient for more intensive efforts.

Statistical Analysis. The analysis of the resulting clinical and genotypic data is itself complex, and refinements of current and development of new methods are constantly becoming available. The type(s) of analyses chosen depend on the type of trait, the type of families to collect, the likely underlying genetic model, and the potential power of the dataset. The various methods of analysis include genetic model-dependent (Chapter 10) and genetic model-independent (Chapter 11) linkage analyses and association studies (Chapter 12). These are not mutually exclusive approaches, and the best overall design may include multiple techniques. In addition, the choice of method may depend on the stage of the discovery process. For example, linkage analysis may be used in the initial genomic screening, but the detailed analysis of specific genomic regions may use association studies. Association studies, in particular, are being further developed and extended, in response to the plethora of SNPs that have been identified over the past few years.

There has been substantial discussion of what constitute a "significant" result from a genomic screen, with widely disparate opinions about the appropriate p-value(s) to be used. This is discussed in more detail in Chapters 10–14. Two points are worth highlighting here. The first is that the p-value(s) chosen depend on the goal of the study. If the goal is to assure that a result is most likely true, then a stringent value, perhaps arrived at through a Bonferroni or false discovery rate approach (Hochberg and Benjamini, 1990), may be appropriate. If, as is genomic screening, the goal is to weed out most of the genome and identify a subset of promising results, then a more liberal statistical criterion, tolerating a higher false-positive rate to maintain power, may be appropriate. The second is to keep in mind that the goal of any disease gene discovery project is finding the susceptibility variant, not adhering to strict statistical theory.

Bioinformatics. There have been tremendous advances in bioinformatics tools over the past few years. It is critically important that the increasingly large amounts of clinical, family history, and genotypic data be stored in well-designed and maintained databases. It is no longer adequate to maintain data simply in static files or spreadsheets. In addition, the massive amount of molecular genetic, sequence, comparative genomic, biochemical pathway, and other data now available in the public domain requires familiarity with numerous different bioinformatic tools to interrogate the public databases and analyze these data. This is discussed in more detail in Chapter 15.

Variation Detection. Once a single gene or associated haplotype has been implicated, it is necessary to examine it for variations that are specifically associated with the trait. For single-gene disorders, this has been a laborious but straightforward process of identifying (usually through sequencing) an extremely rare mutation that usually severely disrupts the normal function of the gene. However, for common and genetically complex traits, the variation may not be rare (in fact, it may be a common polymorphism), and there may be no apparent deleterious function on the gene in question. If this is the case, it becomes difficult to "know" that the right variation has been pinpointed. Clues can be taken from the strength of any allelic association (Chapter 12) found, but the ultimate proof comes with testing the function of the gene in biological systems.

Follow-Up

Replication. The literature is rife with initial reports of allelic or genotypic associations that cannot be replicated at all or are replicated in only a small minority of follow-up studies. Especially for genetic effects that may be relatively modest, the current gold standard is to observe the same effect in a second independent dataset. Increasingly studies now include such a dataset as part of the overall design. Independence does not necessarily require collection by a different group of researchers (although this is certainly possible); rather, it means using families and data not used for the initial exploration. It should be noted that this is often called a "replication" dataset, although the goal is not replication, but rather the testing of the specific hypothesis that there is linkage and/or association to one or more defined polymorphisms. Such datasets may have the same structure as the first dataset (e.g., multiplex, simplex, case–control), but often studies mix these types of datasets. It is particularly powerful if a linkage in multiplex families leads to identification of an association in those families that is then confirmed in an independent case–control dataset.

Functional Studies. While most disease gene discovery efforts have claimed success based on finding rare mutations in a gene, this is, strictly, not sufficient evidence. More conclusive is evidence arising from biological systems (e.g., cultured cells, animal models, or human trials) that the trait can be either induced by introduction of the mutation or ameliorated by blocking the action of the abnormal gene. In genetically complex traits, where the responsible variation may be a common polymorphism, it is even more critical that such evidence be found before success is declared.

Tests in biological systems can be of several types. Perhaps the most common is to test the action of the gene in mice. With transgenic mice the proposed mutation is introduced into the germline of the mouse, and the resulting offspring are examined for evidence of the abnormal phenotype. With knockout mice the action of the gene in question is eliminated and the offspring are examined for evidence of an abnormal phenotype. Similar experiments may be performed in cultured cells, where the control of the mutation or knockout process is generally easier. The cellular phenotype may be difficult to discern, however, and its relationship to the overt trait phenotype in humans may be remote.

Define Interactions. Another critical component in the dissection of a genetically complex disease is an understanding of potential interactions between the genes that underlie the trait and between genes and other risk factors (usually environmental). This is perhaps the least well developed of any of the steps, as it is only now becoming possible to identify and examine more than one gene (and/or risk factor) for complex diseases. This step also requires integration of the techniques used in both genetics and epidemiology, a process that is just now developing. Current approaches to examining interactions are discussed in Chapter 14.

KEYS TO A SUCCESSFUL STUDY

The feasibility of dissecting a genetically complex trait depends on a number of different variables. These include the strength of the genetic component, the number of genes involved and the magnitude of their individual effects, and the consistency of diagnosis, together with the frequency of phenocopies and the amount, type, and power of the available family material. Although every component of a disease gene discovery study is important, there are two overriding keys to success. These points, often overlooked in the rush to initiate studies, must be carefully considered if the study is not later to grind to a halt while these matters are belatedly addressed.

Foster Interaction of Necessary Expertise

To appropriately carry out any disease gene discovery study, one must use techniques from four different fields of expertise (Fig. 16.3). These fields are clinical diagnosis, molecular genetics, statistical genetics, and epidemiology. The first provides the necessary diagnostic and patient ascertainment skills needed to define the phenotype and help collect samples. The second provides the genotyping and functional analysis skills necessary to help locate and identify the gene(s) of interest. The third provides the statistical and analytical framework for the proper design of the study and the analysis of the generated data. The fourth

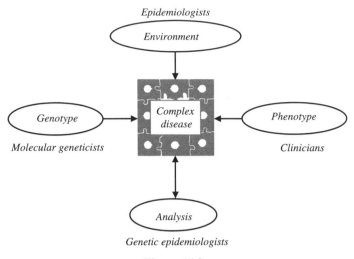

Figure 16.3.

provides expertise to incorporate environmental variables and apply the results at the population level.

During the 21-year history of disease gene discovery primarily for single-gene disorders, the limited number of possible underlying genetic modes of action allowed the development of a general paradigm that could be implemented by researchers having expertise in only one of these fields. Appropriate and perhaps only occasional consultation with experts in the other fields was all that was necessary. With common and genetically complex traits no generally applicable paradigm is possible because the underlying genetic modes of action are unique to each trait. Thus experts in each of these fields must be intimately involved in all aspects of the study. Even with this expertise in place, it is essential that the study not be divided into separate parts with little interaction between the various researchers. For example, genetic epidemiologists should be involved in the discussion of the clinical phenotype to determine the effect of potential changes to the definition of the phenotype on the power to detect any underlying genes.

Develop Careful Study Design

It may seem self-evident that a careful study design is necessary for a successful study. However, it is not enough to decide on a general design of "collect sibpairs, genotype, analyze using sibpair statistical methods." Each step in the process requires substantial thought, and the decisions made for one step will have implications for the other. Much as engineers and architects have to carefully ferret out unintended side effects of any change in their designs, lest a catastrophic failure ensue, researchers must consider carefully all aspects of the experimental design lest they doom themselves to making inappropriate conclusions based on inadequately obtained and interpreted results.

REFERENCES

Barker D, Wright E, Nguyen K, Cannon L, Fain P, Goldgar D, Bishop DT, Carey J, Baty B, Kivlin J, Willard H, Waye JS, Greig G, Leinwand L, Nakamura Y, O'Connell P, Leppert M, Lalouel J-M, White R, Skolnick M (1987): Gene for von Recklinghausen neurofibromatosis is in the pericentromeric region of chromosome 17. Science 236(4805):1100–1102.

Bernstein F (1931): Zur grundlegung der chromosomentheorie der vererbung beim menschen. Z Abst Vererb 57:113–138.

Botstein D, Risch N (2003): Discovering genotypes underlying human phenotypes: Past successes for Mendelian disease, future approaches for complex disease. Nat Genet 33(Suppl): 228–237.

Botstein D, White RL, Skolnick M, Davis RW (1980): Construction of a genetic linkage map in man using restriction fragment length polymorphisms. Am J Hum Genet 32(3):314–331.

Elston RC, Stewart J (1971): A general model for the genetic analysis of pedigree data. Hum Hered 21:523–542.

Fisher, RA (1935a): The detection of linkage with dominant abnormalities. Ann Eugen 6:187–201.

Fisher, RA (1935b): The detection of linkage with recessive abnormalities. Ann Eugen 6: 339–351.

Ford CE, Hamerton JL (1956): The chromosomes of man. Nature 178 (4541):1020–1023.

Goate A, Chartier-Harlin MC, Mullan M, Brown J, Crawford F, Fidani I., Giuffra L, Haynes A, Irving N, James L, Mant R, Newton P, Rooke K, Roques P, Talbot C, Pericak-Vance MA, Roses A, Williamson R, Rossor M, Owen M, Hardy J (1991): Segregation of a missense mutation in the amyloid precursor protein gene with familial Alzheimer's disease. Nature 33:53–56.

Gusella JF, Wexler NS, Conneally MP, Naylor SL, Anderson MA, Tanzi RE, Watkins PC, Ottina K, Wallace MR, Sajaguchi AY, Young AB, Shoulson I, Bonilla E, Martin JB (1983): A polymorphic DNA marker genetically linked to Huntington's disease. Nature 306(5940):234–238.

Haldane JBS, Smith CAB (1947): A new estimate of the linkage between the genes for color blindness and hemophilia in man. Ann Eugen 14:10–31.

Hochberg Y, Benjamini Y (1990): More powerful procedures for multiple significance testing. Stat Med 9(7):811–818.

John S, Shephard N, Liu G, Zeggini E, Cao M, Chen W, Vasavda N, Mills T, Barton A, Hinks A, Eyre S, Jones KW, Ollier W, Silman A, Gibson N, Worthington J, Kennedy GC (2004): Whole-genome scan, in a complex disease, using 11,245 single-nucleotide polymorphisms: Comparison with microsatellites. Am J Hum Genet 75(1):54–64.

Kong A, Gudbjartsson DF, Sainz J, Jonsdottir GM, Gudjonsson SA, Richardsson B, Sigurdardottir S, Barnard J, Hallbeck B, Masson G, Shlien A, Palsson ST, Frigge ML, Thorgeirsson TE, Gulcher JR, Stefansson K (2002): A high-resolution recombination map of the human genome. Nat Genet 31(3):241–247.

Lange K, Elston RC (1975): Extension to pedigree analysis. I. Likelihood calculations for simple and complex pedigrees. Hum Hered 25:95–105.

Lee WC (2004): Case-control association studies with matching and genomic controlling. Genet Epidemiol 27(1):1–13.

Levy-Lahad E, Wasco W, Poorkaj P, Romano DM, Oshima J, Pettingell WH, Yu CE, Jondro PD, Schmidt SD, Wang K, Crowley AC, Fu YH, Guenette SY, Galas D, Nemens E, Wijsman EM, Bird TD, Schellenberg GD, Tanzi RE (1995): Candidate gene for the chromosome 1 familial Alzheimer's disease locus. Science 269:973–977.

McKhann G, Drachman G, Folstein M (1984): Clinical diagnosis of Alzheimer's disease: Report of the NINCDS-ADRDA Work Group under the auspices of the department of health and human services task force on Alzheimer's disease. Neurology 34:939–944.

Morton NE (1955): Sequential tests for the detection of linkage. Am J Hum Genet 7:277–318.

Ott J (1974): Estimation of the recombination fraction in human pedigrees: Efficient computation of the likelihood for human linkage studies. Am J Hum Genet 26:588–597.

Risch N, Merikangas K (1996): The future of genetic studies of complex human disorders. Science 273(5281):1516–1517.

Rogaev EI, Sherrington R, Rogaeva EA, Levesque G, Ikeda M, Liang Y, Chi H, Lin C, Holman K, Tsuda T, Mar L, Sorbi S, Nacmias B, Piacentini S, Amaducci L, Chumakov I, Cohen D, Lannfelt L, Fraser PE, Rommens JM, St George-Hyslop PH (1995): Familial Alzheimer's disease in kindreds with missense mutations in a gene on chromosome 1 related to the Alzheimer's disease type 3 gene. Nature 376(6543):775–778.

Rouleau GA, Wertelecki W, Haines JL, Hobbs WJ, Trofatter JA, Seizinger BR, Martuza RL, Superneau DW, Conneally PM, Gusella JF (1987): Genetic linkage of bilateral acoustic neurofibromatosis to a DNA marker on chromosome 22. Nature 329(6136):246–248.

Sadovnick AD, Ebers GC, Dyment DA, Risch NJ (1996): Evidence for genetic basis of multiple sclerosis. Lancet 347(1728):1730.

Sherrington R, Rogaev EI, Liang Y, Rogaeva EA, Levesque G, Ikeda M, Chi H, Lin C, Li G, Holman K, Tsuda T, Mar L, Foncin J-F, Bruni AC, Montesi MP, Sorbi S, Rainero I, Pinessi L, Nee L, Chumakov I, Pollen D, Brookes A, Sanseau P, Polinsky RJ, Wasco W, DaSilva HAR, Haines JL, Pericak-Vance MA, Tanzi RE, Roses AD, Rommens JM, St George-Hyslop PH (1995): Cloning of a gene bearing missense mutations in early-onset familial Alzheimer's disease. Nature 375:754–760.

Tjio JH, Levan A (1956): The chromosome number of man. Hereditas 42:1–6.

Tsui LC, Buchwald M, Barker D, Braman JC, Knowlton R, Schumm JW, Eiberg H, Mohr J, Kennedy D, Plavsi N, Zsiga M, Markiewicz D, Akots G, Brown V, Helms C, Gravius T, Parker C, Rediker K, Donis-Keller H (1985): Cystic fibrosis locus defined by a genetically linked polymorphic DNA marker. Science 230(4729):1054–1057.

Wald A (1947): Sequential Analysis. New York: John Wiley & Sons.

Watson JD, Crick FH (1953): Molecular structure of nucleic acids; a structure for deoxyribose nucleic acid. Nature 171(4356):737–738.

INDEX

Achondroplasia, point mutations, 17

Acrylamide, marker separation, manual or nonsequencer, 175

Additional medical service needs, family participation studies follow-up, 135–136

Adoption and twin studies, genetic component determination, 103–105

Advanced parametric linkage analysis. See Parametric linkage analysis (advanced)

Affected pedigree member (APM) analysis, nonparametric linkage analysis methods, 301–302

Affected relative pairs (nonparametric linkage analysis methods), 301–311

  affected pedigree member (APM) analysis, 301–302

  likelihood (parametric LOD score) analysis, 307–311

  nonparametric linkage (NPL) analysis, 304–307

  power analysis and experimental design considerations, 311–314

  quantitative traits, 314–318

  SimIBD analysis, 303–304

Affected sibling pairs (ASPs):

  nonparametric linkage analysis methods, 295–301

    identity by descent, 296–301

    identity by state, 295–296

  sample size and power, discrete traits, complex disease, 367–368

Affecteds-only analysis, parametric linkage analysis (advanced), complex traits, 274–275

Alleles:

  genetic linkage analysis, 9–10

  inheritance patterns in Mendelian disease, 13–14, 15–16

Allelic association:

  causes of, linkage disequilibrium, 331–334

  measures of, linkage disequilibrium, 330–331

Allelic heterogeneity, disease phenotype definition, 64

Americans with Disabilities Act (ADA), genetic discrimination, 140

Amyotrophic lateral sclerosis (ALS), point mutations, 17

Anonymous data, identifying data versus, family participation studies, 123–124

Artificial neural networks, serial analysis of gene expression (SAGE), 205–206

Ascertainment:

  family participation studies, 124–131

    informed consent, 128–131

    recruitment, 124–128

  genetic component determination study design, 94–99

    bias, 97–99

    controls, 97

    extended families, 96–97

    relative pairs, 95–96

    single family member, 94–95

Ascertainment bias, genetic component determination study design, 97–99. See also Bias

Genetic Analysis of Complex Diseases, Second Edition, Edited by Jonathan L. Haines and Margaret Pericak-Vance
Copyright © 2006 John Wiley & Sons, Inc.

Association analysis, 335–353. *See also* Genetic linkage disequilibrium
complex genetic interactions detection, 406–415
case-control design, 406–409
case-only studies, 409–410
cohort studies, 410–412
family-based case-control studies, 412
higher order interactions and data reduction, 413–414
genomic screening, 347–348
haplotype data analysis, 345–346
overview, 329–330
sample size and power, 376–383
case-control study design, 380–381
DNA pooling, 381
generally, 376–378
genomic screening strategies, 381–382
SIMLA program, 382–383
transmission/disequilibrium tests, quantitative traits, 380
transmission/disequilibrium tests for discrete traits, 378–379
special populations, 348–349
tests for association, 335–345
case-control tests, 335–340
family-based tests, 340–345
quantitative traits, 347

Bacterial clone maps, physical mapping, bioinformatics, 429–430
Becker muscular dystrophy, deletion/insertion mutations, 18
Bias:
ascertainment, genetic component determination study design, 97–99
complex genetic interactions, detection, 414–415
confounding bias, case-control tests, 339–340
Bioinformatics, 423–454. *See also* Internet resources
candidate genes identification by convergence, 439–440
complex traits gene identification study, 463–464

de novo annotation of genes, 440–444
gene function, 442–444
generally, 440–441
molecular mechanisms of disease, 442
online resources, 441–442
software, 441
genome mapping, 424–434
cytogenetic mapping, 425
generally, 424–425
genetic mapping, 425–427
physical mapping, 428–432
public data repositories and genome browsers, 432–434
radiation hybrid mapping, 427–428
human transcriptome, 444–446
model organisms, 438–439
overview, 423–424
single-nucleotide polymorphisms (SNPs), 434–438
discovery, 435–436
generally, 434–435
resources, 437–438
use, 436–437
Biological sample collection, 153–166
DNA extraction and processing, 157–162
blood, 157
buccal brushes, 160–161
dried blood cards, 161
fixed tissue, 161
quantitation, 157–159
tissue culture, 159–160
whole-genome amplification, 161–162
goals, 153
informed consent/security, 164
sample management, 162–164
types, 153–156
buccal samples, 155–156
dried blood, 156
tissue samples, 156
venipuncture (blood), 153–155
Biological sample management, biological sample collection, 162–164
Blood:
DNA extraction and processing, 157
dried, biological sample collection, 156

venipuncture, biological sample
  collection, 153–155
Buccal brushes, DNA extraction and
  processing, 160–161
Buccal samples, biological sample
  collection, 155–156

Cancer research, serial analysis of gene
  expression (SAGE) applications, 210
Candidate gene prioritization, serial
  analysis of gene expression (SAGE)
  applications, 211
Candidate genes identification, by
  convergence, bioinformatics, 439–440
Case-control studies and design:
  association analysis, complex genetic
    interactions detection, 406–409
  complex traits gene identification study,
    461
  family-based, association analysis,
    complex genetic interactions
    detection, 412
Case-control tests, tests for association,
  335–340
Case-only studies, association analysis,
  complex genetic interactions
  detection, 409–410
Causative concept, susceptibility concept
  versus, 19, 23
Cell reproduction, genetic linkage analysis,
  23–26
Chance, complex inheritance, disease
  phenotype definition, 70–71
Children. See Minor children
Chromosomal rearrangements, in
  mapping, disease phenotype
  definition, 74
Chromosomes, genetic linkage analysis,
  10–13
Classification objects, serial analysis of
  gene expression (SAGE), 204–205
Classification systems, disease phenotype
  definition, 71–72
Classifier evaluation, serial analysis of
  gene expression (SAGE), 206
Clinical Laboratories Improvement
  Amendments of 1988 (CLIA88),
  information release guidelines,
  137–138

Clustering (serial analysis of gene
  expression (SAGE)), 201–204
  conceptualization from, 204
  fuzzy, 203
  generally, 201
  hierarchical, 201–202
  implementation strategies, 204
  partitional, 202
  self-organizing map, 202–203
  stability, 203–204
  validity, 203
Cognitive impairment, informed consent,
  130–131
Cohort studies, association analysis,
  complex genetic interactions
  detection, 410–412
Complete penetrance, sample size and
  power, 386–391. See also Penetrance
Complex genetic disorders:
  analysis, disease phenotype definition,
    75–82
  mapping techniques, genetic linkage
    analysis, 45
  sample size and power, 365–376
    discrete traits, 367–373
    generally, 365–367
    quantitative traits, 373–376
  serial analysis of gene expression
    (SAGE) applications,
    210–211
Complex genetic interactions,
  397–421
  detection, 401–415
    association analysis, 406–415
      case-control design, 406–409
      case-only studies, 409–410
      cohort studies, 410–412
      family-based case-control studies,
        412
      higher order interactions and data
        reduction, 413–414
    biases, 414–415
    generally, 401
    linkage analysis, 402–406
      multivariable regression models,
        405–406
      stratified and conditional, 403–404
      two-locus models, 404–405
    segregation analysis, 402

Complex genetic interactions (*Continued*)
  evidence for, 398–401
    gene-environment interaction, 400–401
    gene-gene interaction (epistasis),
      399–400
    generally, 398
    genetic heterogeneity, 398–399
  overview, 397
Complex inheritance (disease phenotype
    definition), 67–71
  chance, 70–71
  environment, 70
  polygenic and multifactorial models,
    67–69
Complex traits:
  mapping applications, nonparametric
    linkage analysis, 318–319
  parametric linkage analysis (advanced),
    273–277
    affecteds-only analysis, 274–275
    generally, 273–274
    heterogeneity LOD score, 275–276
    maximized maximum LOD score, 275
    MFLINK, 276–277
Complex traits gene identification study,
    455–467. *See also* Study design
  components of, 457–465
    analysis, 463–464
    design development, 460–462
    follow-up, 464–465
    generally, 457–458
    phenotype, 459–460
  overview, 455–456
  requirements for, 465–466
Computer simulation methods, Mendelian
    disease, sample size and power,
    359–363
Conditional linkage analysis, complex
    genetic interactions detection,
    403–404
Confidentiality. *See* Patient confidentiality;
    Security
Confounding bias, case-control tests,
    339–340. *See also* Bias
Consent. *See* Informed consent
Contact maintenance, family participation
    studies follow-up, 137
Contiguous gene mutation, chromosomal
    rearrangements in mapping, 74

Controls. *See also* Case-control studies and
    design; Case-control tests; Family-
    based case-control studies; Quality
    control
  ascertainment, genetic component
    determination study design, 97
  family-based tests, tests for association,
    343–345
  scoring error control, parametric linkage
    analysis, 265–266
  sibling controls, family-based tests, tests
    for association, 343–345
  transcriptional control pathways, serial
    analysis of gene expression (SAGE)
    applications, 209
Convergence, candidate genes
    identification by, bioinformatics,
    439–440
Cosegregation approach, genetic
    component determination, 100
Cystic fibrosis, deletion/insertion
    mutations, 18
Cytogenetic mapping, bioinformatics, 425

Databases, family participation studies
    recruitment, 127
Data collection (family participation
    studies), 131–135
  diagnosis confirmation, 131–132
  field studies, 132–133, 134–135
  special issues, 133, 135
Data integrity, information management,
    228, 230–231
Data management (genotyping methods),
    186–188
  genotype integrity, 187
  objectivity, 187
  quality control, 188
  scoring, 187
  standards, 187–188
Data manipulation software, pedigree
    plotting and, information
    management, 234
Data matrix visualization, serial analysis of
    gene expression (SAGE), 208
Data preparation, serial analysis of gene
    expression (SAGE), 197
Data reduction, higher order interactions
    and, association analysis, 413–414

Decision trees, serial analysis of gene
  expression (SAGE), 206
Deletion/insertion mutations, genetic
  changes, 17–18
Denaturing high-pressure liquid
  chromatography (DHPLC),
  single-nucleotide
  polymorphisms (SNPs)
  detection, 186
Dendrograms, data visualization, serial
  analysis of gene expression (SAGE),
  207–208
De novo annotation of genes, 440–444
  gene function, 442–444
  generally, 440–441
  molecular mechanisms of disease, 442
  online resources, 441–442
  software, 441
Dentatorubropallidoluysian atrophy,
  trinucleotide repeats, 22
Descent, identity by, nonparametric
  linkage analysis, 286–289
Detection methods, 178–180
  fluorescence, 179–180
  radioactive, 178
  silver staining, 178–179
Diagnosis confirmation, family
  participation studies, 131–132
Diamond v. Chakrabarty,
  DNA ownership, 142
Differential expression analysis, serial
  analysis of gene expression (SAGE),
  208
Digenic causation, complex inheritance,
  disease phenotype definition, 69
Dimension reduction (serial analysis of
  gene expression (SAGE)), 198–200
  dimensionality curse, 199
  multidimensional scaling, 199–200
  principal-component analysis, 199
Dinucleotide repeats, sources, 169–170
Disaster recovery, information
  management, 230–231
Discrete traits:
  complex disease, sample size and power,
    367–373
  transmission/disequilibrium tests,
    association analysis, sample size
    and power, 378–379

Discrimination, genetic, family
  participation studies, follow-up,
  139–141
Disease allele frequency, misspecified
  model parameter effects,
  parametric linkage analysis, 261–262
Disease gene discovery, mapping
  techniques, genetic linkage analysis,
  32–41
Disease gene location, mapping
  techniques, genetic linkage analysis,
  41–45
Disease penetrances, misspecified model
  parameter effects, parametric linkage
  analysis, 263–264
Disease phenotype definition, 51–89
  chromosomal rearrangements in
    mapping, 74
  classification systems, 71–72
  complex genetic disorder analysis,
    75–82
  complex inheritance, 67–71
    chance, 70–71
    environment, 70
    polygenic and multifactorial models,
      67–69
  exceptions to Mendelian patterns, 52–64
    generally, 52–53
    genomic imprinting, 61–63
    incomplete penetrance and variable
      expressivity, 58–61
    mitochondrial inheritance, 56–58
    mosaicism, 55–56
    phenocopies and environmentally
      related effects, 63–64
    pseudodominant transmission of a
      dominant, 54–55
    pseudodominant transmission of a
      recessive, 53–54
  heterogeneity, 64–67
    genetic, 64–65
    phenotypic, 65–67
  nonsyndromic phenotypes, 72
  overview, 51–52
  qualitative (discontinuous) and
    qualitative (continuous) traits, 74–75
  resources for, 82
  syndromic phenotypes, 72–73
  unknown causes, 73–74

DNA:
structure of, genetic linkage analysis, 5–8
unstable, novel mutation mechanisms, genetic changes, 18–19, 20–22
DNA array ("chip"), single-nucleotide polymorphisms (SNPs) detection, 181
DNA banking, family participation studies follow-up, 141–142
DNA extraction and processing, 157–162
blood, 157
buccal brushes, 160–161
dried blood cards, 161
fixed tissue, 161
quantitation, 157–159
tissue culture, 159–160
whole-genome amplification, 161–162
DNA pooling:
association analysis, sample size and power, 381
homoxygosity mapping and, marker separation, 177–178
Dried blood, biological sample collection, 156
Dried blood cards, DNA extraction and processing, 161
Duchenne muscular dystrophy, deletion/ insertion mutations, 18

Ensembl Project, 433–434
Environment, disease phenotype definition, complex inheritance, 70
Environmentally related effects, phenocopies and, disease phenotype definition, exceptions to Mendelian patterns, 63–64
Environment-gene interaction, complex genetic interactions, 400–401
Epistasis (gene-gene interaction), complex genetic interactions, 399–400
Equal Employment Opportunity Commission (EEOC), genetic discrimination, 140
Euclidean distance, serial analysis of gene expression (SAGE), 200
Experimental design considerations, power analysis and, nonparametric linkage analysis methods, 311–314

Expression data matrix, serial analysis of gene expression (SAGE), 198
Expressivity, variable, incomplete penetrance and, disease phenotype definition, 58–61
Extended families, ascertainment, genetic component determination study design, 96–97
Extreme concordant pairs, sample size and power, quantitative traits, complex disease, 376
Extreme discordant pairs, sample size and power, quantitative traits, complex disease, 375–376

Familial aggregation approach, genetic component determination, 100–103
Familiality measures (nonparametric linkage analysis), 289–295
qualitative traits, 289–293
quantitative traits, 293–294
Family-based case-control studies, association analysis, complex genetic interactions detection, 412
Family-based design, complex traits gene identification study, 461
Family-based tests, tests for association, 340–345
Family history approach approach, genetic component determination, 101–102
Family members, family participation studies recruitment, 128
Family participation studies, 117–151
ascertainment, 124–131
informed consent, 128–131
recruitment, 124–128
data collection, 131–135
diagnosis confirmation, 131–132
field studies, 132–133, 134–135
special issues, 133, 135
follow-up, 135–142
additional medical services needs, 135–136
contact maintenance, 137
DNA banking, 141–142
genetic discrimination, 139–141
information release guidelines, 137–139
minor children, 139
recontact duty, 136–137

future trends, 142
overview, 117–118
preparations for, 118–124
   confidentiality, 118–119
   family studies director requirement, 119–121
   human subjects, 122–124
sample form for, 142–147
Family studies director requirement, family participation studies, 119–121
Field studies, family participation studies, data collection, 132–133, 134–135
Fixed tissue, DNA extraction and processing, 161
Fixing, marker separation, manual or nonsequencer, 175
Fluorescence, detection methods, 179–180
Fluorescence resonance energy transfer (FRET), Taqman, single-nucleotide polymorphisms (SNPs) detection, 182–184
Fluorescent allele static scanning technique, detection methods, 179
Fluorescent polarization, single-nucleotide polymorphisms (SNPs) detection, 182
Follow-up, complex traits gene identification study, 464–465
Fragile site F, trinucleotide repeats, 20
Fragile site mental retardation-2, trinucleotide repeats, 20
Fragile site 16q22, trinucleotide repeats, 21
Fragile X syndrome:
   ascertainment bias, genetic component determination study design, 99
   trinucleotide repeats, 20
Fuzzy clustering, serial analysis of gene expression (SAGE), 203

Gel formation, marker separation, manual or nonsequencer, 175
Gene annotation:
   within linkage intervals, serial analysis of gene expression (SAGE) applications, 211–212
   serial analysis of gene expression (SAGE) applications, 209
Gene-environment interaction, complex genetic interactions, 400–401

Gene expression profiling data analysis. See Serial analysis of gene expression (SAGE)
Gene function, de novo annotation of genes, 442–444
Gene-gene interaction (epistasis), complex genetic interactions, 399–400
Genetic architecture, quantitative trait linkage analysis, 238–240
Genetic changes, genetic linkage analysis, 14, 17–19
Genetic component determination, 91–115
   cosegregation approach, 100
   familial aggregation approach, 100–103
   family history approach, 101–102
   heritability approach, 106–108
   overview, 91–92
   recurrence risk in relatives approach, 105–106
   segregation analysis approach, 108–109
   study design, 92–99
      ascertainment, 94–99
      bias, 97–99
      controls, 97
      extended families, 96–97
      relative pairs, 95–96
      single family member, 94–95
      generally, 92–93
      population selection, 93–94
   twin and adoption studies, 103–105
Genetic discrimination, family participation studies, follow-up, 139–141
Genetic heterogeneity:
   complex genetic interactions, 398–399
   disease phenotype definition, 64–65
   parametric linkage analysis (advanced), 266–269
Genetic linkage analysis, 1–49. See also Nonparametric linkage analysis; Parametric linkage analysis (advanced)
   alleles, 9–10
   chromosomes, 10–13
   complex genetic interactions detection, 402–406
      multivariable regression models, 405–406
      stratified and conditional, 403–404
      two-locus models, 404–405

Genetic linkage analysis (*Continued*)
DNA structure, 5–8
genetic changes, 14, 17–19
deletion/insertion mutations, 17–18
novel mechanisms, 18–19
point mutations, 14, 17
historical perspective, 2–5
Hardy–Weinberg equilibrium, 5, 6
segregation, 2–5
inheritance patterns in Mendelian
disease, 13–14, 15–16
mapping techniques, 26–45
complex disease, 45
disease gene discovery, 32–41
disease gene location, 41–45
generally, 26, 27, 28
genetic mapping, 29–31
meiotic breakpoint mapping, 31–32
pedigree, information content of, 41
physical mapping, 26–29
mitosis and meiosis, 23–26
overview, 1–2
susceptibility versus causative genes,
19, 23
Genetic linkage disequilibrium, 329–334.
See also Association analysis
allelic association:
causes of, 331–334
measures of, 330–331
genetic mapping using, 334
overview, 329–330
Genetic mapping. *See* Mapping
Genome browsers, public data repositories
and, bioinformatics, 432–434
Genomic imprinting, disease phenotype
definition, exceptions to
Mendelian patterns, 61–63
Genomics and bioinformatics. *See*
Bioinformatics
Genomic screening:
association analysis, 347–348
sample size and power, 381–382
complex disease, discrete traits, 371–373
Genotype integrity, genotyping methods,
data management, 187
Genotyping methods, 167–192
data management, 186–188
genotype integrity, 187
objectivity, 187

quality control, 188
scoring, 187
standards, 187–188
detection methods, 178–180
fluorescence, 179–180
radioactive, 178
silver staining, 178–179
historical review, 167–169
restriction fragment length
polymorphisms (RFLPs), 167–168
short tandem repeats (STRs) or
microsatellites, 168
single-nucleotide polymorphisms
(SNPs), 168–169
variable number of tandem repeat
markers (VNTRs), 168
marker separation, 175–178
DNA pooling and homoxygosity
mapping, 177–178
loading variants, 176–177
manual or nonsequencer, 175
polymerase chain reaction (PCR),
171–174
optimization of laboratory, 171–172
optimization of reagents, 172–173
poor results, remedies for, 173–174
single-nucleotide polymorphisms
(SNPs) detection, 181–186
denaturing high-pressure liquid
chromatography (DHPLC), 186
DNA array ("chip"), 181
fluorescent polarization, 182
invader and PCR-invader assays,
184–186
matrix-assisted laser desorption/
ionization time-of-flight
spectrometry (MALDI-TOF), 184
oligonucleotide ligation assay, 181–182
pyrosequencing, 184
single-base-pair extension, 184
single-strand conformational
polymorphism (SSCP), 186
Taqman, 182–184
sources, 169–171
microsatellites, 169–170
restriction fragment length
polymorphisms (RFLPs), 169
single-nucleotide polymorphisms
(SNPs), 171

Germline mosaicism, disease phenotype
definition, exceptions to
Mendelian patterns, 56

Haplotype data analysis, association
analysis, 345–346
Hardware requirements, information
management, 225–226.
See also Software
Hardy–Weinberg equilibrium, genetic
linkage analysis, 5, 6
Haseman–Elston regression, quantitative
trait linkage analysis, 240–242
Haw River syndrome, trinucleotide
repeats, 22
Health Insurance Portability and
Accountability Act of 1996 (HIPAA),
genetic discrimination, 140–141
Heritability approach, genetic component
determination, 106–108
Heterogeneity, disease phenotype
definition, 64–67
Heterogeneity LOD score, parametric
linkage analysis (advanced), complex
traits, 275–276. *See also* Logarithm
of the odds (LOD) score
Heteroplasmy, mitochondrial inheritance,
disease phenotype definition, 56–57
Hierarchical clustering, serial analysis of
gene expression (SAGE), 201–202
Higher order interactions, data reduction
and, association analysis, complex
genetic interactions detection,
413–414
Homoxygosity mapping, DNA pooling
and, marker separation, 177–178
Human leukocyte antigen (HLA),
susceptibility versus causative genes,
19, 23
Human subjects, family participation
studies, 122–124
Human transcriptome, bioinformatics,
444–446
Huntington's disease:
novel mutation mechanisms, 19
trinucleotide repeats, 21

Identifying data, anonymous data versus,
family participation studies, 123–124

Identity by descent, nonparametric linkage
analysis, 286–289
Identity by state, nonparametric linkage
analysis, 286–289
Implementation:
information management, 227–231
serial analysis of gene expression
(SAGE), clustering, 204
Incomplete penetrance, variable
expressivity and, 58–61
exceptions to Mendelian patterns,
58–61
Information content, of pedigrees,
Mendelian disease, sample size and
power, 358–359
Information management, 219–235
hardware and software requirements,
225–226
implementation, 227–231
conversion, 227–229
data integrity, 228, 230–231
performance tuning, 228
information flow, 222–223
model development, 223–225
overview, 219
pedigree plotting and data manipulation
software, 234
planning, 220–222
security, 231–233
patient confidentiality, 233
system, 233
transmission, 231–233
user interfaces, 231
Information release guidelines, family
participation studies follow-up,
137–139
Informed consent:
biological sample collection, 164
family participation studies, 128–131
Inheritance mode, misspecified model
parameter effects, parametric linkage
analysis, 262–263
Inheritance patterns, Mendelian disease,
genetic linkage analysis, 13–14,
15–16
Insertion/deletion mutations, genetic
changes, 17–18
Institutional databases, family participation
studies recruitment, 127

Institutional Review Board (IRB), family participation studies, 122–124

Interference, genetic mapping, genetic linkage analysis, 30–31

Internet, family participation studies recruitment, 127

Internet resources. *See also* Bioinformatics de novo annotation of genes, 441–442
human transcriptome, 445–446
physical mapping, bioinformatics, 430–431

Invader and PCR-invader assays, single-nucleotide polymorphisms (SNPs) detection, 184–186

Kennedy spinal and bulbar muscular atrophy, trinucleotide repeats, 21

K-nearest neighbor, serial analysis of gene expression (SAGE), supervised machine learning, 205

Laboratory Information Management System (LIMS), biological sample collection, 162

Library analysis, serial analysis of gene expression (SAGE), 195–196

Likelihood (parametric LOD score) analysis, affected relative pairs, nonparametric linkage analysis methods, 307–311

Linear discriminant analysis, serial analysis of gene expression (SAGE), supervised machine learning, 205

Linkage analysis. *See* Genetic linkage analysis

Linkage disequilibrium. *See* Genetic linkage disequilibrium

Loading variants, marker separation, 176–177

Locus heterogeneity, disease phenotype definition, 64–65

Logarithm of the odds (LOD) score. *See also* Likelihood (parametric LOD score) analysis; Nonparametric linkage analysis; Parametric linkage analysis (advanced)

power assessments, definitions for, 363–365
sample size and power, LOD score results example, 385–386

Machado–Joseph disease:
novel mutation mechanisms, 19
trinucleotide repeats, 22

Manual genotyping, marker separation, 175

Mapping:
bioinformatics, 425–427
chromosomal rearrangements in, disease phenotype definition, 74
complex traits, nonparametric linkage analysis, 318–319
genetic linkage analysis, 29–31
linkage disequilibrium, 334
nonparametric linkage analysis, affected relative pairs, quantitative traits, 316–318

Mapping techniques (genetic linkage analysis), 26–45
complex disease, 45
disease gene discovery, 32–41
disease gene location, 41–45
generally, 26, 27, 28
genetic mapping, 29–31
meiotic breakpoint mapping, 31–32
pedigree, information content of, 41
physical mapping, 26–29

Marker allele frequency, misspecified model parameter effects, parametric linkage analysis, 264–265

Marker separation, 175–178
DNA pooling and homoxygosity mapping, 177–178
loading variants, 176–177
manual or nonsequencer, 175

Matrix, expression data matrix, serial analysis of gene expression (SAGE), 198

Matrix-assisted laser desorption/ionization time-of-flight spectrometry (MALDI-TOF), single-nucleotide polymorphisms (SNPs) detection, 184

Maximized maximum LOD score, parametric linkage analysis (advanced), complex traits, 275

Medical clinics, family participation
studies recruitment, 127–128
Medical services needs, family
participation studies follow-up,
135–136
Meiosis, mitosis and, genetic linkage
analysis, 23–26
Meiotic breakpoint mapping, genetic
linkage analysis, 31–32
Mendel, Gregor, 2–5
Mendelian disease:
inheritance patterns in, genetic
linkage analysis, 13–14, 15–16
sample size and power, 358–365
computer simulation methods,
359–363
definitions for power assessments,
363–365
information content of pedigrees,
358–359
MFLINK, parametric linkage analysis
(advanced), complex traits, 276–277
Microarray analysis, serial analysis of gene
expression (SAGE), 196–197
Microsatellites:
historical review, 168
sources, 169–170
Minor children:
family participation studies, follow-up,
139
informed consent, family participation
studies, 130
Misspecified model parameter effects
(parametric linkage analysis),
260–265
disease allele frequency, 261–262
disease penetrances, 263–264
generally, 260–261
inheritance mode, 262–263
marker allele frequency, 264–265
Mitochondrial inheritance, disease
phenotype definition, exceptions to
Mendelian patterns, 56–58
Mitosis, meiosis and, genetic linkage
analysis, 23–26
Model organisms, bioinformatics,
438–439
Molecular mechanisms, of disease, de novo
annotation of genes, 442

Monogenic predisposition, complex
inheritance, disease phenotype
definition, 68–69
Monte Carlo simulation, sample size and
power, 384–385
*Moore v. Regents of University of
California*, DNA ownership, 142
Mosaicism, disease phenotype definition,
exceptions to Mendelian
patterns, 55–56
Multidimensional scaling:
data visualization, serial analysis of gene
expression (SAGE), 207
serial analysis of gene expression
(SAGE), dimension reduction,
199–200
Multifactorial models, polygenic
models and, complex inheritance,
disease phenotype definition,
67–69
Multiloading, marker separation, 176
Multiplexing, marker separation,
176–177
Multipoint analysis:
nonparametric linkage analysis, affected
relative pairs, quantitative traits,
315–316
parametric linkage analysis (advanced),
269–273
Multipoint IBD method (MIM),
quantitative trait linkage
analysis, 242–243
Multivariable regression models, complex
genetic interactions detection, linkage
analysis, 405–406
Muscular dystrophy, Duchenne and
Becker, deletion/insertion
mutations, 18
Myotonic dystrophy:
novel mutation mechanisms, 19
trinucleotide repeats, 22

National Center for Biotechnology
Information (NCBI), 432–433
Network regulation analysis, serial
analysis of gene expression (SAGE),
208–209
Neurofibromatosis, deletion/insertion
mutations, 17–18

Nonparametric linkage analysis, 283–328. *See also* Genetic linkage analysis
affected relative pairs, nonparametric linkage analysis methods, 304–307
affected relative pairs methods, power analysis and experimental design considerations, 311–314
applications for mapping complex traits, 318–319
basic concepts, summarized, 295
familiality measures, 289–295
  qualitative traits, 289–293
  quantitative traits, 293–294
historical perspective, 284–285
methods, 295–318
  affected relative pairs, 301–311
    affected pedigree member (APM) analysis, 301–302
    likelihood (parametric LOD score) analysis, 307–311
    nonparametric linkage (NPL) analysis, 304–307
    power analysis and experimental design considerations, 311–314
    quantitative traits, 314–318
    SimIBD analysis, 303–304
  affected sibling pairs (ASPs), 295–301
    identity by descent, 296–301
    identity by state, 295–296
  generally, 295
overview, 283–284
quantitative trait linkage analysis, 246
software availability, 322
state/descent identification, 286–289
weighted pairwise correlation (WPC), 319–322
Nonsequencer genotyping, marker separation, 175
Nonsyndromic phenotypes, disease phenotype definition, 72
Novel mutation mechanisms, genetic changes, 18–19, 20–22

Objectivity, genotyping methods, data management, 187
Oligogenic causation, complex inheritance, disease phenotype definition, 69
Oligonucleotide ligation assay, single-nucleotide polymorphisms (SNPs) detection, 181–182
Online resources. *See also* Bioinformatics
de novo annotation of genes, 441–442
human transcriptome, 445–446
physical mapping, bioinformatics, 430–431

Parametric linkage analysis (advanced), 255–281. *See also* Genetic linkage analysis
complex traits, 273–277
  affecteds-only analysis, 274–275
  generally, 273–274
  heterogeneity LOD score, 275–276
  maximized maximum LOD score, 275
  MFLINK, 276–277
genetic heterogeneity, 266–269
misspecified model parameter effects, 260–265
  disease allele frequency, 261–262
  disease penetrances, 263–264
  generally, 260–261
  inheritance mode, 262–263
  marker allele frequency, 264–265
multipoint analysis, 269–273
overview, 255–256
scoring error control, 265–266
two-point analysis, 256–260
Partitional clustering, serial analysis of gene expression (SAGE), 202
Pathogenetic sequence, nonsyndromic phenotypes, disease phenotype definition, 72
Patient confidentiality:
family participation studies, 118–119
information management security, 233
Pattern recognition analysis, serial analysis of gene expression (SAGE), 208
PCR-invader assays, single-nucleotide polymorphisms (SNPs) detection, 184–186
Pearson coefficient, serial analysis of gene expression (SAGE), 200–201
Pedigree, information content of, mapping techniques, genetic linkage analysis, 41

Pedigree plotting, data manipulation software and, information management, 234

Penetrance:
  complete, sample size and power, 386–391
  disease penetrances, misspecified model parameter effects, 263–264
  incomplete, variable expressivity and, disease phenotype definition, 58–61
  reduced penetrance, sample size and power, 391–393

Performance tuning, information management, 228

Phenocopies, environmentally related effects and, disease phenotype definition, 63–64

Phenotype, complex traits gene identification study, 459–460

Phenotypic heterogeneity, disease phenotype definition, 65–67

Physical mapping:
  bioinformatics, 428–432
  genetic linkage analysis, 26–29

Planning, information management, 220–222

Point mutations, genetic changes, 14, 17

Polygenic models, multifactorial models and, complex inheritance, 67–69

Polymerase chain reaction (PCR), 171–174
  optimization of laboratory, 171–172
  optimization of reagents, 172–173

Population selection, genetic component determination study design, 93–94

Power analysis, experimental design considerations and, affected relative pairs, nonparametric linkage analysis methods, 311–314. *See also* Sample size and power

Principal-component analysis:
  data visualization, serial analysis of gene expression (SAGE), 207
  serial analysis of gene expression (SAGE), dimension reduction, 199

Pseudodominant transmission:
  of a dominant, disease phenotype definition, exceptions to Mendelian patterns, 54–55
  of a recessive, disease phenotype definition, exceptions to Mendelian patterns, 53–54

Psychological risk, information release guidelines, family participation studies follow-up, 139

Public data repositories, genome browsers and, bioinformatics, 432–434

Pyrosequencing, single-nucleotide polymorphisms (SNPs) detection, 184

Qualitative traits:
  familiality measures, nonparametric linkage analysis, 289–293
  nonparametric linkage analysis, affected relative pairs methods, 311–314
  qualitative traits and, disease phenotype definition, 74–75

Quality control:
  genotyping methods, data management, 188
  serial analysis of gene expression (SAGE), data preparation, 197

Quantitation, DNA extraction and processing, 157–159

Quantitative trait(s):
  familiality measures, nonparametric linkage analysis, 293–294
  nonparametric linkage analysis, affected relative pairs, 314–318
  sample size and power, complex disease, 373–376
  tests for association, 347

Quantitative trait linkage analysis, 237–253
  future directions, 247–249
  genetic architecture, 238–240
  Haseman–Elston regression, 240–242
  multipoint IBD method (MIM), 242–243
  nonparametric methods, 246
  overview, 237–238
  study design, 240
  variance component linkage analysis, 243–246

Radiation hybrid mapping, bioinformatics, 427–428
Radioactive detection methods, 178
Reagents, optimization of, polymerase chain reaction (PCR), 172–173
Real-time scanning, fluorescence detection methods, 179
Recontact duty, family participation studies follow-up, 136–137
Recruitment, family participation studies ascertainment, 124–128
Recurrence risk, in relatives, genetic component determination approach, 105–106
Reduced penetrance, sample size and power, 391–393
Referrals, family participation studies recruitment, 125, 126, 127
Rehabilitation Act of 1973, revised 1992, genetic discrimination, 140
Relative pairs, ascertainment, genetic component determination study design, 95–96
Relatives, recurrence risk in, genetic component determination approach, 105–106
Research databases, family participation studies recruitment, 127
Restriction fragment length polymorphisms (RFLPs):
historical review, 167–168
sources, 169
Roe v. Wade, genetic discrimination, family participation studies follow-up, 139–140

Sample collection. *See* Biological sample collection
Sample size and power, 355–396
association analysis, 376–383
case-control study design, 380–381
DNA pooling, 381
generally, 376–378
genomic screening strategies, 381–382
SIMLA program, 382–383
transmission/disequilibrium tests:
discrete traits, 378–379
quantitative traits, 380

complete penetrance, 386–391
complex disease, 365–376
discrete traits, 367–373
generally, 365–367
quantitative traits, 373–376
LOD score results example, 385–386
Mendelian disease, 358–365
computer simulation methods, 359–363
definitions for power assessments, 363–365
information content of pedigrees, 358–359
Monte Carlo simulation, 384–385
overview, 355–358
reduced penetrance, 391–393
Scanning technologies, fluorescence detection methods, 179–180
Scoring, genotyping methods, data management, 187
Security:
biological sample collection, 164
information management, 231–233
patient confidentiality, 233
system, 233
transmission, 231–233
Segregation, genetic linkage analysis, 2–5
Segregation abnormality, meiosis, genetic linkage analysis, 25–26
Segregation analysis:
complex genetic interactions detection, 402
genetic component determination, 108–109
Self-organizing map, serial analysis of gene expression (SAGE), clustering, 202–203
Serial analysis of gene expression (SAGE), 193–217
applications, 209–212
cancer research, 210
candidate gene prioritization, 211
complex genetic disorders, 210–211
gene annotation, 209
gene annotation within linkage intervals, 211–212
transcriptional control pathways, 209
transcriptional factor binding sites, 209–210

clustering, 201–204
conceptualization from, 204
fuzzy, 203
generally, 201
hierarchical, 201–202
implementation strategies, 204
partitional, 202
self-organizing map, 202–203
stability, 203–204
validity, 203
data preparation, 197
development of, 194–195
dimension reduction, 198–200
dimensionality curse, 199
multidimensional scaling, 199–200
principal-component analysis, 199
expression data matrix, 198
library analysis, 195–196
microarray analysis, 196–197
other types of analysis, 208–209
overview, 193–194
similarity measurements, 200–201
Euclidean distance, 200
Pearson coefficient, 200–201
supervised machine learning, 204–206
artificial neural networks, 205–206
classification objects, 204–205
classifier evaluation, 206
decision trees, 206
k-nearest neighbor, 205
linear discriminant analysis, 205
support vector machines, 206
visualization, 207–208
Short tandem repeats (STRs), historical
review, 168
Sibling controls, family-based tests, tests
for association, 343–345
Sickle cell anemia, point mutations, 17
Silver staining, detection methods,
178–179
SimIBD analysis, affected relative pairs,
nonparametric linkage analysis
methods, 303–304
Similarity measurements (serial analysis of
gene expression (SAGE)), 200–201
Euclidean distance, 200
Pearson coefficient, 200–201
SIMLA program, sample size and
power, association analysis, 382–383

Single-base-pair extension, single-
nucleotide polymorphisms (SNPs)
detection, 184
Single family member, ascertainment,
genetic component
determination study design,
94–95
Single-nucleotide polymorphisms (SNPs):
bioinformatics, 434–438
discovery, 435–436
generally, 434–435
resources, 437–438
use, 436–437
detection, 181–186
denaturing high-pressure liquid
chromatography (DHPLC),
186
DNA array ("chip"), 181
fluorescent polarization, 182
invader and PCR-invader assays,
184–186
matrix-assisted laser desorption/
ionization time-of-flight
spectrometry (MALDI-TOF),
184
oligonucleotide ligation assay,
181–182
pyrosequencing, 184
single-base-pair extension, 184
single-strand conformational
polymorphism (SSCP), 186
Taqman, 182–184
historical review, 168–169
sources, 171
Single-strand conformational
polymorphism (SSCP), single-
nucleotide polymorphisms (SNPs)
detection, 186
Sister chromatids, genetic linkage
analysis, 10
Software. See also Hardware requirements
data manipulation software, pedigree
plotting and, information
management, 234
de novo annotation of genes, 441
information management, 225–226
nonparametric linkage analysis, 322
radiation hybrid mapping,
bioinformatics, 428

Somatic mosaicism, disease phenotype definition, exceptions to Mendelian patterns, 55–56

Special populations, association analysis, 348–349

Spinocerebellar ataxia, trinucleotide repeats, 21, 22

Stability, serial analysis of gene expression (SAGE), clustering, 203–204

Standards, genotyping methods, data management, 187–188

State, identity by, nonparametric linkage analysis, 286–289

Statement of Confidentiality, family participation studies, 119

Stratified linkage analysis, complex genetic interactions detection, 403–404

Study design. *See also* Complex traits gene identification study

  case-control studies and design:

    association analysis, complex genetic interactions detection, 406–409

    complex traits gene identification study, 461

    family-based, association analysis, complex genetic interactions detection, 412

  family-based, complex traits gene identification study, 461

  genetic component determination study design, 94–99

    bias, 97–99

    controls, 97

    extended families, 96–97

    relative pairs, 95–96

    single family member, 94–95

  quantitative trait linkage analysis, 240

Supervised machine learning (serial analysis of gene expression (SAGE)), 204–206

  artificial neural networks, 205–206

  classification objects, 204–205

  classifier evaluation, 206

  decision trees, 206

  k-nearest neighbor, 205

  linear discriminant analysis, 205

  support vector machines, 206

Support groups, family participation studies recruitment, 125, 126

Support vector machines, serial analysis of gene expression (SAGE), supervised machine learning, 206

Supreme Court (U.S.):

  DNA ownership, family participation studies follow-up, 142

  genetic discrimination, family participation studies follow-up, 139–140

Susceptibility concept, causative concept versus, genetic linkage analysis, 19, 23

Syndromes:

  causes of, 74

  disease phenotype definition, unknown causes, 73–74

Syndromic phenotypes, disease phenotype definition, 72–73

System security, information management, 233

Taqman, single-nucleotide polymorphisms (SNPs) detection, 182–184

Tests for association, 335–345

  case-control tests, 335–340

  family-based tests, 340–345

  quantitative traits, 347

Tetranucleotide repeats, sources, 170

Tissue culture, DNA extraction and processing, 159–160

Tissue samples, biological sample collection, 156

Transcriptional control pathways, serial analysis of gene expression (SAGE) applications, 209

Transcriptional factor binding sites, serial analysis of gene expression (SAGE) applications, 209–210

Translocations, chromosomal rearrangements in mapping, disease phenotype definition, 74

Transmission/disequilibrium tests:

  discrete traits, association analysis, sample size and power, 378–379

  family-based tests, tests for association, 341–343

quantitative traits, association analysis, sample size and power, 380

Transmission security, information management, 231–233

Trinucleotide repeats:
novel mutation mechanisms, genetic changes, 18–19, 20–22
sources, 170

Twin and adoption studies, genetic component determination, 103–105

Two-locus linkage analysis, complex genetic interactions detection, 404–405

Two-point analysis, parametric linkage analysis (advanced), 256–260

UCSC Genome Browser, 433

Unaffected sibling, sample size and power, discrete traits, complex disease, 368–369

Unaffected sibling controls, family-based tests, tests for association, 343–345

Uniparental disomy, genomic imprinting, disease phenotype definition, exceptions to Mendelian patterns, 61, 63

United States Supreme Court:
DNA ownership, family participation studies follow-up, 142
genetic discrimination, family participation studies follow-up, 139–140

Unstable DNA, novel mutation mechanisms, genetic changes, 18–19, 20–22

User interfaces, information management, 231

Validity, clustering, serial analysis of gene expression (SAGE), 203

Variable expressivity, incomplete penetrance and, disease phenotype definition, exceptions to Mendelian patterns, 58–61

Variable number of tandem repeat markers (VNTRs), historical review, 168

Variance component linkage analysis, quantitative trait linkage analysis, 243–246

Venipuncture (blood), biological sample collection, 153–155

Visualization, serial analysis of gene expression (SAGE), 207–208

Vulnerable populations, informed consent, family participation studies, 130–131

Weighted pairwise correlation (WPC), nonparametric linkage analysis, 319–322

Whole-genome amplification, DNA extraction and processing, 161–162

Whole-genome shotgun sequencing, physical mapping, bioinformatics, 431–432

YAC maps, physical mapping, bioinformatics, 429